# METHODS IN CELL BIOLOGY

VOLUME 27

*Echinoderm Gametes and Embryos*

*Series Editor*

LESLIE WILSON

*Department of Biological Sciences*
*University of California*
*Santa Barbara, California*

ASCB

# METHODS IN CELL BIOLOGY

*Prepared under the Auspices of the American Society for Cell Biology*

## VOLUME 27
*Echinoderm Gametes and Embryos*

*Edited by*
THOMAS E. SCHROEDER
FRIDAY HARBOR LABORATORIES
UNIVERSITY OF WASHINGTON
FRIDAY HARBOR, WASHINGTON

1986

ACADEMIC PRESS, INC.
**Harcourt Brace Jovanovich, Publishers**
Orlando San Diego New York
Austin Boston London Sydney
Tokyo Toronto

ACADEMIC PRESS, INC.
Orlando, Florida 32887

*United Kingdom Edition published by*
ACADEMIC PRESS INC. (LONDON) LTD.
24–28 Oval Road, London NW1 7DX

LIBRARY OF CONGRESS CATALOG CARD NUMBER: 64-14220

ISBN 0-12-564127-3

PRINTED IN THE UNITED STATES OF AMERICA

87 88 89    9 8 7 6 5 4 3 2

# CONTENTS

## 4. *Ion Measurements in Sea Urchin Sperm*
*Robert Schackmann*



## 5. *Gonad-Stimulating and Maturation-Inducing Substance*
*Hiroko Shirai*



## 6. *Techniques for Observing Living Gametes and Embryos*
*Douglas A. Lutz and Shinya Inoué*



## 7. *Egg and Embryonic Extracellular Coats: Isolation and Purification*
*Peggy J. Weidman and Erica S. Kay*

## 11. *Tubulin-Containing Structures*
*Kathy A. Suprenant*



## 12. *Isolation of Embryonic Cilia and Sperm Flagella*
*R. E. Stephens*



## 13. *Components of the Actin-Based Cytoskeleton*
*R. E. Kane*



## 14. *Cytology and Immunocytochemistry*
*Patricia J. Harris*

# CONTRIBUTORS

*Numbers in parentheses indicate the pages on which the authors' contributions begin.*

CHARLES J. BROKAW, Biology Division, California Institute of Technology, Pasadena, California 91125 (41)

ROBERT D. BURKE, Department of Biology, University of Victoria, Victoria, British Columbia, Canada V8W 2Y2 (295)

ANITA M. COLIN, Department of Zoology, University of Washington, Seattle, Washington 98195 (395)

MICHAEL V. DANILCHIK, Department of Molecular Biology, University of California, Berkeley, California 94720 (175)

ALLAN W. GIBSON, Department of Biology, University of Victoria, Victoria, British Columbia, Canada V8W 2Y2 (295)

GEORGE R. GREEN, Department of Biology, Amherst College, Amherst, Massachusetts 01002 (153)

PATRICIA J. HARRIS, Department of Biology, University of Oregon, Eugene, Oregon 97403 (243)

MERRILL B. HILLE, Department of Zoology, University of Washington, Seattle, Washington 98195 (175)

YUKIO HIRAMOTO, Biological Laboratory, Tokyo Institute of Technology, Tokyo 152, Japan, and Department of Cell Biology, National Institute for Basic Biology, Okazaki 444, Japan (435, 443)

SHINYA INOUÉ, Marine Biological Laboratory, Woods Hole, Massachusetts 02543 (89)

R. E. KANE, Pacific Biomedical Research Center, Kewalo Marine Laboratory, University of Hawaii, Honolulu, Hawaii 96813 (229)

ERICA S. KAY, Department of Biochemistry, University of Washington, Seattle, Washington 98195 (111)

WILLIAM H. KINSEY, Department of Anatomy, University of Miami School of Medicine, Miami, Florida 33101 (139)

TAKEO KISHIMOTO, Department of Developmental Biology, National Institute for Basic Biology, Okazaki 444, Japan (379)

TETSUYA KOMINAMI, Department of Biology, Faculty of Science, Ehime University, Matsuyama 790, Japan (325)

PATRICK S. LEAHY, Kerckhoff Marine Laboratory, California Institute of Technology, Corona del Mar, California 92625 (1)

DOUGLAS A. LUTZ, Department of Anatomy—Histology, University of Toronto, Toronto, Ontario, Canada M5S 1A8 (89)

YOSHIHIKO K. MARUYAMA, Department of Zoology, Faculty of Science, Kyoto University, Kyoto 606, Japan (325)

DAVID R. MCCLAY, Department of Zoology, Duke University, Durham, North Carolina 27706 (309)

IZUMI MITA-MIYAZAWA, Department of Zoology, Faculty of Science, Kyoto University, Kyoto 606, Japan (325)

JOHN B. MORRILL, Division of Natural Sciences, New College of the University of South Florida, Sarasota, Florida 34243 (263)

SHIN-ICHI NEMOTO, Tateyama Marine Laboratory, Ochanomizu University, Tateyama 294-03, Japan (325)

Dominic L. Poccia, Department of Biology, Amherst College, Amherst, Massachusetts 01002 (153)

R. Rappaport, Department of Biological Sciences, Union College, Schenectady, New York 12308, and The Mount Desert Island Biological Laboratory, Salsbury Cove, Maine 04672 (345)

Kiyoshi Sano, Developmental Biology Laboratory, Department of Plastic Surgery, Kitasato University, Sagamihara, Kanagawa 228, Japan (407)

Robert Schackmann, Department of Biochemistry, University of Washington, Seattle, Washington 98195 (57)

G. G. Sekirina, Experimental Station ''Gorki Lininskiye'' of the Institute of Molecular Biology and Genetics, Lenin Academy of Agricultural Sciences, Moscow, USSR (359)

Hiroko Shirai, Division of Developmental Biology, Laboratory of Reproductive Biology, National Institute for Basic Biology, Okazaki 444, Japan (73)

M. N. Skoblina, Institute of Developmental Biology, USSR Academy of Sciences, Moscow, USSR (359)

R. E. Stephens, Marine Biological Laboratory, Woods Hole, Massachusetts 02543 (217)

Kathy A. Suprenant, Department of Physiology and Cell Biology, University of Kansas, Lawrence, Kansas 66045 (189)

Victor D. Vacquier, Marine Biology Research Division, Scripps Institute of Oceanography, University of California, San Diego, La Jolla, California 92093 (15)

S. G. Vassetzky, Institute of Developmental Biology, USSR Academy of Sciences, Moscow, USSR (359)

Peggy J. Weidman, Department of Biochemistry, University of Washington, Seattle, Washington 98195 (111)

Kenya Yamamoto,[1] Department of Zoology, Faculty of Science, Kyoto University, Kyoto 606, Japan (325)

Mitsuki Yoneda, Department of Zoology, Faculty of Science, Kyoto University, Kyoto 606, Japan (421, 443)

[1]*Present address:* Department of Biology, Faculty of General Education, Gifu University, Gifu 501-11, Japan.

# PREFACE

This volume contains a sample of methods for probing the cell biological world of sperm, eggs, and embryos of echinoderms. Each chapter is a self-contained practical guide offering technical details and advice for carrying out a particular kind of investigation. It complements the vast literature of scientific research on echinoderm gametes and embryos where the practical details of experimental protocols are often incompletely explained. Despite the respected, century-long tradition of experimentation upon these gametes and embryos, the content of this volume is distinctly contemporary, reflecting the technical wizardry, advanced and specialized knowledge, and dynamic inquisitiveness of the present times.

The contributors to this volume demonstrate that cellular phenomena, although occurring below the detection limits of our unaided senses, are distinct from purely molecular events. Indeed, one of the volume's main achievements is to reaffirm the vitality of the *cellular* level of biological investigation in spite of the fashionable trend to treat cells as sources of molecular "reagents." The chapters provide access to the intimate world of the living (or "nearly living") cell by outlining devices to extend our sensory acuity, by describing ways of perfecting experimental and analytical skills, and by emphasizing the importance of judiciously choosing the particular cell that is most suitable for a particular inquiry.

In the cell biology laboratory, the selection of research material is too often dictated by the pragmatic imperatives of the availability or abundance of gametes, and as a consequence only a mere handful of echinoderms have been studied in detail. By ignoring the comparative biology of echinoderms, we miss a whole dimension of understanding, and certain distortions set in. Too many students seem to learn that sea urchins, often of one particular species, are the "ideal" echinoderms. Most fail to understand that sea urchin sperm cells have uncharacteristically puny acrosomal processes; that spawned sea urchin eggs are fully postmeiotic and are enveloped in vanishingly thin vitelline layers, unlike virtually all other organisms; or that the 3-tiered, 16-cell sea urchin embryo, with micromeres that are fated to form primary mesenchyme and spicule-forming tissue, is unlike any other embryo, including other echinoderm embryos.

Because there are about 6100 existing species in the Phylum Echinodermata (2200 starfish, 2000 brittle stars and basket stars, 1200 sea cucumbers, 900 sea urchins, and 800 sea lilies and feather stars), choosing the appropriate cell to study is compounded by the wonderful diversity of the parent organisms. Every echinoderm is exotic and distinctive, despite being deuterostomes like ourselves and being familiar encounters at the seashore, in souvenir shops, in fossil fields, and even on our dinner plate. What other organisms have a pentaradial adult body plan that emerges from a bilateral larva, effector organ systems and sensory structures that are dispersed and decentralized, a hard endoskeleton assembled from many small units that are hinged together for overall pliability, connective tissue that is both contractile and instantaneously dissoluble, a nervous system that has no organized synaptic connections, and cells that never have more than one cilium?

This volume, while emphasizing methods for dealing with the gametes and embryos of sea urchins and starfish (due to our presently limited perspectives), repeatedly points out the benefits of comparative studies between species. A future volume of this type will hopefully show that our appreciation of all echinoderms will have kept pace with the anticipated progress of technology in cell biology.

THOMAS E. SCHROEDER

# Chapter 1

# *Laboratory Culture of Strongylocentrotus purpuratus Adults, Embryos, and Larvae*

PATRICK S. LEAHY

*Kerckhoff Marine Laboratory*
*California Institute of Technology*
*Corona del Mar, California*

---



## I. Introduction

The sea urchin embryo has served as both a classical and modern model for the study of early development. At the molecular level this is now the best characterized embryonic system (Davidson *et al.*, 1982). The usefulness of sea urchin embryos for investigations into the mechanism of early development depends in part on the many advantageous practical features of this material, as well as on the broad base of classical data on cellular aspects of development. Gametes can be obtained year round for some species; other species can be obtained in the gravid condition in alternate seasons. Large numbers of fully mature eggs and sperm are produced by individual organisms, and *in vitro* fertilization yields highly synchronous embryonic development with a minimum of treatment. The ready availability of quantities of embryos in excess of $10^8$ provides material for quantitative studies of macromolecules. Such studies are greatly facilitated by the permeability of these embryos to nucleic acid and protein precursors and the ease of preparing several embryonic cell types or subcellular fractions. In addition, several specific experimental techniques have been developed for molecular studies of sea urchin embryos, including methods for *in situ* hybridization (Angerer and Angerer, 1981), cytological procedures for immunofluorescence (McClay and Fink, 1982), and methods for introduction by microinjection of cloned genes into the egg (McMahon *et al.*, 1985; Flytzanis *et al.*, 1985).

Of the many species of sea urchins utilized in biological research today the most widely used in North America is probably the purple sea urchin, *Strongylocentrotus purpuratus*. Here we describe our methods for collecting, maintaining, and culturing this particular species at our facility at the California Institute of Technology's Kerckhoff Marine Laboratory. I will present current techniques used in raising these animals through their entire life cycle, including shedding gametes from gravid animals, culturing embryos, rearing larvae, and rearing juveniles. Since the larvae must feed in order to complete development, I include procedures for raising algae as a food source for the larvae. We then consider practical procedures for induction of the fragile postmetamorphic stages and raising juveniles to adults. The general protocol for rearing purple sea urchins will in most cases apply to other species as well.

Sea urchins have been used in biological research since the nineteenth century (for reviews see Harvey, 1956; Giudice, 1973). In recent years Hinegardner and his associates have simplified the once arduous task of raising sea urchins from egg to egg in the laboratory. Our procedures are modifications of the basic protocol that has emerged from their studies (Hinegardner, 1969; Hinegardner and Rocha Tuzzi, 1981), and with them a greater than 80% success rate may be obtained.

## II. Methods for Management of Adult Sea Urchins

*S. purpuratus* can be collected along the west coast of North America by scuba diving or by gathering animals at low tide in rocky areas where they abound. This particular species is usually found in intertidal and adjacent subtidal regions, but there are rare populations residing in deeper water below the normal thermoclines in habitats where there occurs a minimum of seasonal fluctuations in physical conditions (Leahy *et al.*, 1981). The advantage with collecting these "deep-water" sea urchins is that in this environment gravid animals are found at all times of the year, whereas the intertidal and near-subtidal sea urchins are fertile for only a sharply defined fraction of the year (see comparison of subtidal and deep-water animals in Leahy *et al.*, 1981).

A critical variable in maintaining adult sea urchins in the laboratory is the number of animals per unit volume of seawater and the amount of oxygen required for adequate maintenance. We find that in a running seawater system such as ours we can maintain approximately one "deep-water" sea urchin of average size per liter of water in a tank where the rate of seawater exchange is one tank volume per 1.5 hours. The initial adaptation period may last as long as 2 or 3 weeks. The animals and their medium should be watched very closely during this time. If the water becomes unusually cloudy it should be changed, and any sea urchins found to be dying or dead should be removed immediately.

Even though sea urchins may appear to be invincible with their tough, calcareous spines and test, they are very easily injured if mishandled. Whenever possible, the collector should quickly and lightly twist the animal from its substratum, before it has time to react and fasten itself more tightly to the substrate, in order to avoid loss of numerous tubefeet. This is probably the most common injury incurred when collecting and often can result in the animal's death, usually 2 or 3 weeks after its capture. Under certain conditions, such as heavy turbulence, that cause the animals to attach themselves very firmly, a blunt instrument such as an abalone knife must be used to dislodge them.

Care must also be exercised when packing and transporting the animals. Plastic ice chests or styrofoam containers can be used with a layer of newspaper on the bottom to further protect tubefeet and to soak up any excess water. Seawater serves as a safe and effective coolant when frozen in 1-liter polyethylene bottles, which sea urchin spines cannot puncture. The animals should be placed in a single layer between either seaweed (preferably kelp, e.g., *Macrocytis pyrifera*) or seawater-moistened newspaper. Animals that are spawning should be discarded. Several ice bottles per container are then layered on top to keep the specimens cold. During transit purple sea urchins should be kept at or slightly below the temperature of the water from which they were collected. If gravid animals are subjected to prolonged periods out of the water between collecting and packing, especially on a warm day, they will often spawn either during handling or while in transit. After packing, every effort should be made to get the animals to the aquarium quickly, with a minimum of rough handling. The animals will survive transit conditions for up to 24 hours if packed properly. Other species may require different packing procedures. For example, *Lytechinus pictus,* which is strictly a subtidal sea urchin, should never be packed "dry."

Shipping purple sea urchins and many other species over long distances to inland laboratories can be done successfully using these packing techniques. Before shipping it is wise to keep the animals in a holding tank for a few days to make sure they are adapting well and are healthy enough to make the journey. The travel time should be minimized at both ends, e.g., they should be packed just prior to shipping and met at their destination immediately upon arrival. Many species, including *S. purpuratus,* can be obtained through commercial sources as well. When unpacking do so carefully and be sure not to put a spawning animal in an aquarium with the others or every specimen in the tank may follow suit. It is wise to have several tanks ready just in case an unforeseen accident occurs.

By studying the behavior of sea urchins in the wild we have learned much about their habits and environmental needs. The animals we collect from below the thermocline spend most of their lives at 11–13°C. They are found in groups beneath large rocks away from direct sunlight where they trap and consume drifting blades of kelp. At Kerckhoff Marine Laboratory we have utilized our *in situ* observations to devise a seawater system which closely simulates natural conditions (see Leahy *et al.*, 1978, for a detailed structural description). Our tanks are kept at ~11°C with abundant aeration, an acceptable water exchange rate; because we collect the animals from under rocks, our aquaria are kept away from any direct lighting. We find that we can repeatedly obtain gametes from these animals over several years under these conditions. The animals are spawned every 4 months and 30–50% of the population in each group yield gametes at an average of 3–4 million eggs per spawning female. This is an advance of major usefulness as it removes one of the few disadvantages of these animals for research, the seasonality of gamete supply in all easily accessible (i.e., intertidal and subtidal) populations. Figure 1 demonstrates the *nonseasonal* yield of eggs in one set of laboratory animals over a period of several years. It shows that this is not simply a matter of storage *in vivo* of mature eggs, but rather of ongoing oogenesis that in the cultured population is not calendrically synchronized (Leahy *et al.*, 1981). Thus vitellogenic oocytes as well as mature eggs continue to be produced at all seasons. Except for a few stock specimens, the males are generally discarded. Our facility, which is near the ocean, enables us to operate an open aquarium system by pumping seawater into the building where it is filtered, chilled, and passed through the aquaria housing the sea urchins.

Our system consists of 15 large Plexiglas tanks, mounted in tiers on a steel A-frame structure. Each tank is capable of housing up to 200 average-size *S. purpuratus*. The tanks are 163 × 102 cm in area and 19 cm deep. Each is aerated separately with spargers controlled by individual valves attached to a compressed air line. The main cooling reservoir from which water flows into the whole system is also heavily aerated. When the tanks are fully loaded the demand for dissolved oxygen is high, and mortality may result from lack of aeration. This is true despite the large surface-to-volume ratio of the water in the tanks. The animals live on a perforated Plexiglas tray, under 15 cm of running, chilled seawater. Each tank is surrounded by a thick layer of styrofoam insulation, including the lids, and is equipped with individual drains. Water does not normally flow from one tank to another, a useful feature for isolating populations of sea urchins that may start to spawn or develop a contagious pathology. Each animal tray is seated within a second tray. Feces and other debris fall through the

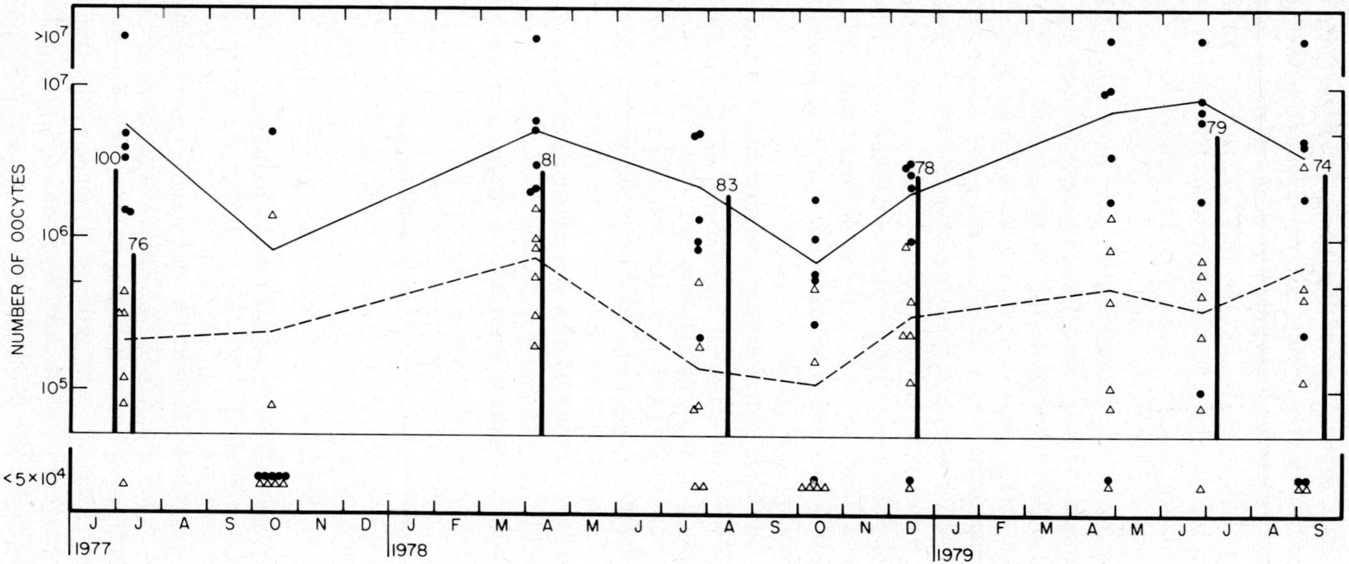


FIG. 1. Egg yield and ovarian occyte content of a typical experimental group held in the *S. purpuratus* maintenance system. Three-hundred and forty "deep-water" sea urchins collected in July 1979 were injected with KCl and 86% spawned. Seventy-six of the 137 spawning females yielded $\geq 10^6$ eggs each, and these animals constituted the experimental group. Individual egg (●) and late vitellogenic oocyte (△) yields were plotted periodically with solid lines and dashed lines connecting the average yields, respectively. Sample animals yielding $<5 \times 10^4$ are grouped below the bottom coordinate. The vertical bars indicate the average yield for those females which spawned when testing the group for fecundity. The number above each bar indicates the percentage of animals yielding gametes when tested. From Leahy *et al.* (1981).

perforations in the Plexiglas animal trays to the undertrays. These undertrays may be removed and cleaned when necessary. This arrangement permits transferring the Plexiglas trays containing the animals to clean undertrays, without removing the animals from the trays. We feel that this is a critical factor in the long-term survival of sea urchins in the maintenance system. The animals adhere tightly to the Plexiglas surface, and it is often difficult to remove them without damage to tubefeet.

The seawater supplied to the trays containing the animals is pumped from nearby Newport Bay through a series of sand filters into a storage tank and then into a reservoir where it is chilled and maintained at approximately 11°C. This water is then distributed to each of the 15 aquaria at a rate of 0.5 gallons per tank per minute. The water then drains into a sump tank through a storage tank for kelp, and finally back into the bay. If for one reason or another the seawater supply fails, i.e., either the incoming seawater pumps or the chiller mechanism malfunctions, recirculation pumps commence operation automatically. Recirculation is controlled by a pump and float valve in the sump tank. Auxiliary compressors are mounted for use in case of emergency, and these maintain the 11°C temperature in the recirculation mode of operation.

Sea urchins in the maintenance system are fed approximately one large kelp (*M. pyrifera*) blade per five animals three times a week. No other food source is required to maintain animals in a healthy condition for years. The kelp is collected weekly and is stored in a running seawater tank associated with the culture system. Individual sea urchin holding tanks are cleaned thoroughly with fresh water between uses. No detergents are used. Between major cleanings, feces and other debris are removed from the trays beneath the holding tanks by means of a suction hose twice a week.

In most laboratories where seawater is not readily available, a closed system must be used. Such a system should be made of nontoxic, nonmetallic materials such as acrylic or glass and equipped with a reliable filtration system. Acceptable salinity and pH levels should be maintained along with adequate aeration. A review of such procedures can be found in Hinegardner *et al.* (1981).

## III. Embryo Culture and Larval Growth

For most species of sea urchins, including *S. purpuratus,* spawning can be induced by injecting the animal with 0.5 *M* KCl. Before injecting, it is wise to rinse the animals thoroughly with filtered seawater to flush from their surfaces any microorganisms which may otherwise settle among the eggs and later contaminate cultures. An average-sized sea urchin (80 g wet wt) is given a 0.5-ml injection of KCl with a 23–26 gauge needle through the peristomial membrane and into the vicinity of the gonads. This amount will vary slightly with the size of the animal. Several injections and vigorous shaking are often required to induce spawning from all five gonopores. When only a few gametes are needed it is usually possible to stimulate only one or two gonads by strategically injecting the animal in one location with less than a full dose of KCl. This, of course, allows the sea urchin to be used again without waiting for a full recovery for regeneration of the gametes.

To collect eggs, female sea urchins are inverted over beakers of filtered seawater which are cooled in a bath of crushed ice. Purple sea urchins are capable of producing anywhere from a few to several million eggs per spawning. Huge animals of this species very rarely produce large amounts of eggs. Sperm is collected "dry" directly from the spawning animal by using a Pasteur pipet. The sperm should then be stored in a test tube on ice until needed. While shedding, the animals should be kept moist and must be kept out of water for no more than 30–40 minutes. When these simple precautions are followed mortality from spawning is negligible, and the animal can be spawned over and over again. In our experience this procedure may be repeated on given animals without noticeable harm for periods of at least several years.

If several females are needed a sample of the eggs of each should first be checked for fertilizability before the eggs are pooled. Occasionally sea urchins will shed immature or deformed oocytes, especially in the first few weeks of the new season. These gametes should be discarded. Even though it is very rare, hermaphrodites that shed fertilized eggs are also occasionally encountered. After spawning the eggs are filtered through 100 μm Nitex (Tetko, Inc., Elmsford, NY) mesh net which is affixed to the bottom of a small acrylic cylinder to exclude spines and other debris. *S. purpuratus* eggs are approximately 80 μm in diameter, and thus 100-μm netting is ideal for this purpose. The eggs are then allowed to settle, and the supernatant seawater is aspirated off. They are washed in this manner three times using Millipore-filtered seawater. When fully settled a volume of 1 ml contains approximately 1,000,000 eggs.

To fertilize the washed eggs it is best to dilute one drop of dry sperm with 5 ml of seawater and to add a fraction of this solution to the eggs. Two milliliters of this sperm suspension is usually sufficient to fertilize $4 \times 10^7$ eggs in 4 liters of seawater. Care should be taken not to have more than 10–20 sperm around each egg. Polyspermy can result from too many sperm. The fertilization membrane will appear within 90 seconds at 16°C and the embryos should develop synchronously. By 24 hours after fertilization the blastulae hatch and become ciliated free-swimming organisms. After gastrulation the larvae develop elongated arms supported by calcareous skeletal rods. The arms are lined with ciliated bands and are used for motility and to gather food. The embryos are cultured in 0.45-μm Millipore-filtered seawater and until the larvae begin to feed 40 mg/liter of the antibiotic, gentamicin sulfate, is added to the medium. We grow embryos in culture chambers which are partially submerged in a water bath to maintain a constant 16°C. These chambers can be made of practically any type of glass or plastic cylindrical containers. We use 2- or 4-liter Nalgene plastic jars depending on the number of embryos we are culturing. The embryos are gently stirred by a plastic paddle attached to a Plexiglas rod operated by a 20-rpm electric clock motor which is mounted on top of the culture chamber. If a large number of embryos is needed for an experiment, i.e., over $10^3$ per ml, aeration of the culture and use of a 60-rpm motor are indicated.

Another method of maintaining the embryos and larvae in suspension is to stir them gently in beakers or flasks on an orbital shaker in a culture room at 16°C. This technique is especially useful when culturing small numbers of embryos. For example, 10 larvae can be reared through metamorphosis in 100 ml of seawater in a 250-ml flask. Parafilm with several air holes or sponge stoppers are used to seal these containers to reduce evaporation. When using such a shaker the speed should be no higher than is necessary to keep the embryos or larvae in suspension.

When raising sea urchins through to the adult stage only a relatively small number of larvae can be dealt with effectively. A feasible schedule for an experienced technician handling all aspects of the operation would be one of starting a new culture of approximately 200 *S. purpuratus* every month. That is, after the first year the laboratory would have to be equipped for maintaining 200 embryos, 200 swimming larvae, 200 metamorphosing larvae, 200 juveniles, plus 1600 young adults of various ages. The water in the larval culture should be changed once a week or more often if it becomes dirty or contaminated. Aside from the stirring no additional aeration is needed as long as the water is changed regularly. Transferring several hundred or more larvae one by one to a clean culture chamber can be tedious and time consuming. One way to shorten this procedure is to collect and transfer the larvae using mesh netting. We construct such a filtering device by cutting away the bottom of a 400-ml plastic beaker and attaching in its place 100-μm Nitex mesh net. Secure the mesh with rubberbands covered over tightly by parafilm. This filter is then placed inside a slightly larger beaker which is half full of seawater. While holding the filter off the bottom of the outer container slowly pour the culture containing larvae into this inner chamber where they are trapped. This procedure is done over a sink to catch the overflowing waste water. Aggregates of dead algal cells which foul the medium are not removed by this method. These aggregates must be removed from the culture with a pipet, being careful to avoid any larvae. Another method of changing the larvae to fresh seawater is by use of a siphon. A siphon can be made by cutting out the bottom of a 50-ml plastic test tube and attaching the mesh net as described above. Tubing with an inside diameter of $\frac{1}{8}$ in. and a wall thickness of $\frac{1}{16}$ in. is threaded through a one-hole stopper at the end opposite the mesh. The tubing should end approximately 1 cm from the mesh, and a clamp is used to control the water flow. When the water level is near the bottom of the container the larvae are then transferred to a clean culture chamber. If culturing a small number of larvae on an orbital shaker it is usually best to use a pipet to transfer the animals to a beaker or flask containing clean water.

Until larvae reach the pluteus stage they can be grown in fairly dense populations, up to $10^4$/ml with sufficient agitation and aeration, but as they begin to feed and grow these numbers are greatly reduced depending on the water exchange rate. We begin to feed *S. purpuratus* on the fifth day after fertilization (at 16°C). At this time we reduce the culture to three larvae/ml. After 14 days the number is reduced to one larva/ml and at 21 days one larva/3 ml. By the time the advanced larvae are ready to metamorphose, usually in 6–8 weeks, the density has been reduced to one larva/10 ml.

The larvae are fed a diet of *Rhodamonas lens,* a single-celled golden brown algae which is cultivated in the laboratory. Larvae are fed daily and given as many cells as they can consume in a 24-hour period. During the first 2 weeks we feed the larvae 3000 *Rhodamonas*/ml of seawater, and during the last 2 weeks 6000 cells/ml. Larvae should be checked with a dissecting microscope for the presence of algae in the gut, and a sample can be taken from the medium for a quick count of *Rhodamonas* in culture. A small flashlight will reveal a faint red cloudiness when the algal concentration is optimal. Overfeeding will foul the medium. Underfeeding will cause slow development and possible degeneration of the larvae.

The algae are cultivated in Provasoli D medium with biotin added to the formula (D. Chapman, personal communication). The medium consists of an artificial seawater base supplemented with vitamins and trace elements. The following is the recipe for 1 liter of this water-enrichment medium:

| | |
|---|---|
| NaCl | 18 g |
| KCl | 0.6 g |
| $NaNO_3$ | 0.5 g |
| $MgSO_4 \cdot 7H_2O$ | 5.0 g |
| Tris base | 1.0 g |
| $CaCl_2 \cdot 2H_2O$ | 277 mg |
| $K_2HPO_4$ | 300 mg |
| $FeCl_3 \cdot 6H_2O$ | 0.5 mg |
| P-II metal mixture | 30 ml |
| Thiamine | 1.0 mg |
| Vitamin $B_{12}$ | 3.0 μg |
| Biotin | 0.04 mg |
| $dH_2O$ | 1000 ml |

P-II metal mixture

| | |
|---|---|
| $Na_2$ EDTA | 1.0 g |
| $H_3BO_3$ | 1.14 g |
| $FeCl_3 \cdot 6H_2O$ | 48 mg |
| $MnCl_2 \cdot 4H_2O$ | 144 mg |
| $ZnCl_2$ | 10.4 mg |
| $CoCl_2 \cdot 6H_2O$ | 4.3 mg |
| $dH_2O$ | 1000 ml |

The chemicals are added to 950 ml of distilled water in sequence as shown. After adding the Tris base to the solution the pH is adjusted to 7.6–7.8 with 1 *N*

HCl. To prevent mold, the P-II metal mixture is kept refrigerated. A mix of the vitamins can be stored frozen in convenient aliquots.

Before inoculating, the enrichment medium should be autoclaved in a container in which the algae will be cultured. We use 100–125 ml of media in 250-ml glass flasks with sponge stoppers. Ten to 20 ml of stock algae is added to these flasks after they cool to the proper temperature. The vitamin mix is added at this time. The cultures grow well between 16–20°C in front of fluorescent lights. Within 4–6 weeks the *Rhodamonas lens* will grow to a dense concentration and appear as a dark reddish brown color. New cultures can be inoculated using this culture as the new stock. Aeration is not necessary as long as the *Rhodamonas* culture is stirred by hand vigorously each day.

The *Rhodamonas* culture medium is toxic to the sea urchin larvae and must be eliminated before feeding. The algae and medium are separated by centrifuging at 200 *g* for 5 minutes. The medium is poured off and the algae resuspended in 50 ml of Millipore-filtered seawater, an amount convenient for counting. We count the cells by adding 0.1 ml of this new suspension to 50 ml of filtered seawater and count a sample of this using a Fuchs–Rosenthal counting chamber. The number of cells needed for a given culture can then be calculated.

*Rhodamonas* can be purchased through the Department of Botany, University of Texas at Austin.

## IV. Metamorphosis and Maintenance of Juveniles

Metamorphosis can be induced in large cultures simply by stopping the paddles and allowing the advanced larvae to settle to the bottom of their culture chamber. With proper care, metamorphosis may occur in 6–8 weeks, when the larvae have achieved maximum growth and development. Because some larvae seem to be more aggressive feeders than others, developmental rate varies slightly among them. Twenty-four hours after settling, the animals which have not yet attached to the bottom of the container are poured into a new culture chamber where they are once again stirred and fed for a few days until the settling process is begun again. The seawater is, of course, replaced in the first chamber. This is repeated until all the larvae have started metamorphosis. The microbial film that develops on the bottom of the containers may be necessary to induce metamorphosis.

Development is usually more synchronous when culturing a small number of animals which can be handled on a one-to-one basis throughout the process. Metamorphosis usually begins within 1 hour after allowing the advanced larvae to settle. The animals will produce five tubefeet, a series of fragile spines, and three pedicellariae, and then remain attached to the bottom for 5–6 days when the metamorphic process is complete. They do not feed during this time.

At the end of the metamorphic period the juvenile sea urchins are transferred to plastic Petri dishes in which diatoms have been cultured. These dishcs house the juveniles for about 10 days or until they are able to graze upon algae-covered rocks or other natural food sources. The diatom we use is of the genus *Nitschia,* a surface-dweller which multiplies rapidly and sticks to the bottom of the plastic dishes. These diatoms are grown in a modification of the water enrichment medium of Guillard and Ryther (1962) and are easily cultured in the laboratory. One liter of this enrichment medium consists of the following ingredients:

| | |
|---|---|
| 3.0 ml | 3% sodium nitrate ($Na_2NO_2$) |
| 1.0 ml | 1% sodium phosphate ($NaH_2PO_4 \cdot 8H_2O$) |
| 1.5 ml | 3% sodium metasilicate ($Na_2MoO_4 \cdot 9H_2O$) |
| 1.0 ml | trace metal solution |
| 2.0 ml | vitamin mixture |
| 1000 ml | Millipore-filtered seawater |

To make 1 liter of the trace metal solution add the following compounds to 1000 ml of distilled water:

| | |
|---|---|
| 6.5 g | $FeCl_3 \cdot 6H_2O$ |
| 8.25 g | $Na_2$ EDTA (disodium ethylenedinitrilotetraacetate) |
| 0.044 g | zinc sulfate ($ZnSO_4 \cdot 7H_2O$) |
| 0.020 g | cobalt chloride ($CoCl_2 \cdot 6H_2O$) |
| 0.360 g | manganese chloride ($MnCl_2 \cdot 4H_2O$) |
| 0.013 g | sodium molybdate ($Na_2MoO_4 \cdot 2H_2O$) |

Because mold can grow in this solution it should be kept refrigerated. The vitamin mixture is made up of the following:

| | |
|---|---|
| 1.0 mg | biotin |
| 0.1 mg | vitamin $B_{12}$ |
| 20.0 mg | thiamin HCl |
| 100 ml | distilled $H_2O$ |

This solution can be stored frozen in small plastic vials and used as needed. The medium is autoclaved before use to maintain sterile conditions for both stock cultures and for growing the diatoms in the Petri dishes. *Nitschia* grows well at 15–18°C in front of fluorescent lights.

The medium is poured 10 mm deep in a 140 × 23 mm plastic Petri dish, which is inoculated with enough diatoms to create a rich film on the bottom of the dish in 3 or 4 days. Approximately $1.5 \times 10^7$ cells are used when inoculating 125 ml of medium per dish. New plates can be inoculated using the old ones as a stock. Since the medium is toxic to the sea urchins it is poured off and replaced with filtered seawater before the juveniles are transferred to it. To move the animals from the culture chamber to the diatom dishes they are given a gentle nudge with a rubber policeman to unfasten their tubefeet and are picked up with a pipet. This must be done very quickly so that the animals will not have time to adhere to the inside of the pipet during transfer. Approximately 20 newly metamorphosed sea urchins can be reared in each dish. Every other day half the seawater is siphoned from the dish and fresh seawater is then added. When the diatoms are nearly all consumed, usually within 5–6 days, the young sea urchins are transferred to new dishes to graze once again until the diatoms are consumed. The juvenile sea urchins are now strong enough to feed upon a variety of microscopic seaweeds commonly found on algae-covered rocks near our facility. We collect small (5–10 cm in diameter), flat rocks in areas where *S. purpuratus* are plentiful. Ten to 20 juveniles, depending on the size of the rock, are transferred to this new environment in an aquarium maintained at 16°C and equipped with a reliable filtration system. Fluorescent lighting is used to keep the algae healthy. A month later, while they are still grazing upon the rocks, we begin feeding these juveniles small pieces of *Ulva* or sea lettuce. *Ulva* has a leaflike thallus that is two cells thick, and it is easy for the young sea urchins to consume it. At about 3 months after metamorphosis bits of kelp are introduced into their diet.

*S. purpuratus* will grow to approximately 22 mm in diameter during their first year and will begin to manufacture gametes as early as 11 months after fertilization. However, the males tend to mature sexually ahead of the females. At 3 years they have attained a diameter of 45–55 mm and can be considered adults. We have found that 2- to 3-year old laboratory-raised sea urchins provide copious eggs that develop perfectly, and we anticipate increasing reliance on wholly cultured animals as the source of gametes for research.

## ACKNOWLEDGMENTS

I thank Drs. Eric H. Davidson and Andrew Cameron for critically reviewing the manuscript. This work was supported by NIH Grant HD-05753.

## REFERENCES

Angerer, L. M., and Angerer, R. C. (1981). *Nucleic Acids Res.* **9,** 2819–2840.

Davidson, E. H., Hough-Evans, B. R., and Britten, R. J. (1982). *Science* **217,** 17–26.

Flytzanis, C. N., McMahon, A. P., Hough-Evans, B. R., Katula, K. S., Britten, R. J., and Davidson, E. H. (1985). *Dev. Biol.* **108,** 431–442.

Giudice, G. (1973). "Developmental Biology of the Sea Urchin Embryo." Academic Press, New York.

Guillard, R. R. L., and Ryther, J. H. (1962). *Can. J. Microbiol.* **8,** 229–239.

Harvey, E. B. (1956). "The American Arbacia and Other Sea Urchins." Princeton Univ. Press, Princeton, New Jersey.

Hinegardner, R. T. (1969). *Biol. Bull. (Woods Hole, Mass.)* **137,** 465–475.

Hinegardner, R. T., and Rocha Tuzzi, M. M. (1981). *In* "Laboratory Animal Management. Marine Invertebrates," pp. 291–302. National Academy Press, Washington, D.C.

Hinegardner, R. T., Atz, J. W., Fay, R. C., Fingerman, M., Josephson, R. K., Meinkoth, N. A., Miller, J. W., and Rice, M. E. (1981). *In* "Laboratory Animal Management. Marine Invertebrates," pp. 13–36. National Academy Press, Washington, D.C.

Leahy, P. S., Tutschulte, T. C., Britten, R. J., and Davidson, E. H. (1978). *J. Exp. Zool.* **204,** 369–380.

Leahy, P. S., Hough-Evans, B. R., Britten, R. J., and Davidson, E. H. (1981). *J. Exp. Zool.* **215,** 7–22.

McClay, D. R., and Fink, R. D. (1982). *Dev. Biol.* **92,** 285–293.

McMahon, A. P., Flytzanis, C. N., Hough-Evans, B. R., Katula, K. S., Britten, R. J., and Davidson, E. H. (1985). *Dev. Biol.* **108,** 420–430.

# Chapter 2

# *Handling, Labeling, and Fractionating Sea Urchin Spermatozoa*

VICTOR D. VACQUIER

*Marine Biology Research Division*
*Scripps Institution of Oceanography*
*University of California, San Diego*
*La Jolla, California*



## I. Introduction

Sea urchin spermatozoa are a favorite material for studying the mechanism of flagellar motility. This work has focused on the mechanism of microtubule sliding and the regulation of the dynein ATPase complex (Brokaw, this volume; Bell and Gibbons, 1983). During the past 5 years new avenues of research have opened regarding the role of the sperm plasma membrane in the initiation of motility, respiration, and induction of the acrosome reaction. These profound physiological "activations" occur when sperm are released into seawater and also when they contact components of the extracellular matrix of the egg known collectively as the egg jelly layer (Garbers *et al.*, 1983, 1985; SeGall and Lennarz, 1979, 1981). The fluxes of $Ca^{2+}$, $Na^{+}$, $H^{+}$, and $K^{+}$ across the sperm plasma membrane underlie these activations (Bibring *et al.*, 1984; Christen *et al.*, 1982, 1983; Johnson *et al.*, 1983; Lee *et al.*, 1983; Schackmann and Shapiro, 1981; Schackmann *et al.*, 1978, 1981, 1984; Shapiro, 1984). The movement of ions across the sperm membrane is currently under intense study, the ultimate aim being the isolation and reconstitution of the ion channels in artificial phospholipid bilayers.

The interaction of egg jelly with the sperm plasma membrane is an excellent example in which the extracellular matrix of one cell (the egg) interacts with specific plasma membrane proteins of another cell (the sperm), resulting in transmembrane signal transduction (ion channel activity). The consequences are both biochemical (changes in cyclic nucleotides metabolism and respiration, reviewed by Garbers *et al.*, 1985) and morphological (the acrosome reaction, Summers *et al.*, 1975). Sea urchin sperm are ideal for studies of the mechanism of signal transduction across the plasma membrane. They can be obtained as pure cells in quantity with ease and at low cost. They have an extraordinarily large amount of membrane surface area in relation to their volume. They do not synthesize protein and possess only four major plasma membrane proteins accessible to $^{125}I$ labeling *in vivo* (Cross, 1983; Lopo and Vacquier, 1980; Podell *et al.*, 1984). Ion fluxes across the sperm membrane are easily measured (Schackmann, this volume) and, when treated with egg jelly, the cells undergo the acrosome reaction with great synchrony in a time span of seconds. These most highly differentiated cells are specialized for two major functions: motility and exocytosis. It is the exocytosis of the acrosomal granule at the anterior apex of the cell (Fig. 1) that exposes the sperm membrane destined to fuse with the membrane of the egg (Summers *et al.*, 1975).

Methods for handling and fractionating sea urchin sperm, and also a wealth of useful information about adult sea urchins, gametes, and embryos have been published (Giudice, 1973; Harvey, 1956; Hinegardner, 1967, 1975; Osanai, 1975; Stearns, 1974; Tyler and Tyler, 1966). Of these works, Tyler and Tyler (1966) contains the most complete description of the experimental handling of sea urchin sperm. Most research on sea urchin sperm in North America utilizes three species, *Arbacia punctulata, Lytechinus pictus,* and *Strongylocentrotus purpuratus.* Additional species are listed in Giudice (1973), Harvey (1956), and Hinegardner (1975). This article describes methods we employ in our research on the biochemistry of sperm–egg interaction.

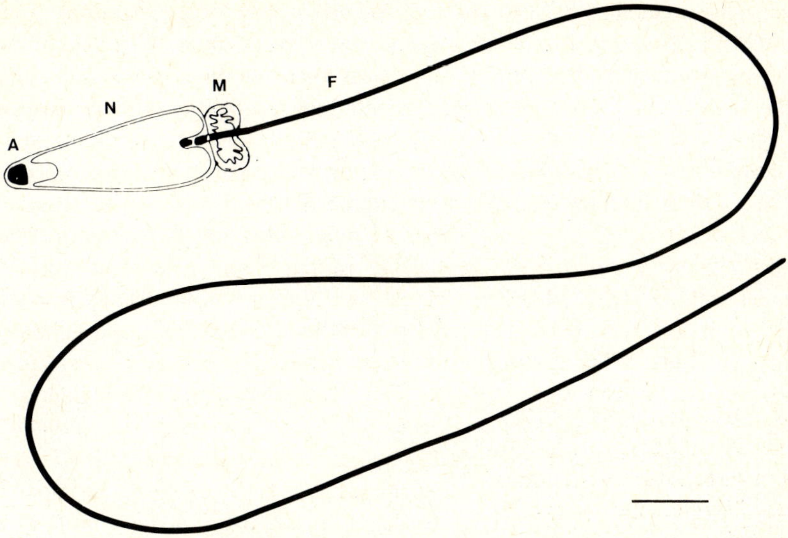


FIG. 1. Drawing of a generalized sea urchin spermatozoan with dimensions taken from Tyler and Tyler (1966). The head consists of an anterior phase-contrast dense acrosome granule (A) containing the adhesive protein bindin, the nucleus (N) which has both an anterior fossa (containing profilamentous actin) and a posterior fossa (containing a pair of centrioles), and a midpiece (M) containing a single giant mitochondrion. The tail, or flagellum (F), is 50 × 0.1 μm consisting of an axonemal complex covered by a plasma membrane. The entire cell is covered by a continuous plasma membrane. The bar is 1 μm.

## II. Seasonality, Quantity, and Methods of Collection

Gametes of *S. purpuratus* are most readily available during winter. In northern California the reproductive season for intertidal specimens is from the first week of December until the second week of April. Animals collected from mid-April to December are devoid of gametes. In southern California the season for intertidal *S. purpuratus* can extend from December to mid-June. Animals collected below the thermocline by scuba divers have gametes throughout the year (Leahy *et al.*, 1981). The summer months are optimal for *A. punctulata* and *L. pictus* gametes, with the peak for the latter in San Diego being in August when the ocean temperature reaches 23°C. During this period *L. pictus* will spontaneously spawn in aquaria open to the sea, but if kept well fed will attain full gravidity again in 2 to 3 weeks.

Of these three species, sperm of *S. purpuratus* are obtained in greatest quantity, large males (8 cm test diameter) yielding 10 ml of undiluted semen ("dry sperm") at a fairly consistent concentration of $4 \times 10^{10}$ spermatozoa/ml. *L. pictus* and *A. punctulata* males are much smaller, yielding a maximum of 1 ml of semen/male at about $1 \times 10^4$ cells/ml. The advantages of *S. purpuratus* are the quantity of cells and their relative toughness to physical abuse. The disadvantage of *S. purpuratus* is that we find it impractical to spawn the males in a manner in which they survive so that they can be cultured for repeated spawning. However, if one has sufficient quantities of 11°C seawater and can wait several months for the completion of gametogenesis, *S. purpuratus* adults can be repeatedly spawned and maintained in continuous culture (Leahy *et al.*, 1978). Regarding tolerance of the cells to physical abuse, *S. purpuratus* and *A. punctulata* sperm can be repeatedly pelleted at 1500 *g* (15 minutes) and resuspended in fresh seawater without appreciable lysis. However, *L. pictus* sperm usually lyse when similarly treated, lysis being determined by release of the DNA and gelation of the tube contents. When a 20-μl drop of a fresh suspension of *L. pictus* is placed on a slide and a coverslip is dropped on, virtually all the cells undergo the acrosome reaction. This does not occur with *S. purpuratus* or *A. punctulata* sperm. Although the quantity of sperm obtained from *A. punctulata* and *L. pictus* is small, their great advantage is that the adults can be cultured and spawned repeatedly at monthly intervals. We routinely maintain 100 *A. punctulata* or 200 *L. pictus* adults in 150 gallon (20°C) closed Instant Ocean aquaria (Aquarium Systems, Inc., Eastlake, OH). The aquaria are subdivided into separate compartments with perforated plastic partitions, each compartment containing animals spawned during 1 week. The adults are fed on brown kelp, especially *Macrocystis pyrifera,* care being taken not to overfeed which may pollute the seawater. Urchin feces and plant debris are siphoned from the bottom each week. The culturing of *L. pictus* in closed systems has been described previously (Hinegardner, 1975). Both *A. punctulata* and *L. pictus* held in aquaria open to the sea and fed *M. pyrifera* maintain gametes throughout the year. To spawn both *L. pictus* and *A. punctulata* so that they survive, we inject not more than 0.3 ml of 0.5 *M* KCl with a 25-gauge syringe needle inserted just beneath the peristomal membrane. The animals are placed inverted on a beaker filled with seawater and their surfaces kept wet using a squirt bottle (''wet'' gamete collection). Alternatively, the semen is pipetted directly from the gonopores and placed undiluted on ice (''dry'' collection). For an adult to survive forced spawning the outer surface must never become dry. After spawning the males are held for 1 hour in a small continuously flowing aquarium to allow completion of spawning and washing away of all seminal material. *S. purpuratus* adults are spawned by injection of 2 to 10 ml KCl.

Depending on what is to be done with the sperm cells these are the two ways to collect the semen. ''Wet'' sperm collection is simple and appropriate where sperm is used for isolation of the plasma membrane or other cellular components. However, this is a poor method for collecting sperm for ion flux measurements or acrosome reaction induction because dilution into seawater induces the $Na^+/H^+$ exchange that elevates the internal pH and activates respiration and motility (Bibring *et al.*, 1984; Christen *et al.*, 1982, 1983; Lee, 1984a; Lee *et al.*, 1983; Schackmann *et al.*, 1981, 1984). "Wet" collection also increases the amount of protease activity associated with the sperm (Resing *et al.*, 1985). For physiological studies of intact sperm, it is best to collect semen "dry," placing it in capped 1.5-ml microfuge tubes packed in ice. Stored in this manner, sperm will remain capable of fertilization for several days. However, even when stored as described the percentage of sperm capable of undergoing the acrosome reaction decreases with time after spawning. The cryopreservation of sea urchin sperm has been reported (Asahina and Takahashi, 1979).

## III. Artificial Seawaters

The formulations of three artificial seawaters routinely used for sea urchin gametes appear in Table I. Similar recipes can also be found in Cavanaugh (1964), Hinegardner (1967), and Tyler and Tyler (1966). High $K^+$ seawater is prepared by substituting $K^+$ and $Na^+$. All natural seawater is passed through a 0.45-μm Millipore filter before use. Additional buffering capacity can be obtained by addition of extra $NaHCO_3$ or 5 to 10 m*M* Tris [tris(hydroxymethyl)aminomethane] or HEPES (*N*-2-hydroxyethyl-piperazine-*N*′-2-ethanesulfonic acid), adjusted to pH 7.8–8.0. However, some workers warn that Tris can be inhibitory to the acrosome reaction (M. Gould and S. Podell, personal communications), but neither organic buffer inhibits the respiratory increase occurring when sperm are treated with the ethanol-soluble fraction of egg jelly (Vacquier, unpublished).

## IV. Preparing a Pure Sperm Suspension

### A. Composition of Semen and Removal of Pigmented Phagocytes

Semen can be separated into cells and seminal plasma by centrifugation at 40,000 *g*. The ionic composition of seminal plasma is given in Table II. In *S. purpuratus,* semen is roughly 70% cells and 30% plasma. A high $CO_2$ content in the seminal plasma is thought to depress sperm pH, keeping motility and respiration turned off (Christen *et al.*, 1982, 1983; Johnson *et al.*, 1983; Lee *et al.*, 1983). Although the great majority of cells in semen are mature spermatozoa, a resident population of highly phagocytic, darkly pigmented cells is always present. These cells are most probably responsible for the maintenance of the gonad as an axenic environment (Manahan *et al.*, 1983). These pigmented cells, al-

TABLE I

FORMULAE FOR ARTIFICIAL SEAWATERS (MILLIMOLES PER LITER)[a]

| | NaCl | KCl | $CaCl_2$ | $MgCl_2$ | $MgSO_4$ | $KHCO_3$ or $NaHCO_3$ | Choline chloride |
|---|---|---|---|---|---|---|---|
| Artificial seawater | 454 | 9.7 | 9.6 | 26.7 | 28.9 | 2.5 | 0.0 |
| Calcium-free seawater | 469 | 9.7 | 0.0 | 26.7 | 28.9 | 2.5 | 0.0 |
| Sodium-free seawater | 0.0 | 7.2 | 9.6 | 26.7 | 28.9 | 2.5 | 454 |

[a] For additional buffering capacity add 5–10 m*M* $NaHCO_3$, HEPES, or Tris and adjust to pH 7.8.

though representing only 1/100 or 1/1000 of the cellular volume of semen, exhibit extremely high proteoesterase activity and must be separated from the sperm before isolation of sperm components (Resing *et al.*, 1985).

This small population of contaminating cells is easily removed by suspension of dry semen in at least 50 vol of seawater and centrifugation in a swinging bucket rotor at 200 *g* for 15 minutes. The pigmented cells sediment, forming a tightly cohesive pellet in a 50-ml conical tube. The sperm suspension is decanted into a clean tube and the centrifugation is repeated. To wash the sperm from the seminal plasma, the supernatant sperm suspension is decanted into a 50-ml tube which is centrifuged for 15 minutes at 1500 *g* in a swinging bucket rotor. The supernatant is removed by aspiration and the cells are resuspended in chilled seawater using a Teflon spatula. This method works well for *S. purpuratus* and *A. punctulata* but not for *L. pictus* sperm because of their greater fragility.

TABLE II

COMPOSITION OF SEA URCHIN SEMINAL PLASMA (MILLIMOLAR)

| | *L. pictus*[a] | | |
|---|---|---|---|
| | Spawned by injecting with air | Spawned by KCl injection | *S. purpuratus*[b] (Spawned by KCl injection) |
| $Na^+$ | 440 | 520 | 373 |
| $K^+$ | 19 | 23 | 52 |
| $Ca^{2+}$ | 9.7 | 9.7 | 5.2 |
| $Mg^{2+}$ | 48 | 46 | 42 |
| $SO_4^{2-}$ | — | — | 0.24 |
| $Zn^{?+}$ | | — | 0.52 |
| $PO_4^{3-}$ | — | — | 0.25 |

[a] From Johnson *et al.* (1983).
[b] From Gache and Vacquier (unpublished).

### B. Method for Washing Sperm

A simple, effective technique has been devised to wash sperm of all three species rapidly and completely without pelleting the cells. A three-step Ficoll–seawater gradient is layered in a 50-ml soft plastic centrifuge tube (28.6 × 103.7 mm): 5 ml of 35% Ficoll (Sigma Chemical Co., St. Louis, MO) in seawater on the bottom, 5 ml 20% Ficoll–seawater as the middle layer, and 25 ml 10% Ficoll–seawater at the top. Up to 10 ml of sperm suspension is layered on top and (for *S. purpuratus*) the tube is centrifuged for 18 minutes at 5000 rpm (4100 *g*) in a Sorvall HB-4 rotor (for *L. pictus* and *A. punctulata* sperm longer times and higher forces are necessary, or the 25-ml step is made with 7% Ficoll). The cells are washed as they sediment through the 10% step and band in the 20% step. They are perfectly viable as judged by fertilizability when diluted into fresh seawater. Results of such a wash, with radiolabel added to the sperm, are shown in Fig. 2.

## V. The Effect of Storage on Respiration and Fertilizability

Concern about the possible deleterious effects of long-duration storage at various temperatures led us to investigate these parameters on dry *S. purpuratus* sperm sealed in microfuge tubes. The data (Fig. 3) show that storage for 23 hours at 0°C has little deleterious effect on sperm respiration when the semen is later diluted into seawater at 20°C, whereas storage at 17 and 24°C is definitely bad for preservation of the respiratory activation of the cells.

A loss in fertilizing capacity of *S. purpuratus* sperm roughly parallels the loss in respiratory ability. The fertilizing capacity of sperm stored at 17°C for 5.5 hours and at 0°C for 5.9 hours was compared by fertilizing *S. purpuratus* eggs with serial dilutions of sperm. Attaining 50% fertilization (as judged by fertilization envelope or hyaline layer formation of eggs) required three times more sperm stored at 17°C than sperm stored at 0°C. Heavy metal chelators, amino acids, and peptides also prolong life of sperm (Johnson and Epel, 1983; Tyler and Tyler, 1966).

## VI. Determining Sperm Concentration and Quantitating Fertilizability

A simple method for determining sperm concentration is to measure the turbidity of a sperm suspension with a spectrophotometer and relating the value to a

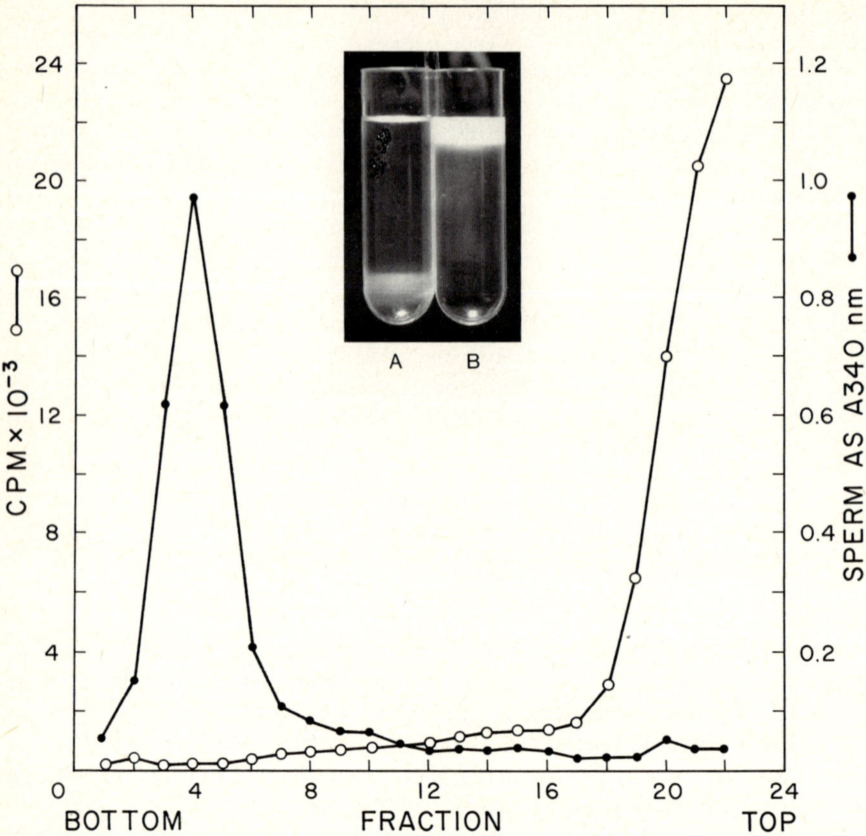


FIG. 2. One-step procedure for washing *S. purpuratus* sperm done without pelleting the cells. A three-layer Ficoll–seawater step gradient (35, 20 and 10%) is prepared in a 50-ml tube. Five milliliters of a fresh sperm suspension containing [$^{125}$I]methyltyrosine cyclic AMP (as a nonbinding label) was layered on top of the 10% step, and the tube was centrifuged for 20 minutes in a swinging bucket rotor at 4100 *g*. The photograph shows a tube before (B) and after (A) centrifugation. After centrifugation the tube was punctured and 2-ml fractions were collected from the bottom. The absorbance ($A_{340}$) and radioactivity in the fractions were determined. The cells were completely separated from the extracellular label.

standard curve constructed by hemocytometer counts. We read turbidity at 340 nm where sensitivity is greater than at longer wavelengths (Vacquier and Payne, 1973). Since the turbidity is a measure of light scattering and absorption a different standard curve must be constructed for each spectrophotometer. Suspensions of motile sperm are easily read. However, because of their high axial ratio, immotile sperm with straightened flagella tend to orient, producing large variations in $A_{340}$ nm. This problem can be overcome by rapid mixing in the cuvette, or by averaging the high and low readings (Vacquier and Payne, 1973).

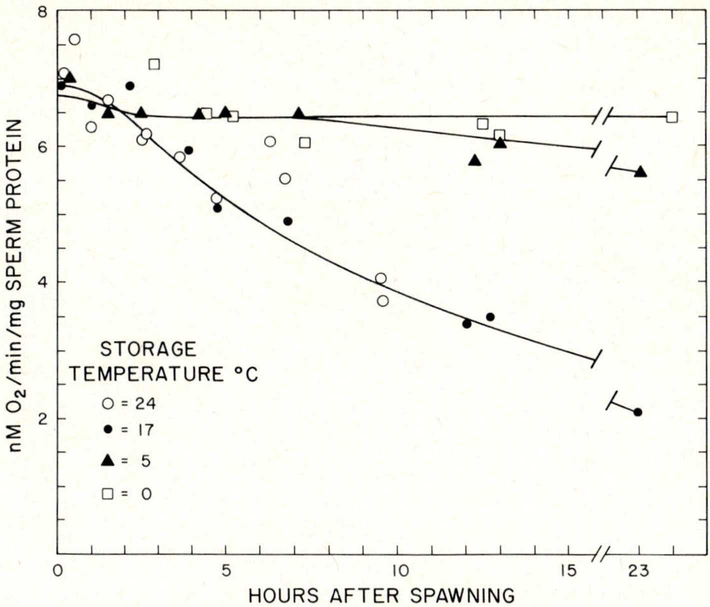


FIG. 3. Effect of storage time and temperature on dry sperm. *S. purpuratus* semen was collected "dry" and stored at 0, 5, 17, and 24°C in sealed microfuge tubes. For each assay, 50 μl of semen was added to 2.4 ml of stirring seawater buffered with 10 m*M* HEPES, pH 7.8, at 20°C in a 3-ml chamber containing an oxygen electrode. Data were recorded for 7 minutes, and data shown are the average rates for $O_2$ utilization between minutes 1 and 6 of each run. Storage at 0°C for 23 hours has no effect on the respiratory potential of sperm.

Three simple methods utilizing eggs are available for studying the effects of enhancers or inhibitors of gamete interaction on fertilization. The first method involves determination of the percentage fertilization of eggs exposed to various sperm concentrations. Sperm or egg suspensions are treated with the agent under investigation at a reasonable temperature (not greater than 20°C). The agent is either washed out or retained with the cells. The sperm are diluted in a series of tubes, each dilution being a 50% decrease in cell concentration. If plastic pipet tips are used, they should be trimmed to an opening of at least 1 mm diameter. This is essential for both sperm and eggs since the shear of passing through narrow tips is injurious to the cells. A measured volume of egg suspension is added to each tube and after gentle mixing the tubes are incubated for the desired time. An equal volume of 2% formaldehyde or glutaraldehyde in seawater is added and a minimum of 200 eggs per tube is scored microscopically; fertilized eggs are determined by the presence of a fertilization envelope, a hyaline layer, or cleavage. A control of untreated gametes is run in parallel. The concentration of sperm needed to obtain 50% fertilization is determined from a plot of the data. After determining the sperm concentration that yields a value between 100 and

0%, a more accurate determination can be made by repeating the experiment using 10–25% reductions in sperm concentration per dilution.

A second method is to quantitate the binding of sperm to eggs, as previously described in detail (Vacquier and Payne, 1973; Vacquier *et al.*, 1979). The only additional improvement to this method to report here is the inclusion of 10 m*M* HEPES (pH 7.8–8.0) to all media (see below). The "Pronase" method for quantitating the number of sperm bound per egg indicates that a finite number of sperm receptor sites might exist on the surface of the egg vitelline layer (Vacquier and Payne, 1973).

The third method is to quantitate the number of sperm pronuclei in the egg cytoplasm. After treating the gametes with the agent under study, sperm and eggs are mixed together and 10 minutes later an equal volume 1% formaldehyde in seawater is added. After 30 minutes the eggs are washed from the fixed sperm by gentle hand centrifugation and then treated 10 minutes with 1 mg protease/ml seawater. The protease digestion removes sperm fixed to the egg surface. After washing in fresh seawater, the eggs are fixed for at least 30 minutes (overnight is better) in fresh Carnoy's fixative (3 parts ethanol:1 part acetic acid), with two changes. The pellet of eggs is then resuspended in 60% acetic acid and observed by phase-contrast microscopy. Sperm pronuclei appear as phase-dense spheres in the egg cytoplasm (Fig. 4). The living eggs should be fixed before pronuclear decondensation, the timing of which will vary with species and temperature. A sample of eggs not exposed to sperm must always be prepared, since eggs occasionally contain phase-dense inclusions resembling sperm pronuclei. Drops of eggs in acetic acid are placed on a slide. Broken pieces of coverslips are added which act as spacers, a clean coverslip is added, and the eggs are viewed at 400× magnification. By carefully focusing through the cytoplasm the number of sperm pronuclei present in each egg is determined.

## VII. Detecting and Quantitating the Acrosome Reaction

The natural eliciter of the sea urchin sperm acrosome reaction is the egg jelly coat. Preparations of egg jelly are obtained by lowering the pH of a suspension of fresh eggs by dropwise addition of 0.1 *N* HCl and stirring by hand (not with a magnetic stir bar) while measuring pH with a combination pH electrode. When the pH is 5.0, the eggs are stirred 2 minutes and then the pH is returned to 7.8 with 0.1 *N* NaOH. The eggs are allowed to settle and the supernatant is removed by pipet and centrifuged by hand to remove remaining eggs and large debris. The resulting supernatant is then clarified by centrifugation for 30 minutes at 20,000 *g*. Quantitation of the jelly solution is usually done on the basis of its fucose content (Dubois *et al.*, 1956). Characterization of sea urchin egg jelly has been reported (SeGall and Lennarz, 1979, 1981; Garbers *et al.*, 1983).

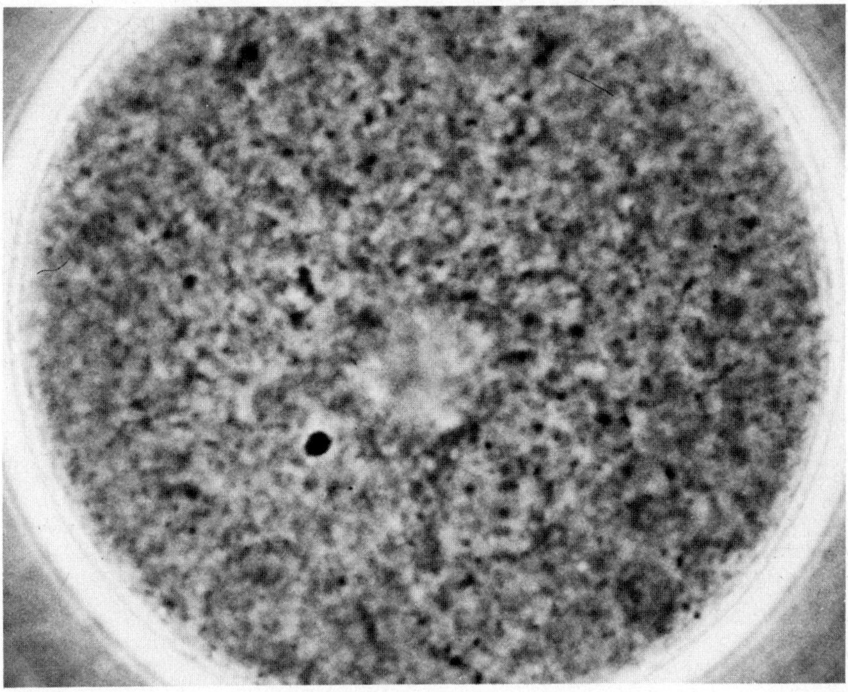

FIG. 4. The phase-dense sperm pronucleus before decondensation in an egg (*S. purpuratus*). The clear area in the egg center is the egg pronucleus. The egg is approximately 80 μm in diameter.

To induce the acrosome reaction with egg jelly, ionophores, or other reagents, fresh sperm are obtained "dry" and stored on ice in microfuge tubes. Immediately before use the sperm are diluted into seawater to the desired concentration and aliquots added to tubes containing 50% dilutions of the agent to be tested. After 5 minutes an equal volume of 5% glutaraldehyde in seawater is added. A sample of untreated sperm should be included for determining the frequency of spontaneous acrosome reaction. It is important to note that if buffer is omitted the respiration of concentrated sperm suspensions will lower the pH of the seawater to about 7.2, which is inhibitory to respiration, motility, and the acrosome reaction (Schackmann and Shapiro, 1981). The rate of decrease in the pH of natural seawater as a function of sperm concentration appears in Fig. 5. We find that seawater buffered with 10 m*M* HEPES, pH 7.8–8.0, will maintain that pH with a sperm suspension of $2 \times 10^8$ cells/ml for at least 10 minutes. If organic buffers are undesirable, adding extra $NaHCO_3$ will greatly increase the buffering capacity of seawater (compare Figs. 5 and 6).

Induction of the acrosome reaction by ionophores is accomplished by mixing a dilution of fresh sperm with the ionophore dissolved in an organic solvent. For

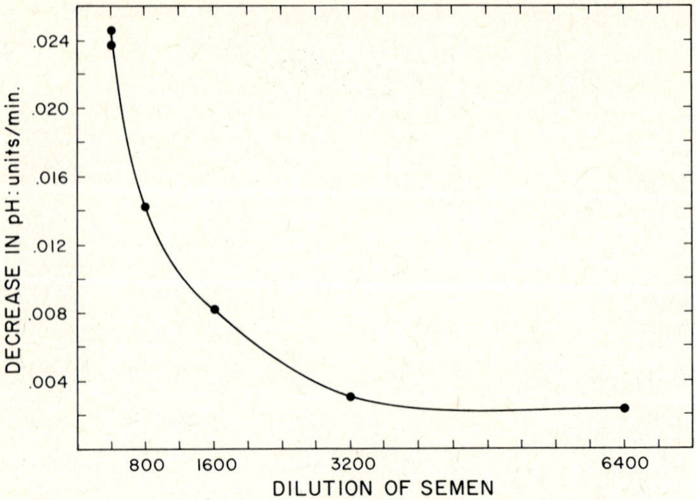


FIG. 5. Rate of decrease in pH of natural seawater as a function of diluting *S. purpuratus* sperm. "Dry" sperm were diluted into natural seawater (20°C) and the pH was recorded for 10 minutes. A 3200-fold dilution of dry sperm equals $1.25 \times 10^7$ cells/ml.

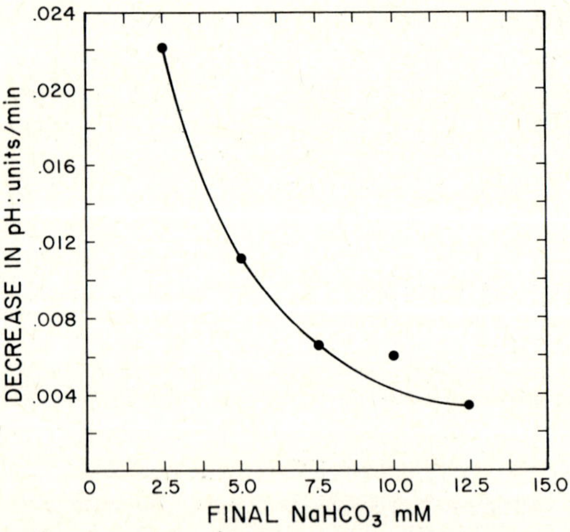


FIG. 6. Effect of additional $NaHCO_3$ on the stability of the pH of *S. purpuratus* sperm suspensions. Five milliliters natural seawater plus various concentrations of $NaHCO_3$ (pH 7.8) were placed on a magnetic stir plate and a pH electrode was immersed in the medium. From a 1:10 dilution of "dry" sperm into natural seawater 0.125 ml was added and the decrease in pH was recorded for 6 minutes. The final dilution of dry sperm was 400-fold, or $1 \times 10^8$ cells/ml.

example, to induce almost 100% acrosome reaction in *S. purpuratus* sperm with the $Ca^{2+}/H^{+}$ carrying ionophore A23187 we dissolve 2 mg of ionophore in 1 ml of dimethyl sulfoxide. A fresh sperm suspension is prepared and 99 vol is rapidly added to 1 vol of dimethyl sulfoxide–ionophore solution. A control with dimethyl sulfoxide alone is run in parallel. The final ionophore concentration is thus approximately 38 $\mu M$ (Podell and Vacquier, 1984a; Fig. 7). The acrosome reaction can also be induced by elevation of the pH of a sperm suspension to pH 9.1–9.4 using dilute $NH_4OH$ (Tilney *et al.*, 1973).

After the sperm have been fixed in suspension several methods are available for determining the frequency of acrosome reaction. For determination by scanning electron microscopy the cells are transferred to salt-free medium and critical point dried on a support. After coating with metal, the cells are examined for the

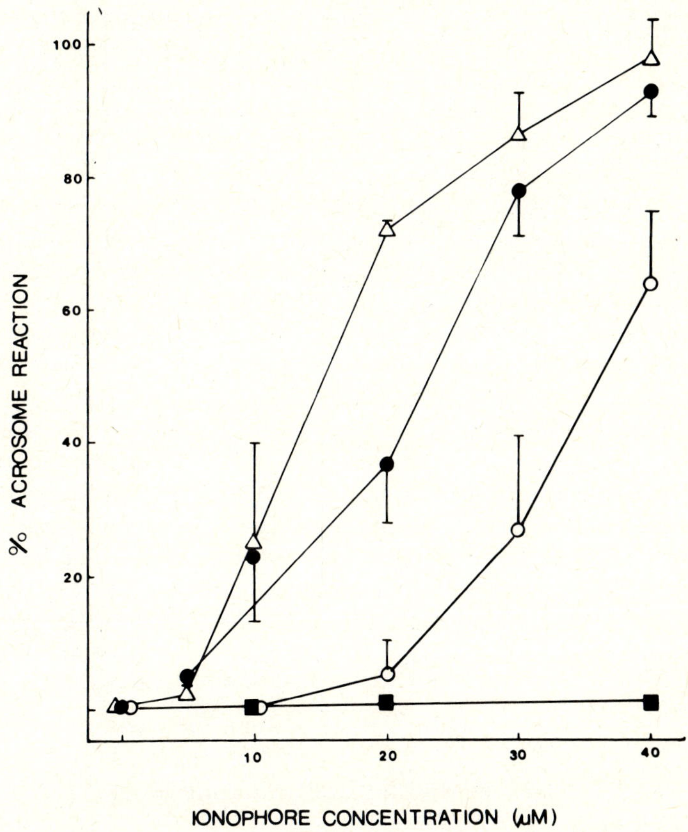


FIG. 7. The induction of acrosome reactions in *S. purpuratus* sperm by ionophores. Nigericin (△), A23187 (○), monensin (■), A23187 plus monensin (●). Each point is the average of three experiments. (From Podell and Vacquier, 1984a.)

presence of the acrosomal process protruding as a narrow rod about 1 μm in length from the anterior apex of the cell (Collins and Epel, 1977; Schackmann *et al.*, 1978).

For detecting the acrosome reaction by transmission electron microscopy the fixed sperm are picked up on a coated grid, the grid is blotted, desiccated, and placed directly in the microscope for observation of the acrosome process (Gregg and Metz, 1976; Kato and Sugiyama, 1969; Saling *et al.*, 1982; Talbot *et al.*, 1976).

We find that frequency of acrosome reaction is easily determined in *S. purpuratus* and *L. pictus* sperm by phase-contrast microscopy at approximately 1200× magnification. A drop of fixed cells is placed on a clean slide and a clean cover slip is added. The slide is held on edge against tissue for 15 seconds in order to drain and inverted onto a tissue. A tissue is placed on top of the slide and squashing pressure is applied with a thumb, being careful that the cover slip and slide do not slip across each other. The sperm squash is then turned cover slip up, immersion oil is added, and the preparation is viewed by phase-contrast microscopy. Occurrence of the acrosome reaction in *S. purpuratus* and *L. pictus* sperm is distinguished by loss of the phase-dense acrosomal granule and extension of the acrosome process.

In some species, such as *A. punctulata,* it is indeed difficult to score accurately the acrosome reaction by phase-contrast microscopy. An additional procedure has been developed involving the reaction of the fixed sperm with rhodamine-conjugated rabbit antibody to the protein bindin, the major constituent of the acrosome granule. The fixed sperm are transferred to a blocking solution, then treated with the antibody and finally washed and viewed with epifluorescence. Sperm that have undergone the acrosome reaction, which exposes the bindin, show a brilliant localization of fluorescence at their anterior tip, whereas untreated sperm are completely nonfluorescent (Fig. 8; Vacquier, 1984). This rhodamine-conjugated antibody cross-reacts with all echinoid sperm bindin yet tested (Moy and Vacquier, 1979). If a fluorescence microscope is not available, horseradish peroxidase-conjugated rabbit anti-bindin can be used (Fig. 8; Vacquier, 1984). When scoring acrosome reactions quantitatively it is important to have a disinterested colleague code the tubes to prevent bias by the experimenter. The samples are scored "blind" and decoded after all samples have been observed.

## VIII. Labeling Sperm Components

### A. Radioiodination of Membrane Glycoproteins

Techniques have been devised to radioiodinate sperm using iodogen (Pierce Chemical Co., Rockford, ILL) and lactoperoxidase (Lopo and Vacquier, 1980;

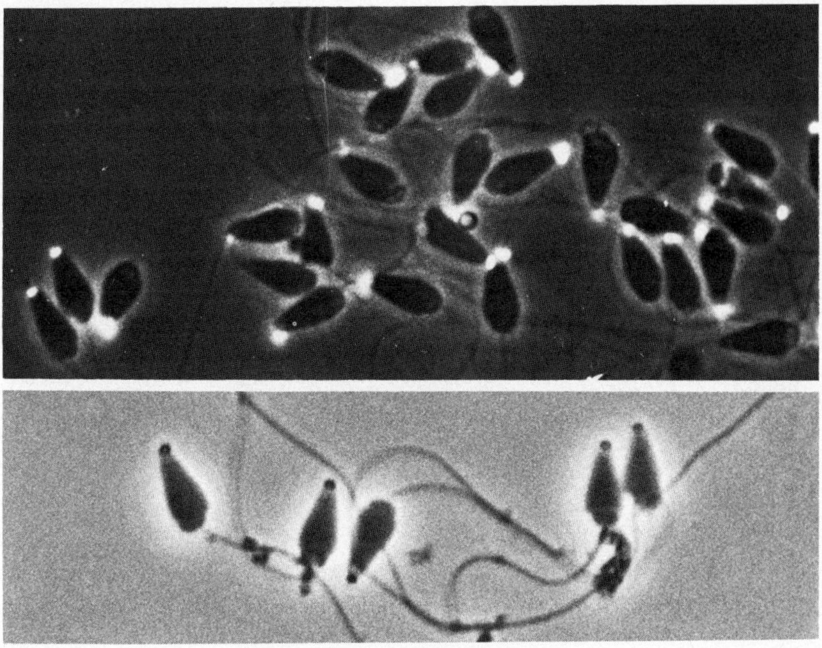

FIG. 8. Scoring acrosome reactions of *S. purpuratus* sperm using rhodamine-conjugated (upper panel) or peroxidase-conjugated (lower panel) rabbit antibody to bindin. The acrosome-reacted sperm show intense labeling with the antibody at their anterior tips. Approximately 3700×. (From Vacquier, 1984.)

Podell *et al.*, 1984). If care is taken, the cells will survive without a decrease in their ability to undergo the acrosome reaction. Two interesting findings emerged from this study: first, a great percentage of the $^{125}$I is incorporated into phospholipids, and second, only four glycoproteins are labeled to an appreciable extent. We have refined the method for labeling sea urchin sperm plasma membrane glycoproteins using a new commercial reagent called Iodobeads (Pierce Chemical Co.). To surface iodinate sperm the pigmented phagocytes are first removed as described in Section IV,A. The sperm are then suspended at $4 \times 10^8$ cells/ml in seawater containing 50 m*M* Tris, pH 7.5. To 10 ml of sperm suspension is added 0.5–2.0 mCi [$^{125}$I]sodium iodide, and after mixing, two Iodobeads previously washed for 2 minutes in Tris–seawater are added. Two notes of caution must be included here: first, the $^{125}$I must be added before the Iodobeads, and second, the Iodobeads must be washed thoroughly in an excess of buffered seawater before adding to the cells. Failure to follow this exact procedure will result in dead sperm with little protein-incorporated label. The sperm surface glycoproteins that label under *in vivo* conditions will run in completely denaturing acrylamide gels with $M_r$ values of approximately 210,000, 140,000, 80,000, and 60,000 (Podell *et al.*, 1984; Fig. 9). Similar labeling patterns have been obtained by Cross (1983). When the plasma membranes are isolated from iodinated sperm the specific activities of these bands are enriched (Fig. 9). If the cells lyse either before or during the iodination, the tubulin bands at the 50,000- to 55,000-Da region of the gel will show heavy incorporation of label. *S. pur-*

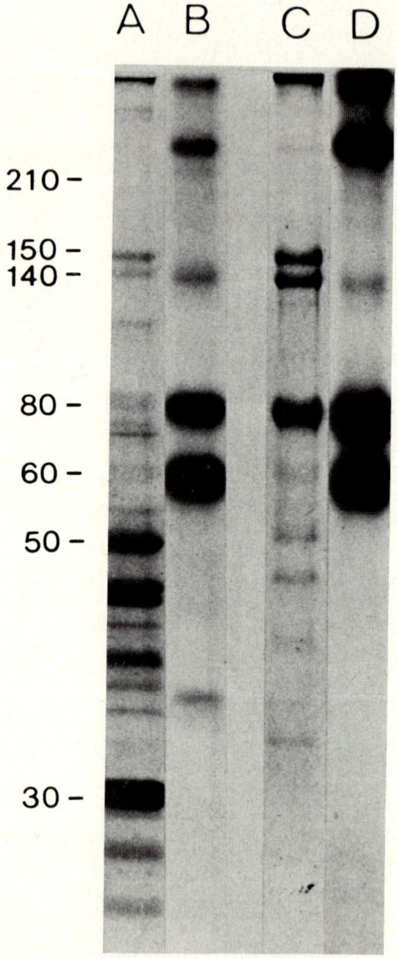


FIG. 9. Radioiodination of *L. pictus* sperm and the isolation of the labeled proteins with the plasma membranes. (A) Coomassie Blue-stained whole sperm (50 μg) labeled *in vivo;* (B) autoradiograph of (A) showing labeled bands at $M_r$ 210,000+, 140,000, 80,000, and 60,000; (C) Coomassie Blue-stained isolated sperm membranes showing the major proteins at $M_r$ 150,000, 140,000, and 80,000; (D) autoradiograph of (C), showing enrichment of the four labeled bands in this membrane fraction. (From Podell *et al.*, 1984.)

*puratus* and *A. punctulata* sperm are relatively easier than *L. pictus* to radiolabel while maintaining viability. *L. pictus, S. purpuratus,* and *S. franciscanus* sperm show essentially the same pattern of labeling of these four plasma membrane glycoproteins. The only differences found with *A. punctulata* are that all four bands are of slightly higher apparent molecular weight and the band just above the 210,000 region is not as intensely labeled.

## B. Labeling with [$^{125}$I]Diiodofluorescein Isothiocyanate (IFC)

This probe was first synthesized by Gabel and Shapiro (1978). It was subsequently used for the isolation by immunoabsorption of labeled sperm components using antibodies to IFC (Gundersen and Shapiro, 1984a) and to study the fate of IFC-labeled sperm components in the embryo (Gundersen and Shapiro, 1984b). About 10% of this probe is incorporated into proteins of greater than 15,000 Da that are sensitive to protease digestion of sperm. The remaining label is in low-molecular-weight components that are not phospholipids (Gundersen and Shapiro, 1984b). The advantages of this probe as a membrane label are that it is both radioactive and fluorescent, and specific antibodies to IFC make possible the isolation of the labeled components.

## C. Labeling the Surface with Lectins and Antibodies

Of all lectins tested, we found wheat germ agglutinin to be the only potent agglutinin of *S. purpuratus, A. punctulata,* and *S. franciscanus* sperm, whereas *L. pictus* sperm were not agglutinated (Podell and Vacquier, 1984a). Wheat germ agglutinin was also the only lectin, when rhodamine labeled, that stained the cells brilliantly fluorescent. The methods for reacting sperm with fluorescent lectins, determining agglutination of cells by wheat germ agglutinin, and the inhibition of the acrosome reaction and its associated $Ca^{2+}$ influx and $H^+$ efflux by wheat germ agglutinin have been described in detail (Podell and Vacquier, 1984a). A note of caution to add is that wheat germ agglutinin rapidly loses activity when dissolved in seawater; even if stored at −70°C this lectin is ineffective as an agglutinin after a few days. We also found that wheat germ agglutinin, isolated by us by the method of Vretblad (1976), stored as a lyophilized powder at 4°C, and made up fresh each day in seawater, had the highest potency as an agglutinin and inhibitor of the acrosome reaction. Lectin blots of isolated sperm membrane proteins indicate that the major membrane protein reacting with wheat germ agglutinin is the 210,000+ Da component (Podell and Vacquier, 1984a). However, the conditions used for lectin blots select for the highest binding affinities and minor reacting membrane proteins, or ones of much lower affinity for wheat germ agglutinin might not be detected.

Living sperm are reacted with anti-sperm immunoglobulins by simply dialyzing the antibody (usually dissolved in phosphate-buffered saline) first against 0.5 *M* NaCl, 10 m*M* HEPES, pH 7.5, and then against seawater. This is to avoid the formation of calcium phosphate precipitate that develops when phosphate-buffered saline is dialyzed directly against seawater. Most anti-sperm antibodies react well in seawater. Assessment of antibody reaction with the sperm surface can be by agglutination, as with wheat germ agglutinin, or by binding of a second antibody containing a fluorescent label. After washing, the quantity of second antibody bound to the cells is determined using a fluorescence spectrophotometer, as described in detail by Trimmer *et al.* (1985). If a fluorescence spectrophotometer is not available, a second method to quantitate anti-sperm immunoglobulins binding to sperm is by a second antibody conjugated to horseradish peroxidase, as described in detail by Podell and Vacquier (1984b).

We find that both polyclonal and monoclonal antibodies to sea urchin sperm react equally well with fixed and living cells. When using living cells, 10 m*M* sodium azide is added to immobilize them during the washes. Sodium azide is a completely reversible inhibitor of sperm motility, even after exposure to 50 m*M*. For all immunological work such as ELISA assays for hybridomas that are positive to the sperm surface (Trimmer *et al.*, 1985) and immunofluorescent localization of antigens (Podell and Vacquier, 1984b; Trimmer *et al.*, 1985), sperm are fixed in seawater containing 3% formaldehyde and 0.1% glutaraldehyde, after Tokuyasu and Singer (1976). Monoclonal antibodies, directed against the 210,000+ Da *S. purpuratus* sperm membrane protein, stain living sperm in patches of fluorescence on the flagellum, whereas they stain fixed sperm evenly along the entire flagellum (Trimmer *et al.*, 1985).

## D. Labeling Sperm with Radiophosphate

The sperm of *A. punctulata* will take up $^{32}PO_4^{3-}$ from seawater and incorporate it into ATP. Dry sperm are diluted roughly 100-fold ($10^8$ cells/ml), $^{32}PO_4^{3-}$ is added to 0.5 mCi/ml, and the sperm is incubated at 15–21°C. The total sperm protein labels to equilibrium in about 2–3 hours. Sodium dodecyl sulfate (SDS) gel analysis and autoradiography show that the major labeled protein has an $M_r$ of 160,000. Upon treatment of $^{32}P$-labeled sperm with egg jelly this protein loses all its radiophosphate and its $M_r$ shifts to 150,000 (Ward and Vacquier, 1983). Subsequent work shows that this phenomenon is not causally linked to the acrosome reaction (Ward *et al.*, 1985b); rather, the dephosphorylation is caused by a peptide of known sequence named ''resact'' that is present in egg jelly (Suzuki *et al.*, 1984). The $M_r$ 160,000/150,000 protein in the sperm is guanylate cyclase and its dephosphorylation is associated with a 38-fold decrease in specific activity (Ward *et al.*, 1985a). All of the methods involved in the study of this phenomenon can be found in the articles cited above. The guanylate cyclase of sea urchin sperm is present only in the membrane-bound "particulate" form, at the highest levels known for any cell. The enzyme has been purified from sperm of *L. pictus* (Garbers, 1976), *S. purpuratus* (Radany *et al.*, 1983), and *A. punctulata* (Ward *et al.*, 1985a).

Other minor proteins also incorporate $^{32}PO_4^{3-}$ when *A. punctulata* sperm are labeled *in vivo*. Bands at $M_r$ values of about 37,000 and 50,000 are also found, the latter possibly being the flagellar phosphoprotein called axokinin (Tash *et al.*, 1984).

Our standard method to prepare sperm samples for SDS–polyacrylamide gel electrophoresis is as follows. Five microliters of dry *S. purpuratus* sperm (500 μg of protein) are suspended in 2 ml seawater. Fresh 100% trichloroacetic acid (TCA) is added to 10% final concentration. The tube is vortexed and put in ice for 30 minutes. The fixed cells are then sedimented by centrifugation for 10 minutes at 3000 *g*. The supernatant is removed by aspiration and 3 ml 90% acetone (10% water) is added. The pellet is completely resuspended by vortexing, which may be aided by a metal spatula. After 10 minutes the cells are sedimented as above, the supernatant is removed, and the pellet is completely resuspended in 3 ml 90% acetone. The TCA denatures degradative enzymes and fragments the DNA. The acetone washes remove the TCA. Care is taken to wash the tube walls with acetone to completely remove all of the TCA. The second acetone pellet is then completely dissolved in 1 ml of 10% SDS (Sigma, St. Louis, MO, Stock No. L5750) by vortexing, heating, and using a spatula or 2-mm glass beads to help break up the insoluble precipitate. Fifteen-milliliter Corex centrifuge tubes work best for this entire procedure. After complete solubilization (usually overnight) 25-μl samples are removed for protein determination (Lowry *et al.*, 1951). The tube contents are then precipitated by the addition of 3 ml 100% acetone and after centrifugation the pellet is resuspended in sample buffer (Laemmli, 1970) at 1 μg/μl for Coomassie Blue-stained gels or 0.1 μg/μl for silver-stained gels. After complete solubilization in the SDS and mercaptoethanol-containing buffer the samples are boiled 3 minutes.

## IX. Isolation of Sperm Components

### A. Separation of Head from Flagellum

The isolation of flagellar axonemes is found in Stephens (this volume), and a review on axonemal dynein has been recently published (Bell and Gibbons, 1983). We have worked out several methods for separating heads from flagella. The methods all involve shearing the cells in a blender or tissue grinder and then separating the heavier heads from the lighter flagella by differential centrifugation. A blender will rapidly cut the tails into small pieces while leaving the heads fairly intact. The quality of isolated *S. purpuratus* heads has been assessed by scanning electron microscopy. When the isolated heads are added to acid-dejellied eggs, a low percentage of fertilization occurs. However, if soluble egg jelly is added to induce the acrosome reaction, a high percentage of fertilization occurs (Vacquier, 1979). The amount of blending must be carefully controlled, since damage to the heads occurs with time. The sperm suspension (0–4°C) is blended for 30-second intervals and the amount of sperm heads with remaining flagella is assessed by phase-contrast microscopy. The flagella-less heads are sedimented in a swinging bucket rotor at 1000 *g* for 20 minutes. The supernatant is removed and the heads are resuspended in fresh seawater. The heads are washed in this manner until flagellar fragments are completely removed (Vacquier, 1979).

The least injurious method to separate the heads from flagella is by homogenization in seawater in a 40-ml Wheaton tissue grinder with a round-bottom glass mortar and tight Teflon pestle driven up and down by hand. The amount of head and tail separation is assessed by phase-contrast microscopy. This method yields long, relatively intact flagella. For *A. punctulata,* heads (nucleus and mitochondrion) are separated by sedimentation at 1000 *g* for 10 minutes. The supernatant containing flagella is retained in another tube and the pellet containing heads is resuspended and washed two additional times. The flagella are isolated by three successive centrifugations for 15 minutes at 4000 *g;* each time the supernatant is removed and recentrifuged while the pellet of broken nuclei and mitochondrial midpieces is discarded. The third supernatant after the 4000 *g* centrifugation is then recentrifuged 15 minutes at 10,000 *g* to sediment the flagella. The purity of each fraction, as determined by SDS gels and silver staining, is shown in Fig. 10 (Ward and Vacquier, 1983).

## B. Isolation of Components of the Sperm Head

The midpiece of sea urchin sperm is a single large mitochondrion (Longo and Anderson, 1969). A method to isolate sperm mitochondria has been described (Rinaldi *et al.*, 1983). First, the heads are isolated from the tails in a Tris–sucrose–EDTA buffer. The heads are homogenized with greater force and the nuclei are sedimented at 500 *g*. A second centrifugation at 11,000 *g* removes fragments of nuclei. The mitochondria are then sedimented at 14,000 *g*. Electron microscopy shows the pellet of mitochondria to be fairly pure and excellent for isolating mitochondrial DNA with essentially no contamination by nuclear DNA (Rinaldi *et al.*, 1983).

The centrioles of both starfish (Kuriyama and Kanatani, 1981) and sea urchin (Ishikawa *et al.*, 1977) sperm have been isolated. For sea urchin sperm sonication in 1 *M* hexylene glycol was used to disrupt the cell, and differential centrifugation separated a fraction rich in centrioles.

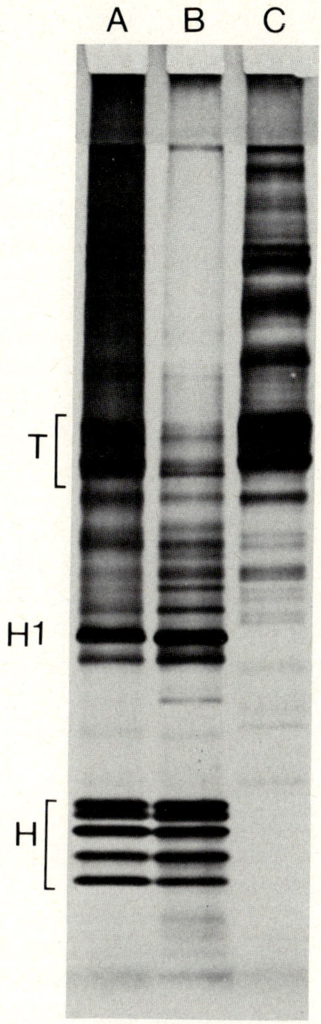


FIG. 10. Separation of head from flagellum as demonstrated for *A. punctulata* sperm by silver-stained SDS gels (7.5 μg protein/lane). (A) Whole sperm; (B) heads; (C) flagella. Note that the core histones (H) and histone $H_1$ are not found in the flagellar fraction. T, Tubulin. (From Ward and Vacquier, 1983.)

Demembranated acrosomal granules containing the protein called bindin, which acts as the adhesive between sperm and egg, were first isolated from *S. purpuratus* sperm. Sperm were suspended in calcium-free seawater (see Section III) and agitated in Triton X-100, which dissolved the membranes and released the acrosomal granules. Acrosomal granules were separated from broken flagella by filtration through spun glass fiber (glass wool), the granules passing through while the flagellar fragments became trapped. SDS gel electrophoresis of the isolated granules showed them to be composed of one major protein (bindin) with an $M_r$ of 30,500 (Vacquier and Moy, 1977, 1978; Bellet *et al.*, 1977). Attempts to repeat these original isolations of very pure acrosomal granules were hampered by contamination with tubulin. The problem appeared to be caused by the type of glass fiber. Spun glass fiber used in the original isolation was of unknown age and manufacture, and that available today is not as effective in separating axonemal fragments from acrosomal granules. For this reason, a new method involving the isolation of a fraction enriched in acrosomal granules and the purification of bindin from this fraction by DEAE–cellulose chromatography was developed (Vacquier, 1983). A complete review of the work on bindin and discussion of the isolation technique can be found in Vacquier (1983). Our interest in bindin is that it is species specific in its interaction with eggs (Glabe and Lennarz, 1979) and that it may also be a mediator of membrane fusion (Glabe, 1985a,b).

The nuclei of *S. purpuratus* sperm have been isolated by homogenization in seawater containing 0.1% Triton X-100 and a protease inhibitor. The final step is the centrifugation of nuclei through 1.5 *M* sucrose (Kunkle, 1984). Isolated nuclei retain their conoid shape even when the DNA is enzymatically digested and the histones are extracted. The conclusion of Kunkle's study is that a fairly resilient nuclear matrix must be responsible for maintaining the shape of the sperm nucleus.

## C. Isolation of the Plasma Membrane

The lipid composition of sea urchin sperm has been previously reported (Kozhina *et al.*, 1978). The first method to isolate plasma membrane from sea urchin sperm was presented by Gray and Drummond (1976), their primary objective being to study the adenylate cyclase. Cross (1983) refined this procedure and presented the first detailed characterization of the membrane fraction obtained from isolated flagella. Since both the above methods are laborious and result in low yield, we developed a simple method to isolate a plasma membrane fraction in bulk quantity that has been slightly revised since our recent publication (Podell *et al.*, 1984). The pigmented phagocytes are removed from *S. purpuratus* semen and the sperm are pelleted at 1500 *g* and resuspended in 25 vol of natural seawater. An equal volume of natural seawater (15°C) containing 20 mg/100 ml of both penicillin and streptomycin, 20 m*M* benzamidine, and 40 m*M* Tris (the pH adjusted to 9.1 with HCl) is gently stirred into the sperm suspension. The suspension is then incubated at 15°C for 5 to 15 hours. It is important not to agitate or stir the suspension since this leads to tubulin contamination. After incubation the suspension is centrifuged 30 minutes in a swinging bucket rotor at

7000 *g*. The supernatant is carefully removed and centrifuged as before. The second 7000 *g* supernatant is then centrifuged at 40,000 *g* for 30 minutes, which sediments sperm membrane vesicles. The supernatant is removed and the pellet is resuspended in one-tenth the original volume in seawater containing 5 m*M* benzamidine, 10 m*M* HEPES, or Tris buffer at pH 7.8. The preparation is centrifuged 30 minutes at 7000 *g*. The supernatant is removed and the centrifugation repeated. The sperm membrane vesicles are collected from the second 7000 *g* supernatant at 40,000 *g* for 30 minutes. The preparation is enriched for adenylate cyclase activity. The sperm membrane vesicles are microscopically free of broken axonemes and fragmented sperm heads and appear to be in the form of sealed right-side-out vesicles that show species-specific affinity to eggs (Podell *et al.*, 1984).

This is a very advantageous method to isolate 10- to 100-mg quantities of membrane. As revealed by silver staining of SDS gels (Fig. 11), this membraneous fraction consists of a small number of major proteins. This preparative method of isolating sperm membrane vesicles has been used as the starting material for the purification of the $M_r$ 80,000 and 210,000+ sperm membrane glycoproteins (Podell and Vacquier, 1985). One problem with the above method is that the 15°C of incubation (which is necessary to obtain the vesicles) might be deleterious to potentially interesting enzymes. A nitrogen bomb cavitation method at 2°C was therefore developed to isolate a plasma membrane fraction from *A. punctulata* sperm (Ward *et al.*, 1985a), based on one developed for the isolation of plasma membrane from mammalian sperm (Gillis *et al.*, 1978). *A. punctulata* sperm are washed in cavitation buffer {480 m*M* NaCl, 10 m*M* $MgCl_2$, 10 m*M* KCl, 10 m*M* benzamidine, 0.5 m*M* phenylmethylsulfonyl fluoride and 10 m*M* MES [(*N*-morpholino)ethanesulfonic acid]} at pH 6.0 and resuspended in 10 vol of the same. A nitrogen atmosphere at 400 psi is applied, the valve is opened, and the sperm is cavitated into a tube. The cavitate is centrifuged for 10 minutes at 8000 *g*, and the resulting supernatant is centrifuged for 30 minutes at 11,000 *g*. Membranes are pelleted from the second supernatant by centrifugation for 45 minutes at 75,000 *g*. Membranes obtained by this procedure appear essentially identical by SDS gel analysis to the sperm membrane vesicles previously described (Figs. 9 and 11).

There is no doubt that the major proteins of the sea urchin sperm membrane ($M_r$ values of 210,000+, 160,000, 140,000, 80,000, and 60,000) will be intensely studied in the coming years with the hope of understanding the ion channel regulation governing sperm function. Recent papers report the presence in sperm tail membranes of an $Na^+/H^+$ exchanger that is sensitive to membrane potential (Lee, 1984b), the identification of an $M_r$ 77,000 protein from sperm membranes that is a receptor for the peptide called speract from egg jelly (Dangott and Garbers, 1984), and the increase in $Ca^{2+}$ flux into isolated sperm membrane vesicles in response to egg jelly (Darszon *et al.*, 1984).

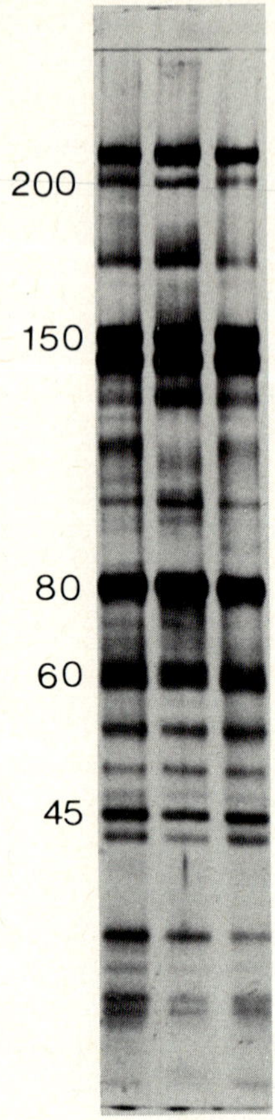


FIG. 11. Silver-stained gel (4 μg/lane) of isolated *S. purpuratus* sperm membrane vesicles. Comparison to Fig. 9, lane (C), shows many more protein bands are detectable by silver staining. This pattern of bands on SDS gels is consistently observed.

## ACKNOWLEDGMENTS

I thank and am indebted to those who have worked with me in attempting to understand the physiology of these most fascinating cells. Without their work this article would not have been possible. They are Charles G. Glabe, Alina C. Lopo, Gary W. Moy, Brigitte Brandriff, Neal F. Bellet, Gregory S. Kopf, Sheila B. Podell, Gary E. Ward, Donald C. Porter, and James S. Trimmer. The research was supported by NIH Grant HD 12986.

## REFERENCES

Asahina, E., and Takahashi, T. (1979). *Dev. Growth Differ.* **21,** 423–430.

Bell, C. W., and Gibbons, I. R. (1983). *In* "Muscle and Nonmuscle Motility" (A. Stracher, ed.), Vol. 2, pp. 1–34. Academic Press, New York.

Bellet, N. F., Vacquier, J. P., and Vacquier, V. D. (1977). *Biochem. Biophys. Res. Commun.* **79,** 159–165.

Bibring, T., Baxandall, J., and Harter, C. C. (1984). *Dev. Biol.* **101,** 425–435.

Cavanaugh, G. M. (1964). *In* "Formulae and Methods V, Marine Biological Laboratory Chemical Room," pp. 51–56. Marine Biological Laboratory, Woods Hole, Massachusetts.

Christen, R., Schackmann, R. W., and Shapiro, B. M. (1982). *J. Biol. Chem.* **257,** 14881–14890.

Christen, R., Schackmann, R. W., and Shapiro, B. M. (1983). *J. Biol. Chem.* **258,** 5392–5399.

Collins, F., and Epel, D. (1977). *Exp. Cell Res.* **106,** 211–222.

Cross, N. L. (1983). *J. Cell Sci.* **59,** 13–25.

Dangott, L. J., and Garbers, D. L. (1984). *J. Biol. Chem.* **259,** 13712–13716.

Darszon, A., Gould, M., De La Torre, L., and Vargas, I. (1984). *Eur. J. Biochem.* **144,** 515–522.

Dubois, M., Gilies, K. A., Hamilton, J. K., Rebers, P. A., and Smith, F. (1956). *Anal. Chem.* **28,** 350–356.

Gabel, C. A., and Shapiro, B. M. (1978). *Anal. Biochem.* **86,** 396–406.

Garbers, D. L. (1976). *J. Biol. Chem.* **251,** 4071–4077.

Garbers, D. L., Kopf, G. S., Tubb, D. J., and Olson, G. (1983). *Biol. Reprod.* **29,** 1211–1220.

Garbers, D. L., Bentley, J. K., Dangott, L. J., Ramarao, C. S., Shimomura, H., Suzuki, N., and Thorpe, D. (1985). *In* "The Molecular and Cellular Biology of Fertilization" (J. L. Hedrick, ed.). Plenum, New York.

Gillis, G., Peterson, R., Russell, L., Hook, L., and Freund, M. (1978). *Prep. Biochem.* **8,** 363–378.

Giudice, G. (1973). "The Developmental Biology of the Sea Urchin." Academic Press, New York.

Glabe, C. G. (1985a). *J. Cell Biol.* **100,** 794–799.

Glabe, C. G. (1985b). *J. Cell Biol.* **100,** 800–806.

Glabe, C. G., and Lennarz, W. J. (1979). *J. Cell Biol.* **83,** 595–604.

Gray, G. P., and Drummond, G. I. (1976). *Arch. Biochem. Biophys.* **172,** 31–38.

Gregg, K. W., and Metz, C. B. (1976). *Biol. Reprod.* **14,** 405–415.

Gundersen, G. G., and Shapiro, B. M. (1984a). *Biochim. Biophys. Acta* **799,** 68–79.

Gundersen, G. G., and Shapiro, B. M. (1984b). *J. Cell Biol.* **99,** 1343–1353.

Harvey, E. B. (1956). "The American Arbacia and Other Sea Urchins." Princeton Univ. Press, Princeton, New Jersey.

Hinegardner, R. T. (1967). *In* "Methods in Developmental Biology" (F. H. Wilt and N. K. Wessells, eds.), pp. 139–156. Crowell, New York.

Hinegardner, R. T. (1975). *In* "The Sea Urchin Embryo, Biochemistry and Morphogenesis" (G. Czihak, ed.), pp. 10–22. Springer-Verlag, Berlin.

Ishikawa, M., Ohta, T., and Kato, K. H. (1977). *Acta Embryol. Exp.* **1,** 21–32.

Johnson, C. H., and Epel, D. (1983). *J. Exp. Zool.* **226,** 431–440.

Johnson, C. H., Clapper, D. L., Winkler, M. M., Lee, H. C., and Epel, D. (1983). *Dev. Biol.* **98,** 493–501.

Kato, K. H., and Sugiyama, M. (1983). *Dev. Growth Differ.* **25,** 153–163.

Kozhina, V. P., Terekhova, T. A., and Svetashev,V. I. (1978). *Dev. Biol.* **62,** 512–517.

Kuriyama, R., and Kanatani, H. (1981). *J. Cell Sci.* **49,** 33–49.

Laemmli, U. K. (1970). *Nature (London)* **227,** 680–685.

Leahy, P. S., Tutschulte, T. C., Britten, R. J., and Davidson, E. H. (1978). *J. Exp. Zool.* **204,** 369–380.

Leahy, P. S., Hough-Evans, B., Britten, R. J., and Davidson, E. H. (1981). *J. Exp. Zool.* **215,** 7–22.

Lee, H. C. (1984a). *J. Biol. Chem.* **259,** 4957–4963.

Lee, H. C. (1984b). *J. Biol. Chem.* **259,** 15315–15319.

Lee, H. C., Johnson, C., and Epel, D. (1983). *Dev. Biol.* **95,** 31–45.

Longo, F. J., and Anderson, E. (1969). *J. Ultrastruct. Res.* **27,** 486–509.

Lopo, A. C., and Vacquier, V. D. (1980). *Dev. Biol.* **76,** 15–25.

Lowry, O. H., Rosebrough, N. J., Farr, A. L., and Randall, R. J. (1951). *J. Biol. Chem.* **193,** 265–275.
Manahan, D. T., Davis, J. P., and Stephens, G. C. (1983). *Science* **220,** 204–206.
Moy, G. W., and Vacquier, V. D. (1979). *Curr. Top. Dev. Biol.* **13,** 31–44.
Osanai, K. (1975). *In* "The Sea Urchin Embryo, Biochemistry and Morphogenesis" (G. Czihak, ed.), pp. 26–36. Springer-Verlag, Berlin.
Podell, S. B., and Vacquier, V. D. (1984a). *J. Cell Biol.* **99,** 1598–1604.
Podell, S. B., and Vacquier, V. D. (1984b). *Exp. Cell Res.* **155,** 467–476.
Podell, S. B., and Vacquier, V. D. (1985). *J. Biol. Chem.* **260,** 2715–2718.
Podell, S. B., Moy, G. W., and Vacquier, V. D. (1984). *Biochim. Biophys. Acta* **778,** 25–37.
Radany, E. W., Gerzer, R., and Garbers, D. L. (1983). *J. Biol. Chem.* **258,** 8346–8351.
Resing, K., Green, J. D., and Walsh, K. A. (1985). *Dev. Biol.* **107,** 87–93.
Rinaldi, A. M., Salcher-Cillari, I., Comito, L., and Carra, E. (1983). *Cell Biol. Int. Rep.* **7,** 915–921.
Saling, P. M., Eckberg, W. R., and Metz, C. B. (1982). *J. Exp. Zool.* **221,** 93–103.
Schackmann, R. W., and Shapiro, B. M. (1981). *Dev. Biol.* **81,** 145–154.
Schackmann, R. W., Eddy, E. M., and Shapiro, B. M. (1978). *Dev. Biol.* **65,** 483–495.
Schackmann, R. W., Christen, R., and Shapiro, B. M. (1981). *Proc. Natl. Acad. Sci. U.S.A.* **78,** 6066–6070.
Schackmann, R. W., Christen, R., and Shapiro, B. M. (1984). *J. Biol. Chem.* **259,** 13914–13922.
SeGall, G. K., and Lennarz, W. J. (1979). *Dev. Biol.* **71,** 33–49.
SeGall, G. K., and Lennarz, W. J. (1981). *Dev. Biol.* **86,** 87–93.
Shapiro, B. M. (1984). *Ciba Found. Symp.* **103,** 86–99.
Stearns, L. W. (1974). "Sea Urchin Development, Cellular and Molecular Aspects." Dowden, Hutchinson, & Ross, Stroudsburg, Pennsylvania.
Summers, R. G., Hylander, B. L., Colwin, L. H., and Colwin, A. L. (1975). *Am. Zool.* **15,** 523–551.
Suzuki, N., Shimomura, H., Radany, E. W., Ramarao, C. S., Ward, G. W., Bentley, J. K., and Garbers, D. L. (1984). *J. Biol. Chem.* **259,** 14874–14879.
Talbot, P., Summers, R. G., Hylander, B. H., Keough, E. M., and Franklin, L. E. (1976). *J. Exp. Zool.* **198,** 383–392.
Tash, J. S., Kakar, S. S., and Means, A. R. (1984). *Cell* **38,** 551–559.
Tilney, L. G., Hatano, S., Ishikawa, H., and Mooseker, M. S. (1973). *J. Cell. Biol.* **59,** 109–126.
Tokuyasu, K. T., and Singer, S. J. (1976). *J. Cell Biol.* **71,** 894–906.
Trimmer, J. S., Trowbridge, I. S., and Vacquier, V. D. (1985). *Cell* **40,** 697–703.
Tyler, A., and Tyler, B. S. (1966). *In* "The Physiology of Echinodermata" (R. A. Boolootian, ed.), pp. 639–682. Wiley, New York.
Vacquier, V. D. (1979). *Dev. Growth Differ.* **21,** 61–69.
Vacquier, V. D. (1983). *Anal. Biochem.* **129,** 497–501.
Vacquier, V. D. (1984). *Exp. Cell Res.* **153,** 281–286.
Vacquier, V. D., and Moy, G. W. (1977). *Proc. Natl. Acad. Sci. U.S.A.* **74,** 2456–2460.
Vacquier, V. D., and Moy, G. W. (1978). *In* "Cell Reproduction" (D. M. Prescott, E. R. Dirksen, and C. F. Fox, eds.). Academic Press, New York.
Vacquier, V. D., and Payne, J. E. (1973). *Exp. Cell Res.* **82,** 227–235.
Vacquier, V. D., Brandriff, B., and Glabe, C. G. (1979). *Dev. Growth Differ.* **21,** 47–60.
Vretblad, P. (1976). *Biochim. Biophys. Acta* **434,** 169–176.
Ward, G. E., and Vacquier, V. D. (1983). *Proc. Natl. Acad. Sci. U.S.A.* **80,** 5578–5582.
Ward, G. E., Garbers, D. L., and Vacquier, V. D. (1985a). *Science* **227,** 768–770.
Ward, G. E., Moy, G. W., and Vacquier, V. D. (1985b). *In* "The Molecular and Cellular Biology of Fertilization" (J. L. Hedrick, ed.). Plenum, New York.

# Chapter 3

## *Sperm Motility*

CHARLES J. BROKAW

*Division of Biology*
*California Institute of Technology*
*Pasadena, California*



## I. Experimental Preparations

### A. Sperm Collection

Sea urchins are normally induced to shed gametes by injection of a small quantity of 0.5 *M* KCl solution through the soft tissues around the mouth into the perivisceral cavity. Spermatozoa are collected from the aboral surface of the urchin with a Pasteur pipet (''dry spermatozoa'') or by the time-honored method of inverting the urchin on top of a full beaker of seawater, so that the shed sperm drops to the bottom of the beaker. Starfish spermatozoa are generally obtained by dissecting out a portion of gonad and macerating it gently in a small volume of seawater, followed by filtering and centrifuging the suspension (Mabuchi *et al.*, 1976); this method can also be used with sea urchins.

Concentrated spermatozoa can usually be stored on ice for up to 4 hours before use, although spermatozoa of *Tripneustes gratilla* are reported to remain in better condition with storage at room temperature (Gibbons, 1982). With some species, improved viability during storage may be obtained if the spermatozoa are slightly diluted with 0.5 *M* NaCl or seawater. Spermatozoa of *Strongylocentrotus purpuratus* and *Colobocentrotus atratus* have been reported to remain in good condition during storage for several days. ''Dry spermatozoa'' are normally too thick to pipet accurately; for quantitative work the sperm suspension must be diluted with 1 to 2 vol of seawater or 0.5 *M* NaCl to obtain a suspension that is thin enough to pipet accurately. Spermatozoa used for observations of motility are not ordinarily washed before use, because it is difficult to avoid breakage of the sperm flagella during resuspension after centrifugation.

For some work, it is desirable to standardize the quantity of spermatozoa used. This is most easily done by measuring the optical density of diluted spermatozoa and adjusting the concentration of the concentrated stock sperm suspension by adding 0.5 *M* NaCl or seawater. The optical density measurements can be made by adding 10 μl of sperm suspension to 5.0 ml of 0.5 *M* NaCl solution and measuring the optical density at 540 nm (Brokaw and Simonick, 1977) or by measuring the optical density at 550 nm using a sperm sample diluted with 1% tricholoroacetic acid (Gibbons, 1982). These sperm concentrations can be calibrated by measurements of total protein concentration and/or by sperm counts.

Sperm counts are subject to two sources of rather large errors: (1) with some sperm samples, many of the spermatozoa may be found in large clumps, which may be completely overlooked when small aliquots of highly diluted spermatozoa are counted, and (2) spermatozoa may stick to glass surfaces when a sample is placed in a counting chamber, resulting in a highly nonuniform distribution of spermatozoa in the chamber. The following procedure (Gibbons and Fronk, 1972; Brokaw and Simonick, 1977) avoids the second source of error, but careful inspection of the sperm sample at high sperm concentrations is also necessary, to confirm the absence of sperm clumps. A 50-μl portion of concentrated sperm suspension is mixed with 450 μl of 5% formaldehyde in 0.5 *M* NaCl. After 20–30 minutes, the fixed spermatozoa are diluted 1:40,000 with distilled water (two successive 1:200 dilutions). A small drop of the final dilution (5–15 μl) is placed on a previously weighed microscope slide, and the slide is immediately reweighed to determine the volume of the drop. The drop is evaporated to dryness with warm air, and *all* of the spermatozoa in the drop are counted, using a phase-contrast microscope with a 25× objective. At least three drops should be counted from each diluted sample.

Concentrated, ''dry,'' sperm samples from *S. purpuratus* typically contain 4–10 × $10^{10}$ sperm/ml. A suspension diluted with 0.5 *M* NaCl so that a 1:500 dilution has an optical density of 0.22 contains about 2 × $10^{10}$ sperm/ml.

## B. Live Spermatozoa

Spermatozoa from some species of sea urchins (e.g., *S. purpuratus*) become motile as soon as they are diluted into a large volume of seawater. Natural seawater is normally filtered with an 0.45-μm Millipore filter before use in experiments on sperm motility. However, modified seawater solutions containing EDTA (usually 0.5 m*M*) to chelate toxic heavy metal ions (Tyler, 1953) and substances such as polyvinylpyrollidinone (PVP K-30, average $M_r$ 40,000; Matheson, Coleman, and Bell #8069; usually 1% wt/vol) to minimize adhesion to glass surfaces have often been used to prolong the useful life of the spermatozoa under the conditions used for microscopic observations. The pH must be readjusted to 8.1 after addition of EDTA, even if a neutralized stock solution of EDTA is used. In other echinoderm species, the conditions for activation of spermatozoan motility may be more stringent and may be favored by addition of other agents to the seawater. Histidine (10 m*M*) and urea (0.1 *M*) have been found to be useful (Fujii *et al.*, 1955; Brokaw, 1965). The cardinal rule is that one should not assume that spermatozoa diluted with pure seawater will immediately show optimal motility. Unless microscopic observation indicates that all of the spermatozoa are swimming uniformly, with bending waves that maintain normal amplitude along the full length of the flagellum, experimentation to improve the suspending medium is appropriate, as is examining sperm samples from different animals. For some experiments, especially when variations in the ionic environment are required, artificial seawater solutions may be preferable to the use of natural seawater. A very simple artificial seawater solution that has been used successfully in my laboratory contains 470 m*M* NaCl, 10 m*M* KCl, 54 m*M* $MgSO_4$, 10 m*M* $CaCl_2$, and 10 m*M* Tris buffer, with the pH adjusted to 8.1 with HCl. Recipes used by other workers include: 454 m*M* NaCl, 9.7 m*M* KCl, 24.9 m*M* $MgCl_2$, 9.6 m*M* $CaCl_2$, 27.1 m*M* $MgSO_4$, 4.4 m*M* $NaHCO_3$, and 10 m*M* Tris buffer (Hansbrough and Garbers, 1981); 493 m*M* NaCl, 10 m*M* KCl, 27 m*M* $MgCl_2$, 10 m*M* $CaCl_2$, 28 m*M* $MgSO_4$, 2.5 m*M* $NaHCO_3$, and 10 m*M* HEPES buffer (pH 8.0) (Johnson *et al.*, 1983). Most of these recipes are derived from the "Woods Hole" artificial seawater of Cavanaugh (1956).

## C. Demembranated Spermatozoa

Although early workers reported some success studying ATP-reactivated motility of spermatozoa from sea urchins and other echinoderms by using glycerol-permeabilized spermatozoa (Kinoshita, 1958; Brokaw, 1966; Brokaw and Josslin, 1973), the modern era of work with ATP-reactivated sperm motility began with the report by Gibbons and Gibbons (1972) on the preparation of fully demembranated sperm flagella with the nonionic detergent, Triton X-100. Previous reviews of these techniques have been presented by Goldstein (1974) and by Gibbons (1982). Since 1972 there have been three major modifications in the original techniques, based on (1) recognition of the role of the calcium ion concentration during demembranation in controlling the symmetry of the subsequent reactivated motility (Brokaw *et al.*, 1974); (2) recognition that acetate or other organic anions provide a more favorable reactivation environment than chloride ions (Gibbons *et al.*, 1982); and (3) recognition that incubation with cAMP may sometimes be required in order to obtain fully activated motility under reactivation conditions (Ishiguro *et al.*, 1982; Brokaw, 1985a). Additionally, some more recent workers have substituted Nonidet P-40 for Triton X-100, with equivalent results.

Most of this work has been carried out with spermatozoa from one of four Hawaiian or Californian sea urchin species, but Japanese species have been studied using the same procedures. Current procedures for these four species will be described below.

## 1. General Considerations

The use of high-quality deionized and/or glass-distilled water, and scrupulously clean glassware, is essential. When poor results are obtained, these factors are the first to suspect. Demembranated preparations are more sensitive to high temperatures than are live spermatozoa. The use of laboratory air conditioning, temperature-controlled microscope stages, and heat filtering of the microscope illumination may be essential. Spermatozoa in demembranation solutions deteriorate rather rapidly, especially if millimolar calcium is included; standard procedure is to make a fresh preparation of demembranated spermatozoa for each sample of reactivated spermatozoa that is observed. Solutions containing dithiothreitol (DTT) should be freshly made on the day of use. The presence of oxidizing impurities in commercial preparations of Triton X-100 has been reported. These are probably neutralized by the use of DTT in the solutions, but a purification procedure is described by Ashani and Catravas (1980).

Two types of demembranated sperm preparations are distinguished (Brokaw *et al.*, 1974; Gibbons and Gibbons, 1980). Spermatozoa demembranated in the continuous presence of EGTA to maintain low $Ca^{2+}$ concentrations are "potentially asymmetric," because they swim in circles with asymmetric bending waves in the usual EGTA-containing reactivation solutions. "Potentially symmetric" spermatozoa are obtained by exposing the demembranated spermatozoa to millimolar $Ca^{2+}$ concentrations in the presence of Triton X-100; these spermatozoa swim in nearly straight paths, with nearly symmetrical bending waves, in the same reactivating solutions.

In the following recipes, "Tris buffer" refers to either Tris hydroxymethyl aminomethane, in base form, with the solution pH lowered to the desired value by the addition of HCl, or to a mixture of Tris base and Tris HCl that will give the desired pH with minimal adjustment.

## 2. *Tripneustes gratilla* (Gibbons, 1982)

These urchins are obtained from warm-water environments, and experimentation is usually done successfully at a room temperature of 21–23°C. Spermatozoa are diluted with 1 to 5 vol of seawater containing 0.1 to 0.2 m*M* EDTA (pH adjusted to 8.3) and kept at room temperature for no more than 2 to 4 hours.

Demembranation solution (for "potentially symmetric" spermatozoa): 10 m*M* Tris buffer, 1 m*M* DTT, 0.1 m*M* EGTA, 4 m*M* $CaCl_2$, and 0.04% Triton X-100, at pH 8.1.

Reactivation solution: 0.15 *M* KCl, 10 m*M* Tris buffer, 2 m*M* $MgSO_4$, 1 m*M* DTT, 0.1 m*M* EGTA, 1 m*M* ATP, and 2% polyethylene glycol (PEG 20,000; Fisher Scientific Co., Pittsburgh, PA; #P-162) at pH 8.1. In more recent work, the KCl has been replaced with potassium acetate at 0.15 *M* (Gibbons *et al.*, 1982) or 0.20 *M* (Gibbons and Gibbons, 1984).

Twenty-five microliters of the diluted sperm suspension is mixed with 300 μl of demembranation solution. After 20–30 seconds, an aliquot of the demembranated sperm suspension is diluted 50- to 250-fold with reactivation solution. Note that because of the seawater added with the spermatozoa, the demembranating mixture contains approximately 4 m*M* $Mg^{2+}$ and 40 m*M* NaCl.

### 3. *Colobocentrotus atratus* (Gibbons and Gibbons, 1972, 1974, 1979)

These spermatozoa are also prepared and observed at 21–23°C. The same procedures described for *Tripneustes* spermatozoa are followed, but the optimized solutions for *Colobocentrotus* spermatozoa were originally given as:

Demembranation solution: 0.15 *M* KCl, 2 m*M* Tris buffer, 2 m*M* $MgSO_4$, 0.5 m*M* EDTA, 1 m*M* DTT, 0.04% Triton X-100, at pH 8.1.

Reactivation solution: 0.15 *M* KCl, 20 m*M* Tris buffer, 2 m*M* $MgSO_4$, 0.5 m*M* EDTA, 5 m*M* DTT, 1 m*M* ATP, and 2% PEG, at pH 8.1.

This demembranation solution lacks calcium because the motility of demembranated *Colobocentrotus* spermatozoa appears to be relatively independent of calcium concentration during demembranation and sufficient calcium is supplied by the seawater used to dilute the spermatozoa. Modifications used to obtain "rigor wave" spermatozoa are described by Gibbons and Gibbons (1974).

This species is important because the outer dynein arms are relatively easily removed by brief extraction with 0.5 *M* KCl or NaCl, leaving the sperm flagella in a state that can be reactivated with subnormal beat frequency or augmented by exogenous dynein (Gibbons and Gibbons, 1973, 1976, 1979). With this species, partially demembranated sperm flagella can be obtained by using a lower Triton X-100 concentration (Brokaw and Gibbons, 1973).

### 4. *Strongylocentrotus purpuratus* (Brokaw *et al.*, 1974)

These demembranated spermatozoa show rapid deterioration at temperatures greater than 21–22°C. In my laboratory, they are routinely prepared and examined at 18°C.

Demembranation solution: 0.15 *M* KCl, 2 m*M* Tris, 2 m*M* $MgSO_4$, 0.5 m*M* EDTA, 2 m*M* DTT, 2–5 m*M* $CaCl_2$, 0.04% Triton X-100, pH 8.2.

Reactivation solution: 0.15 *M* KCl, 20 m*M* Tris, 1.5 m*M* $MgSO_4$, 2 m*M* EGTA, 2 m*M* DTT, 1.2 m*M* ATP, 2% PEG, pH 8.2.

Concentrated spermatozoa are diluted 1:1 with 0.5 *M* NaCl for storage on ice. Ten microliters of these spermatozoa is added to 0.5 ml of demembranation solution. After 30 seconds, these spermatozoa are diluted 1:100 into reactivation solution. This procedure produces "potentially symmetric" spermatozoa. Methods for preparation of "rigor wave" spermatozoa from this species and from *Lytechinus* are described by Penningroth *et al.* (1982).

### 5. *Lytechinus pictus* (Brokaw, 1985a)

These spermatozoa are not quite as temperature sensitive as *S. purpuratus,* but 18°C is also recommended. These spermatozoa require incubation with cAMP for full activity.

Demembranation solution: 0.25 *M* KCl, 10 m*M* Tris buffer, 2 m*M* $MgSO_4$, 1 m*M* EGTA, 1 m*M* DTT, 0.04% Triton X-100, pH 8.2.

Activation solution: 0.05 *M* KCl, 10 m*M* Tris buffer, 2 m*M* $MgSO_4$, 1 m*M* EGTA, 1 m*M* DTT, 1 m*M* ATP, 10 μ*M* cAMP, pH 8.2.

Reactivation solution: 0.25 *M* potassium acetate, 20 m*M* Tris buffer, 3 m*M* $MgSO_4$, 1 m*M* EGTA, 1 m*M* DTT, up to 1 m*M* ATP, 0.05% PEG, pH 8.2.

Concentrated spermatozoa are used without dilution. One microliter of sperm is added to 100 μl demembranation solution. After 30 seconds, 300 μl of activation solution is added. After incubation for 2 to 3 minutes, the demembranated spermatozoa are diluted 1:100 into reactivation solution. This procedure yields "potentially asymmetric" spermatozoa. "Potentially symmetric" spermatozoa are prepared by adding 5 μl of 0.2 *M* $CaCl_2$ at the end of the incubation with cAMP and waiting an additional 20 or 30 seconds before dilution into reactivation solution.

### 6. Calculation of Solution Compositions

The algorithms given in Table I are useful for calculating recipes for reactivation solutions for experiments in which it is desired to change the concentration of one component, such as substrate ($MgATP^{2-}$) or $Ca^{2+}$, without changing the concentrations of other species in the solutions. The free $Ca^{2+}$ concentration becomes very sensitive to errors in total Ca content when the total Ca content approaches the EGTA content; or when EDTA is used, when the total Ca content is small. With 2 m*M* EGTA or EDTA at pH 8, EGTA should be used for solutions with free $Ca^{2+}$ concentration less than $4 \times 10^{-7}$ *M* and EDTA should be used for solutions with free $Ca^{2+}$ above this value. An accurate $CaCl_2$ stock

TABLE I

ALGORITHMS FOR CALCULATION OF SOLUTION COMPOSITIONS[a]

| | Equilibrium constants | | |
|---|---|---|---|
| | EGTA | | EDTA |
| $K_1$ ($H{:}E^{4-}$) | $10^{9.5}$ | | $10^{10.2}$ |
| $K_2$ ($H{:}HE^{3-}$) | $10^{8.9}$ | | $10^{6.2}$ |
| $K_3$ ($Mg{:}E^{4-}$) | $10^{5.2}$ | | $10^{8.6}$ |
| $K_4$ ($Mg{:}HE^{3-}$) | $10^{3.4}$ | | $10^{2.8}$ |
| $K_5$ ($Ca{:}E^{4-}$) | $10^{11}$ | | $10^{10.7}$ |
| $K_6$ ($Ca{:}HE^{3-}$) | $10^{5}$ | | $10^{4}$ |
| $K_7$ ($Mg{:}ATP^{4-}$) | | $10^{4.0}$ | |
| $K_8$ ($Ca{:}ATP^{4-}$) | | $10^{3.6}$ | |
| $K_9$ ($H{:}ATP^{4-}$) | | $10^{6.9}$ | |

Let $E$ = total concentration of EGTA or EDTA
Let $H$ = $10^{-\mathrm{pH}}$
Let $C_1$ = desired concentration of $MgATP^{2-}$
Let $C_2$ = desired concentration of $Mg^{2+}$
Let $C_3$ = desired concentration of $Ca^{2+}$

$C_4$ = (MgE species) = $EB_2/(B_1 + B_2 + B_3)$
$C_5$ = (CaE species) = $EB_3/(B_1 + B_2 + B_3)$
where $B_1 = 1 + HK_1 + HHK_1K_2$
$B_2 = C_2(K_3 + HK_1K_4)$
$B_3 = C_3(K_5 + HK_1K_6)$
$C_6$ = ($ATP^{4-}$) = $C_1/(C_2K_7)$
$C_7$ = ($Ca{:}ATP^{2-}$) = $C_3C_6K_8$
$C_8$ = ($H{:}ATP^{3-}$) = $HC_6K_9$

Total ATP required = $C_1 + C_6 + C_7 + C_8$
Total Mg required = $C_1 + C_2 + C_4$
Total Ca required = $C_3 + C_5 + C_7$

[a] Equilibrium constants have been collated from work of Durham (1983), Garfinkel and Garfinkel (1984), and Sillen and Martel (1964).

solution can be made by weighing out anhydrous $CaCO_3$ and neutralizing the solution with HCl. Accurate EGTA and CaEGTA stock solutions require calibration by pH titrations, as described by Miller and Smith (1984). Allowance must be made for the Ca added with the demembranated spermatozoa when calculating recipes for reactivation solutions.

Some of the constants used in Table I are "apparent equilibrium constants" for equilibria in which "free ATP" (shown as $ATP^{4-}$) includes both $ATP^{4-}$

and $KATP^{3-}$, and "free Mg" (shown as $Mg^{2+}$) includes both $Mg^{2+}$ and $MgCl^{1+}$ or magnesium acetate. These constants were originally obtained in solutions containing 0.1 *M* KCl, and are only approximately correct for solutions containing 0.15 to 0.25 *M* KCl or potassium acetate. For precise work an expanded algorithm that explicitly considers the formation of $KATP^{3-}$ and $MgCl^{1+}$ or magnesium acetate should be used (Garfinkel and Garfinkel, 1984).

## II. Observation and Measurement

### A. Microscopy

Microscopic observation of spermatozoa swimming in the interior of a microscope slide chamber, away from surfaces, is difficult. Spermatozoa quickly leave this part of the preparation and accumulate near surfaces, either sticking to the surface or swimming close to the surface with the plane of flagellar bending parallel to the surface. When swimming in this manner, the sperm flagellum is relatively easily observed and photographed, and individual spermatozoa remain observable in the plane of focus for extended periods. Most of the available data on sperm motility have been obtained from spermatozoa swimming near surfaces. Proximity to a surface will modify the hydrodynamic resistances encountered by a motile spermatozoon (Katz and Blake, 1975), but because the distance between the sperm flagellum and the surface is seldom known accurately, the magnitude of this hydrodynamic effect is difficult to estimate.

Sticking of the spermatozoa to glass surfaces can sometimes be minimized by the use of PVP or PEG in the suspending media and by coating the surfaces with Formvar (Gibbons, 1982), siliconizing preparations such as Siliclad or AquaSil (catalog #42799, Pierce Chemical Co., Rockford, IL), or egg white (expose the surface to concentrated raw egg white, pour off the egg white, rinse with the solution that will be used to suspend the spermatozoa, and use without drying the surface).

#### 1. Sperm Progression

A relatively easy photographic method for measuring the motility of spermatozoa swimming next to the surface of a microscope slide or at the upper surface of an open drop is the "dark-field-track method" (Rothschild and Swann, 1949). Using a dark-field condenser, such as the Zeiss 0.7–0.85 NA condenser (#48-55-07), and relatively low magnification (a 10× or lower magnification objective), spermatozoa are photographed with exposure durations of 0.5 or 1.0 second, so that the sperm head leaves a record of its path on the film.

Either continuous or stroboscopic illumination can be used. Examples of photographs obtained with this procedure can be seen in Gibbons and Gibbons (1972). This procedure can provide information about swimming velocity, the curvature of the sperm path, and the number of motile spermatozoa in a sample.

## 2. Flagellar Bending

*a. Illumination for Observation and Photographic Recording.* Stroboscopic illumination, using light flashes with durations less than 100 μsec, has become recognized as the method of choice for observing rapidly beating flagella such as sea urchin sperm tail flagella. The principal commercial source for suitable stroboscopic equipment is the Chadwick Helmuth Co., El Monte, CA. This company currently manufactures a basic strobe power supply, model 236. However, the older model 136 is advantageous because the circuitry allows operation above its rated wattage for short bursts during brief photographic exposures. The model 271 point source microscope lamp is usable over a wide range of intensities and flash frequencies, including single flash operation. However, for operation at reduced intensities and frequencies of 100 Hz or greater, a more efficient lamp, model LX150F, is available from ILC Technology, Sunnyvale, CA. This lamp requires a special interface, such as the model 276V Trigger Module, available from Chadwick Helmuth. The output of these lamps is relatively cool, but a simple filter of heat-absorbing glass (e.g., #E4010, Edmund Scientific Co., Barrington, NJ) can be used to further reduce heating of the preparations.

A high-quality oscillator, such as the General Radio model 1310B, with a stable output that is not sensitive to the radio frequency noise from a nearby strobe lamp, is required to provide trigger pulses to the strobe power supply. For precise quantitative work, a digital frequency counter should be available to calibrate the oscillator output frequencies.

*b. Microscope Optics.* I have found three combinations to be particularly useful:

1. For low-resolution examination of spermatozoa swimming at the upper surface of an open drop: Zeiss dry dark-field condenser, 0.8–0.95 NA (#48-55-15); Zeiss 16×, 0.40 NA. Neofluar objective (#46-05-20); 4 or 5× eyepiece for photography on 35-mm film.

2. For medium-resolution examination of spermatozoa under a cover glass: Zeiss dark-field condenser, as in (1); Zeiss 40×, 1.0 NA apochromat objective with iris diaphragm to reduce the aperture (#46-17-46); 3.15 or 4× eyepiece magnification for photography.

3. For highest resolution: Zeiss ultra-dark-field condenser (#48-55-01); Zeiss

100×, 1.3 NA apochromat objective with iris diaphragm to reduce the aperture (#46-19-46); 2.5× eyepiece magnification for photography.

Many boxes of microscope slides from commercial suppliers have etched surfaces and are unsuitable for high-resolution dark-field microscopy. When a box with clean surfaces is found, it should be treasured, and the slides should be washed and reused until they become obviously scratched.

*c. Photographic Recording.* Any ordinary 35-mm photomicrographic camera set-up can be used successfully to obtain photographic records, and it is especially convenient to use multiple flash photography to record several images of a flagellum on each frame. Four or five images per frame is the useful maximum, and this number can be limited by selecting an appropriate shutter speed, or, more exactly, by controlling the strobe trigger pulses. To obtain more extensive image sequences, moving film photography is useful. A 35-mm oscilloscope recording camera, such as the model C4R from Grass Instrument Co., Quincy, MA, is ideal. The frame width is reduced to 5 to 10 mm by a mask close to the film plane. The shutter and lens supplied with the camera are replaced with an adapter to rigidly couple the camera to a Zeiss photomicrography attachment. The relay in the camera can be used to switch between a low-frequency oscillator to trigger strobe flashes for observation and a high-frequency oscillator for photography. With this arrangement, high-resolution photographs at flash rates of up to at least 600 Hz are obtainable.

I routinely use Kodak Tri-X 35-mm film, either developed with Acufine (Acufine, Inc., Chicago, IL) to give approximately 1600 ASA or with Perfection XR-1 developer (Perfection Photographic Products, Los Angeles, CA) to give approximately 6400 ASA.

*d. Image Analysis.* New methods of computer-assisted image analysis are needed to deal with the large amounts of detailed information about flagellar bending that can be obtained with these photographic techniques. A method for extracting bending wave parameters from digitized low-resolution images has been described recently (Brokaw, 1985b), and several research groups are currently exploring methods for computer-assisted analysis of high-resolution images.

## 3. Micromanipulation and Stiffness Measurements

The papers of Okuno and Hiramoto (1976, 1979) and Okuno (1980) should be consulted for these techniques.

## 4. Microtubule Sliding

Summers and Gibbons (1971) introduced the use of trypsin-digested axonemes as a preparation for studying ATP-dependent sliding of axonemal microtubules.

Techniques using trypsin have been summarized by Gibbons (1982). More recently, elastase has been suggested to be a more specific protease for this purpose (Brokaw, 1980), but elastase has been found to be most useful in work with axonemes that are immobilized with polylysine, since polylysine is not digested by elastase. A useful method using elastase is described by Asai and Brokaw (1980): A small portion of concentrated sperm suspension is placed in a test tube and homogenized with a vortex mixer for about 30 seconds. The duration of mixing should be determined by experimentation so that there are no intact spermatozoa, yet the axonemal fragments are not too short. A 10-μl portion of these broken spermatozoa is mixed with 0.5 ml of demembranation solution (Section I,C,4) and 1.5 μg of elastase (Calbiochem, San Diego, CA, #324689; or Sigma, St. Louis, MO, E-0127) and incubated for approximately 15 minutes at 18°C. The mixture is then diluted with 2.5 ml of ATP-free reactivation solution containing 5 μg/ml of chicken ovoinhibitor (Sigma, T-1886). These elastase-digested axonemes can be stored at 0°C for several hours. They are then diluted with an equal volume of reactivation solution, with or without ATP, for observation. To watch the disintegration of axonemal fragments by ATP-dependent microtuble sliding, the usual procedure is to place the axonemal suspension, without ATP, on a microscope slide under a cover glass, and add a drop of reactivation solution containing ATP to the edge of the cover glass; disintegration can be observed as the ATP diffuses into the preparation.

Quantitative measurements of the rate of microtubule sliding during the ATP-induced disintegration of protease-digested axonemes and of the force generated during this sliding have been described (Yano and Miki-Noumura, 1980; Kamimura and Takahashi, 1981; Takahashi *et al.*, 1982). For these measurements, the axonemal fragments are usually exposed simultaneously to ATP and either trypsin or elastase, and the measurement is made as soon as the digestion has proceeded to the point where sliding begins.

## B. Other Methods for Assaying Motility

Although photographic methods can provide very detailed information about flagellar bending, they sample relatively small numbers of spermatozoa in a population. It would be desirable to have a method that measured average motility parameters of a larger sperm population and of a population of spermatozoa that was not confined to the vicinity of a surface. Such a method would be particularly useful for comparing motility with parameters such as metabolic rate.

One very simple method was introduced by Branham (1966). This method relies on the observation that live spermatozoa sediment more rapidly than dead spermatozoa in a mild centrifugal field, presumably because of geotactic orientation of the swimming spermatozoa. In Branham's method, 15-ml samples of suspensions of sea urchin spermatozoa at a concentration of approximately $2 \times 10^7$ spermatozoa/ml were centrifuged at 200 *g* for 20 minutes in a conical centrifuge tube, using a swinging bucket rotor. After centrifugation, 5 ml of suspension was removed from a point 1 cm above the bottom of the tube. The difference between the optical density of this sample and of a sample of uncentrifuged spermatozoa was taken as a measure of motility. Nelson (1972) modified this technique by centrifuging for 4-minute periods at 120 *g* in a spectrophotometer cuvette and measuring the optical density (OD) at 540 nm before and after centrifugation. Sperm killed with 1% formaldehyde showed little or no change in OD upon centrifugation, while the OD of live sperm suspensions decreased because the spermatozoa swam to the bottom of the tube under the influence of the centrifugal field and accumulated there.

A major limitation of these methods is that the measurement does not distinguish between changes in the ability of the spermatozoa to swim (''swimming velocity''), changes in the helical parameters of the three-dimensional sperm paths, or changes in the ability of the spermatozoa to respond to gravitational forces. These different responses need to be distinguished by microscopic observation of the sperm samples. With this additional information the technique can be a useful tool.

## C. Measurement of Sperm Energy Metabolism

Since 90% or more of sperm energy metabolism is normally coupled to motility, measurements of the oxidative metabolism of live spermatozoa or the rate of ATP dephosphorylation by demembranated spermatozoa are important adjuncts of sperm motility studies. In both cases, measurements should be made on suspensions with relatively low sperm density if it is also desired to measure motility parameters in samples taken from the suspensions during the measurements, without modifying the sperm environment by additional dilution.

### 1. Measurements on Live Spermatozoa

*a. Measurements of Oxygen Consumption.* These measurements typically utilize a polarographic oxygen sensor, obtainable from Yellow Springs Instrument Co., Yellow Springs, OH, or Beckman/Altex, Fullerton, CA. The sample must be isolated from atmospheric oxygen during the measurement. A sample chamber with a Beckman oxygen sensor inserted from the side and the surface of the solution sealed by a layer of mineral oil was designed so that samples could be removed during a measurement and transferred to a microscope slide for examination of motility (Brokaw and Benedict, 1968b). The suspending buffer can be the same as for other observations of motility. The stirring rate must be carefully chosen to avoid sperm breakage while still giving adequate exchange of solution at the electrode surface.

Measurements have been made on *S. purpuratus* spermatozoa suspended at a density of 2–3 $\times$ $10^7$ spermatozoa/ml, with observations of motility on samples taken during the oxygen consumption measurements showing 90% or more of the spermatozoa with normal motility (Brokaw and Benedict, unpublished; results quoted in Brokaw and Gibbons, 1975). The spermatozoa were suspended in artificial seawater augmented with 0.5 m*M* EDTA, 10 m*M* histidine, 0.1 *M* urea, and 1% PVP, at pH 8.1 and 16°C. The mean beat frequency was 31 $\pm$ 1.6 Hz (SD) and the mean respiratory rate was 44 $\pm$ 7 n*M*/min/$10^9$ spermatozoa (28 samples). Spermatozoa that were broken by vortex mixing of the concentrated stock sperm suspension, to eliminate motility, had a respiratory rate of 4 $\pm$ 1.5 n*M*/min/$10^9$ spermatozoa (nine samples), indicating that 90% or more of respiration is coupled to motility. Spermatozoa of *L. pictus* measured under the same conditions gave values that were essentially the same as for *S. purpuratus*. Lower values have been obtained by other workers using more concentrated sperm suspensions without confirming the motility of the spermatozoa during the respiration measurements: 20 n*M*/min/$10^9$ spermatozoa for *Hemicentrotus pulcherrimus* spermatozoa at a density of 1.4 $\times$ $10^9$ spermatozoa/ml in artificial seawater with 25 m*M* HEPES buffer at pH 8.2 (Suzuki *et al.*, 1981); 4.3 n*M*/min/$10^9$ spermatozoa for washed spermatozoa of *S. purpuratus* at a density of 8 $\times$ $10^8$ spermatozoa/ml at 15°C in seawater containing speract, the motility-stimulating peptide from egg jelly coats (Hansbrough and Garbers, 1981). The value of 20 $\mu$*M* per $10^9$ spermatozoa (assuming 5 $\times$ $10^{10}$ spermatozoa/ml of stock sperm suspension) for *S. purpuratus* spermatozoa at a density of 2.5 $\times$ $10^8$ spermatozoa/ml in artificial seawater at 10–12°C is obviously in error unless micromoles are interpreted to be nanomoles (Christen *et al.*, 1983).

*b. pH-Stat Measurements of Acid Production Associated with Metabolism.* The rate of acid production is linearly proportional to oxygen uptake over a considerable range of sperm concentrations. This measurement is somewhat more sensitive than measurements with an oxygen electrode and can be used at lower sperm concentrations. Its sensitivity depends on minimizing the buffer capacity of the solution; added buffers should be avoided, and the buffering capacity can be reduced further by using $CO_2$-free artificial seawater solutions rather than natural seawater solutions. Details of these methods have been given in a paper on *Ciona* spermatozoa (Brokaw and Benedict, 1968b), but the same methods have been used successfully with sea urchin spermatozoa. For *S. purpuratus* spermatozoa, the ratio between $H^+$ produced and $O_2$ consumed is 1.10 at pH 8.1, 16°C rather than the ratio of 1.34 obtained under the same conditions with *Ciona* spermatozoa. If a pH-stat is not available, measurements can be made with an expanded scale pH meter and a potentiometric recorder, by adding known amounts of NaOH solution at intervals to keep the pH within a range of 0.1 pH unit.

## 2. Measurements on Demembranated Spermatozoa

Methods for pH-stat measurements of ATP dephosphorylation in suspensions of demembranated spermatozoa have been described (Brokaw and Benedict, 1968a; Brokaw and Simonick, 1977). These measurements are made in solutions with the same composition as the solutions for observation of reactivated motility, except for the omission of buffer compounds such as Tris and the inclusion of oligomycin (8 μg/ml) to inhibit ATPases other than flagellar dynein ATPase; it may also be useful to reduce the concentration of EGTA or EDTA in order to further reduce the buffer capacity. Measurements can be readily carried out on 10-ml samples containing sperm concentrations of about $2 \times 10^7$ spermatozoa/ml, with typical dephosphorylation rates of 250 n*M*/min/$10^9$ spermatozoa.

ATP dephosphorylation can also be measured with colorimetric assays for inorganic phosphate, but the reactivation solution cannot contain PEG or PVP, as these components interfere with the phosphate assay. Newly developed methods using Malachite Green are as sensitive as the pH-stat method. In my laboratory, the method of Carter and Karl (1982) has been used to measure inorganic phosphate concentrations up to 50 μ*M;* it can be usefully used for ATPase assays with substrate concentrations as low as 50 μ*M*. The reagents are

A. Four volumes of 2 *N* $H_2SO_4$ mixed with 3 vol of 0.1 *M* $Na_2MoO_4$

B. 0.042% (wt/vol) Malachite Green hydrochloride (Sigma M-9636) in 1% polyvinylacetate (Aldrich Chem. Co., Milwaukee, WI, #18933-2). The polyvinylacetate is dissolved by stirring in warm water.

C. 7.8% $H_2SO_4$

These reagents are stable for at least several months; reagent B is normally kept refrigerated. A 10-μl sample of concentrated spermatozoa (adjusted to give OD = 0.21 as described in Section I,A) is mixed with 200 μl of demembranation solution (Section I,C,4). After 30 seconds a 100-μl portion is added to 15 ml of reactivation solution (Section I,C,4) with oligomycin and no PEG, to give a sperm concentration of approximately $7 \times 10^6$/ml. A 1.5-ml aliquot is removed, and added to a tube containing 1.0 ml of reagent A; 0.5 ml of reagent B is added immediately, and the tube contents are mixed with a vortex mixer. Two minutes later, 3.0 ml of reagent C is added to the tube. Subsequent aliquots are removed from the sperm suspension at 2-minute intervals. Five aliquots, spanning an 8-minute period, are normally used to obtain a measurement of the rate of ATP dephosphorylation. The OD is read at 625 nm 40–60 minutes later.

Measurements with radiolabeled ATP provide an extremely sensitive method which can be used with 0.2-ml samples containing approximately $10^7$ spermatozoa/ml and ATP concentrations below 10 μ*M* (Penningroth *et al.*, 1982).

## Acknowledgment

Preparation of this article has been supported in part by a grant from the National Institutes of Health, No. GM 18711.

## References

Asai, D. J., and Brokaw, C. J. (1980). *J. Cell Biol.* **87,** 114–123.
Ashani, Y., and Catravas, G. N. (1980). *Anal. Biochem.* **109,** 55–62.
Branham, J. M. (1966). *Biol. Bull.* **131,** 251–260.
Brokaw, C. J. (1965). *J. Exp. Biol.* **43,** 155–169.
Brokaw, C. J. (1966). *J. Exp. Biol.* **45,** 113–139.
Brokaw, C. J. (1980). *Science* **207,** 1365–1367.
Brokaw, C. J. (1985a). *Ann. N.Y. Acad. Sci.* **438,** 132–141.
Brokaw, C. J. (1985b). *Cell Motil.* **4,** 417–430.
Brokaw, C. J., and Benedict, B. (1968a). *Arch. Biochem. Biophys.* **125,** 770–778.
Brokaw, C. J., and Benedict, B. (1968b). *J. Gen. Physiol.* **52,** 283–299.
Brokaw, C. J., and Gibbons, I. R. (1973). *J. Cell Sci.* **13,** 1–10.
Brokaw, C. J., and Gibbons, I. R. (1975). *In* "Swimming and Flying in Nature" (T. Y. Wu, C. J. Brokaw and C. Brennan, eds.), pp. 89–126. Plenum, New York.
Brokaw, C. J., and Josslin, R. (1973). *J. Exp. Biol.* **59,** 617–628.
Brokaw, C. J., and Simonick, T. F. (1977). *J. Cell Sci.* **23,** 227–241.
Brokaw, C. J., Josslin, R., and Bobrow, L. (1974). *Biochem. Biophys. Res. Commun.* **58,** 795–800.
Carter, S. G., and Karl, D. W. (1982). *J. Biochem. Biophys. Methods* **7,** 7–13.
Cavanaugh, G. M. (1956). "Formulae and Methods V. of the Marine Biological Laboratory Chemical Room." Marine Biological Laboratory, Woods Hole, Massachusetts.
Christen, R., Schackmann, R. W., and Shapiro, B. M. (1983). *J. Biol. Chem.* **258,** 5392–5399.
Durham, A. C. H. (1983). *Cell Calcium* **4,** 33–46.
Fujii, T., Utida, S., and Mizuno, T. (1955). *Nature (London)* **176,** 1068–1069.
Garfinkel, L., and Garfinkel, D. (1984). *Biochemistry* **23,** 3547–3552.
Gibbons, B. H. (1982). *Methods Cell Biol.* **25,** 253–271.
Gibbons, B. H., and Gibbons, I. R. (1972). *J. Cell Biol.* **54,** 75–97.
Gibbons, B. H., and Gibbons, I. R. (1973). *J. Cell Sci.* **13,** 337–357.
Gibbons, B. H., and Gibbons, I. R. (1974). *J. Cell Biol.* **63,** 970–985.
Gibbons, B. H., and Gibbons, I. R. (1976). *Biochem. Biophys. Res. Commun.* **73,** 1–6.
Gibbons, B. H., and Gibbons, I. R. (1979). *J. Biol. Chem.* **254,** 197–201.
Gibbons, B. H., and Gibbons, I. R. (1980). *J. Cell Biol.* **84,** 13–27.
Gibbons, B. H., and Gibbons, I. R. (1984). *Nature (London)* **309,** 560–562.
Gibbons, I. R., and Fronk, E. (1972). *J. Cell Biol.* **54,** 365–381.
Gibbons, I. R., Evans, J. A., and Gibbons, B. H. (1982). *Cell Motil. Suppl.* **1,** 181–184.
Goldstein, S. F. (1974). *In* "Cilia and Flagella" (M. A. Sleigh, ed.), pp. 111–130. Academic Press, London.
Hansbrough, J. R., and Garbers, D. L. (1981). *J. Biol. Chem.* **256,** 2235–2241.
Ishiguro, K., Murofushi, H., and Sakai, H. (1982). *J. Cell Biol.* **92,** 777–782.
Johnson, C. H., Clapper, D. L., Winkler, M. M., Lee, H. C., and Epel, D. (1983). *Dev. Biol.* **98,** 493–501.
Kamimura, S., and Takahashi, K. (1981). *Nature (London)* **293,** 566–568.
Katz, D. F., and Blake, J. R. (1975). *In* "Swimming and Flying in Nature" (T. Y. Wu, C. J. Brokaw, and C. Brennan, eds.), pp. 173–184. Plenum, New York.

Kinoshita, S. (1958). *J. Fac. Sci. U. Tokyo Sect. IV.* **8,** 219–228.
Mabuchi, I., Shimizu, T., and Mabuchi, Y. (1976). *Arch. Biochem. Biophys.* **176,** 564–576.
Miller, D. J., and Smith, G. L. (1984). *Am. J. Physiol.* **246** (*Cell Physiol.* **15**), C160–C166.
Nelson, L. (1972). *Exp. Cell Res.* **74,** 269–274.
Okuno, M. (1980). *J. Cell Biol.* **85,** 712–725.
Okuno, M., and Hiramoto, Y. (1976). *J. Exp. Biol.* **65,** 401–413.
Okuno, M., and Hiramoto, Y. (1979). *J. Exp. Biol.* **79,** 235–243.
Penningroth, S. M., Cheung, A., Olehnik, K., and Koslosky, R. (1982). *J. Cell Biol.* **92,** 733–741.
Rothschild, L., and Swann, M. M. (1949). *J. Exp. Biol.* **26,** 164–176.
Sillen, L. G., and Martell, A. E. (1964). ''Stability Constants of Metal-Ion Complexes.'' Chemical Society, London.
Summers, K. E., and Gibbons, K. R. (1971). *Proc. Natl. Acad. Sci. U.S.A.* **68,** 3092–3096.
Suzuki, N., Nomura, K., Ohtake, H., and Isaka, S. (1981). *Biochem. Biophys. Res. Commun.* **99,** 1238–1244.
Takahashi, K., Shingyojo, C., and Kamimura, S. (1982). *Symp. Soc. Exp. Biol.* **35,** 159–177.
Tyler, A. (1953). *Biol. Bull.* **194,** 224–239.
Yano, Y., and Miki-Noumura, T. (1980). *J. Cell Sci.* **44,** 169–186.

# Chapter 4

# *Ion Measurements in Sea Urchin Sperm*

ROBERT SCHACKMANN

*Department of Biochemistry*
*University of Washington*
*Seattle, Washington*



## I. Introduction

Recent literature, both biochemical and physiological, pays increasing attention to relationships among cellular ion movements, ion-translocating systems, and cellular function. Ion transport, once thought to be specific for tissues containing excitable cells or cells specialized for salt or acid secretion, appears now to be a component of cell responses to stimuli throughout growth and development. $Na^+/H^+$ exchanges, which increase $pH_i$[1] (Johnson *et al.*, 1976; Shen and Steinhardt, 1978; Hesketh *et al.*, 1985; L'Allemain *et al.*, 1984; Moolenaar *et al.*, 1984; Rothenberg *et al.*, 1983), changes in $[Ca^{2+}]_i$ (Ridgway

[1]Abbreviations: $pH_i$, intracellular pH; $pH_e$, extracellular pH; $[Ca^{2+}]_i$, intracellular calcium concentration; Ar, accumulation ratio; TPP, tetraphenylphosphonium; SCN, thiocyanate; DMO, 5,5-dimethyloazolidine-2,4-dione; $P_i$, inorganic phosphate.

*et al.*, 1977; Steinhardt *et al.*, 1977; Hesketh *et al.*, 1985), or changes in the plasma membrane potential (Rothenberg *et al.*, 1982) are among the earliest responses of quiescent eukaryotic cells to stimulation of cell growth. Cellular responses to a variety of hormonal agonists have been linked to both alterations of $[Ca^{2+}]_i$ (Thomas *et al.*, 1984) and $pH_i$ (Moore *et al.*, 1982), and ATP production by oxidative phosphorylation (Boyer *et al.*, 1977) involves pH and potential gradients operating at the subcellular level. These are but a few examples of biological systems in which ions serve as regulatory components of cell biology.

Historically, echinoid sperm were among the first cells used to provide evidence that ionic changes are important to cell activation. It was demonstrated 30 years ago by Jean Dan that the sea urchin sperm underwent an acrosomal reaction when exposed to factors released from the egg (egg jelly). This reaction was defined morphologically to consist of exocytosis of the acrosomal granule and polymerization of actin into the acrosomal rod. Dan (1954) showed that removal of $Ca^{2+}$ from the seawater prevented the acrosomal reaction. Likewise increasing the $pH_e$ to 9.2 spontaneously initiated the acrosomal reaction in the absence of any of the egg factors (Dan, 1952). Thus, ionic alterations were implicated in the activation of the sperm for fertilization. We now know that egg factors that initiate the acrosomal reaction cause both $Ca^{2+}$ influx and increased $pH_i$ (Schackmann *et al.*, 1978, 1981). The $Ca^{2+}$ influx is thought to lead to exocytosis of the acrosomal granule while increased $pH_i$ can cause actin polymerization (Tilney *et al.*, 1978; Schroeder and Christen, 1982). Additionally, $pH_i$ is a primary regulatory component of sea urchin sperm motility and respiration (Christen *et al.*, 1982, 1983a,b; Lee *et al.*, 1983). At low $pH_i$ ($< \sim 7.0$) sperm are immotile. As $pH_i$ is increased to the normal $pH_i$ of $\sim 7.4$, sperm respiration and motility are initiated. These findings underlie the importance for studying ion movements as regulatory components of sperm activation.

This article discusses methods used to measure ion changes important to activation of sea urchin sperm motility and the acrosome reaction. I describe briefly the methods used by several laboratories to measure $pH_i$, ion fluxes, and membrane potentials in sea urchin sperm. It is not my intention to explain in complete detail how to make the ion measurements, but to provide a supplement to published methods which hopefully contains additional insight on their application.

## II. Sperm Preparation and Determination of Intracellular Water Space

### A. Sperm Preparation

Sperm are typically collected from sea urchins by injecting a few milliliters of 0.5 *M* KCl into the coelomic cavity. After initiation of spawning, the animals are inverted and allowed to shed into a Petri dish without seawater ("dry sperm"). Experiments are usually performed at 0.1 to 10% suspensions of "dry sperm" (volume:volume). The supernatant of centrifuged dry sperm has been analyzed by atomic absorption spectroscopy (Johnson *et al.*, 1983) and except for $[K^+]$ (20–30 m*M*), the composition approximates that of seawater. Thus $[Na^+]$ is > 400 m*M* and $[Ca^{2+}]$ is ~10 m*M*. The high $[Na^+]$ necessitates that sperm be washed, if they are to be removed from $Na^+$. A 100-fold dilution contains ~4 m*M* $Na^+$. This is sufficient $Na^+$ to activate sperm motility and respiration (Christen *et al.*, 1982). Washing of sperm is easily accomplished by gentle centrifugation at 1000–1500 *g* for ~10 minutes.

## B. Measurements of Intracellular and Extracellular Water Space

It is convenient to measure uptake and cell content of ions, expressed per unit of cell protein or cell number. However, interpretation of the physiological significance of the ion content often requires an approximation of the intracellular concentration, which requires evaluation of the intracellular water space. The accepted method for evaluating intracellular water consists of a difference measurement between total $[^3H]H_2O$ in a pellet of cells and the pellet volume occupied by molecules impermeant to the plasma membrane (Rottenberg, 1979). Water rapidly penetrates biological membranes and the volume of $[^3H]H_2O$ in a pellet represents the total of intra- and extracellular water space. Typically, sugar molecules fill the role as impermeant, extracellular markers. $[^{14}C]$Inulin, sucrose, and sorbitol have all been used with sea urchin sperm (Schackmann *et al.*, 1981; Lee *et al.*, 1983). Values for the extracellular space are ~0.2 μl/$10^8$ sperm. The intracellular water space is 0.7–0.8 μl/$10^8$ sperm (Schackmann *et al.*, 1981; Lee *et al.*, 1983) and is quite close to the morphologically determined value of 0.72 μl/$10^8$ sperm for the sea urchin *A. punctulata* (Harvey and Anderson, 1943). Sucrose is easier to use than carboxy-inulin as the latter tends to accumulate precipitative radioactive material which can lead to erroneously large volumes for extracellular space (in extreme cases exceeding the total $[^3H]H_2O$ space). However, sucrose is finitely permeant and incubations as an extracellular marker should be kept to only a few minutes. Since the intracellular $H_2O$ space measurement is determined by a difference between experimental values, each with its own inherent experimental error, the total error in the intracellular water content can be larger than that for measurement of uptake of the ions or $pH_i$ probes themselves. Small differences or changes in intracellular volume are also not easily detected. It is clearly important to evaluate the cell and pellet water spaces under each set of experimental conditions used. The accuracy of the calculations of ion concentrations, $pH_i$, or membrane potential are no better than the water space measurements themselves.

## III. pH Measurements Important to Sperm Activation

Initiation of motility and induction of the acrosome reaction are associated with increases in $pH_i$ regardless of the stimulus applied. To be motile the sperm $pH_i$ must be $\geq \sim 7.2$ (Christen *et al.*, 1982, 1983a,b; Lee *et al.*, 1983). For the acrosome reaction to occur the $pH_i$ must be raised to $\sim 7.6$ (Schackmann *et al.*, 1981). However, the experiments first indicating the importance of $pH_i$ increases in sperm activation were not based on direct measurements of $pH_i$, but rather were observations that acid efflux occurred upon activation of either motility or the acrosome reaction. Activation of sperm respiration and motility by dilution (Johnson *et al.*, 1983), by addition of the egg peptide "speract" (Hansbrough and Garbers, 1981; Repaske and Garbers, 1983), or by addition of $Na^+$ to sperm in $Na^+$-depleted seawater (Nishioka and Cross, 1978; Lee, 1984a), all result in acid efflux and elevation of $pH_i$. Induction of the acrosome reaction by either egg jelly or ionophores was likewise found to be accompanied by acid efflux (Schackmann *et al.*, 1978; Tilney *et al.*, 1978).

### A. Measurement of Acid Efflux

Measurement of the acid efflux upon activation of motility or the acrosome reaction provides rapid quantitative information necessary to understand how $H^+$ efflux is coupled to other ion movements such as $Na^+$ entry (see Section IV,B). To measure the efflux a pH electrode is placed in a sperm suspension, agents are added to stimulate the sperm, and changes in the external pH are followed with a recorder. For example, when $Na^+$ is added to a 1% sperm suspension in $Na^+$-free seawater, $H^+$ efflux of $\sim 100$ nmol occurs. The resulting $pH_e$ decrease is easily monitored with most of the better modern pH meters and electrodes. Our laboratory uses a pHM64 Radiometer pH meter with a GK2310-C combination electrode. The output to the recorder must be offset (Madeira, 1975). With seawater buffered with $HCO_3^-$ (Lee, 1984a) or low concentrations of Tris or HEPES ($\sim 1$ m*M*), variation of the sperm concentration allows manipulation of the total external pH change. I try to keep the change to $\sim 0.05$ pH units to approximate measurements at a constant $pH_e$. The response time of the electrode must be reasonably fast as the rapid pH changes due to activation of the acrosome reaction and motility initiation are complete within 15 seconds.

Calculation of acid equivalents released is usually performed by titration with freshly prepared NaOH or KOH. Stock solutions should be standardized against an acid of known concentration since hydroxide reacts with dissolved $CO_2$. Additionally, as sperm respire, acid is produced (Brokaw and Benedict, 1968). This acid release can be sufficiently rapid, particularly at higher temperatures, to make measurement of the activation-associated $pH_e$ changes difficult. The resultant recorder trace can appear as a small rapid change on top of much larger continuing acid efflux. Resolution of the rapid activation-dependent $pH_e$ changes is improved by reducing the temperature or by inhibiting mitochondrial $O_2$ consumption with oligomycin (Hansbrough and Garbers, 1981) or cyanide (Schackmann *et al.*, 1978; Lee, 1984a).

The mixing chamber and electrode must be tested for their ability to respond to the appearance of acid in the extracellular medium. This is most easily performed by addition of a small aliquot of a strong acid such as HCl to a solution with and without sperm. The response of the electrode to the HCl addition should be considerably faster than that obtained upon activation of the sperm. The system we use gives responses about four-fold faster for HCl than to sperm acid release. Faster mixing is necessary for kinetic measurements.

## B. Measurement of $pH_i$

Methods available for measurement of $pH_i$ include uptake of radiolabeled weak acids and bases, $^{31}P$-nuclear magnetic resonance, and incorporation of molecular probes such as carboxyfluorescein which undergo spectral or fluorescent shifts with changes in $pH_i$. Two excellent reviews on $pH_i$ measurements and methods have recently been published (Nuccitelli and Deamer, 1982; Busa and Nuccitelli, 1984). Since each probe for $pH_i$ may be subject to systematic error (examples are given below), the best experimental approach is to apply several methods to evaluate $pH_i$. Agreement between unrelated methods is a strong indication that values determined are reliable.

### 1. Uptake of Labeled Weak Bases and Acids

*a. Uptake of [$^{14}$C]Methylamine and [$^{14}$C]Diethylamine.* For reasons discussed below, uptake of weak bases has served as the primary method for measurement of $pH_i$ in sea urchin sperm. Uptake occurs because the uncharged, unprotonated amine rapidly equilibrates across the plasma membrane and is trapped when protonated within the cell (Schuldiner *et al.*, 1972). Weak bases are concentrated by this mechanism into acidic environments. At high concentrations (10 m*M* is sufficient for sea urchin sperm at pH 8) uptake can increase $pH_i$; at low concentrations it allows for measurement of $pH_i$ as intracellular buffering compensates for the small amount of amine accumulated. If a few specific conditions are satisfied,

$$pH_i = pH_e - \log(Ar) \tag{1}$$

where Ar is the accumulation ratio, the apparent intracellular concentration of the weak base (or other probe molecule) divided by the extracellular concentration. For the simplest of conditions, measurement of $pH_i$ consists only of monitoring $pH_e$ and measuring the amount of weak base accumulated.

Evaluation of uptake is performed by incubation of the sperm, typically a ~100-fold dilution of dry sperm, with sufficient radioactive amine (~0.3 μCi/ml) until a steady state level of uptake is reached. Following rapid separation of the sperm from the reaction medium, radioactivity is counted and the Ar is determined (Schackmann *et al.*, 1981). Several separation techniques have been employed. Centrifugation in a microfuge either directly or by layering an aliquot of the sperm suspension onto silicone oil gives values for $pH_i$ of ~7.4 (Lee *et al.*, 1983; Christen *et al.*, 1982), in agreement with those of other techniques such as $^{31}$P-nuclear magnetic resonance ($^{31}$P-NMR) (Christen *et al.*, 1983b). Assays are facilitated by spinning through silicone oil (Schackmann *et al.*, 1981) as the pellets are then stable throughout the remainder of the experiment and do not require immediate attention. For sperm in artificial seawater (pH 8), $pH_i < pH_e$ (Ar $\geq$ ~3), the amount of radioactive weak base trapped in the extracellular space is therefore <10% of the total in the pellet. Accurate and reproducible measurements are possible because the weak base is concentrated within the sperm.

Filtration has also been used (Repaske and Garbers, 1983) to separate the sperm, but gave a $pH_i$ value of ~7.7. Although filtration with subsequent washing has the advantage of not requiring extracellular water space measurements, if fluxes are fast, appreciable weak base may exit during the washes. Such an effect results in Ar values artificially low and a $pH_i$ too high. This may explain the $pH_i$ value of 7.7.

For Eq. (1) to be valid, (1) the p*K* of the weak base must be much greater than $pH_e$ or $pH_i$, (2) weak bases should not bind to sperm components, and (3) uptake must reach a steady state level determined by the acid base equilibrium of the amine across the plasma membrane. Since the p*K* of both methylamine and diethylamine are above 10, the first condition is satisfied for sperm in seawater at pH 8. Satisfying condition (2) is not simple, as binding can cause a variety of changes in uptake. Because diethylamine uptake shows excellent agreement with $pH_i$ measured by $^{31}$P-NMR, it is unlikely that binding of that amine is significant (Christen *et al.*, 1983b). Uptake of another weak base, 9-aminoacridine, is associated with a high degree of binding, but still can be used as a measure of $pH_i$, as described in Section III,B,1,c. Condition (3) dictates that uptake of weak bases be followed with time for each of the experimental conditions that may alter the uptake rate or level. In particular, it has been shown that alteration of $pH_e$ affects the kinetics of weak base uptake. Decreasing $pH_e$ decreases the concentration of uncharged amine; consequently, it decreases the rate of uptake, for example, of 9-aminoacridine (Christen *et al.*, 1982). Also, it has been calculated that increasing the $\Delta$pH (i.e., increasing Ar) increases the time required for equilibrium to be attained (Christen *et al.*, 1982).

*b. Uptake of [$^{14}$C]DMO.* Uptake of the weak acid dimethyloxazolidine-2,4-dione (DMO) has also been used as a measure of $pH_i$ in sperm (Schackmann *et al.*, 1981). Unlike weak bases, weak acids accumulate into regions of higher pH as the unprotonated acid is charged (Waddell and Butler, 1959). An Ar of ~0.5 for [$^{14}$C]DMO in sperm in artificial seawater (pH 8) gives an apparent $pH_i$ of 7.7. The difference between this value and that for weak bases ($pH_i$ ~7.4) can be explained by uptake of the weak acid into the mitochondrion. Mitochondria are usually basic with respect to the cytoplasm (Hoek *et al.*, 1980). In sea urchin sperm, the single mitochondrion may occupy a volume of up to 30% of the total intracellular space (Harvey and Anderson, 1943; Schackmann *et al.*, 1984), hence it is a substantial fraction of the intracellular volume. If the intramitochondrial space pH is 8.0, then uptake of both diethylamine (Ar ~5) and DMO (Ar ~0.5) is consistent with a value for the cytoplasm $pH_i$ of ~7.3. A complete, detailed analysis of the problem of subcellular compartments of different pH has been provided by Hoek and colleagues (1980). As calculated by Christen and colleagues (1982), at $pH_e$ 8 only ~7% of the weak base partitions into the mitochondrion whereas as much as 80% of the weak acid may be located there. In sea urchin sperm in seawater at pH 8 weak bases therefore provide a more direct indicator of the cytoplasm $pH_i$.

*c. Uptake of 9-Aminoacridine.* An alternative to measurement of $pH_i$ with radioactive weak bases involves measurement of fluorescence quenching of 9-aminoacridine. 9-Aminoacridine is a weak base and accumulates by diffusion of the unprotonated amine (Schuldiner *et al.*, 1972; Christen *et al.*, 1982). It has the property that its fluorescence is highly quenched upon uptake by sperm. Measurement of fluorescence quenching is equivalent to following disappearance of the 9-aminoacridine from the seawater as it is accumulated. Up to 80% of the dye may be taken up, yielding Ar values of 2200 or larger (Schackmann *et al.*, 1981; Christen *et al.*, 1982). If a $pH_i$ value were determined directly from Eq. (1), a $pH_i$ of ~4.7 would be calculated, in complete disagreement with values for other weak bases. However, a careful study revealed that most acridine uptake occurs as a result of the intracellular binding of the dye following accumulation as weak base (Christen *et al.*, 1982). That is, even if $\Delta$pH is completely collapsed with detergents or ionophores, substantial uptake remains. The binding properties of the sperm for 9-aminoacridine were studied and it was found that at an initial external concentration of 1–2 $\mu M$, intracellular binding was far from saturation. As explained in detail in the supplement to the paper by Christen *et al.* (1982), this allows for correction of the binding in a simple manner to give the following equation:

$$pH_i = pH_e - \log \frac{Q}{1-Q} \frac{1-Q^1}{Q^1} \quad (2)$$

where $Q$ is fraction of fluorescence quenched (fraction of 9-aminoacridine taken up) and $Q^1$ is the fraction quenched in the presence of either detergent or ionophores to collapse $\Delta pH$ across the plasma membrane. The unexpected advantage of this method of $pH_i$ measurement is that intracellular water space need not be measured. With this correction $pH_i$ values determined by uptake of 9-aminoacridine agree well with those determined by uptake of [$^{14}C$]diethylamine or with $^{31}P$-NMR (Christen *et al.*, 1982, 1983b).

## 2. $pH_i$ Measurements by $^{31}P$-NMR and with Carboxyfluorescein

Two measurements of sea urchin sperm $pH_i$ by $^{31}P$-NMR have been reported (Christen *et al.*, 1983b; Johnson *et al.*, 1983). Measurement of the chemical shift of the $^{31}P_i$ is directly related to $pH_i$ (for review, see Nuccitelli and Deamer, 1982). The chemical shift of $P_i$ varies with the degree of protonation of $P_i$ and is a good indicator of $pH_i$ in the vicinity of the p$K$ of $P_i$. Measurement of $pH_i$ therefore necessitates a pH calibration of the $P_i$ chemical shift within the sperm. Christen and colleagues (1983b) approached this problem by collapsing the $\Delta pH$ with the ionophores nigericin and monensin so that $pH_i = pH_e$. Varying $pH_e$ and recording the $P_i$ chemical shift allows for a calibration within the intracellular environment. Values determined by $^{31}P$-NMR in sperm are in excellent agreement with those determined by uptake of diethylamine (Christen *et al.*, 1983b). For example, at $pH_e$8 for sperm in artificial seawater, the $pH_i$ was 7.4 by $^{31}P$-NMR, 7.3 by [$^{14}C$]diethylamine accumulation, and 7.4 by 9-aminoacridine uptake (Christen *et al.*, 1982). And for quiescent sperm in $Na^+$-free seawater, $pH_i$ was 6.8 by $^{31}P$-NMR and 6.9 by [$^{14}C$]diethylamine uptake.

A limitation of NMR is the requirement for high cell concentration. A 10% sperm suspension was necessary to yield a satisfactory spectrum in a reasonable amount of time (minutes). This precludes time-dependent resolution of $pH_i$ changes which are complete in seconds. However, $^{31}P$-NMR does provide the only reasonable method of analyzing the $pH_i$ in dry sperm before dilution. A $pH_i$ value of ~7 was obtained, which is sufficiently low to explain sperm immotility in semen (Johnson *et al.*, 1983; Christen *et al.*, 1983b).

Carboxyfluorescein and other optical pH probes have yet to be used as $pH_i$ probes for sea urchin sperm. A detailed report of carboxyfluorescein use for measurement of $pH_i$ in bovine sperm has been published by Babcock (1983) and the interested reader is referred to this article. It is mentioned here because intracellular optical probes provide the possibility of real-time resolution of the $pH_i$ changes important to sperm activation.

For sea urchin sperm, measurement of $pH_i$ by following weak base uptake is the most direct method and can be performed without elaborate equipment. Uptake reflects the $pH_i$ of the apparent "cytoplasm" since it is in agreement with values determined by $^{31}$P-NMR and the bases will not accumulate into the mitochondria. It remains to be seen whether weak base uptake will be useful for evaluation of $pH_i$ in sperm of other species. The large acrosomal region of mammalian sperm has led Meizel and Deamer (1978) to propose that 9-aminoacridine uptake measures the intraacrosomal pH in hamster sperm.

## IV. Measurement of Cation Uptake

In the previous section $pH_i$ was discussed as it is an important regulatory component of sperm motility and the acrosome reaction. $pH_i$ changes in sea urchin sperm correlate with or are linked with other cation movements as well. In particular, a $Na^+/H^+$ exchange is a component of the sperm plasma membrane that can either raise or lower the $pH_i$ (Lee, 1984a,b). Furthermore, induction of the acrosome reaction results in multiple changes in the sperm cation composition (Schackmann *et al.*, 1981; Cantino *et al.*, 1983). $Na^+$ and $Ca^{2+}$ enter the sperm and $K^+$ (as well as $H^+$) exit. Of these $Ca^{2+}$ entry is a crucial requirement for the acrosome reaction regardless of the method used to initiate it.

### A. Measurement of $Ca^{2+}$

#### 1. $^{45}Ca^{2+}$ Uptake

The most frequently used method for measurement of $Ca^{2+}$ entry into sperm consists of following $^{45}Ca^{2+}$ accumulation. The basic methodology is the same as that for monitoring uptake of weak bases (Section III,B,1,a). Isotopic $^{45}Ca^{2+}$ is added to the seawater and incubated with the sperm under the experimental conditions to be tested. Termination of uptake is effected by rapid separation of the sperm from the medium by either centrifugation or filtration. Most experiments have been performed by filtration (Schackmann *et al.*, 1978; Kopf *et al.*, 1984; Kopf and Garbers, 1980). I have found that centrifugation of sperm through silicone fluid works equally well, although background $^{45}Ca^{2+}$ uptake is higher (~4 nmol/$10^8$ sperm) than for filtration (~1 nmol/$10^8$ sperm). Whether this represents rapid $Ca^{2+}$ uptake, $Ca^{2+}$ binding to the surface, or both, is not known. When the acrosome reaction is triggered, increased $^{45}Ca^{2+}$ uptake is initiated and may reach ~20 nmol/$10^8$ sperm (Schackmann *et al.*, 1978). Such an increase is easily resolved by either filtration or centrifugation. Stimulation of $^{45}Ca^{2+}$ entry indicates that enhanced $Ca^{2+}$ influx occurs but does not provide direct evidence that net $Ca^{2+}$ uptake results. In order to show that a net increase occurred, atomic absorption spectroscopy and X-ray microanalysis were performed. I will not deal with the methodology directly but refer the reader to work by Cantino and colleagues (1983). The results obtained by either method show that net $Ca^{2+}$ uptake does correspond to $^{45}Ca^{2+}$ uptake. X-Ray microanalysis also allowed for localization of the $Ca^{2+}$ uptake to the mitochondria as predicted from inhibitor studies (Schackmann and Shapiro, 1981). The slow, long-term accumulation of $Ca^{2+}$ for ~30 minutes after the acrosome reaction is complete represents a complex function of $Ca^{2+}$ entry through the plasma membrane and accumulation by the mitochondrion.

An additional factor limiting the usefulness of $^{45}Ca^{2+}$ uptake is the level of resolution attained. With a background level of 1–4 nmol/$10^8$ sperm, it is extremely difficult to resolve increases of even 0.1 nmol/$10^8$ sperm as were found when the acrosome reaction was initiated in the presence of drugs to inhibit mitochondrial $Ca^{2+}$ uptake (Schackmann and Shapiro, 1981). Yet a change of this magnitude could still represent an extremely large change in $[Ca^{2+}]_i$. A need for measurement of changes in free $[Ca^{2+}]_i$ is clearly indicated.

### 2. Measurement of $[Ca^{2+}]_i$

Tsien *et al.* (1982) and colleagues (Grynkiewicz *et al.*, 1985) have developed several $Ca^{2+}$ indicator dyes which show altered fluorescence upon binding of $Ca^{2+}$. These dyes can be loaded into cells as uncharged molecules if the carboxyl groups involved in binding $Ca^{2+}$ are first esterified. The ester groups are hydrolyzed within cells to generate the native $Ca^{2+}$ indicators. These dyes, "Quin2" (Tsien *et al.*, 1982), "fura-2," and "indo-1" (Grynkiewicz *et al.*, 1985), have dissociation constants of ~100 n*M* and are sufficiently insensitive to $[Mg^{2+}]$ and $[H^+]$ to allow for measurement of $[Ca^{2+}]$ within living cells. As highly charged anions, they do not readily leak from cells. These dyes have only recently begun to be applied to sperm physiology. In an experiment described in Table I, nigericin, a monovalent cation inonophore that triggers the acrosome reaction and stimulates $^{45}Ca^{2+}$ uptake (Schackmann *et al.*, 1978), was found to increase Quin2 fluorescence. The data indicate that $[Ca^{2+}]_i$ increases approximately 8-fold within a minute following addition of the nigericin. Further application of these dyes to sperm physiology will allow for a more complete understanding of $[Ca^{2+}]_i$ changes important to sperm activation.

## B. Measurement of $Na^+$ and $K^+$

The simplest method for measurement of $Na^+$ or $K^+$ uptake is to follow $^{22}Na^+$ or $^{42}K^+$ fluxes. $^{42}K^+$ has a 12-hour half-life and $^{86}Rb^+$ has been used as a $K^+$ substitute (Schackmann *et al.*, 1978). Recent attention has focused on

TABLE I

NIGERICIN INCREASES $[Ca^{2+}]_i$[a]

| Conditions | Fluorescence (arbitrary units) | $[Ca^{2+}]_i$ (units of $K_d$) |
|---|---|---|
| ASW, 5 m*M* $Ca^{2+}$ | 43 | 0.30 |
| + nigericin (60 μ*M*) | 91 | 2.3 |
| ASW, 0 m*M* $Ca^{2+}$ | 41 | 0.27 |
| + nigericin (60 μ*M*) | 40 | 0.25 |
| ASW, 5 m*M* $Ca^{2+}$ | | |
| + ionomycin (30 μ*M*) | 122 | — |
| + ionomycin + 3 m*M* $Mn^{2+}$ | 0 | — |

[a] Sperm (10% dry sperm by volume) of the sea urchin, *S. purpuratus*, were incubated 5 hours in calcium-free seawater, pH 6.8, with 1 m*M* EGTA and 50 μ*M* Quin2-acetoxymethyl ester (Tsien *et al.*, 1982). After washing the sperm and resuspending the pellet as 20% dry sperm, 100 μl was added to 1.9 ml ASW at the indicated $[Ca^{2+}]$. Ionophores were followed by 1-minute incubations prior to fluorescence measurements. $[Ca^{2+}]$ was calculated as $(F - F_{min}/F_{max} - F)K_d$ (Tsien *et al.*, 1982), where $F$ = fluorescence and $F_{max}$ = the fluorescence with 5 m*M* $Ca^{2+}$ and ionomycin. $F_{min} = 0.16\ \Delta F$ where $\Delta F$ is between $F_{max}$ and the $F$ with ionomycin + $Mn^{2+}$ to quench Quin2 fluorescence (Hesketh *et al.*, 1983). $K_d$ is the dissociation constant for the Quin2·$Ca^{2+}$ complex within the sperm.

$^{22}Na^+$ movements because $Na^+/H^+$ exchange has been linked to $pH_i$ changes associated with motility activation (Lee, 1984a; Hansbrough and Garbers, 1981). Uptake of $^{22}Na^+$ is complete within seconds when motility and respiration are stimulated by $Na^+$ addition to sperm in $Na^+$-free seawater (Lee, 1984a; Schackmann, unpublished data). $^{22}Na^+$ entry is rapid when the acrosome reaction (Schackmann and Shapiro, 1981) is triggered and is sufficiently rapid in intact sea urchin sperm tails (Lee, 1984a,b) to preclude a kinetic analysis. These studies of $^{22}Na^+$ uptake were performed by filtration on glass fiber filters (Whatman GF/C). As noted in Section IV,A,1, for $^{45}Ca^{2+}$ uptake, both centrifugation and filtration gave comparable results. However, centrifugation of the sperm through silicone oil gives values of $^{22}Na^+$ uptake approximately two-fold higher than by filtration (Cantino *et al.*, 1983) and these values are in much closer agreement to those determined by X-ray microanalysis. Presumably, when sperm are filtered and washed, some $^{22}Na^+$ exits rapidly. The time taken for $^{22}Na^+$ to equilibrate across the sperm membranes is about the same magnitude as the time necessary for washing the filters and could lead to underestimates for $Na^+$ uptake. Such apparent differences demand careful attention when quantifying $Na^+$ uptake. The recently described $Na^+/H^+$ exchange in isolated tails from sea urchin sperm is thought to be electrically neutral (Lee, 1984b) on the basis of an apparent 1:1 stoichiometry. Since different techniques yield different values for $Na^+$ uptake, arguments for electroneutrality based on filtration measurements alone are unwarranted without careful substantiation.

Even larger differences between data acquired by filtration and centrifugation have been registered for $K^+$ uptake (Cantino *et al.*, 1983). Again, centrifugation yields values in closer agreement to those determined by atomic absorption or X-ray microanalysis.

An additional difficulty exists with $^{22}Na^+$ measurements. Evaluation of ion uptake is difficult if the ion is not concentrated within the sperm; that is if Ar $\leq$ 1. This means that measurement of $K^+$ is quite simple since the $Ar_{K^+} = {\sim}20$ but for sperm in artificial seawater $Ar_{Na^+} = {\sim}0.3$. The amount of $Na^+$ trapped in the extracellular space is at least 50% of the total amount pelleted with the sperm. At normal levels of seawater $Na^+$ both methods, filtration and centrifugation, are experimentally difficult to use as accurate quantitative measurements for $Na^+$ uptake.

## V. Measurement of Membrane Potentials in Sperm

The small size of sea urchin sperm has precluded attempts at measurement of membrane potentials with conventional microelectrodes. Measurements on intact sperm have relied on uptake of the diffusable lipophilic ions [$^{14}$C]thiocyanate$^-$ and [$^3$H]tetraphenylphosphonium$^+$ (TPP$^+$) (Schackmann *et al.*, 1981, 1984). This method was designed to measure membrane potentials in bacteria and in membrane vesicles where electrode technology was not applicable (Lichtshtein *et al.*, 1979; Rottenberg, 1979). If a negative membrane potential exists, the TPP$^+$ cation will accumulate and the thiocyanate anion will be excluded (Ar $<$ 1). We verified this for sperm of the sea urchin *Strongylocentrotus purpuratus*. Additionally, [$^3$H]TPP$^+$ uptake was decreased by increasing seawater [$K^+$], an event that is consistent with the depolarization of the plasma membrane potential by elevated extracellular [$K^+$]. Conversely, [$^{14}$C]thiocyanate uptake increased with increasing seawater [$K^+$]. Qualitatively these data imply that the sea urchin sperm plasma membrane potential is similar to that of many other cells.

Quantitative evaluation of the potentials with these membrane potential probes is more difficult, particularly for TPP$^+$, which largely accumulates into the mitochondrion (Schackmann *et al.*, 1984) by virtue of the large negative mitochondrial potential. We were able to explore the contribution of mitochondrial potential in intact sperm using [$^{14}$C]thiocyanate ([$^{14}$C]SCN) to measure the plasma membrane potential and using [$^3$H]TPP$^+$ to measure a composite potential that is dependent upon the two negative potentials acting in series. Additionally, by making the plasma membrane permeant with digitonin, we were able to generate with an exogenous substrate a mitochondrial potential consistent with our *in vivo* measurements (−155 mV). If corrections are made for the mitochondrial accumulation of $TPP^+$, both $TPP^+$ and SCN yield values of the plasma membrane potential of ~ −30 mV for sperm in artificial seawater. [$^{14}$C]SCN uptake in principle provides a more direct measurement of the plasma membrane potential; but since Ar < 1, the error in these measurements is large.

Measurements of [$^{3}$H]$TPP^+$ and [$^{14}$C]SCN uptake have been made principally by centrifugation through silicone oil (Schackmann *et al.*, 1984). $TPP^+$ was found to have limited solubility in the silicone oil. Diffusion from the sperm pellet was not substantial within 24 hours, but diffusion from the supernate into the silicone oil layer was found to erroneously enhance the value for Ar (supernatant counts decreased). Samples should, therefore, be processed immediately following an experiment.

Although we have used the lipophilic ions to provide the first values for membrane potentials in sea urchin sperm, the methods have serious limitations. The kinetics of $TPP^+$ uptake into sperm are extremely slow, thus only static, long-term charges in potential can be observed. It has proved possible to measure a depolarization of ~30 mV when the acrosome reaction is induced. But if transient changes occurred prior to the net depolarization, we would be unable to detect them. And like uptake of weak bases, the time-course for changes in uptake does not reflect the time-course for changes in $\Delta\psi$, but reflects a complex function of the potentials and sperm permeability to the probe molecules.

Recent advances in patch-clamping small sections of cell membranes (Neher *et al.*, 1978) should ultimately allow more exact, rapid measurements of sperm membrane potentials. In this regard, Liévano and colleagues (1985) have only recently reported making lipid bilayers of sea urchin sperm membranes on a patch electrode. They found $K^+$ channels which could be blocked by tetraethylammonium. This potentially represents an exciting new method of analyzing ion movements through sperm membranes.

## VI. Conclusion

I have briefly attempted to provide the reader interested in measuring ionic changes in sperm with a perspective on how the measurements have been and should be made and the potential limitations inherent with existing techniques. Simple ionic manipulation, regulation of $pH_i$, and [$Ca^{2+}$] are of fundamental importance to sperm physiology. Investigations of ionic changes associated with activation of sea urchin sperm have allowed us to begin to define how these important events are initiated and lead to activation of development. Whether these changes are unique to echinoid sperm or are general characteristics of sperm remains to be shown. Similar ionic changes, increases in $pH_i$, and $[Ca^{2+}]_i$ are known to attend to a host of cell activation events throughout the animal kingdom. The apparent similarity between echinoid sperm activation and activation of other eukaryotic cells raises additional interest in the study of echinoid sperm and the ionic changes important to their activation.

## ACKNOWLEDGMENTS

I wish to thank Merry Peters for typing this manuscript. This work was supported by Grant HD 18670 from the National Institutes of Health.

## REFERENCES

Babcock, D. F. (1983). *J. Biol. Chem.* **258,** 6380–6389.
Bibring, T., Baxandall, J., and Harter, C. C. (1984). *Dev. Biol.* **101,** 425–435.
Boyer, P. D., Chance, B., Ernster, L., Mitchell, P., Racker, E., and Slater, E. C. (1977). *Annu. Rev. Biochem.* **46,** 955–1026.
Brokaw, C. J., and Benedict, B. (1968). *J. Gen. Physiol.* **52,** 283–299.
Busa, W. B., and Nuccitelli, R. (1984). *Am. J. Physiol.* **246,** R409–438.
Cantino, M. E., Schackmann, R. W., and Johnson, D. E. (1983). *J. Exp. Zool.* **226,** 255–268.
Christen, R., Schackmann, R. W., and Shapiro, B. M. (1982). *J. Biol. Chem.* **257,** 14881–14890.
Christen, R., Schackmann, R. W., and Shapiro, B. M. (1983a). *J. Biol. Chem.* **258,** 5392–5399.
Christen, R., Schackmann, R. W., Dahlquist, F. W., and Shapiro, B. M. (1983b). *Exp. Cell Res.* **149,** 289–294.
Dan, J. C. (1952). *Biol. Bull. (Woods Hole)* **103,** 54–56.
Dan, J. C. (1954). *Biol. Bull. (Woods Hole)* **107,** 335–349.
Grynkiewicz, G., Poenie, M., and Tsien, R. Y. (1985). *J. Biol. Chem.* **260,** 3440–3450.
Hansbrough, J. R., and Garbers, D. L. (1981). *J. Biol. Chem.* **256,** 2235–2241.
Harvey, E. B., and Anderson, T. F. (1943). *Biol. Bull. (Woods Hole)* **85,** 151–156.
Hesketh, T. R., Smith, G. A., Moore, J. P., Taylor, M. V., and Metcalfe, J. C. (1983). *J. Biol. Chem.* **258,** 4876–4882.
Hesketh, T. R., Moore, J. P., Morris, J. D. H., Taylor, M. V., Rogers, J., Smith, G. A., and Metcalfe, J. C. (1985). *Nature (London)* **313,** 481–484.
Hoek, J. B., Nicholls, D. G., and Williamson, J. R. (1980). *J. Biol. Chem.* **255,** 1458–1464.
Johnson, C. H., Clapper, D. L., Winkler, M. M., Lee, H. C., and Epel, D. (1983). *Dev. Biol.* **98,** 493–501.
Johnson, J. D., Epel, D., and Paul, M. (1976). *Nature (London)* **262,** 661–664.
Kopf, G. S., and Garbers, D. L. (1980). *Biol. Reprod.* **22,** 1118–1126.
Kopf, G., Lewis, C. A., and Vacquier, V. D. (1984). *J. Biol. Chem.* **259,** 5514–5520.
L'Allemain, G., Paris, S., and Pouysségur, J. (1984). *J. Biol. Chem.* **259,** 5809–5815.
Lee, H. C. (1984a). *J. Biol. Chem.* **259,** 4957–4963.
Lee, H. C. (1984b). *J. Biol. Chem.* **259,** 15313–15319.
Lee, H. C., Johnson, C., and Epel, D. (1983). *Dev. Biol.* **95,** 31–45.
Lichtshtein, D., Kaback, H. R., and Blume, A. J. (1979). *Proc. Natl. Acad. Sci. U.S.A.* **76,** 650–654.
Liévano, A., Sánchez, J., and Darszon, A. (1985). *Biophys. J.* **47,** (2, Pt. 2), 145a.
Madeira, V. M. C. (1975). *Biochem. Biophys. Res. Commun.* **64,** 870–876.
Meizel, S., and Deamer, D. W. (1978). *J. Histochem. Cytochem.* **26,**98–105.

Moolenaar, W. H., Tertoolen, L. G. J., and de Laat, S. W. (1984). *J. Biol. Chem.* **259,** 7563–7569.
Moore, R. A., Fidelman, M. L., Hansen, J. C., and Otis, J. N. (1982). *In* "Intracellular pH: Its Measurement, Regulation, and Utilization in Cellular Functions" (R. Nuccitelli and D. W. Deamers, eds.), pp. 386–416. Liss, New York.
Neher, E., Sakmann, B., and Steinbach, J. H. (1978). *Pflügers Arch. Ges. Physiol.* **375,** 219–228.
Nishioka, D., and Cross, N. (1978). *In* "Cell Reproduction" (E. R. Dirksen, D. Prescott, and D. F. Fox, eds.), pp. 403–413. Academic Press, New York.
Nuccitelli, R., and Deamer, D. W. (1982). "Intracellular pH: Its Measurement, Regulation, and Utilization in Cellular Functions" (R. Nuccitelli and D. W. Deamers, eds.). Liss, New York.
Repaske, D. R., and Garbers, D. L. (1983). *J. Biol. Chem.* **258,** 6025–6029.
Ridgway, E. B., Gilkey, J. C., and Jaffe, L. F. (1977). *Proc. Natl. Acad. Sci. U.S.A.* **74,** 623–627.
Rothenberg, P., Reuss, L., and Glaser, L. (1982). *Proc. Natl. Acad. Sci. U.S.A.* **79,** 7783–7787.
Rothenberg, P., Glaser, L., Schlesinger, P., and Cassel, D. (1983). *J. Biol. Chem.* **258,** 12644–12653.
Rottenberg, H. (1979). *In* "Methods in Enzymology" (S. Fleisher and L. Packer, eds.), Vol. 55, pp. 547–569. Academic Press, New York.
Schackmann, R. W., and Shapiro, B. M. (1981). *Dev. Biol.* **81,** 145–154.
Schackmann, R. W., Eddy, E. M., and Shapiro, B. M. (1978). *Dev. Biol.* **65,** 483–495.
Schackmann, R. W., Christen, R., and Shapiro, B. M. (1981). *Proc. Natl. Acad. Sci. U.S.A.* **78,** 6066–6070.
Schackmann, R. W., Christen, R., and Shapiro, B. M. (1984). *J. Biol. Chem.* **259,** 13914–13922.
Schroeder, T. E., and Christen, R. (1982). *Exp. Cell Res.* **140,** 363–371.
Schuldiner, S., Rottenberg, H., and Avron, M. (1972). *Eur. J. Biochem.* **25,** 64–70.
Shen, S. S., and Steinhardt, R. A. (1978). *Nature (London)* **272,** 253–254.
Steinhardt, R., Zucker, R., and Schatten, G. (1977). *Dev. Biol.* **58,** 185–196.
Thomas, A. P., Alexander, J., and Williamson, J. R. (1984). *J. Biol. Chem.* **259,** 5574–5584.
Tilney, L. G., Kiehart, D. P., Sardet, C., and Tilney, M. (1978). *J. Cell Biol.* **77,** 536–550.
Tsien, R. Y., Pozzan, T., and Rink, T. J. (1982). *J. Cell Biol.* **94,** 325–334.
Waddell, W. J., and Butler, T. C. (1959). *J. Clin. Invest.* **38,** 720–729.

# Chapter 5

# *Gonad-Stimulating and Maturation-Inducing Substance*

HIROKO SHIRAI

*Division of Developmental Biology*
*Laboratory of Reproductive Biology*
*National Institute for Basic Biology*
*Okazaki, Japan*



## I. Introduction

In starfish, the release of gametes is controlled by two endogenous hormones, "gonad-stimulating substance" (GSS) and "maturation-inducing substance" (MIS). Progress in isolating these substances, and in understanding their mechanism of action, now permits the experimenter to manipulate the ripe adult starfish in order to obtain gametes and their accessory cells at will. Before exploring the details of isolating or characterizing GSS and MIS, it may be useful to review the basic biology of spawning and oocyte maturation in starfish (Chaet, 1967; Kanatani, 1973, 1979; Kanatani and Nagahama, 1983).

When ripe female starfish spawn, either naturally or artificially as a result of injection of GSS, "eggs" are released as isolated cells in an intermediate stage of meiotic maturation, usually near first meiotic metaphase (or germinal vesicle breakdown) (Chaet and McConnaughy, 1959; Kanatani, 1964). Spawning, or egg release, is the culmination of several events triggered in various tissues of the starfish. One of the earliest events is the release from nervous tissue of GSS, whose principal action seems to be the stimulation of the follicle cells that surround each full-grown secondary oocyte in the ovary. During storage, each immature occyte is separated from its neighbors and is apparently prevented from moving by these follicle cells. GSS causes these follicle cells to produce MIS, which is the agent that directly causes stored oocytes to resume meiotic maturation (Kanatani and Ohguri, 1966; Kanatani and Shirai, 1969, 1970; Schuetz, 1969a). MIS also indirectly causes the uncoupling of intercellular junctions between follicle cells and oocytes through its action on the oocyte cytoplasm (Kishimoto *et al.*, 1983; Schroeder, 1971).

With respect to spawning itself, some active substance closely associated with the maturing oocytes seems to act on the ovarian walls to stimulate contraction of ovarian muscles. Neither GSS nor MIS is able to make the ovarian wall contract (Shirai *et al.*, 1981). As a result of this train of events, oocytes that have begun their maturation process are freed from their attaching follicle cells and squeezed out through oviducts and gonopores to the surrounding seawater. Thus, in starfish, oocyte maturation is intimately linked with the spawning process.

In this article I shall describe methods for preparing and assaying GSS and MIS. These methods are derived from experience with selected starfish. In principle, if not in specifics, they may be useful for additional species of starfish, as well as for studying GSS and MIS in other classes of echinoderms.

## II. Gonad-Stimulating Substance

### A. Preparation of GSS

#### 1. Nerve Collection

Although all nervous tissues in starfish probably contain GSS, only the radial nerves are easy to dissect and handle for extracting GSS. Radial nerves are located along the ambulacral groove on the oral surface of each arm. To separate radial nerves from a soft-bodied starfish such as *Asterias amurensis,* the oral surface of the arms is dissected by cutting along the rows of tube feet with scissors (Fig. 1). Each ambulacral groove is then sliced with scissors, slightly off center so that the radial nerve remains in one of the halves. Fine forceps are then used to dissect the tissue at the exact midline of the ambulacral groove near the

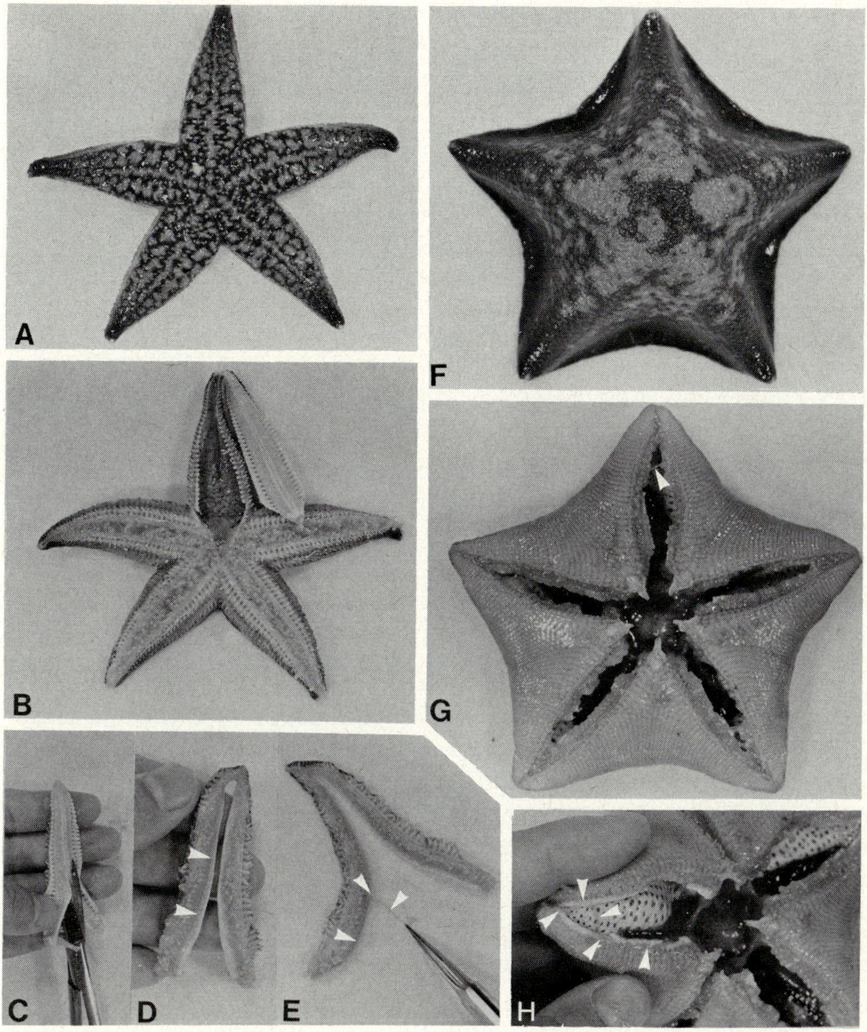


FIG. 1. Nerve collection. (A) Aboral view of adult *Asterias amurensis*. (B) Oral view. One ambulacral zone has been cut out with scissors. (C) An ambulacral zone being sliced along the ambulacral groove slightly off center. (D) A sliced ambulacral zone. A radial nerve is seen on the cut edge of the left half (arrow heads). No radial nerve is present in the right half. (E) A radial nerve being peeled with fine forceps (arrow heads). (F) Aboral view of adult *Asterina pectinifera*. (G) Oral view of (F) after being folded along each ambulacral groove by hand. Each tear is continued distally (toward the tip of the arm). A radial nerve is seen across a tear (arrow head). (H) A radial nerve along with the edges of the same tear (arrow heads).

distal end (tip of the arm) and to grasp the radial nerve. The radial nerve is then firmly peeled in the proximal direction (toward the mouth). The direction of this removal is important, as the tissue surrounding the radial nerve separates more easily distoproximally than in the opposite direction.

In the case of hard-bodied starfish such as *Asterina pectinifera,* animals can be folded along the nerve by hand in the air so that the radial nerves will be exposed between the two sides of a broken ambulacral groove (Fig. 1). The direction of removal of the nerve is opposite to that of *Asterias*.

It is recommended that freshly collected nerves (from one or more animals) be frozen rapidly, because autodigestion of tissue soon begins. Nerves can be kept frozen for a few days, but longer term storage results in a decrease of activity, probably because of oxidization of GSS. For rapid freezing, a mixture of ice and salt is convenient (−20°C obtained by a mixture of ice-NaCl, 3:1; −50°C by ice-$CaCl_2 \cdot 2H_2O$, 1:1). Nerves can also be lyophilized. In the case of the five radial nerves taken from an adult *A. amurensis* (with an arm length of about 8 cm), the weight of lyophilized nerve is about 10 mg, five to six times less than the weight of fresh nerves. The activity (content) of GSS in lyophilized nerves is retained when the samples are stored in a desiccator at room temperature.

## 2. Extraction of GSS

GSS can be extracted from stored nerves (fresh, frozen, or lyophilized) by homogenizing in water or aqueous solutions containing salts. Mechanical disruption of the tissue is important in order to completely extract GSS, which is stored intracellularly in some granular form (de Angelis *et al.*, 1972). If fresh nerves are homogenized using a Potter–Elvehjem glass homogenizer with 0.25 *M* sucrose solution, the granules containing GSS can be liberated without disruption and spun down by centrifugation; this implies a lower recovery rate of GSS in the supernatant of the homogenate, i.e., insufficient extraction of GSS. Freeze-and-thaw is recommended for the complete extraction of GSS from fresh tissue with aqueous solution. A mixture of acetic acid and ethanol (2 *M* acetic acid and pure ethanol, 1:1, v/v) is also effective for GSS extraction, and can be used for nerve storage, fresh nerves being accumulated in this solution in a deep freezer. A Potter–Elvehjem glass homogenizer, Waring blender, or sonicator can be used for homogenization. The homogenate is then centrifuged; the first centrifugation should use a high force, generally 30,000 *g* for 30 minutes.

After extraction with aqueous solutions, heat treatment (boiling 15–30 minutes) is needed to denature digestive enzymes in the extract in order to protect the GSS. To avoid coprecipitation of GSS with denatured proteins, only the supernatant from the first centrifugation should be boiled. To remove the denatured and coagglutinated components from the heat-treated supernatant, a second centrifugation, similar to the first, is recommended. In the case of extraction with acid–ethanol, boiling can be omitted.

Since GSS is insoluble in nonpolar solvents such as chloroform and ether, the crude extract can be delipidized by partitioning with these solvents. An equal volume of nonpolar solvent is added to the crude extract and the mixture is shaken vigorously for about 10 minutes. After allowing the mixture to stand, the aqueous phase can be isolated. The separation of two phases can be accelerated by centrifugation (about 10,000 *g* for 30 minutes). If delipidization is performed on the acid–ethanol extract, the ethanol must first be removed from the extract. The extract is then neutralized with ammonia and concentrated to dryness in a rotary evaporator (below 40°C). An appropriate amount of water is added to dissolve the GSS, and partitioning is then carried out as for the aqueous extract. Since GSS is also insoluble in polar solvents such as ethanol and acetone (over 95%), lyophilized nerves can be delipidized with these polar solvents. The lyophilized nerves are homogenized with the solvents, and the homogenate is then centrifuged. Delipidized powder is obtained by drying the dispersed pellet in air or *in vacuo*.

### 3. Removal of Salts from Crude GSS

A GSS sample without common salts is sometimes needed, for example, to examine GSS action under calcium-free conditions. Gel-permeation and partition chromatography can be used. The schematic elution pattern of GSS in three gels is shown in Fig. 2; Sephadex G-25 and G-50 are suitable for desalting the crude GSS. The conditions and procedures are shown in Table I.

Step-wise partition chromatography on Poragel PN (Waters Assoc., Milford, MA) or silan-bonded gel (spherical porous silica gel; Noumura Chem. Co., Seto, Japan) can also be used. Figure 3 shows the construction of an open column in the elution, and the conditions and procedures are shown in Table II. In partition chromatography, the crude GSS must be delipidized in advance in order to prolong the life of the gels, which are generally expensive. Just after separation from the nonpolar solvent the delipidized aqueous phase generally contains a small amount of the solvent, which affects the retention and elution of GSS on the gels. Therefore the aqueous phase must be concentrated with a rotary evaporator to remove the solvent. In partition chromatography, a large volume of sample can generally be applied at once (GSS from 100 animals/1 g dry gel), because the gel surface is very large (about 200 $m^2$/1 g dry gel). Salt components are eluted with the washes while GSS is retained on the gel (eluant III). For complete elution, care must be taken that air is not pushed into the gel bed.

Other detailed procedures for purification of GSS and for determination of GSS nature are omitted here. The readers will find adequate references elsewhere

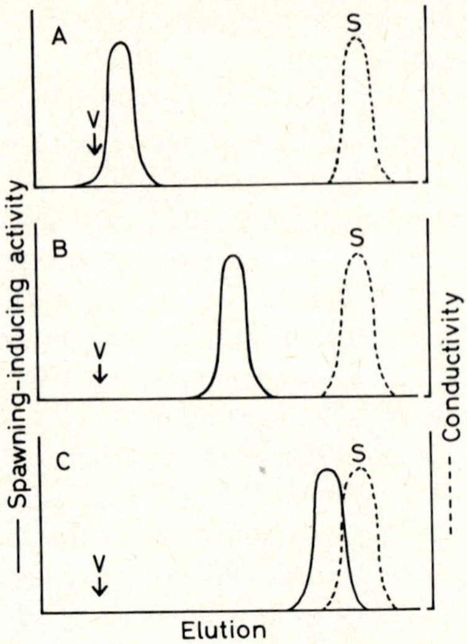


FIG. 2. Schematic elution pattern of GSS in various types of gel-permeation chromatography. (A) Sephadex G-25. The void volume (elution position of Blue Dextran) is shown by V, which is generally at about 45% of the gel bed volume. GSS elutes almost at the void volume, and salt (S) elutes at about 85% of the gel bed volume. (B) Sephadex G-50. The void volume is about 35% of the gel bed volume. GSS elutes about 70% of the gel bed volume, and salts elute at about 90% of the gel bed volume. (C) Sephacryl S-200. The void volume is about 42% of the gel bed volume. GSS elutes about 96% of the gel bed volume, and salts eluted at about 98% of the gel bed volume.

(Kanatani and Noumura, 1962; Kanatani *et al.*, 1971; Schuetz, 1969b; Shirai and Kanatani, 1968). Briefly, the molecular weight of GSS from the two species of starfish is around 2000, according to their behavior by gel-permeation chromatography and dialysis. Their isoelectric points are near pH 4 (*Asterias*, pH 4.6, *Asterina*, pH 4.2). They are fairly hydrophobic and heat stable and are digested by Pronase. *Asterina* GSS can stimulate antibody formation in rabbits. GSS possesses species specificity: *Asterina* GSS stimulates spawning of an *Asterias* ovary but *Asterias* GSS does not stimulate the *Asterina* ovary. From these preliminary checks, GSS is considered to be a small peptide. These directions should help readers to handle, keep, and purify GSS or other peptides resembling GSS.

## B. Bioassay of GSS

As described before, a water extract of nerves can cause spawning in ripe starfish. A concentration of GSS of 10–100 μg dry nerve Eq/ml is needed in the body fluid (coelomic fluid). Since the body fluid of *Asterias amurensis* is gener-

TABLE I

CONDITIONS OF GEL-PERMEATION CHROMATOGRAPHY FOR GSS DESALTING[a]

| | | | |
|---|---|---|---|
| Gel type | Sephadex G-25 (fine) | Sephadex G-50 (fine) | Sephacryl S-200 |
| Gel state at purchase | Dry powder | Dry powder | Wet sludge |
| Initial treatment | Boil the gel with water for 30 minutes to make the gel swell completely | | Not necessary |
| Removal of small particles | For even packing and better separation of GSS from other components, mix the gel with about 10 vol of water by spatula (don't use a stirrer), stand for 5–10 minutes, and discard the water layer containing the small particles by decanting or aspirating. Repeat more than five times | | |
| Equilibration with eluant | Wash the gel in the column with about 20 vol of eluant (0.1–0.5 *M* NaCl solution or 100 m*M* ammonium acetate is recommended) using a flow rate of 5–10 ml/cm$^2$/hr. The column height should be at least 20 times larger than the diameter | | |
| | Elution position (percentage gel bed volume) | | |
| Blue Dextran (for void volume estimation) | 45% | 35% | 42% |
| GSS | 45% | 70% | 96% |
| Salts | 85% | 90% | 98% |
| Gel regeneration | Wash the column with more than two bed volumes of eluant. For standing the column, 0.05% sodium azide solution is used | | |

[a] For 1 ml of GSS sample (about 50 mg dry nerve Eq.), the size column recommended is about 1.5 cm (diameter) × 50 cm (height).

ally 10–20 ml in volume, 50 μl of GSS stock sample (1 mg dry nerve Eq/ml) should be injected. Although spawning will occur when the entire sample is injected into one arm, it is better to inject it into all the arms little by little, in order to achieve an even mixing of GSS in the body fluid.

Since the conditions inside ovaries are somewhat inhibitory to the progression of maturation (due to the existence of inhibitory substances against GSS and MIS, low pH and low oxygen, etc.) (Ikegami, 1976; Ikegami *et al.*, 1967, 1976; Shirai *et al.*, 1982), maturation progresses more quickly in earlier spawned eggs. All the spawned eggs are fertilizable. The most appropriate period for normal fertilization and development is from just after germinal vesicle breakdown to the first polar body formation. During this period, the polyspermy-block mechanism functions well (Miyazaki and Hirai, 1979).

Ripe animals with physiologically normal ovaries can be kept for nearly 2 months without food if the seawater is cooled (1–2°C colder than the habitat temperature, generally 14°C in the case of *Asterias* or *Asterina* at my Institute). Induction of spawning in individuals is difficult due to individual variation and

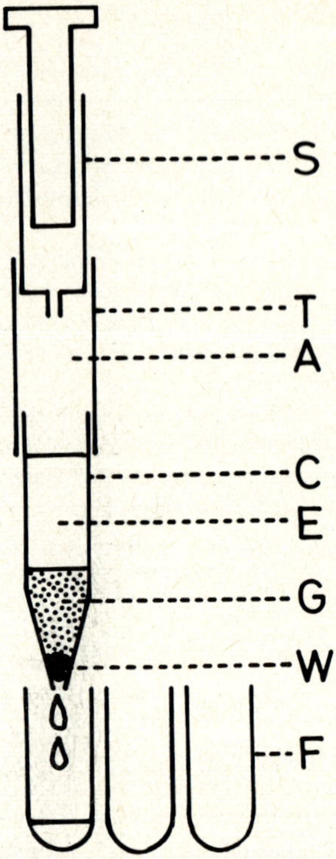


FIG. 3. Construction of an open column for step-wise elution in partition chromatography. A, Air; C, tip of the air-displacement pipet; E, sample or eluant; F, fractions; G, gel; 250 mg dry powder is put in the tip. S, A syringe for air compression; T, tubing to connect the tip and the syringe; W, a small amount of Teflon wool as a stopper. Glass wool can be used, although it tends to adsorb GSS nonspecifically, especially at higher degrees of purification. Sample application and eluant addition are made by deconnecting the tubing from the tip.

the need for large amounts of GSS. Thus a semimicroassay using ovarian fragments has been introduced (Chaet *et al.*, 1964). Since all parts of an ovary taken from one female possess generally similar reactivity, problems of individual variation are overcome. Ovaries can be kept intact in seawater without spawning even after being cut into many pieces (about 5 mm in size), because of adhesion among oocytes and other cells and the absence of high tension in the ovarian wall. Normal reactivity in isolated ovaries can be retained for 1 day in cold medium (about 4°C). Since the ionic conditions of the body fluid are very similar to seawater (Shirai *et al.*, 1982), artificial or filtered seawater with pH adjusted to about 7 can be used as a convenient medium. Unless otherwise stated, the term seawater will be used to describe such pH-adjusted seawater.

The semimicroassay is carried out as follows (Table III): (1) Ovaries are cut

TABLE II

CONDITIONS FOR STEP-WISE PARTITION CHROMATOGRAPHY ON AN OPEN COLUMN

| | | |
|---|---|---|
| Gel type | Poragel PN (particle size, 37–75 μm) 250 mg dry gel | Octadecyl silane-bonded gel (particle size, 30 μm) 250 mg dry gel |
| Prewash | | |
| Eluant I | 100% isopropanol, 1 ml × 10 | |
| Eluant II | 50% isopropanol in 10 m*M* ammonium acetate, 1 ml × 10 | 20% acetonitrile (ACN) in 10% isopropanol and 0.1% trifluoroacetic acid (TFA), 1 ml × 10 |
| Equilibration with eluant III | 10 m*M* ammonium acetate, 1 ml × 10 | 5% acetonitrile in 10% isopropanol and 0.1% TFA, 1 ml × 10 |
| Sample application and wash | Sample pH is adjusted to about 7; eluant III, 1 ml × 5 | Sample pH is adjusted to about 3; eluant III, 1 ml × 5 |
| | Eluted fraction is called breakthrough | |
| GSS elution | Eluant II, 1 ml × 5 | |
| Posttreatment of fractions | | Eluted fraction must be neutralized just after elution to retain GSS |
| For bioassay | Sample can be diluted with seawater directly. Isopropanol does not affect spawning if the concentration is less than 0.1%. If GSS concentration in the eluate is too low, it can be concentrated in advance of the assay with a rotary evaporator | |
| Gel regeneration | Wash with eluant I, 1 ml × 10, and subsequently wash with eluant II, 1 ml × 10 | |
| | For standing the column, eluant I (100% isopropanol) is used | |

into small fragments and washed several times with seawater. Only batches that have not spawned within about 1 hour after dissection can be used for assay. Spawning often occurs, especially late in the breeding season. (2) Four hundred microliters of seawater (initial medium) is placed into the wells of an 80-well acrylic plate (as used for hemagglutination and hemolysis). (3) Part (e.g., 10 μl) of the test sample (e.g., 15 ml) is mixed with the medium in the first well. The first dilution is therefore 41-fold. (4) Two hundred microliters of the first diluted sample is then transferred to the second well, giving a three-fold dilution. The symbol ⤵ indicates this process. (5) Similar three-fold dilutions are continued serially until an appropriate dilution is achieved. A similar serial dilution is performed on a standard GSS solution (assumed to be 1.3 mg dry nerve Eq/ml). (6) Ovarian fragments are immersed in the diluted samples. (7) Observation and evaluation of spawning are made 1–2 hours later.

TABLE III

BIOASSAY OF GSS

| | | | | | |
|---|---|---|---|---|---|
| The standard GSS (1.3 mg dry nerve Eq/ml) | | | | | |
| Initial medium | 400 μl | 400 μl | 400 μl | 400 μl | 400 μl |
| | 10 μl → | 200 μl → | 200 μl → | 200 μl → | 200 μl → |
| Dilution degree | 41 | 41 × $3^1$ | 41 × $3^2$ | 41 × $3^3$ | 41 × $3^4$ |
| Assessment of spawning | +[a] | +[b] | −[c] | − | − |
| The test sample (15 ml) | | | | | |
| Initial medium | 400 μl | 400 μl | 400 μl | 400 μl | 400 μl |
| | 10 μl → | 200 μl → | 200 μl → | 200 μl → | 200 μl → |
| Dilution degree | 41 | 41 × $3^1$ | 41 × $3^2$ | 41 × $3^3$ | 41 × $3^4$ |
| Assessment of spawning | + | + | +[d] | − | − |

[a] +, A large amount of eggs is released.
[b] Minimum dose of GSS for spawning: 1.3 mg/(41 × $3^1$).
[c] −, No eggs are released.
[d] The concentration of GSS at this dilution corresponds to the minimum dose.

The calculation process, shown in Table III, is based on the determination of the minimum dose of GSS for inducing of spawning in a certain batch (ovarian sensitivity). This is generally about 10 μg dry nerve Eq/ml, although extreme variations are often observed early or late in the breeding season. With respect to the serial dilution, by changing the volume of the initial medium or of transferred diluted samples, $1.5^n$-fold, $2^n$-fold, $5^n$-fold, etc., are possible. With a $3^n$-fold dilution, two dilutions will make 9-fold, which is approximately 10-fold, and is therefore useful for a rough and rapid estimation of GSS content.

## III. Maturation-Inducing Substance

### A. Preparation of MIS

#### 1. Biosynthesis of MIS

Since spawned eggs are always mature, it was first considered that GSS not only induces spawning but also oocyte maturation independently. However, results using the semimicroassay in *Asterina* did not support this, since immature oocytes were seen at the edge of the mass of matured oocytes released from the cut end of the ovarian fragment. Since GSS was thought to be evenly distributed in the assay medium, this indicated that GSS does not directly trigger oocyte maturation (Kanatani and Shirai, 1967; Schuetz and Biggers, 1967).

To obtain large amounts of MIS, fresh ovaries (200 mg/ml) are incubated in seawater containing GSS (500 μg dry nerve Eq/ml) for 6 hours (20°C). The medium is filtered through a mesh to remove ovarian walls and then centrifuged to remove eggs (3000 rpm for 10 minutes). A second centrifugation at high force is preferable to obtain a clear supernatant.

MIS production is detectable soon (1–2 minutes) after the addition of GSS into the incubation mixture and is nearly proportional to the added GSS, up to several hundred micrograms of dry nerve Eq/ml (Kanatani and Shirai, 1970, 1971; Shirai, 1972; Shirai *et al.*, 1972). To stop the biochemical reaction, the incubation mixture is boiled for 10 minutes after an appropriate length of time, and the boiled sample is then homogenized. The homogenate is centrifuged (30,000 *g* for 30 minutes) and the supernatant is assayed for MIS. When the ovary is incubated with sufficient GSS at 20°C, the amount of MIS increases for nearly 6 hours.

The target cells of GSS in the ovary are follicle cells (Hirai and Kanatani, 1971; Hirai *et al.*, 1973; Kubota and Kanatani, 1975, 1978). When several hundred micrograms of dry nerve Eq/ml of GSS is used, MIS produced by follicle cells taken from about $10^4$ oocytes/ml can generally be detected without a concentration process. The procedures for obtaining oocytes, oocytes with follicle cells, and follicle cells are described later. MIS is also produced in testes under the influence of GSS and also induces spawning. The target cells in the testis have been identified as the interstitial cells (α-cells) (Cochran and Kanatani, 1975; Kubota *et al.*, 1977). To collect the α-cells, testes are taken from animals, first washed, and then teased in seawater. The contents of a testis are spermatozoa, α-cells, and β-cells (including spermatogonia, spermatocytes, spermatids, and perivisceral–epithelial cells). To remove the fragmented testis walls the suspension is filtered through a mesh and the filtrate is centrifuged at 1000 rpm for 1–2 minutes to spin down the cellular components (other than spermatozoa). To separate α-cells from other kinds of cells, gradient centrifugation is needed several times, each time changing the gradient conditions. α-cells are obtained at the 13% Ficoll layer of the final gradient of 17 ml of 13% Ficoll on 7 ml of 15% Ficoll in seawater.

## 2. Nature of MIS

Since the MIS of starfish has been identified as 1-methyladenine, it can be purchased and used at will. However, for the reader's convenience, several checks that were carried out in advance of the original purification of MIS will be described briefly.

The activity of MIS does not decrease after freeze-and-thaw, is heat resistant (the activity is retained through the boiling treatment), and is dialyzable. These physical properties suggest that MIS is a rather small substance. As to its chemical nature, its behavior on Sephadex G-25 is particular, MIS being eluted after common salts. The tendency of being retained on the gel is more apparent at higher pH (8–9) than at lower pH (7), which permits desalting. When partitioned between water and nonpolar solvents, such as ether, benzene, chloroform, and petroleum ether, MIS remains in the water phase. The properties indicate that MIS is probably positively charged at neutral pH. With respect to its biochemical nature, MIS is resistant to Pronase. Further, incubation of GSS with Pronase, which destroys GSS, does not produce any MIS activity, indicating that MIS is not derived from a peptide and is not a part of GSS.

Based on these observations, MIS has been purified and, in starfish, has been identified as 1-methyladenine (1-MeAde) (Kanatani *et al.*, 1969).

## B. Bioassay of MIS

Since 1-MeAde alone can induce spawning (Kanatani, 1969; Kanatani and Shirai, 1967), the bioassay procedure for spawning can be used if it is known that the test sample does not contain GSS. GSS- and 1-MeAde-induced spawning are indistinguishable, therefore for the separate assay of MIS itself, maturation-inducing activity should be checked.

Ovaries taken from animals are washed with seawater to remove tissue fluid, especially fluid from the stomach, and teased with fine forceps. To release oocytes, a teased fragment is grasped with fine forceps and oocytes with follicle cells are gently scraped away by nipping and sliding another pair of forceps along the fragment. Ovarian growth is generally synchronized and the ovaries are filled with numerous (several million) oocytes at the same physiological stage. In the early part of the breeding season, large numbers of younger (smaller) oocytes are found. Irregular inclusions (cellular and noncellular) are observed, especially in the later part of the breeding season. These contaminants can be removed by repeated washing with seawater. Gentle pipetting then detaches oocytes from follicle cells, and defolliculated oocytes and follicle cells obtained as a suspension. When follicle cells are needed, the supernatant of the suspension can be collected, leaving the oocytes at the bottom. After several washings with seawater, a suspension consisting of oocytes alone is obtained. Follicle cells can also be collected as follows: Oocytes with follicle cells are treated with seawater containing $10^{-7} M$ 1-MeAde, and after maturation, separated follicle cells are collected as described above and washed with seawater by several centrifugations (5000 rpm for 15 minutes) to remove 1-MeAde. Maturing oocytes are obtained simultaneously.

In the case of *Asterias amurensis,* follicle cells begin to produce MIS without GSS when they are transferred to seawater, so-called ''spontaneous maturation'' (Cloud and Schuetz, 1973; Schuetz and Biggers, 1968; Shirai, 1974). However, MIS production by follicle cells is inhibited under calcium-free conditions. Therefore if immature oocytes of *Asterias* are needed, ovaries should be washed with calcium-free seawater for nearly 1 hour, replacing the seawater several times before being teased, so that the oocytes and follicle cells become detached within the ovary. After washing oocytes from contaminating follicle cells with calcium-free seawater, oocytes are transferred to normal seawater. Oocytes prepared in this way from both *Asterina* and *Asterias* can be kept for nearly 1 day at 4°C. Serial dilutions of a test sample and standard solution (generally a 1-MeAde solution is used) are made as for the bioassay of GSS (see Section II,B). To each diluted medium, 200–300 oocytes are transferred; a thin capillary is useful to suck up oocytes with a very small amount of seawater. Filter paper cut into small pieces (about 1 × 10 mm) can also be used to collect oocytes with the aid of fine forceps. Oocytes gently adhering to the piece are rinsed off by gentle agitation. About 1–2 hours later, numbers of mature and immature oocytes are counted under an ordinary microscope (objective 4×), and the maturation rate (%) is calculated (for example, see Table IV).

The calculation procedure is based on the determination of the 1-MeAde concentration which induces 50% maturation in a certain batch (effective dose at 50%, $ED_{50}$) (Shirai, 1974). First, the percentage of matured oocytes in each diluted sample of the standard solution is plotted using semilogarithmic paper (Fig. 4A). In this example, the 50% (net) response level (shown as — - - —) is 53% maturation. The intersection (*l*) between the maturation curve (the dose–response curve) and the 50% line gives an $ED_{50}$ for 1-MeAde of $5 \times 10^{-8}$ *M*.

Next, the percentage of matured oocytes in each diluted test sample is similarly plotted although the direction of the semilogarithmic paper is reversed for easier reading of the dilution degree (Fig. 4B). From the intersection (*m*) be-

TABLE IV

BIOASSAY OF MIS

| | | | | | |
|---|---|---|---|---|---|
| The standard solution (1-MeAde, $10^{-6}$ *M*) | | | | | |
| Initial medium | 400 µl<br>200 µl → | 400 µl<br>200 µl → | 400 µl<br>200 µl → | 400 µl<br>200 µl → | 400 µl<br>200 µl → |
| Dilution degree | $3^1$ | $3^2$ | $3^3$ | $3^4$ | $3^5$ |
| Maturation (%) | 100[a] | 100[b] | 40[c] | 6[d] | 7 |
| The test sample | | | | | |
| Initial medium | 400 µl<br>10 µl → | 400 µl<br>200 µl → | 400 µl<br>200 µl → | 400 µl<br>200 µl → | 400 µl<br>200 µl → |
| Dilution degree | 41 | $41 \times 3^1$ | $41 \times 3^2$ | $41 \times 3^3$ | $41 \times 3^4$ |
| Maturation (%) | 100[e] | 100[f] | 100[g] | 57[h] | 7[i] |
| Seawater alone (control) | | 7%[j]; net 50% response: 100–(100–7)/2=53%[k] | | | |

[a–i] Maturation rate (%) in each diluted sample.

[j] Maturation rate (%) in seawater alone (control).

[k] Maturation rate (%) of net 50% response.

*Note:* This key applies to the corresponding letters in Fig. 4.

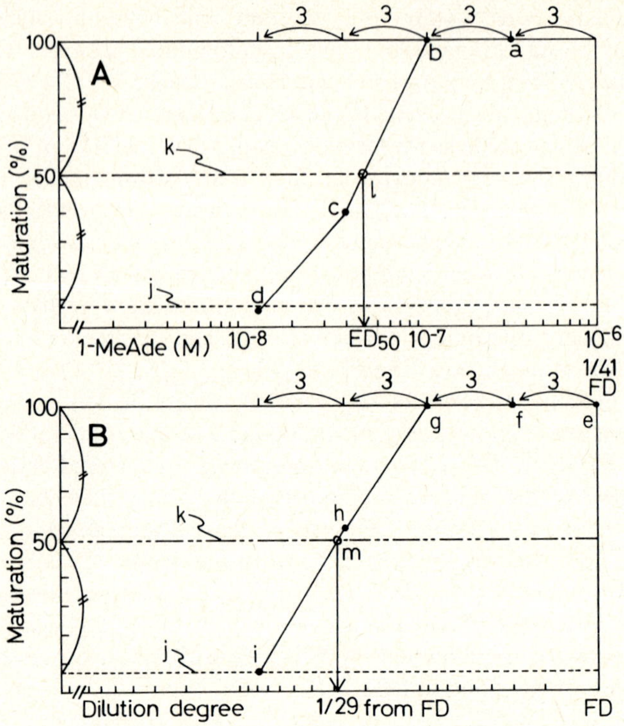


FIG. 4. Bioassay of MIS. (A) A dose–response curve obtained using diluted standard solution (1-MeAde, $10^{-6}$ *M*). (B) A dilution–response curve obtained by diluting a test sample. a–m correspond to a–m in the Table IV, in which *a–i* show the maturation rates in diluted samples, *j* is the maturation rate in seawater alone (control), *k* is the net 50% response level, and *l* and *m* are intersections between the 50% level and dose response curve (or dilution–response curve). FD in (B) signifies the first dilution degree in the test sample (cf. Table IV).

tween the maturation curve of the test sample and the 50% line, the dilution degree corresponding to $ED_{50}$ is here read as 29-fold. Since this corresponds to the concentration of 1-MeAde at $ED_{50}$, $5 \times 10^{-8}$ *M* in this case, the original MIS concentration in the test sample is $5.9 \times 10^{-5}$ *M* (1-MeAde equivalent). Since the molecular weight of 1-MeAde is 149, and the volume of the test sample is known, the weight of MIS in the original test sample can be calculated (as 1-MeAde equivalent).

## IV. Outlook

The chemical aspects of GSS and MIS action are open to future investigation. Artificially synthesized GSS and its active fragments based on the chemical analysis of GSS structure, will be useful for studies of the early processes in GSS action on follicle cells. *In vitro* culture and proliferation of follicle cells (or in the case of male, the interstitial cells) may also be attempted. Radioisotopes of 1-MeAde and insoluble or derivatized 1-MeAde will be useful tools in this aspect (Kanatani and Shirai, 1971). Since 1-MeAde is synthesized in males and exerts certain effects on spermatozoon physiology (Shirai and Kanatani, 1982); spermatozoa, as well as oocytes, possess the receptor for 1-MeAde. Early biochemical processes in 1-MeAde action may be partly common to both oocytes and spermatozoa, and cell membranes of spermatozoa may be useful materials to investigate this aspect.

Studies of GSS and MIS are now at the chemical and subcellular level. Identification and use of GSS and MIS in many other animals will enable more suitable materials to be used, leading to a deeper understanding of spawning and oocyte maturation.

## Acknowledgments

I thank Dr. P. Lumsden for critical reading of the manuscript. I also thank the members of the Laboratory of Reproductive Biology, National Institute for Basic Biology, with whom I have been spending a pleasant time, despite the recent loss of the Late Prof. H. Kanatani. My thanks also to Drs. N. Kondo, K. Imai, M. Isobe, and T. Goto of the Faculty of Agriculture, University of Nagoya, and Dr. P. Bulet for their mutual collaboration through the purification of GSS.

## References

Chaet, A. B. (1967). *Symp. Zool. Soc. London* **20,** 13.

Chaet, A. B., and McConnaughy, R. A. (1959). *Biol. Bull.* **117,** 407.

Chaet, A. B., Andrews, P. M., and Smith, R. H. (1964). *Fed. Proc., Fed. Am. Soc. Exp. Biol.* **23,** 204.

Cloud, J., and Schuetz, A. W. (1973). *Exp. Cell Res.* **79,** 446.

Cochran, R. C., and Kanatani, H. (1975). *Biol. Bull.* **149,** 424.

De Angelis, E., Viglia, A., Watanabe, T., Shirai, H., Kubota, J., and Kanatani, H. (1972). *Annot. Zool. Jn.* **45,** 16.

Hirai, S., and Kanatani, H. (1971). *Exp. Cell Res.* **67,** 224.

Hirai, S., Chida, K., and Kanatani, H. (1973). *Dev. Growth Differ.* **15,** 21.

Ikegami, S. (1976). *J. Exp. Zool.* **198,** 359.

Ikegami, S., Tamura, S., and Kanatani, H. (1967). *Science* **158,** 1052.

Ikegami, S., Kamiya, Y., and Shirai, H. (1976). *Exp. Cell Res.* **103,** 233.

Kanatani, H. (1964). *Science* **146,** 1177.

Kanatani, H. (1969). *Exp. Cell Res.* **57,** 333.

Kanatani, H. (1973). *Int. Rev. Cytol.* **35,** 253.

Kanatani, H. (1979). *In* "Hormones and Evolution" (E. J. W. Barrington, ed.), Vol. 1, p. 273. Academic Press, London.

Kanatani, H., and Nagahama, Y. (1983). *In* "Reproductive Biology of Invertebrates. Vol. I. Oogenesis, Oviposition and Oosorption" (K. G. Adiyodi and R. G. Adiyodi, eds.), p. 611. Wiley, Chichester.

Kanatani, H., and Noumura, T. (1962). *J. Fac. Sci. Univ. Tokyo IV* **9,** 403.

Kanatani, H., and Ohguri, M. (1966). *Biol. Bull.* **104.**
Kanatani, H., and Shirai, H. (1967). *Nature (London)* **216,** 284.
Kanatani, H., and Shirai, H. (1969). *Biol. Bull.* **137,** 297.
Kanatani, H., and Shirai, H. (1970). *Dev. Growth Differ.* **12,** 119.
Kanatani, H., and Shirai, H. (1971). *Dev. Growth Differ.* **13,** 53.
Kanatani, H., Shirai, H., Nakanishi, K., and Kurokawa, T. (1969). *Nature (London)* **211,** 273.
Kanatani, H., Ikegami, S., Shirai, H., Oide, H., and Tamura, S. (1971). *Dev. Growth Differ.* **13,** 151.
Kanatani, H., Shirai, H., and Taguchi, S. (1976). *Dev. Growth Differ.* **18,** 25.
Kishimoto, T., Usui, N., and Kanatani, H. (1983). *Dev. Biol.* **101,** 28.
Kubota, J., and Kanatani, H. (1975). *Science* **187,** 654.
Kubota, J., and Kanatani, H. (1978). *Dev. Growth Differ.* **20,** 349.
Kubota, J., Nakao, K., Shirai, H., and Kanatani, H. (1977). *Exp. Cell Res.* **106,** 63.
Miyazaki, S., and Hirai, S. (1979). *Dev. Biol.* **70,** 327.
Schroeder, P. C. (1971). *Naturwissenschaften* **58,** 270.
Schuetz, A. W. (1969a). *Biol. Bull.* **137,** 524.
Schuetz, A. W. (1969b). *Gen. Comp. Endocrinol.* **12,** 209.
Schuetz, A. W., and Biggers, J. D. (1967). *Exp. Cell Res.* **46,** 624.
Schuetz, A. W., and Biggers, J. D. (1968). *J. Exp. Zool.* **168,** 1.
Shirai, H. (1972). *Exp. Cell Res.* **74,** 124.
Shirai, H. (1974). *Exp. Cell Res.* **87,** 31.
Shirai, H., and Kanatani, H. (1968). *Doubutsugaku Zasshi* **77,** 128.
Shirai, H., and Kanatani, H. (1982). *Zool. Mag.* **91,** 272.
Shirai, H., Kanatani, H., and Taguchi, S. (1972). *Science* **172,** 1366.
Shirai, H., Yoshimoto, Y., and Kanatani, H. (1981). *Biol. Bull.* **161,** 172.
Shirai, H., Ikegami, S., Kanatani, H., and Mohri, H. (1982). *Dev. Growth Differ.* **24,** 419.

# Chapter 6

# *Techniques for Observing Living Gametes and Embryos*

DOUGLAS A. LUTZ

*Department of Anatomy–Histology*
*University of Toronto*
*Toronto, Ontario, Canada*

SHINYA INOUÉ

*Marine Biological Laboratory*
*Woods Hole, Massachusetts*

# I. Introduction

In this article, we describe ways of handling relatively small numbers of echinoid gametes and embryos and of microscopically observing their physiology and development. We emphasize ways to keep the embryos healthy for extended periods of observations. Many of the points described should apply equally well to other types of studies.

For microscopic observations, we prefer cells that are optically clear. Thus, we emphasize the use of *Lytechinus variegatus* and *Echinarachnius parma,* whose eggs are among the clearest that we have found on the United States Atlantic coast.

To obtain healthy embryos, it is important to have healthy adults. Therefore, our discussion starts with brief comments on the adult animals.

# II. Handling and Care of Animals

We have had good success obtaining *L. variegatus* from two sources: Gulf Specimens Co., Inc., Panacea, FL, and Marine Specimens Unlimited, Big Pine Key, FL. Because of the warmer water temperature in the Florida Keys, the urchins from these areas show less of a population-wide breeding season than those from other areas. The advantage is that shipments can be obtained at least 10 months of the year in which a percentage of the animals are ripe (from late November to early January appears to be the period of the year when this is not true). We have generally tried to order animals just prior to the beginning of the breeding season. By warming our holding tanks (25°C) and supplying generous amounts of food, we ripen the animals in the laboratory. Animals that are shipped at the peak of the breeding season are more easily stressed, and this results in a higher mortality rate and spawning of animals during shipping or shortly after arrival.

Upon receipt of the animals, they are slowly acclimated to the tank's temperature (by floating the plastic bags containing the animals in the tank). Once adjusted they are carefully removed from the bags and placed at the bottom of the tank. They should very soon begin to extend their tube feet and raise their spines. An urchin with drooping spines should be watched closely, as this indicates that they are very stressed and may not survive. Soon after arrival they should be supplied with food, and after about a week the temperature in the tank can be slowly elevated. Once the animals have reached maturity (this can be assessed by electric shock-induced shedding, see Section III), the temperature of the tank should be reduced to a holding level (for *L. variegatus* about 18°C).

To keep our urchins year round, we have been using closed system aquaria equipped with under-gravel filtration. Animals are kept at a density of less than one animal per gallon of seawater. One-half of the volume of the tank is replaced at least once a week when we are using natural seawater. Adequate illumination should be provided for the tanks, especially if one feeds the urchins their natural food, to prevent spoilage.

We are fortunate enough to have running seawater available to us year round, but we have also used aritificial seawater. We have been able to keep animals alive and healthy in our closed system aquaria without changing the seawater for over 1 year using "My Sea" artificial seawater (Jamarin Laboratory, Osaka, Japan). This represents a dramatic improvement over previous results with other artificial or even natural seawaters; using other seawaters, successful maintenance of sea urchins has usually required a replenishment of at least one-half tankful with fresh, tempered seawater every 2 weeks. The specific gravity of the "My Sea" seawater was checked once a week and adjusted with deionized water to make up for evaporation. In addition to "My Sea," Jamarin Laboratory also produces a variety of excellent, culture-quality artificial seawaters and chemicals that we now use routinely in our lab, such as "Jamarin U" and $Ca^{2+}$-free and $Ca^{2+}$,$Mg^{2+}$-free mixtures. These are distributed through the Chemical Stockroom of the Marine Biological Laboratory.

We supply our urchins with natural food supplemented with "urchin cakes." The recipe given in Table I is one developed by E. B. Harvey, as modified by B. M. Woodward in our laboratory. It is important to remove excess pieces of food after a day, since they can spoil and contaminate the tanks.

TABLE I

URCHIN CAKE RECIPE[a]

500 ml seawater
500 ml distilled water
30 g agar
Six hard-boiled eggs: Remove shells, dry, and crush them
TetraMin fish flakes (about one-half of a large container)

1. Simmer the agar in the water for 30 minutes
2. Mash eggs through fine screen
3. Add eggs, finely crushed egg shells, and fish flakes to agar. Mix well
4. Pour mixture into several clean Petri dishes
5. Place clean Kimwipes on top of Petri dishes so that they cover the dishes but do not touch the cake mixture, and let cool for several hours
6. Cover dishes and store in the refrigerator
7. For feeding, cut cakes into small pieces and place in urchin tanks

[a] From B. M. Woodward, after E. B. Harvey.

## III. Obtaining Gametes

Gametes can be shed through the gonopores, by stimulating ovarian muscles to contract, using one of several methods. Each has associated advantages and disadvantages that will be outlined in the following discussion.

Many echinoderms can be stimulated electrically. This method has the advantage that the number of gametes shed from a given animal can be finely controlled. Only the gametes required for a given experiment are collected, and the same animal may be used repeatedly. Not all species of echinoids respond to electrical stimulation (e.g., *E. parma*), and an alternative method must be employed in these cases.

Figure 1 illustrates, in schematic form, the electrical stimulator we have used. Line voltage is stepped down using a variable transformer. Current is passed through a low wattage light bulb which provides a resistance (to prevent short circuit) as well as visual indicator for the passage of current through the circuit. A volt meter is connected into the circuit to measure the voltage applied to the animal. Two carbon electrodes (used for carbon coating in electron microscopy) are fixed to the ends of the wires, and the connection is insulated so that seawater will only come into contact with the carbon tip. Plastic tubing or tubing connectors make good insulators as well as being convenient handles for the electrodes.

Species vary in the voltage required to stimulate shedding. For *L. variegatus*, we found that 25 to 30 V is required while for *Arbacia punctulata* only 10 V is

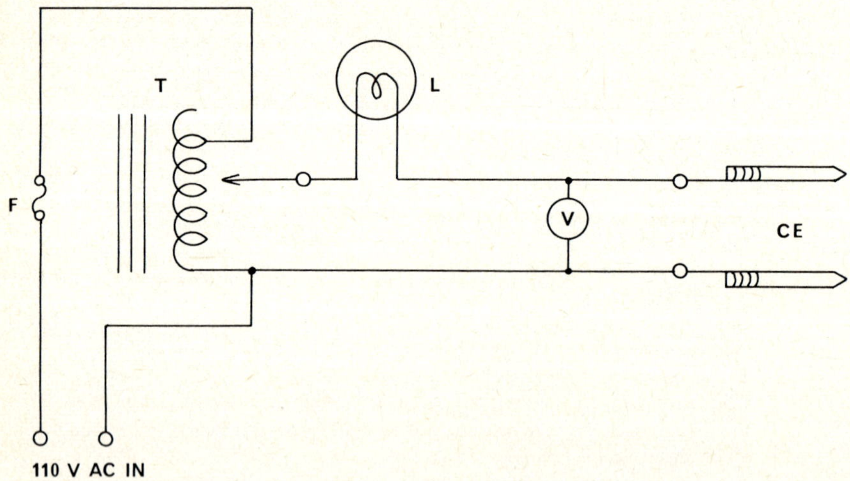


FIG. 1. Circuit diagram for electrical stimulator. F, Fuse; T, variable transformer; L, 120-V 40-W lamp; V, volt meter; CE, carbon electrodes from EM carbon coater.

necessary. If only one species is used in the laboratory, a simplified electrical stimulator may be used consisting of a single output voltage transformer.

For electrical stimulation, the animal is placed in a finger bowl partially filled with seawater, leaving the aboral surface exposed. One electrode is placed in the seawater while the other is lightly touched to the animal's surface 0.5 to 1 cm below a gonopore. In some species the gonopores are quite small, and it is desirable to use a stereomicroscope to observe the placement of the electrode as well as to determine the flow of gametes. We find that the animal is minimally stressed if the stimulation is given in approximately 10-second pulses separated by short pauses rather than in a continuous manner. After several pulses, gametes begin to issue from the gonopore if the animal is gravid. Stimulation is stopped and the gametes are collected from the gonopore region using a mouth-operated capillary tube. Semen is stored "dry" (undiluted) in these tubes while eggs are transferred into a seawater-filled beaker for washing. By using this pulse method of stimulation, problems of overstimulation, resulting in too many gametes being shed, or even complete shedding of the animal, are largely avoided. After collecting a sufficient number of gametes and making sure the animal is no longer shedding, the aboral surface is rinsed with seawater and the animal is returned to its tank. Male and female animals are held in separate tanks to prevent spawning.

As mentioned previously, not all species respond to electrical stimulation. In these cases one can inject 0.1 to 1 ml of a 0.5 *M* KCl or a 10 m*M* acetylcholine solution to induce shedding. A small-gauge needle should be used to inject the stimulant through the peristomal membrane surrounding the mouth into the body cavity. KCl injection usually causes complete shedding of the animal, but a small amount of acetylcholine can be injected instead to release a limited number of gametes. Thus, gametes can be collected from a single animal several times by repeated injections of small volumes of acetylcholine solution.

After injection of either stimulant, the animal is placed aboral side down onto a seawater-filled beaker whose diameter is smaller than the animal. Streams of gametes are then observed to come from the gonopores. Males are removed from the beaker and concentrated semen is collected "dry" from the aboral surface of the animal, as previously described. Eggs are allowed to settle to the bottom of the beaker. A steady stream of individual eggs usually indicates that the animal is ripe and the eggs will be healthy and fertilizable. A discontinuous stream or clumping of eggs as they fall from the gonopore may indicate problems. A small sample of eggs collected from the bottom of the beaker or directly from the gonopore should be checked for germinal vesicle stages or deteriorating eggs. Overstimulation will result in a mixture of "good" and "bad" eggs, and thus injection volumes should be empirically adjusted to provide a minimum dose for shedding.

Eggs should be thoroughly washed to remove debris and any toxic secretions before fertilization is attempted. Washing eggs is accomplished by allowing them to settle to the bottom of a beaker by gravity, decanting the seawater above, and resuspending the eggs in fresh seawater. This process is repeated two to three times. Debris which is heavier than the eggs can be removed by allowing it to settle, then decanting the suspended eggs into a clean beaker. Alternately, one can carefully aspirate the debris from the bottom using a Pasteur pipet. Some eggs withstand washing by centrifugation, but others do not tolerate such rough handling well.

Washed eggs may be stored after shedding. The length of time one can store a batch of eggs and still retain good fertilizability depends on a number of factors. In our experience, storage of eggs in "Jamarin U" artificial seawater increases the storage time by at least a factor of two over other artificial seawaters we have tried. Storing the eggs as a monolayer covered by about 1 cm of seawater appears to be better than maintaining the eggs in more crowded conditions with a thicker covering layer of seawater. Chilling the eggs slightly (5–10°C below tank temperature) also lengthens storage time of warm-water species although lowering the temperature even further, in our experience, decreases their viability and the percentage fertilization.

Semen may be conveniently stored for 1 to 2 days in sealed capillary tubes. Many investigators place sea urchin semen at 4°C for storage, but we found that

TABLE II

DISHWASHING PROCEDURE

| |
|---|
| 1. Never mix other laboratory glassware with glassware for living material. "Living" glassware includes culture dishes, beakers for collection and washing eggs, and glassware used to make up and store seawaters |
| 2. Rinse dishes thoroughly after use in tap water, and leave soaking in tap water. Never allow soaking dishes to be exposed to air: Dishes must be completely covered with water |
| 3. Before starting to wash dishes, be sure you can complete all the steps up to (7). Dishes may be left to soak at this stage |
| 4. Soak dishes in hot detergent solution (Alconox, Micro, or equivalent) for 30 minutes. Do not allow the dishes to stay in the detergent solution longer than 30 minutes. Add only enough detergent to the wash water to make it sudsy: Too much detergent requires additional rinsing |
| 5. Brush each dish thoroughly inside and out |
| 6. Rinse every part of each dish at least 10 times with running tap water. A clean dish begins to "squeak" in your fingers at this point. Check that each dish is visibly clean before proceeding to the next step |
| 7. Transfer tap water-rinsed dishes into a tub filled with tap-distilled water. Dishes may be held at this stage for an extended period if they are completely covered |
| 8. Rinse every part of each dish 10 times in tap-distilled water and immediately rinse every part of the same dish 5 times in glass-distilled water |
| 9. Invert dishes on layers of clean Kimwipes laid on a clean surface, cover with clean Kimwipes, and allow to dry overnight |
| 10. Cover beakers, flasks, graduates, etc., with Parafilm or aluminum foil. Smaller items are most conveniently stored in plastic bags |

sperm, like eggs, retain their viability longer if we stored them 5–10°C cooler than the animal's holding temperature.

## IV. "Biocleaning" Glassware

We have developed and used the procedure outlined in Table II to clean our culture glass or plastic ware. Depending upon local water conditions, one may be able to reduce the number of rinses at each step. Those listed here are guidelines which have been safely used under a variety of laboratory conditions, and modifications must be evaluated by individual investigators. The soaking steps are carried out in small plastic wash tubs reserved for this purpose.

Biocleaning procedures for cover glasses and slides (Table III) are similar to the one for dishes. Again, the number of rinses at each step may be modified depending upon local water quality. We prefer to store both slides and cover glasses in 80% ethanol, removing them with clean forceps, and drying them with four thicknesses of Kimwipe (to avoid transfer of oils from fingertips to slides) just before use. This lessens the possibility of contamination during storage.

## V. Fertilization and Culturing

Methods to fertilize echinoderm eggs have been described in great detail in a number of reviews (Harvey, 1956; Hinegardner, 1975). In the following section,

TABLE III

"BIOCLEANING" SLIDES AND COVER SLIPS

1. Fill 500-ml beaker with hot tap water and enough detergent to make the solution sudsy
2. Drop 30–50 cover glasses or slides *individually* into the beaker. Try to position the slides in a criss-cross arrangement in the beaker
3. Soak for 15–30 minutes
4. Place beaker in ultrasonic washer for 3–5 minutes
5. Rinse many times in tap water to remove the majority of the detergent
6. Using stainless steel forceps, transfer slides or cover glasses individually to another beaker containing tap-distilled water. Repeat this individual transfer rinse 10 times. Sonicate beaker for 3–5 minutes before the last transfer
7. Individually transfer slides or cover glasses into a beaker containing glass-distilled water. Repeat this 10 times. Sonicate beaker for 3–5 minutes before the last transfer
8. Individually transfer slides or cover glasses into bioclean storage jars filled with 80% ethanol for storage
9. Before use, remove slide or cover glass from alcohol using clean forceps and wipe dry with a Kimwipe folded twice (four thicknesses). Do not rub

we briefly outline the procedure we use, adding a number of hints that improve the success of fertilization and the viability of our cultures.

Eggs are placed in a culture dish so that they cover, at most, one-third of the bottom area. Overcrowding results in poorer fertilization success and higher mortality during culture. Concentrated "dry" semen is diluted into seawater before adding to the eggs. First, push out a small amount of sperm from the storage capillary and discard. This semen contains a large number of nonviable sperm due to evaporation. A small drop ($< 20\ \mu l$) is added to 2–3 ml of seawater (dilution $>$ 1:100) and is gently mixed. A sample of the diluted sperm should be checked under the microscope to make sure that the sperm are actively motile. If a majority of the sperm in the sample is nonmotile, make another dilution from a different batch of sperm until most sperm are swimming.

To fertilize the eggs, add one to three drops of the diluted sperm to the dish and gently mix the culture for approximately 10 seconds. Do not mix by vigorously sucking in and out of a pipet. Avoid adding too much sperm since the eggs become polyspermic. Also, the excess sperm quickly die and encourage bacterial and protozoan growth. Within 1 minute after the addition of sperm, the eggs should be surrounded by a fertilization envelope which fully develops 4–6 minutes after the addition of sperm. Fertilization success may be assessed by scanning the culture dish and evaluating the percentage of eggs in the culture displaying fertilization envelopes. For experiments, we generally use cultures with >95% fertilization success. After insemination, let the eggs settle, aspirate off the supernatant seawater, and add fresh seawater to remove excess sperm.

For a given species, the degree of elevation of the fertilization envelope from the egg surface is a good indicator of egg condition. As the eggs age after shedding, their fertilization envelopes elevate progressively less. This is particularly easy to see in those species whose fertilization envelopes are normally quite high (e.g., *L. variegatus, E. parma*). Another indicator of declining egg condition is the increased concentration of sperm required to inseminate an older batch of eggs.

## A. Removal of Fertilization Envelopes

It is often desirable or necessary to remove the fertilization envelope from the eggs. A number of techniques have been published in the literature, and we will list a few in this section. When first starting, we recommend that several techniques be tried before making a decision on which technique to use routinely. Species and investigator differences both play a role in the ease and success of these methods.

While the fertilization envelope expands to its maximum extent in the first minute or two following insemination, it is quite soft and fragile. Shaking the eggs in a test tube during this period of time (approximately the first 4 minutes following fertilization—during the next 4–6 minutes the fertilization envelope becomes progressively tougher and more difficult to remove) breaks open some of the fertilization envelopes. Often, however, the agitation of the eggs cause some to lyse, it is difficult to use with small numbers of eggs, and the percentage removal is quite variable.

Another mechanical method for fertilization envelope removal which is widely used is passing the newly fertilized egg suspension through Nitex, a nylon screen material which comes in a wide variety of mesh sizes (Tetko Inc., Precision Woven Screening Media, Elmsford, NY). Useful mesh size depends on the diameter of the egg and should be about 10 μm smaller than the average egg diameter. Too small a mesh size results in fragmentation of the eggs while too large a mesh size allows eggs with their fertilization envelopes to pass through. Removal of fertilization envelopes is accomplished by gently pouring the egg suspension (about 4 minutes following insemination) through a small piece of Nitex, which is fastened to the bottom of a short plastic tube, into a culture dish. Success with this technique depends on determining the mesh size of the Nytex, the time after fertilization to begin pouring, and the right pouring rate. The filter should be well rinsed with running tap water followed by distilled water after use to keep it free from debris. This method works well with both small and large numbers of eggs: For best results the area of the Nitex should be reduced for small numbers of eggs and increased for larger cultures.

Removal of fertilization envelopes by shearing through Nitex screen can be facilitated by fertilizing the eggs in seawater containing 1 m*M* 3-amino-1,2,4-triazole (Showman and Foerder, 1979). This peroxidase inhibitor blocks fertilization envelope hardening as long as it is present during the first 5 minutes following insemination. The fertilization envelope remains soft and may be removed at any point in development up to hatching. In addition, "stickiness" and clumping of fertilization envelope-free embryos was reduced by letting the embryos establish their initial hyaline layer (approximately 1 hour after insemination) prior to demembranation.

We routinely use an isosmotic solution of glycerol for fertilization envelope removal (Salmon and Segal, 1980). After insemination and once the fertilization envelopes have begun to elevate, the eggs are gently pelleted using a hand centrifuge (about 30 seconds). The seawater supernatant is aspirated off and the eggs are resuspended in 10 to 50 vol of "wash buffer" (1 *M* glycerol, 5 m*M* Tris, pH 8.0–8.2). After the eggs have been in the wash buffer for 30–60 seconds, they are gently pelleted and washed by centrifugation with at least two changes of seawater. In our hands, this method is very effective and reproducible and is not injurious to the subsequent development of the embryos. Another advantage of the method is that it works equally well with eggs producing high or low fertilization envelopes.

### B. Handling Small Numbers of Eggs

In order to load our manipulation or observation chambers, we must be able to select, pick up, transfer, and dispense a small number of gametes. In addition, eggs that have had their fertilization envelopes removed have a tendency to stick to glass pipets. We have overcome these problems by making suitably sized and tapered tips of polyethylene tubing (Intramedic polyethylene tubing, Clay Adams, Inc., Parsippany, NJ). To taper the tubing, a small microburner is constructed from an 18-gauge syringe needle placed into the end of a convenient length of tubing with a screw-type tubing clamp used to regulate the gas flow. It takes very little heat to soften the tubing, and unlike glass, the polyethylene tubing has a tendency to contract when heated. The tubing is pulled slowly and steadily to avoid tearing, and this is most easily accomplished by continued mild heating. When the desired taper and end diameter have been reached, remove the tubing from the heat but maintain tension on the tubing until it has cooled (the polyethylene turns from clear to translucent).

Tips pulled from ''PE-50'' tubing fit onto the ends of most microliter syringe needles. By using a 10-μl syringe, a fine degree of control can be exercised. Because the polyethylene is translucent, eggs can be easily seen in the lumen of the tubing. ''PE-160'' tubing is useful for making tips that fit on the ends of standard Pasteur pipets. As long as the tips are well rinsed after each use, they can be reused many times before they need to be replaced.

## VI. Observation Chambers

### A. Microdrop Preparation

Temporary slides can be made by placing a small sample of eggs in seawater on a bioclean slide, covering the drop with a cover slip whose edges have a thin layer of vaseline or silicon high-vacuum grease. The grease seals the preparation to prevent evaporation and also supports the weight of the cover slip. These preparations are useful only for short-term observation since they do not allow for gas exchange and the cells very soon become anoxic. For long-term observations of individual embryos, one of us (S.I.) developed what we call a microdrop preparation. Using this chamber, one can follow the development of individual embryos from just after fertilization through gastrulation to late pluteus stages.

Figure 2 illustrates the construction of this chamber. A bioclean slide and cover slip must be used. A spacer is cut from filter paper slightly smaller than the size of the cover slip, and a hole is punched or cut from the center. The thickness of the spacer used depends upon the diameter of the egg under observation. One

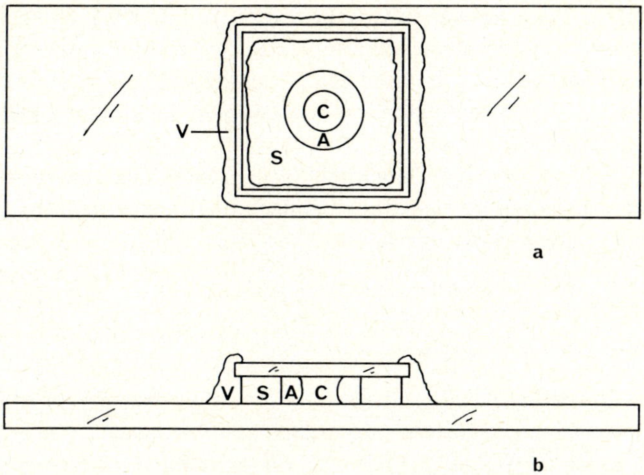


FIG. 2. Microdrop observation chamber, in top (a) and side (b) view. V, Valap sealing wax; S, filter paper spacer; A, air space; C, droplet of culture medium containing cells.

wishes to hold each egg in place without overly compressing it. The spacer must also be made from an absorbent material since its second function is to provide a reservoir of seawater to saturate the gas phase of the chamber. If you measure the thickness of your dry spacer material, remember that the thickness approximately doubles when it is wet.

It is important to construct and seal the chamber as quickly as possible to limit evaporation of the small egg-containing droplet. To this end, we make our preparations in the following sequence. First, place the bioclean slide and cover glass on a support made from applicator sticks. This keeps the slide and cover glass away from the bench top where they will quickly pick up dirt. Next, place the dry filter paper spacer on the center of the slide. Third, draw fresh seawater into a clean Pasteur pipet, and also load a sample of eggs into the polyethylene tubing tip (about 10 eggs in a 20-μl drop) of a microliter syringe. Then, wet the spacer with the fresh seawater, quickly place the drop of eggs in the center of the spacer hole, and gently place the cover slip on top of the spacer. The drop must remain surrounded by air and must not touch the spacer. Finally, the cover slip is held in position while the edges are completely sealed with Tackiwax or valap (a 1:1:1 mixture of vaseline, lanolin, and paraffin or beeswax, simmered on low heat for 2 hours). It is a good idea to check the sealing job using a stereomicroscope and to patch any holes quickly, since even a small pinhole will lead to the preparation drying out.

Instead of placing the drop in the center of the hole on the slide, one can place the drop on the center of the cover slip, then carefully lower the slide with wet spacer on top of the cover slip. The adhesion of the cover slip to the spacer/slide is firm enough to allow the whole preparation to be lifted, inverted, and then sealed. Either method may be used for assembling the preparation. Its success depends upon the speed with which it is assembled, and on the bioclean quality of the components.

As mentioned previously, ''Jamarin U'' rather than filtered seawater is recommended for suspending the eggs for extended culture. It is the only medium that has allowed *L. variegatus* zygotes to grow into full plutei in the microcrop preparation.

## B. Perfusion Chambers

Figure 3 illustrates the construction of a wedge-shaped perfusion chamber designed for use on an upright microscope stand, modeled after the chamber devised by Inoué and Tilney (1982) used on an inverted microscope stand. The wedge-shaped spacer is an improvement over the more familiar parallel spacer in two ways: The cells are trapped in the wedge, and the perfusate is forced to spread out, providing a uniform exchange of media throughout the entire chamber. This chamber permits high-resolution observation of cells while allowing for rapid and complete exchange of media.

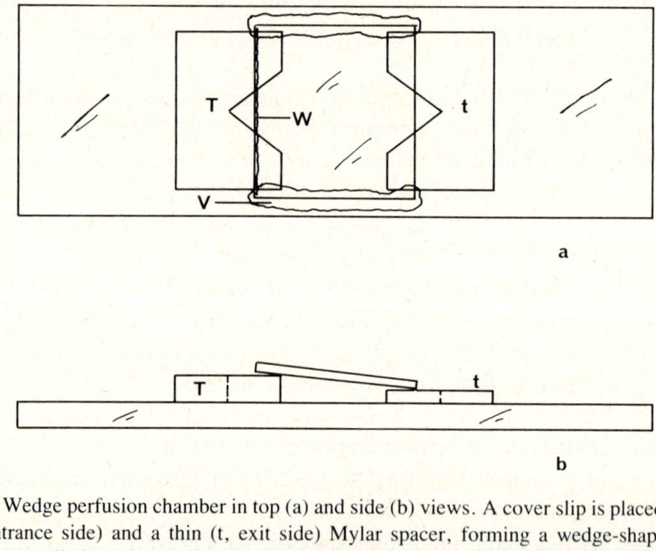


FIG. 3. Wedge perfusion chamber in top (a) and side (b) views. A cover slip is placed on top of a thick (T, entrance side) and a thin (t, exit side) Mylar spacer, forming a wedge-shaped chamber space. The cover slip is sealed to the slide on two sides with valap (V). A drop of perfusate is placed at the opening of the thick side of the chamber and is drawn through the chamber by a filter paper wick placed at the opening of the thin side. A grease pencil line (W) drawn on the entrance edge of the cover slip helps to prevent the perfusate from flooding the cover slip surface.

The spacers are cut from Mylar sheets (sample sheets may be obtained by contacting the Products Division, E. I. DuPont, East Orange, NJ) of different thicknesses, or from Scotch brand Magic tape (3M Co., St. Paul, MN; one tape layer is approximately 60 $\mu$m thick). A cover slip is placed on top of the spacers and sealed on two sides with valap. To help prevent drops of perfusate from flooding the cover slip surface, a grease pencil line is drawn on the entrance edge. Tapered filter paper wicks are used to draw solution through the chamber. Flow rate through the chamber can be crudely adjusted by changing the angle of taper of the wicks, and by using different filter paper types (e.g., Whatman #4—medium to fast; Whatman #501—slow to medium).

## C. Egg White-Coated Perfusion Chambers

Although the wedge perfusion chamber was ideal for experiments with eggs, in studies of sperm flagellar motility (Schweitzer *et al.*, 1983), we observed that live sperm of some species, as well as detergent-extracted sperm models, quickly became stuck to the glass surfaces along their entire length. To overcome this problem, a coating procedure was developed to lessen the ''stickiness'' of the preparation. Both slides and cover slips were coated with fresh egg white by immersing them for 5–10 minutes. After this time, the slides/cover slips were removed from the egg white, one side was wiped clean, and they were stored in a high-humidity environment. Perfusion chambers were assembled as described previously, with the egg white being wiped away from all areas of the slide except those included in the wedge space. After sealing the chamber with valap, the interior was rinsed with seawater (live sperm) or buffer (sperm models), and the chamber was stored in a moist container in this condition.

Sperm, either models or living, preferentially attach by their midpieces, leaving the flagella free for observation. Although the image is degraded slightly because of the unevenness of the egg white coating, surprisingly high-resolution images can be obtained using this chamber design.

## D. Manipulation Chambers

The design and construction of a chamber, designed by G. W. Ellis and used extensively in our laboratory, which allows easy access to the eggs by micropipets or microtools, has already been described in detail by Kiehart (1982). Eggs in this chamber may be observed using high-resolution and high-extinction microscopy modes for extended periods of time without affecting the subsequent development of the embryos. An alternative chamber configuration, and the use of polyanionic coatings of chamber surfaces, is described here.

Rather than the parallel chamber space formed by the coverslip fragment and the support cover slip described by Kiehart (1982), we have used a chamber in which the cover slip fragment and the support overslip form a wedge-profiled space, from front to back (Fig. 4), as described by Hiramoto (1956; also see Kishimoto, this volume). This is accomplished by supporting the fragment only at its front corners by two small tape or Mylar spacers. Using this design, the cells need not be as greatly flattened to hold them in place as in the parallel design.

For micromanipulation work, one of us (D.A.L.) found that neither chamber design held the eggs firmly enough for movement of the microneedle inside the cell. To overcome this problem the surfaces of both the fragment and the support cover slip, in contact with the cells, were coated with either poly-L-lysine (Mazia *et al.*, 1975) or protamine sulfate (Steinhardt *et al.*, 1971) prior to chamber assembly (Table IV). I found species differences in how well these agents worked and recommend that other investigators try each polycation. For example, I found that demembranated *L. variegatus* eggs adhered more tightly to protamine sulfate-coated cover slips than to poly-L-lysine-coated ones. In addition, the embryos grown on the poly-L-lysine-coated cover slips showed tendencies of postgastrulation abnormalities, while eggs grown on protamine sulfate-coated cover slips did not.

## VII. Considerations for Microscopic Observation of Development of Single Embryos

### A. Separation of Blastomeres

The study of later cleavage stages can be difficult, even in clear eggs, due to the interference produced by out-of-focus blastomeres above or below the cell of

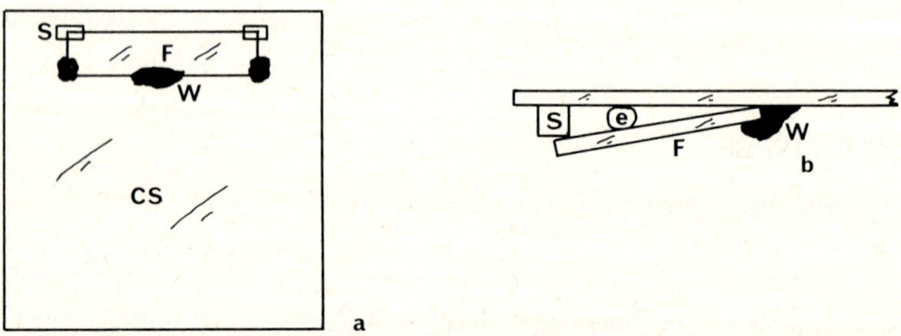


FIG. 4. Wedge manipulation chamber. (a) A cover glass fragment (F) is supported at the front by two small pieces of tape or Mylar (S) while it rests on the main cover slip (CS) at the back. It is positioned 1–2 mm from the front edge of the main cover slip, and is held in place by 3 dabs of tackiwax (W). For additional details of chamber and support slide construction see Kiehart (1982). (b) An egg (e) in the chamber in side view.

TABLE IV

PREPARING COVER SLIPS FOR CELL ADHESION[a]

| | |
|---|---|
| 1. | Biocleaned cover slips are immersed in concentrated nitric acid for 30 minutes |
| 2. | Rinse well with plenty of glass-distilled water |
| 3. | Store in glass-distilled water |
| 4. | Remove the cover slip from water and blot off excess water by touching edge of cover slip to filter paper |
| 5. | Flood surface of cover slip with several drops of 1 mg/ml poly-L-lysine in glass-distilled water (approximate $M_r$ 220,000; Sigma Chemical Co., St. Louis, MO) or 1% protamine sulfate in glass-distilled water (Gr. A, Calbiochem-Behring, La Jolla, CA). Aliquots of both solutions are kept frozen until use |
| 6. | Leave solution on cover slip 2–5 minutes |
| 7. | Rinse cover slip with glass-distilled water |
| 8. | Blot excess water from cover slip edge with filter paper and air dry |

[a] Steps 1–3 may be done ahead of time; steps 4–8 are done *the day of use*.

interest. In a study of the nuclear migration and mitotic apparatus positioning prior to the unequal fourth division in vegetal blastomeres, we used one-half and one-quarter embryos produced by separating blastomeres following first and second division.

Following insemination, fertilization envelopes are removed by one of the previously described methods, and the embryos are cultured in $Ca^{2+}$-free artificial seawater. Blastomeres are separated from each other after the completion of first (one-half embryos) or second division (one-quarter embryos) by carefully pushing them apart with a fine glass needle. The needle is made by pulling a glass capillary tube in a microburner flame. Once the blastomeres are separated, they are collected and further cultured in calcium-containing artificial seawater. In seawater containing calcium, the hyaline layer reforms between the blastomeres of the subsequent divisions, "glueing" them together.

## B. Light Source

For long-term observation of dividing embryos, standard microscope light sources must be appropriately filtered before reaching the specimen. Quartz–halogen, tungsten, mercury, and especially xenon lamps all emit infrared radiations which can heat the specimen and damage optical components (particularly sensitive are polarizing sheet filters). One should insert reliable heat-cut filters between the illuminator and the mirror reflecting the light to the polarizer and condensor (note that differential interference contrast optics also use polarizers).

Short-wavelength UV irradiation can cause cell damage and even death over a short period of time, and the addition of a UV-cut filter may be required if one is using a UV-rich source (e.g., mercury or xenon). Use of an interference filter for a longer wavelength, in combination with the glass lenses in the illuminating optical path, should sufficiently attenuate the UV. However, even wavelengths in the visible blue range can have noticeable detrimental effects on ciliary and flagellar motility as well as on cell division, depending on the species.

The best wavelength to select for illuminating the specimen is 546-nm green or 577/579 yellow light. Not only do they appear to be harmless to cellular activities but they have several other advantages as well. The mercury arc lamp has strong emission lines at these wavelengths. Optical components of the microscope are best corrected in this wavelength range, especially important when using achromatic objective lenses, and our eye has its peak sensitivity in green light (for photopic vision). It is worth the expense to obtain a high quality, 546-nm *narrow band-pass* interference filter for living cell observations. Many microscope manufacturers supply them or one may obtain them from manufacturers of optical components (e.g., Baird-Atomic, Inc., Cambridge, MA; Ealing Optics Corp., South Natick, MA). Upon receipt, it is a good idea to check the transmission spectrum of your filter with a spectrophotometer to make sure that it meets the published specifications. Although transmission of the rest of the visible spectrum would be expected to be quite low, they often transmit a surprising amount of the infrared spectrum, again stressing the need for a high-quality heat-cut filter to be placed directly following the illuminator.

## C. Controlling Temperature during Observation

Many species have optimum development temperatures that are below normal room temperatures (e.g., *E. parma*, 16°C; *L. pictus*, 18°C), and therefore some mechanism must be employed to keep the temperature of the embryos under observation near optimum and preferably constant. One simple method is to observe the cultures for no longer than a minute or so at a time, after which they are immediately returned to a low-temperature incubator. However, if long-duration observation is required, another solution must be found.

The entire microscope can be placed in a temperature-regulated room. This is the easiest solution, although investigator comfort begins to be a problem at lower temperatures. The microscope should be placed in the room before the temperature is decreased or it should be insulated so that the temperature change is gradual. A sharp temperature drop can cause stress and strain in the glass and metal components, which can affect their performance. We have also directed the filtered output from an air conditioner, by channeling the cool air through a flexible duct, close to the stage of the microscope. This method has worked quite well in lowering the stage temperature up to 10°C below ambient. Troyer (1974) has described a thermoelectric, temperature-regulated stage providing for either heating or cooling, and we have developed a special slide for rapidly regulating the specimen chamber temperature (Inoué *et al.*, 1975).

## D. Microscope Use and Cleaning

A detailed description of the proper use of the light microscope is beyond the scope of this article and we direct you to several other sources for this information. Slayter (1970), Spencer (1982), and Inoué (1986) contain details of image information and practical instructions for microscope alignment and use. Also, the microscope manufacturers, in addition to instructions for use of their own equipment, often have available excellent reference material on different image-forming systems, interpretation of the image, and theoretical aspects of image formation.

Cleanliness is of the utmost importance when using any microscope to its full potential. All optical components, from the illuminator to the oculars or camera, must be kept free of dust, oil, and fingerprints. Dust may be removed by using a soft brush (used only for this purpose), a blower bulb, or a can of compressed air. Commercial cans of compressed air often expel droplets of oil or propellant that are difficult to remove, and they must be used with caution. Oily residues and fingerprints can sometimes be removed by moist breath and a soft piece of lens tissue (hard lens tissues are not recommended as they can scratch the coatings present on lens surfaces). Do not rub the lens surface even with lens tissue. Rather, gently wipe, working from the center of the lens to the edge. Stronger organic solvents, such as ethanol or benzene, may have to be employed for stubborn oil contamination. Never put the solvent directly on the lens surface. Instead, place a drop onto the tip of a rolled piece of lens tissue and gently wipe the surface with the wet tip (see Inoué, 1985, for further details). Slides, when not on the microscope stage, should be supported above the bench top surface. This is easily accomplished by resting the slide on two wooden applicator sticks positioned under each end, and covering the cover slip area with a small Petri dish cover.

## E. Types and Selection of Optics

Because of the generation of ''phase halos'' and the optical noise in the image plane produced by out-of-focus objects, the use of phase-contrast microscopy has slowly declined in recent years in favor of differential interference contrast (DIC) microscopy. However, in certain circumstances it may still be the method of choice. In studies of flagellar motility in living sperm and reactivated detergent models, we found that the phase-contrast image was of higher contrast and more easily recorded than the DIC image. In order to get the most from the phase-contrast microscope it is important to align precisely the phase annulus of the condenser with the phase ring in the objective lens. This is accomplished by observing the objective aperture using a telescope and moving the condenser annulus until it is superimposed over the dark phase ring of the objective. It is especially important to keep the preparation as clean as possible since out-of-focus images of dirt add greatly to the optical noise seen in the image, obscuring specimen detail.

Polarized light microscopy has been used to study the assembly, function, and disassembly of the mitotic spindle during cell division (review Inoué, 1981a) and the morphogenesis of the biocrystalline, spicule skeleton of the pluteus larva (Okazaki and Inoué, 1976). Contrast is produced by the interaction of plane-polarized light with an anisotropic alignment of molecules in the specimen. Using polarized light microscopy, one can quantitatively determine molecular arrangements inside living cells and observe how they change during the cell cycle or during differentiation.

Theoretical and practical aspects of polarized light microscopy are described in the references cited at the beginning of this section and in Bennett (1950). For information on the particular problems associated with biological polarized light microscopy and methods used to overcome them, see Swann and Mitchison (1950), Inoué and Dan (1951), Inoué (1961, 1981b, 1986), and Allen *et al.* (1981a).

A DIC microscope is a type of polarizing interference microscope. The shaded image that one observes is a result of the rate of change of the refractive index experienced by two beams of light separated (sheared) by a distance at or slightly less than the resolving power of the particular objective used. It gives good contrast to fine specimen detail when that structure is oriented perpendicular to the shear direction (the direction that the light appears to be coming from to give rise to the shaded image). In addition, depth of field discrimination is very good, and the specimen may be "optically sectioned."

There are two commercially available DIC systems: the Smith design manufactured by Leitz, and the Nomarski design manufactured by most other microscope companies. The difference between the two designs lies in the physical placement of the Wollaston prisms (the beam splitters). In the Smith system, the Wollastons are located at the condenser and objective aperture planes. This means that prisms are mounted in the objective lenses and these lenses are used exclusively for DIC. The prism design has been slightly modified in the Nomarski system such that the prism does not physically reside within the objective. Thus, the lenses may be used for other types of microscopy.

Fluorescence microscopy techniques are becoming increasingly valuable tools in the study of cell physiology and development. Vital fluorescent dyes have been found which bind or associate with specific locations or structures within the cell (DNA—Hoechst dyes, Latt and Statten, 1976). The fate of a microinjected molecule can be determined in the living cell if the molecule has a fluorophore group attached (Wang *et al.*, 1982; Kreis and Birchmeir, 1982; Salmon *et al.*, 1984). As discussed in the next section, certain fluorescent dyes can be used as cell markers in embryonic cell lineage studies.

Any standard microscope can be easily and inexpensively converted into a fluorescence microscope by inserting an appropriate excitation filter before the specimen, and an appropriate barrier filter after the specimen. While this arrangement is simple and relatively cheap, and high numerical aperature objectives can be used for the efficient collection of the fluorescent light, the separation of the strong excitation light from the weak fluorescent light by the barrier filter is difficult to accomplish, even with the best quality filters.

A dark-field condenser may be used to keep the illuminating excitation light from entering the objective, providing a low background level of light. However, objectives of rather low numerical aperture must be used in conjunction with the dark-field condensers, and the amount of fluorescent light collected by the objective is reduced.

An incident light or epiillumination system helps to solve these problems. In this case both excitation and emitted wavelengths are well separated by a dichroic beam splitter. The objective is also the condenser and high numerical aperture objectives may be used for efficient collection of the fluorescent light. This system is, however, the most expensive.

Selection of objective lenses for fluorescence work is a compromise between magnification (intensity is related to the inverse of the magnification squared) and numerical aperture (intensity is related to numerical aperture to the fourth power for epifluorescence). Many improvements in objective lens design for fluorescence microscopy have been made by the various manufacturers. They have begun to maximize numerical aperture for both low and high magnifications. The use of fluorite and other special optical glass elements has improved the transmission properties of these lenses down to 350 nm. Various immersion lenses have been produced for use with or without coverglasses to limit light loss from reflections at air–glass interfaces.

## F. Use of Video for Data Recording

Recent advances in video technology, and the increased availability of high-quality, moderately priced equipment has led many investigators to use closed-circuit television systems in conjunction with a variety of light microscopical techniques. Besides the convenience of these systems for data recording, video imaging has many advantages for the researcher studying cell structure and physiology. In the following section, we will briefly point out some of the advantages of video. Further specific information may be obtained by consulting Allen *et al.* (1981a,b) and Inoué (1981b, 1986).

A dramatic increase in image contrast is obtained using video cameras. This contrast enhancement can be automatically set by the camera, or with many of the cameras now available, the operator has the option of manual control. A standard video camera records one field of information, each one-sixtieth of a second, and therefore time resolution is quite good. Swimming gametes and embryos can be clearly recorded. Using intensified video cameras, very low light levels can be detected; this is especially useful in fluorescence techniques. This great sensitivity is accompanied by a decrease in camera resolution due to increased noise levels, and a compromise must be made between resolution and sensitivity.

Camera output can be recorded by a video tape recorder. Two tape formats are commonly used: $\frac{1}{2}$-in. VHS or $\frac{3}{4}$-in. U-Matic. Higher resolution can be recorded on the $\frac{3}{4}$-in. format, although the tape and the recorder itself are more expensive. All machines enable one to make an audio recording at the same time as the video, and this greatly facilitates commenting on an experiment in progress. Time-lapse recorders are available, a feature very useful to investigators of cell motility and development. However, at the time of this writing, only three machines are commercially available with this capability and all are $\frac{1}{2}$-in. VHS format (Panasonic model #AG6010; Panasonic model #AG6050 for higher resolution; Gyyr model #TLC2051).

A variety of analog (and recently also digital) devices can be connected to a video system. These peripherals can be used for a variety of functions such as contrast enhancement, noise reduction, and quantitative measurements. Using a special-effects generator one can record such things as an oscilloscope trace along with the microscope image (e.g., see Inoué, 1986).

## VIII. Iontophoretic Injection of Lucifer Yellow CH to Study Embryonic Cell Lineage

In collaboration with R. I. Woodruff, we have been investigating the distribution and developmental fate of the progeny of the micromeres that are formed at the fourth division in sea urchin embryos (Woodruff *et al.*, 1982). For this study we have injected the fluorescent marker dye Lucifer Yellow CH into the blastomeres iontophoretically, using a simple, inexpensive circuit (Fig. 5). Lucifer Yellow quickly binds to cytoplasmic components and is nontoxic in low concentrations making it useful as an *in vivo* lineage marker.

Using a SIT video camera, we were able to follow individual injected embryos from fourth division until midpluteus stages and observe the fate of the injected cell and its progeny at various stages in between. Even at the later stages, we were able to clearly identify the descendants of a single, microinjected cell.

### Acknowledgments

The preparation of this article was supported, in part, by Grants NIH R01 GM31617, NSF PCM 8216301 (to S.I.), NSERC Postgraduate Fellowship and MRC Postdoctoral Fellowship (to D.A.L.).

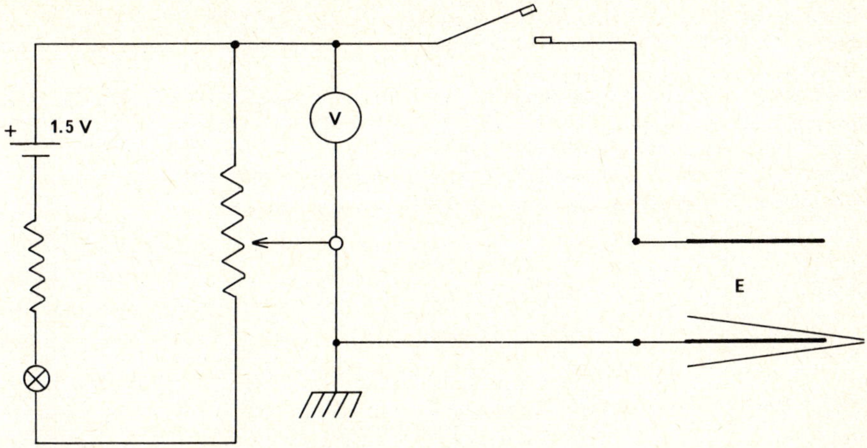


FIG. 5. Circuit diagram of iontophoretic injection apparatus. V, Volt meter; E, Ag–AgCl electrodes.

## REFERENCES

Allen, R. D., Travis, J. L., Allen, N. S., and Yilmaz, H. (1981a). *Cell Motil.* **1,** 275–289.

Allen, R. D., Allen, N. S., and Travis, J. L. (1981b). *Cell Motil.* **1,** 291–302.

Bennett, H. S. (1950). *In* "Handbook of Microscopial Technique" (C. E. McClung, ed.), pp. 591–677. Hoeber, New York.

Harvey, E. B. (1956). "The American Arbacia and Other Sea Urchins." Princeton Univ. Press, Princeton, New Jersey.

Hinegardner, R. (1975). *In* "The Sea Urchin Embryo. Biochemistry and Morphogenesis" (G. Czihak, ed.), pp. 10–25. Springer-Verlag, New York.

Hiramoto, Y. (1956). *Exp. Cell Res.* **11,** 630–636.

Inoué, S. (1961). *In* "Encyclopedia of Microscopy" (G. L. Clarke, ed.), pp. 480–485. Reinhold, New York.

Inoué, S. (1981a). *J. Cell Biol.* **91,** 131s–147s.

Inoué, S. (1981b). *J. Cell Biol.* **89,** 346–356.

Inoué, S. (1986). "Video Microscopy." Plenum, New York.

Inoué, S., and Dan, K. (1951). *J. Morphol.* **89,** 423–456.

Inoué, S., and Tilney, L. G. (1982). *J. Cell Biol.* **93,** 812–819.

Inoué, S., Fuseler, J., Salmon, E. D., and Ellis, G. U. (1975). *Biophys. J.* **15,** 725–744.

Kiehart, D. P. (1982). *Methods Cell Biol.* **25B,** 13–31.

Kreis, T., and Birchmeir, W. (1982). *Int. Rev. Cytol.* **75,** 209–227.

Latt, S. A., and Statten, G. (1976). *J. Histochem. Cytochem.* **24,** 24–33.

Mazia, D., Schatten, G., and Sale, W. (1975). *J. Cell Biol.* **66,** 198–200.

Okazaki, K., and Inoué, S. (1976). *Dev. Growth Differ.* **18,** 413–434.

Salmon, E. D., and Segal, R. R. (1980). *J. Cell Biol.* **83,** 355–365.

Salmon, E. D., Saxton, W. M., Leslie, R. J., Karow, M. L., and McIntosh, J. R. (1984). *J. Cell Biol.* **99,** 2157–2164.

Schweitzer, P. A., Otter, T., Inoué, S., and Burgos, M. (1983). *J. Cell Biol.* **97,** 15a.

Showman, R. M., and Foerder, C. A. (1979). *Exp. Cell Res.* **120,** 253–255.

Slayter, E. M. (1970). "Optical Methods in Biology." Wiley, New York.

Spencer, M. (1982). "Fundamentals of Light Microscopy." Cambridge Univ. Press, London and New York.

Steinhardt, R. A., Lundin, L., and Mazia, D. (1971). *Proc. Natl. Acad. Sci. U.S.A.* **68,** 2426–2430.

Swann, M. M., and Mitchison, J. M. (1950). *J. Exp. Biol.* **28,** 434–444.

Troyer, D. (1974). *J. Microsc.* **102,** 215–218.

Wang, Yu-Li, Heiple, J., and Taylor, D. L. (1982). *Methods Cell Biol.* **24B,** 1–11.

Woodruff, R. I., Lutz, D. A., and Inoué, S. (1982). *Biol. Bull.* **163,** 379a.

# Chapter 7

# *Egg and Embryonic Extracellular Coats: Isolation and Purification*

PEGGY J. WEIDMAN AND ERICA S. KAY

*Department of Biochemistry*
*University of Washington*
*Seattle, Washington*

The extracellular coats of the sea urchin embryo known as the fertilization membrane and the hyaline layer have been the subjects of intense fascination and investigation for over a century. The term "fertilization membrane" was coined in 1847 by Derbès to describe the translucent, tough envelope assembled around the egg after penetration of a spermatozoon (for review see Kay and Shapiro, 1985). We now know that this "membrane" is actually a protective glycoprotein coat assembled, under the influence of divalent cations, from the egg glycocalyx (the vitelline layer) and proteins secreted from the egg cortical granules at fertilization. Within 30 seconds of fertilization, the vitelline envelope is elevated and the highly organized structure, stabilized by ovoperoxidase-catalyzed dityrosine crosslinks, is finished only minutes later. This remarkable example of extracellular morphogenesis is one of the few eukaryotic self-assembly processes that is amenable to dissection and analysis with current biochemical technology. Although the simultaneous formation of the hyaline layer may be less spectacular than fertilization membrane assembly, it is of no less interest as a subject of investigation. This thick, amorphous layer is a relatively simple extracellular matrix and is known to be involved in maintaining the integrity of the embryo.

This article focuses primarily on the methods available for analysis of fertilization membrane assembly, namely the isolation of precursors, individual components, and intermediates of the assembled fertilization membrane. Isolation of hyalin, the major protein component of the hyaline layer, readily follows as a useful by-product of these methods. All of the methods described pertain (primarily) to *Strongylocentrotus purpuratus* eggs and embryos, the species for which fertilization membrane assembly has been most studied. Due to the plethora of techniques available, we have chosen to depart somewhat from the goal of this series, i.e., to present in detail descriptions of all relevant methods. Instead, we have endeavored to give a critical review of the methods that we feel are valuable for investigators interested in examining modifications in egg surface coats after fertilization.

Figure 1 presents a schematic diagram of the several fertilization membrane components that will be discussed and the relationships between them. Each section of the text includes a brief description of an experimental method, with references to the original publications. The applicability, advantages, and disadvantages of the method are also discussed, including any helpful information that we have gleaned from our experience. Methods with which we are most familiar have been elaborated upon. In the section dealing with the purification of proteins from fertilization product, we have attempted to demonstrate how the

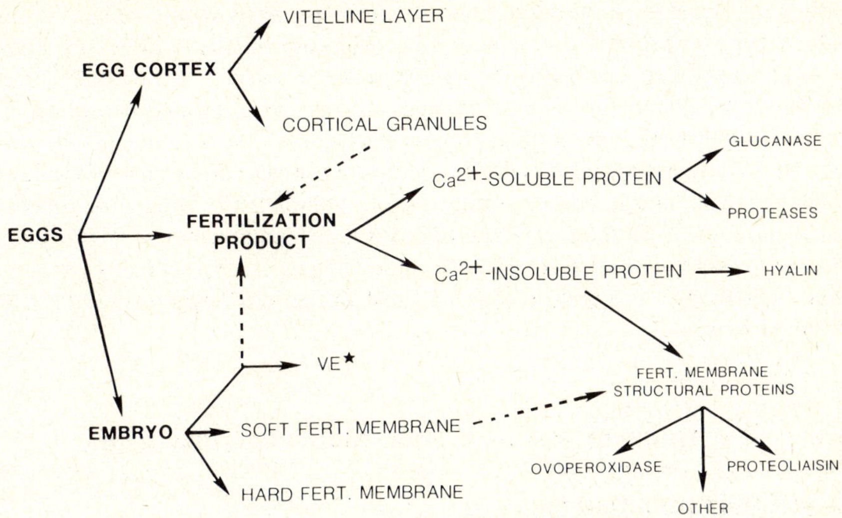


FIG. 1. An overview of the various fertilization membrane components and intermediates that can be isolated from sea urchin eggs and the relationships between them.

protocols for isolating individual proteins might be integrated to purify several components from the same starting material. We hope that, in this way, we have not only shared our knowledge about current methods in this field but also imparted sufficient information to allow investigators to modify these methods in accordance with their own requirements.

# I. Preparation of Gametes

Prior to beginning most isolation procedures, it is necessary to wash the eggs and remove the egg jelly. Debris such as spines should first be removed by passing the egg suspension through several layers of cheesecloth or through 100- to 150-μm mesh nylon, which also removes some of the egg jelly. Lowering the pH of the egg suspension to 5.5–6.0 for 5–10 minutes, then raising it to pH 7.5–8.0 again, is the most common method for removing egg jelly. We prefer to use concentrated (1 *M*) solutions of weak acids and bases such as MES [2-(*N*-morpholino)ethanesulfonic acid] and Tris rather than dilute HCl or NaOH (0.1 *M*) to prevent accidental pH overshooting, which causes egg lysis. Short-chain aliphatic acids, such as acetic acid, should not be used since they may also trigger parthenogenic activation (Section III,C). Treatment with EGTA artificial seawater (Section II,A) or dithiothreitol (Section V,A) also removes egg jelly.

During these and most other procedures, egg suspensions should be stirred, but only enough to prevent egg settling since vigorous stirring causes egg lysis.

After removal of egg jelly, the eggs should be washed two to three times before use by simply suspending the eggs in fresh seawater and decanting the wash after they settle. The settling process can be very time consuming. The only way we have found to speed up the process is to use very dilute suspensions and containers with a large bottom surface area to volume ratio. Siphoning (instead of decanting) is a particularly useful alternative for large volumes; this removes more of the wash without disturbing the settled egg layer. The settled egg volume after "dejellying" is approximately half that of the initial egg suspension with hydrated jelly.

## II. Isolation of Fertilization Membrane Precursors

### A. Isolation of Egg Cortices

The precursors of the fertilization membrane are located in the egg cortex, which is essentially the plasma membrane with the vitelline layer and cortical granules attached to its outer and inner aspect, respectively. Although egg cortices have been used primarily to study the exocytotic apparatus, they are also useful as intermediates in the isolation of either vitelline layers or cortical granules, as illustrated in Fig. 2. Two basic approaches for isolating cortices have been described: One method produces cortices attached to a substratum (cortical lawns; Vacquier, 1975, 1976) while the other yields cortices in suspension (Decker and Lennarz, 1979; Detering *et al.*, 1977). Several $Ca^{2+}$-free, isotonic media containing 10–25 m*M* EGTA can be used for isolation. These include artificial seawater (Vacquier, 1975; Decker and Lennarz, 1979; Detering *et al.*, 1977), isotonic KCl (Sasaki and Epel, 1983; Haggerty and Jackson, 1983), and isotonic glycine media (Sasaki and Epel, 1983; Moy *et al.*, 1983). These different media were developed primarily to allow $Ca^{2+}$-mediated exocytosis to be elicited in isolated cortices. In terms of isolating fertilization membrane precursors, no preference has been reported, although glycine has been observed to disturb normal cortical granule attachment to the plasma membrane and KCl appears to extract some proteins from the cortex during isolation (Sasaki and Epel, 1983). The egg cortex is most stable at low temperatures, so all manipulations should be conducted on ice.

#### 1. Cortical "Lawns"

To prepare cortical lawns, a dejellied egg suspension is added to a plastic tissue culture dish coated with protamine sulfate. After several minutes, unbound

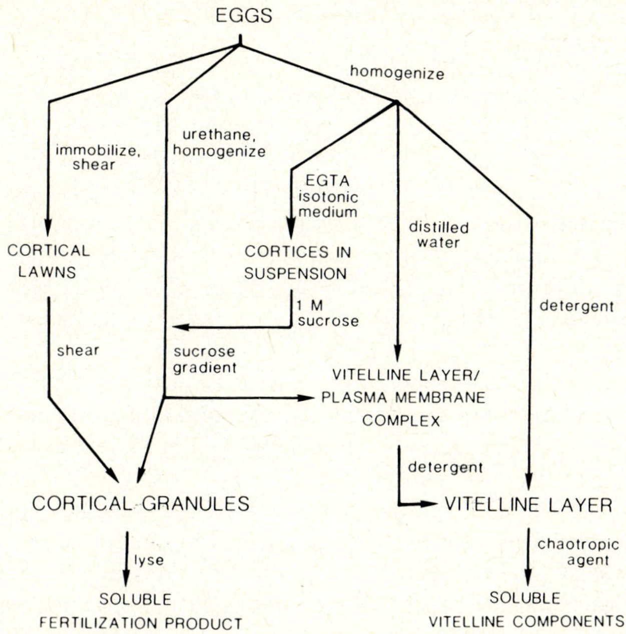


FIG. 2. Approaches to the isolation of fertilization membrane precursors from sea urchin eggs. (See Section II for a description of methods.)

eggs are washed away with EGTA-isolation medium and bound eggs are sheared by a jet of cold medium from a squirt bottle. The egg cytoplasm is then gently washed away with additional isolation medium, leaving the egg cortex firmly attached to the dish.

### 2. Cortices in Suspension

To prepare suspensions of egg cortices, eggs are washed with, then suspended in, 9 vol of EGTA isolation medium. This treatment removes the egg jelly as well as $Ca^{2+}$ (Section I). The egg suspension is homogenized with a Dounce homogenizer in aliquots (3–5 ml) until no whole eggs remain. The homogenate is then diluted 10-fold into isolation medium and centrifuged at low speed to collect the cortices. We prefer hand centrifugation in conical tubes, so that any unbroken cells and nuclei are packed in the bottom of the tube and the egg cortices are loosely layered on top. To avoid fragmenting the cortices, only the supernatant should be removed with a pipet; the cortices can then be resuspended with gentle agitation and decanted into another conical tube containing isolation medium. This process is repeated until the wash supernatant is clear.

## B. Isolation of Cortical Granules

Cortical granules can be isolated from cortical lawns (Kopf *et al.*, 1982), suspensions of cortices (Decker and Kinsey, 1983), or, more directly, from urethane-treated eggs (Hylander and Summers, 1981; see Fig. 2). In the first method, cortical granules are detached from cortical lawns by further shearing with a special isolation buffer (Kopf *et al.*, 1982). We have no experience with this method so the reader is referred to the original publication for details. One limitation of this method is that only small amounts of material can be processed at one time; however, it has the major advantage of minimizing contamination from the vitelline layer, since this layer is secured to the substratum during isolation.

Vitelline layer contamination becomes a major difficulty when isolating cortical granules from cortex suspensions. In the published procedure (Decker and Kinsey, 1983), cortical granules are dislodged from the plasma membrane by lightly homogenizing a suspension of cortices in 1 *M* sucrose and centrifuging to remove the vitelline layer–plasma membrane complexes. The cortical granules suspended in the supernatant are further purified by sucrose density centrifugation. In our hands, mild homogenization in sucrose releases only a small fraction of the cortical granules. More vigorous treatment increases the yield but also produces small vitelline layer–plasma membrane fragments that cannot be easily separated from the cortical granules.

In the third cortical granule isolation technique (Hylander and Summers, 1981), intact eggs are treated with local anesthetics and then centrifuged; this treatment dislodges the cortical granules from the plasma membrane. After homogenization, cortical granules readily separate from the plasma membrane–vitelline layer complex and cytoplasmic components during sucrose density gradient centrifugation. Unfortunately, the morphology of these isolated cortical granules suggests that certain components are lost during isolation. This may be a consequence of the anesthetic treatment (direct exposure of cortices to high concentrations of urethane dissolves the cortical granules, Decker and Kinsey, 1983); it is not known whether the use of a stabilizing isolation medium could overcome this problem.

## C. Isolation of Vitelline Layers

Isolation of fertilization membrane precursors from the vitelline layer has been, and remains, a difficult problem in the analysis of fertilization membrane assembly. The vitelline layer is a glycocalyx that surrounds the egg and is covalently attached to the plasma membrane. There are no procedures available as yet for extracting only vitelline layer components from intact eggs. Thus, isolated, intact vitelline layers may contain some components that are functionally part of the plasma membrane and not precursors of the fertilization membrane. Vitelline layer/plasma membrane complexes (egg ''ghosts'') derived from the various cortical granule isolation techniques (Section II,B) can be used in the isolation of vitelline layers (see also Kinsey *et al.*, 1980). Alternatively, ghosts can be prepared by lysing eggs in distilled water (Schmell *et al.*, 1977). In either case, it is advisable to include protease and ovoperoxidase inhibitors in the isolation medium (Rossignol *et al.*, 1984). To obtain the insoluble vitelline matrix, these ghosts are next extracted with nonionic detergent (usually Triton X-100) to remove the plasma membrane. Vitelline matrices have also been prepared by homogenizing eggs directly in Triton X-100 medium and collecting the insoluble residue (Glabe and Vacquier, 1977; Niman *et al.*, 1984). Presumably, extraction of isolated egg cortices would work as well.

Because of the extreme insolubility of the vitelline matrix, the isolation of individual vitelline components is more problematic. Isolated vitelline layers can be solubilized under severe denaturing conditions to yield a minimum of 25 individual peptides when analyzed by two-dimensional SDS electrophoresis (Niman *et al.*, 1984). Only a small fraction of the proteins in this preparation are accessible to iodination in the intact egg, or appear to have analogs in uncrosslinked, assembled fertilization membranes (Weidman, unpublished observation), in contrast to what might be expected for egg surface components that are precursors of the fertilization membrane. Vitelline layers can be solubilized under less drastic conditions using chaotropic agents such as guanidinium thiocyanate (Rossignol *et al.*, 1984). Purification of these soluble proteins is further complicated by the fact that many appear to be covalently linked together by intermolecular disulfide bonds (Epel *et al.*, 1970). Even after reduction of these bonds, some of the components, including the egg receptor for sperm, have apparent molecular weights in excess of 1,000,000 and exhibit other characteristics commonly associated with proteoglycans (Rossignol *et al.*, 1984). We have observed that when soft fertilization membranes (Section IV,B), prepared from eggs with $^{125}$I-labeled vitelline layers, are solubilized under denaturing conditions, all but one of the labeled proteins fail to enter even a low percentage stacking gel during SDS electrophoresis (Kay, 1985). This suggests that the proteoglycan-like proteins of the vitelline layer may also be precursors of the fertilization membrane. With the exception of the egg surface receptor for sperm, no vitelline components have been functionally identified or purified to date.

## III. Fertilization and Parthenogenic Activation of Eggs

There are many ways to induce cortical granule exocytosis and fertilization membrane assembly in sea urchin eggs. The three most commonly used methods include fertilization, parthenogenic activation with the $Ca^{2+}$ ionophore A23187, and parthenogenic activation with organic acids such as acetic and butyric acids. Each method has distinct features that determine its usefulness for a given type of isolation.

## A. Fertilization

The easiest and most inexpensive technique is, of course, fertilization. Sperm are generally collected "dry" for this purpose and diluted approximately 100-fold into seawater immediately prior to use. To inseminate, this suspension is simply added to the egg suspension with gentle stirring. With suspensions containing up to 15% eggs (vol/vol), a sperm-to-egg ratio of 10:1 or 20:1 is usually adequate. The amount of sperm suspension to use is estimated by assuming a concentration of $5 \times 10^{10}$ sperm/ml dry semen and $10^6$ eggs/ml settled eggs. As the egg concentration is increased above 20%, fertilization efficiency decreases and adding excess sperm does not necessarily result in a proportional increase in fertilization. The composition of the fertilization medium and treatments that alter the egg surface may prevent fertilization. For example, fertilization cannot occur in $Ca^{2+}$-free artificial seawater, and reduced fertilization efficiencies are observed when the egg vitelline layer is disrupted prior to fertilization. It is also important to consider how sperm contaminants can be separated from the desired final product. This method is most frequently used in the isolation of fertilization membrane intermediates that can be easily washed to remove sperm.

## B. Ionophore-Induced Activation

The $Ca^{2+}$ ionophore A23187 has been described as a "universal egg activator" (Steinhardt *et al.*, 1974). Very low (10–50 $\mu M$) concentrations induce cortical granule exocytosis in suspensions containing as much as 25% eggs. External $Ca^{2+}$ is not required for exocytosis, making this an attractive method for activating eggs in $Ca^{2+}$-free seawater. A23187 is insoluble in water and is usually added to egg suspensions as a 5 m$M$ solution in dimethylsulfoxide (Steinhardt and Epel, 1974). The major disadvantage of this method is the expense (and toxicity) of A23187, making it impractical for use with very large-scale preparations.

## C. Activation with Organic Acids

The parthenogenic activation of sea urchin eggs by organic acids was investigated over 70 years ago by Jacques Loeb (1913) with butyric acid being the most extensively studied activator. The method has been recently revived in a modified form (Deits *et al.*, 1984) with acetic acid replacing its more malodorous cousin. In this method, a 20% suspension of dejellied eggs (vol/vol) is gently stirred while adding 1 *M* acetic acid to a final concentration of 5.5 m*M*. The pH at this point should be in the range of 4.1–4.3; however, in our experience, it is the acetic acid concentration rather than pH which is most important. The length of incubation required for optimal activation generally falls between 2 and 4 minutes, and should be determined in advance of each activation with trial activations using small aliquots of the egg suspension. After the acetic acid incubation, the pH is elevated above 5.5 by the addition of concentrated weak base such as Tris (see Section I). Within a minute or so of elevating the pH, cortical granule exocytosis occurs. At pH greater than about 7.2, a normal fertilization membrane forms. If the pH is held between 5.5 and 6.5, fertilization membrane assembly is inhibited, although occasionally a very thin membrane first appears, then disappears again, presumably collapsing back onto the egg surface. It is not clear whether divalent cations are required for acetic acid activation, since exposure of eggs to low pH in the absence of divalent cations usually results in substantial egg lysis. Acetic acid activation is the method of choice for isolating fertilization membrane components on a preparative scale, since very large (2 liter) and dense (up to 25% vol/vol) egg suspensions can be activated reproducibly.

The success of all of these methods ultimately depends on the viability (i.e., fertilizability) of the gametes used. Egg viability diminishes fairly rapidly after spawning (1 day) but can be maintained as long as 48 hours if the eggs are stored in thin layers at 4°C. It is always advisable to test egg viability and determine optimal fertilization or activation conditions with minisamples before proceeding with a full-scale preparation. This is particularly important when activating in the presence of reagents whose effects on fertilization/exocytosis are unknown. Fertilization efficiency is assessed as the percentage of eggs exhibiting fertilization membranes. When fertilization membrane assembly is prevented, fertilization efficiency can be determined by comparing the amount of ovoperoxidase, a cortical granule marker enzyme, found in the medium after activation with the total amount of ovoperoxidase in a homogenate of the same egg suspension.

## IV. Isolation of Fertilization Membrane Intermediates

Following fertilization, the components exocytosed from the cortical granules and the vitelline layer combine to form the nascent structure that ultimately matures into a hard fertilization membrane. This hard fertilization membrane has been successfully isolated from embryos and studied *in vitro* (Foerder and Shapiro, 1977, Hall, 1978; Carroll and Baginski, 1978), as have two earlier intermediates in the assembly process (Carroll and Endress, 1982; Kay *et al.*,

1982). One such intermediate, VE*, is obtained by preventing divalent cation-directed assembly of cortical granule components onto the elevating vitelline layer. The other intermediate (termed a soft fertilization membrane) is isolated by blocking the ovoperoxidase-catalyzed crosslinking step to prevent hardening. These isolated structures provide a means for functionally dissecting the assembly process in a fashion analogous to the genetic dissection of phage head assembly using mutants. In all three cases, the isolation procedure is essentially the same: After fertilizing eggs under conditions appropriate for producing the intermediate of interest, the resulting fertilization membrane is stripped from the embryo and purified by repeated low-speed centrifugation. Presumably, any stable intermediate could be isolated using this approach.

## A. Isolation of VE*

Eggs fertilized in the absence of divalent cations form thin, fragile fertilization membranes, designated VE*s (Carroll and Endress, 1982). Since sperm require calcium for the acrosome reaction, fertilization must be initiated in normal seawater. When the eggs are diluted 5- to 40-fold into divalent cation-free seawater (575 m*M* NaCl, 10 m*M* KCl, 5 m*M* EGTA, and 10 m*M* Tris, pH 8) at 15 seconds postinsemination, cortical granule exocytosis occurs normally but fertilization membrane assembly is blocked, leaving only the vitelline scaffold surrounding the embryo. After several washes to remove sperm and soluble fertilization product (see Section V,C), this intermediate fertilization membrane can be stripped from the embryo. This is accomplished by subjecting a dense embryo suspension to one or two passes of a Dounce homogenizer, gentle vortexing, or passage through 53-μm Nitex mesh (see Section IV,B). Hand centrifugation provides the most convenient method for removing the embryos, leaving the VE* still in suspension. VE*s are then collected by gentle centrifugation (300 *g* for 10 minutes) and further purified by washing with divalent cation-free seawater until the supernatant is clear and no contamination is detected microscopically. Preparations troubled by egg lysis may require a sucrose gradient centrifugation step; several types of gradients have been reported (Carroll and Endress, 1982) but we prefer a simple spin (1300 *g* for 15 minutes) through 50% sucrose. The major difficulties with VE* isolation stem from the fragility of this fertilization membrane intermediate. VE*s are easily fragmented during isolation and purification; even gentle washes to remove sperm occasionally wash away the envelopes, decreasing yields profoundly. Considering these difficulties, it might prove more economical to investigate ionophore-induced activation of eggs (Section III,B) directly in divalent cation-free seawater and stripping the envelopes prior to completing the washing steps. This modification might also increase the potential scale of this procedure that, as described here, is optimal for only small egg volumes (< 20 ml).

## B. Isolation of Soft Fertilization Membranes

Fertilization in seawater containing 2 m*M* 3-amino-1,2,4-triazole blocks ovoperoxidase-catalyzed hardening but permits normal assembly of the fertilization membrane. Once aminotriazole is removed, these membranes can be crosslinked *in vitro* by the addition of $H_2O_2$ (Kay *et al.*, 1982). We recommend the Nitex (nylon mesh) method for stripping these soft fertilization membranes from the embryos. A convenient apparatus for this purpose is a 30-ml plastic syringe with the end sliced off and 53-μm Nitex mesh held in place over the cut end with a tight-fitting ring cut from a 50-ml plastic syringe. Aliquots (ca. 30 ml) of a 50% embryo suspension are then gently forced through the mesh with the syringe plunger. Embryo lysis is minimized if hyalin, which tends to clog the pores, is washed from the mesh with deionized water between each batch of embryos. In this way, up to 250 ml of packed embryos can be efficiently processed at one time. Removal of the embryos, collection, and purification of the soft fertilization membranes are accomplished as previously described for VE*s. After low-speed centrifugation, soft fertilization membranes form a loose, easily suspended pellet. Egg debris, when present, tends to form a more cohesive pellet that can often be removed after gentle resuspension of the soft fertilization membranes, although with some sacrifice in soft fertilization membrane yields. Soft fertilization membranes are hardier than VE*s; however, they may be fragmented by overly vigorous vortexing. The composition of these highly purified membranes remains unchanged during storage at −70°C, as long as aminotriazole is present. There is, however, some recent evidence suggesting that proteolytic events associated with assembly may be delayed as long as 1 hour in aminotriazole-containing seawater (Justice and Carroll, 1984; Kay, unpublished observations). Thus, the composition of soft fertilization membranes immediately after isolation may differ from the more stable form obtained after complete purification. This raises the possibility of using protease inhibitors to isolate another stable fertilization membrane intermediate, analogous to the soft fertilization membrane, for analyzing the role of proteases in fertilization membrane assembly.

## C. Isolation of Hard Fertilization Membranes

The highly crosslinked, "hard" fertilization membrane that forms under standard conditions of fertilization can be isolated by a simple modification of the procedures presented for soft fertilization membranes and VE*s. The major difference is that hard fertilization membranes, as the name implies, cannot be stripped from the intact embryo easily. Consequently, the embryos must be vigorously homogenized either directly in seawater (Kay *et al.*, 1982) or in low-salt buffers to help lyse the embryos (Hall, 1978; Foerder and Shapiro, 1977).

Complete disruption of the embryos is not necessary to release the fertilization membrane; any intact embryos are readily removed by hand centrifugation. Further purification employs the same techniques previously described (Sections IV,A and IV,B).

## V. Isolation of Soluble Fertilization Product

The term "fertilization product" has been used to describe the proteins released from eggs at fertilization (Ishihara, 1964; Carroll and Epel, 1975). In general, these proteins fall into three categories: the $Ca^{2+}$-insoluble structural proteins of the fertilization membrane, hyalin (also $Ca^{2+}$ insoluble) and the $Ca^{2+}$-soluble proteins that are involved in the extracellular events occurring at fertilization but are not incorporated into the embryonic coats. Under normal conditions of fertilization, most of the released protein associates with the embryo in either the fertilization membrane or hyaline layer. Several methods have been devised for obtaining these proteins in soluble form. The protein composition of fertilization product obtained by most methods is qualitatively and quantitatively similar. The methods differ primarily in economy of scale and the ease with which the fertilization product can subsequently be fractionated into the three major protein classes. The various techniques for preparing fertilization product are summarized for comparison in Table I. Since exocytosed protein represents less than 10% of the total egg protein, fertilization product is preferred over egg homogenates as starting material for the isolation and purification of fertilization membrane proteins and hyalin.

### A. Disruption of the Vitelline Layer and Parthenogenic Activation with Acetic Acid

As described above, most of the protein released at fertilization is incorporated into the insoluble embryonic coats. In this method, fertilization membrane assembly and $Ca^{2+}$-dependent precipitation of the exocytosed protein are prevented by the combined effects of disrupting the vitelline layer prior to fertilization and maintaining a low pH during exocytosis (Section III,C). The vitelline layer can be disrupted with either reducing agents such as dithiothreitol (Epel *et al.*, 1970) or protease such as trypsin (Epel, 1970). Neither method completely removes the vitelline layer; however, sperm receptors on the egg surface are depleted by these treatments. Parthenogenic activation thus provides more efficient activation and higher yields of fertilization product proteins than does fertilization.

Of the two methods for disrupting the vitelline layer, we prefer treatment with

TABLE I

A COMPARISON OF METHODS FOR OBTAINING FERTILIZATION PRODUCT

| | Soluble fertilization product | | | | |
|---|---|---|---|---|---|
| | Composition | | | | |
| | $Ca^{2+}$-soluble proteins | Hyalin | Fertilization membrane structural proteins | Protein concentration[a] (mg/ml) | Section |
| Eggs with disrupted vitelline layers, activated at[b] | | | | | |
| pH 8.0 (−benzamidine) | + | − | − | <<0.5 | |
| pH 8.0 + benzamidine | + | − | + | 0.5–1.0 | V,A,2 |
| pH 6.0 (−benzamidine)[c] | + | + | + | >1.0 | V,A,1 |
| pH 6.0 + benzamidine[d] | + | + | + | >1.0 | V,A,3 |
| Eggs with intact vitelline layers, fertilized in[e] | | | | | |
| Normal seawater | + | − | − | <<0.5 | V,B |
| Normal seawater, diluted into EGTA–seawater[c] | + | + | + | <0.5 | V,C |
| Glycine–seawater[c] | + | + | + | 0.5–1.0 | V,D |
| Soft fertilization membranes diluted into | | | | | |
| Divalent cation-free buffer | − | − | + | 0.5–1.0 | V,E |

[a] Protein concentrations are only approximate and are based on an initial egg concentration of 20% (vol/vol) for activation and 15% (vol/vol) for fertilization.

[b] Up to 4 liters of a 20% egg suspension (vol/vol) can be activated with acetic acid.

[c] These preparations can be fractionated into $Ca^{2+}$-soluble and -insoluble fractions. Complete fractionation of the $Ca^{2+}$-insoluble fraction into hyalin and fertilization membrane structure proteins is more difficult to obtain.

[d] This preparation can also be fractionated into $Ca^{2+}$-soluble and -insoluble fractions. Hyalin and the fertilization membrane structural proteins can be easily separated.

[e] Fertilization is generally optimal for up to 1-liter suspensions containing 15% or less eggs (vol/vol).

dithiothreitol, since residual protease contamination of fertilization product is difficult to prevent. For this procedure, it is not necessary to dejelly the eggs first since dithiothreitol treatment also removes egg jelly (Section I). After the eggs have been filtered through cheesecloth and allowed to settle once, as much seawater as possible is removed. An equal volume of 20 m*M* dithiothreitol in seawater at pH 9.4 is added to this dense egg suspension. After gentle stirring for 5–7 minutes, the suspension is diluted 7- to 10-fold with seawater (pH 7.5–8.0). Following two more washes with the same volume of seawater, an accurate measure of settled egg volume is made in preparation for activation with acetic acid as described in Section III,C. The major disadvantage of dithiothreitol is the fairly substantial amount of time and seawater required to wash it away thoroughly prior to activating the eggs.

Activation of dithiothreitol-treated eggs requires only that the pH be elevated above 5.5 after acetic acid treatment. Thus $Ca^{2+}$-dependent precipitation of fertilization product proteins is easily prevented by inducing exocytosis at pH 6.0 (see Section VI,A). If the protease inhibitor, benzamidine, is added during exocytosis, $Ca^{2+}$-dependent precipitation of the structural proteins of the fertilization membrane is inhibited at all pH values (Weidman *et al.*, 1985). The combined use of pH and benzamidine makes it possible to control the composition of the soluble fertilization product obtained (Table I).

### 1. Soluble Fertilization Product Containing a Mixture of Fertilization Membrane Structural Proteins and Hyalin

To obtain fertilization product containing essentially all of the protein released at fertilization, the pH is elevated to 6.0 following acetic acid treatment (Section III,C). This pH adjustment is accomplished by the addition of 1.0 *M* MES, pH 8.5, to a final concentration of 10 m*M*. After gentle stirring for 7–10 minutes, the eggs are allowed to settle and the fertilization product is decanted. When the pH of this fertilization product is increased to 7.5–8.0, a mixed precipitate of fertilization membrane structural proteins and hyalin forms (Bryan, 1970a; see Section VI,A).

### 2. Soluble Fertilization Product Containing Fertilization Membrane Structural Proteins Relatively Free of Hyalin

In this method, benzamidine is employed to specifically inhibit precipitation of the fertilization membrane structural proteins (Weidman *et al.*, 1985). After acetic acid treatment, 0.5 *M* benzamidine is added to a final concentration of 10 m*M*. Concomitantly, the pH of the stirred egg suspension is raised directly to

7.5–8.0 with 1 *M* Tris, pH 8.5 (final concentration = ca. 20 m*M*). Within 2–3 minutes of increasing the pH, the egg suspension will take on a "curdled" appearance due to the polymerization of hyalin and resultant clumping of the eggs. After the eggs are settled and the fertilization product has been decanted, any residual hyalin is removed by centrifugation. Hyalin comprises at least half of the protein released at fertilization (Bryan, 1970b), so this procedure constitutes a significant purification step in the isolation of fertilization membrane structural proteins.

#### 3. Fertilization Product Yielding the Fertilization Membrane Structural Proteins and Hyalin in Separate Fractions

By combining the techniques used in methods 1 and 2, relatively pure preparations of both hyalin and fertilization membrane structural proteins can be obtained. After acetic acid treatment, benzamidine and MES are added (to 10 m*M*) and the pH is adjusted to 6.0–6.5. Under these conditions, all of the exocytosed proteins remain soluble. After the eggs are removed, the pH of the fertilization product is increased to 7.5–8.0 by the addition of 1 *M* Tris, pH 8.5, to selectively precipitate the hyalin. Once the hyalin is collected by centrifugation, the resulting soluble fertilization product contains the fertilization membrane structural proteins and $Ca^{2+}$-soluble proteins.

### B. $Ca^{2+}$-Soluble Proteins Released from Untreated, Fertilized Eggs

When eggs with intact vitelline layers are fertilized in seawater (pH 7.5–8.0), only a fraction of the protein released at fertilization remains soluble after fertilization membrane assembly and hyalin polymerization. The proteases (Vacquier *et al.*, 1972), the β-1,3-glucanohydrolase (Epel *et al.*, 1969), and a small amount of the ovoperoxidase (Foerder and Shapiro, 1977; Hall, 1978) released at fertilization are the only known components of this $Ca^{2+}$-soluble fraction. There is probably very little glucanase in the $Ca^{2+}$-soluble fraction obtained by this method because this enzyme appears to be confined to the perivitelline space (space between the fertilization membrane and hyaline layer) after normal fertilization. This fertilization product is easily obtained by fertilizing eggs, collecting the fertilization medium, and removing the sperm by high-speed centrifugation. Although one chymotrypsin-like protease has been isolated from this fertilization product (Fodor *et al.*, 1975), it has not been used in other isolation procedures. This is due largely to the very low protein content of this fertilization product; attempts at concentration by ultrafiltration often result in the formation of a gelatinous mass that is difficult to work with.

## C. Dilution of Eggs into EGTA-Containing Seawater after Fertilization

Since fertilization membrane assembly and hyalin polymerization require divalent cations, eggs diluted into EGTA-containing seawater after fertilization release most of the exocytosed protein as soluble fertilization product. This procedure is virtually the same as that described previously for the isolation of VE* (Carroll and Endress, 1982; Section IV,A). After fertilization and dilution into EGTA–artificial seawater, the eggs are allowed to settle and the supernatant containing soluble fertilization product is collected (Baginski *et al.*, 1982). The major advantage of this method is that it is not necessary to disrupt the vitelline layer prior to fertilization, thereby eliminating the time-consuming washes required to remove disrupting agents and remnants of the vitelline layer prior to fertilization (see Section V,A). A disadvantage of the method is that very large-scale preparations require enormous quantities of EGTA artificial seawater, at least four times the volume of the fertilized eggs suspension, which must be $<$ 20% eggs (vol/vol) for efficient fertilization. Although we have not investigated this possibility, activation with A23187 directly in EGTA–artificial seawater may be a useful alternative with this method (see also Section IV,A).

## D. Fertilization in the Presence of Amino Acids

Dilution of embryos into isotonic glycine media immediately after fertilization is a technique used to prevent fertilization membrane assembly and to remove hyalin in order to investigate the role of the hyaline layer in embryonic development (Kane, 1973). Although the effects of glycine were initially attributed to the lack of divalent cations in the medium, it has been subsequently found that fertilization directly in glycine–artificial seawater (with divalent cations) also has similar effects (Hall, 1978). In addition to solubilizing hyalin, glycine solubilizes the fertilization membrane crosslinking enzyme, ovoperoxidase, and other exocytosed proteins that are presumably components of the fertilization membrane as well (Hall, 1978). The mechanism appears to be largely nonspecific, since other amino acids, sucrose, and glycerol all appear to have a similar effect. In this method, an equal volume of 1.1 *M* glycine (pH 8.0) is added to a 10–20% suspension of eggs and the eggs are fertilized while stirring. After 10 minutes, the eggs are settled and the soluble fertilization product is decanted. Although the method has not been used for the isolation and purification of fertilization membrane components, it remains a potentially useful technique since it is not necessary to remove the vitelline layer prior to fertilization (Section V,A) or to dilute the egg suspension after fertilization (Section V,C).

## E. Soluble Proteins Released from Soft Fertilization Membranes

Isolated soft fertilization membranes (see Section IV,B) require divalent cations in the external medium to maintain their structural integrity (Kay *et al.*, 1982). When these membrane intermediates are diluted into divalent cation-free medium, proteins are released into solution. These soluble fertilization membrane proteins are obtained by lightly centrifuging soft fertilization membranes (300 *g* for 10 minutes) and resuspending them in dilute buffer (20 m*M*) at pH 7–8. The residual fertilization membrane, termed a "wraith" (Kay *et al.*, 1982), is then removed by filtration or centrifugation. Although it is impractical to isolate fertilization membrane components on a preparative scale using this method, these proteins are useful standards for identifying fertilization membrane proteins and for analyzing the interactions of these proteins in the assembly process (Weidman *et al.*, 1985; see Section VIII).

# VI. Methods for Concentrating Fertilization Product

Prior to the isolation of proteins from fertilization product, the soluble fertilization product must be concentrated and desalted. Several techniques commonly used include $Ca^{2+}$-mediated precipitation, ammonium sulfate precipitation, and ultrafiltration. The first two methods are most efficient when the protein concentration of the fertilization product is greater than 0.5 mg/ml. Ultrafiltration is required for more dilute solutions.

## A. $Ca^{2+}$-Dependent Precipitation

As described in Section V,A,1, the $Ca^{2+}$-insoluble fraction of fertilization product (the fertilization membrane structural proteins and hyalin) will form a precipitate in normal seawater at pH 7.5–8.0 (Bryan, 1970a,b). This $Ca^{2+}$-dependent precipitation also functions as a purification step since the fertilization product proteases, the glucanase, and other contaminants arising from egg lysis do not precipitate. The method is impractical for the large-scale purification of ovoperoxidase, since only 50–80% of this enzyme is usually precipitated by this procedure. Although the fertilization membrane structural proteins form a paracrystalline precipitate and hyalin forms an amorphous polymer, this mixed precipitate is difficult to fractionate. A crude separation is possible after low-speed centrifugation in conical tubes, where hyalin is found predominantly in the bottom of the tube and the fertilization membrane structural proteins are enriched in the less cohesive, upper layer of the pellet (Bryan, 1970a). Both polymers can be resolubilized by suspension and dialysis in 20 m*M* buffer at neutral pH, and reprecipitated by adding $CaCl_2$ to 5–10 m*M*. We have found that precipitation of the fertilization product structural proteins by this method is not quantitative, even at high protein concentrations (i.e., only 80% precipitation is typically obtained when the protein concentration is 2–3 mg/ml), so that purification by several cycles of $Ca^{2+}$ precipitation significantly reduces recovery of these proteins.

$Ca^{2+}$-dependent precipitation of fertilization product proteins is inhibited to varying degrees by the presence of EGTA, glycine, or benzamidine in the fertilization product (Sections V,A; V,C; V,D). EGTA prevents the precipitation of both fertilization membrane structural proteins and hyalin; however, this inhibition can be overcome by increasing the concentration of free $Ca^{2+}$ to greater than 5 m*M*. Dialysis against buffered seawater or buffer (ca. 20–30 m*M* Tris, neutral pH) containing 5–10 m*M* $CaCl_2$ avoids the extreme decrease in pH that occurs upon the addition of $Ca^{2+}$ to solutions of EGTA. Failure to control the pH can result in isoelectric precipitation and denaturation of the fertilization product protein. We have found that exposure of the fertilization membrane structural proteins to concentrations of EGTA exceeding about 10 m*M* irreversibly depresses the $Ca^{2+}$-dependent precipitation of these proteins, even after removal of EGTA. Repeated exposure to lower concentrations of EGTA, for example during several cycles of precipitation and solubilization, also compromises yields. Since solubilization of the $Ca^{2+}$-dependent precipitate occurs readily in dilute buffer at neutral pH without EGTA (Bryan, 1970a,b), we suggest that its use be kept to a minimum when concentrating proteins by this method. The precipitation of both hyalin and fertilization membrane structural proteins is also inhibited in fertilization product that contains glycine. Although the mechanism of inhibition is unknown, it is apparently reversible since both classes of proteins can be precipitated with $Ca^{2+}$ after removal of glycine. In fertilization product containing benzamidine, hyalin precipitates normally at pH values $> 7.0$ (see Section V,A,2), but precipitation of the fertilization membrane structural proteins is selectively inhibited unless excess $CaCl_2$ (20–30 m*M* free $Ca^{2+}$) is added or benzamidine is removed (Section VII,D). This and other observations concerning the various effects of benzamidine on fertilization product proteins (Weidman *et al.*, 1985) suggest that this compound may be inhibiting precipitation by competing for divalent cations binding sites on fertilization membrane structural proteins.

## B. Ammonium Sulfate Precipitation

Ammonium sulfate precipitation can be used to either precipitate quantitatively or to fractionate partially fertilization product. At low ammonium sulfate concentrations (7–15% saturating), approximately 75% of the fertilization product protein precipitates, including 100% of the proteoliaisin and 60–80% of the ovoperoxidase. Quantitative precipitation can be achieved at higher concentrations (30–50% saturating), making this a practical alternative for rapidly concentrating large volumes (> 2 liters) of fertilization product. Ammonium sulfate precipitation can also be used to rescue preparations having contaminants due to egg lysis, since most cytoplasmic egg proteins do not precipitate, even at 50% saturation (Weidman, unpublished observation). Ammonium sulfate-precipitated proteins retain their sensitivity to precipitation with $Ca^{2+}$ and higher yields of $Ca^{2+}$-insoluble protein are obtained if the fertilization product is first concentrated by ammonium sulfate precipitation. When ammonium sulfate precipitation is not followed by a $Ca^{2+}$ precipitation step, precautions should be taken to inhibit proteolytic activity. Benzamidine, 10 m*M*, works well for this purpose but may interfere in subsequent purification procedures (Weidman *et al.*, 1985; see Section VII,D). An inhibitor mixture containing 1–5 μg/ml each of soybean trypsin inhibitor, pepstatin, and leupeptin provides an effective substitute, as long as care is taken to replenish these inhibitors after dialysis steps. We have found that phenylmethylsulfonylfluoride is not satisfactory for this purpose, possibly due to its relative instability in aqueous solutions.

### C. Ultrafiltration

In the event of incomplete egg activation that produces dilute fertilization product (<0.5 mg/ml protein), preliminary concentration by ultrafiltrations will increase the efficiency of subsequent precipitation steps. Several filtration apparatuses are available, ranging from centrifugal filtration devices for small samples (2–25 ml), pressure filtration cells (10–500 ml capacity), and tangential flow dialysis systems for volumes greater than 500 ml. With large volumes, concentration can require a great deal of time, making it prudent to use antibacterial agents and protease inhibitors. (Azide should be avoided for this purpose as it is also a potent, irreversible inhibitor of ovoperoxidase.) In general, ultrafiltration is not the method of choice for concentrating large-scale preparations.

## VII. Purification of Components from Fertilization Product

In this section, rather than reiterate the details of published protocols, we have integrated the various purification schemes (Fig. 3) to illustrate how more than one component may be isolated from the same starting material. Readers are referred to the original, published protocols for specific details of individual

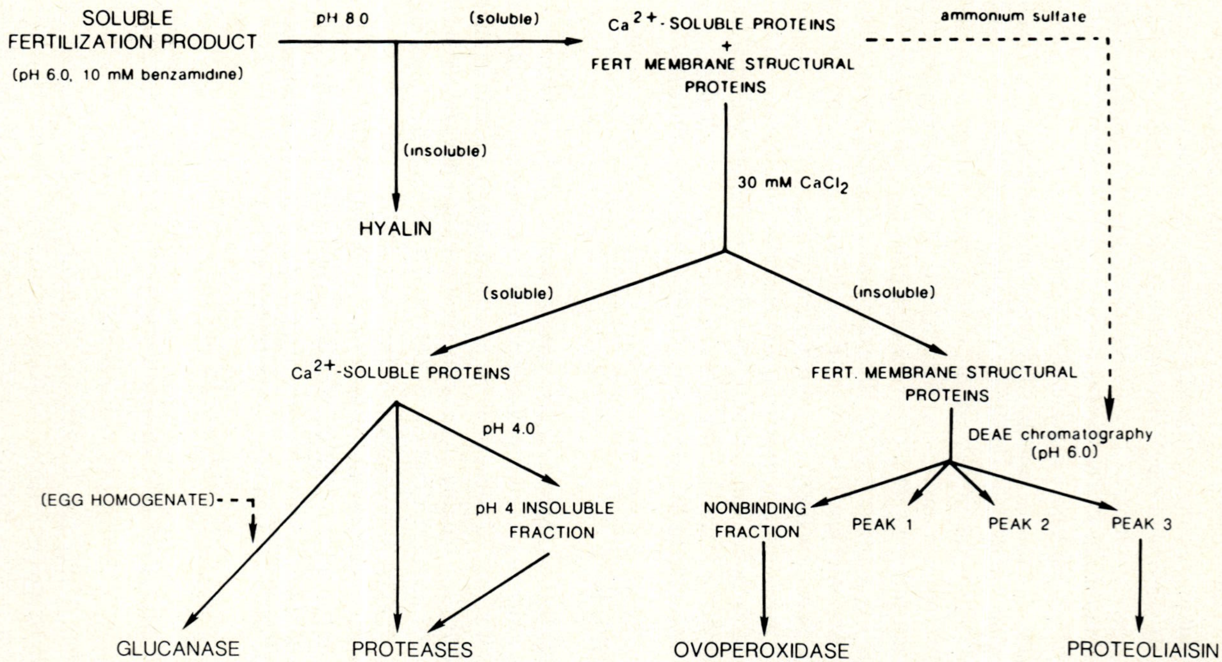


FIG. 3. The purification of multiple components from a single preparation of fertilization product. (See Section VII for a description of methods.)

purification techniques. In this scheme, the starting material is fertilization product obtained as described in Section V,A,3. This fertilization product, in seawater, 10 m*M* MES, pH 6.0, and 10 m*M* benzamidine, contains all of the exocytosed cortical granule proteins in soluble form.

## A. Purification of Hyalin

To precipitate hyalin specifically, the pH of the fertilization product is increased to 7.5–8.0 by the addition of concentrated Tris buffer (see Section VI,A). Precipitation should be complete by 1–2 hours, at which time the hyalin is collected by centrifugation and purified by 2–3 cycles of precipitation and solubilization (Kane, 1970). Hyalin is solubilized by dialysis against 10–30 m*M* Tris, pH 7.5–8.0, and precipitated by addition of 10 m*M* $CaCl_2$. Although exposure to EGTA does not appear to interfere with the subsequent $Ca^{2+}$-mediated precipitation of hyalin, addition of EGTA to a hyalin suspension does not accelerate the process of solubilization and can lead to irreversible isoelectric precipitation (Section VI,A). Dialysis against EGTA-containing buffers offers no advantage over dialysis against buffer alone, since EGTA must be removed before $Ca^{2+}$ can be added back.

## B. Purification of the Fertilization Product Proteases

To obtain the protease-enriched, $Ca^{2+}$-soluble fraction of fertilization product, the fertilization membrane structural proteins are precipitated by increasing the $Ca^{2+}$ concentration of the soluble fertilization product remaining after the removal of hyalin to ca. 20 m*M* (Section VI,A). After removal of the $Ca^{2+}$-dependent precipitate by centrifugation, the proteases then are precipitated from the soluble fraction at pH 4.0 by gradually adding 1 *M* sodium acetate, pH 4.0 (Carroll, 1976). After solubilization, the pH 4.0-insoluble material can be further fractionated by ammonium sulfate precipitation (Lois *et al.*, 1981; Carroll *et al.*, 1982). Alternatively, the proteases can be isolated by soybean trypsin inhibitor affinity chromatography (Fodor *et al.*, 1975; Alliegro and Schuel, 1983) or *p*-aminobenzamidine affinity chromatography (Carroll and Epel, 1975). After years of investigation by several different laboratories, there is still no consensus as to the actual number or molecular weight(s) of the fertilization product proteases. Whether this is due to various amounts of autoproteolysis occurring with different purification methods or molecular polymorphism remains to be established.

## C. Purification of β-1,3-Glucanohydrolase

The $Ca^{2+}$-soluble fraction of the fertilization product also contains the exoglucanase activity released at fertilization. It can be isolated by a combination of DEAE-ion exchange chromatography and affinity chromatography on gluconolactone–Sepharose (Muchmore *et al.*, 1969; Talbot and Vacquier, 1982). Various studies using specific glucanase inhibitors have failed to elucidate a role for this enzyme in fertilization membrane assembly. Since only a fraction of the total egg glucanase is released at fertilization, isolation of this component directly from egg homogenates rather than from fertilization product results in higher yields (Talbot and Vacquier, 1982).

## D. Purification of Ovoperoxidase

During the $Ca^{2+}$-precipitation step that follows hyalin removal, ovoperoxidase is distributed between both the $Ca^{2+}$-soluble fraction, which contains the proteases and glucanase, and the $Ca^{2+}$-insoluble fraction, which contains primarily fertilization membrane structural proteins. Although the enzyme can be isolated from either fraction, the enzyme in the $Ca^{2+}$-soluble fraction loses its activity very quickly, probably as a result of proteolytic digestion, and is difficult to recover. Higher yields of ovoperoxidase can be obtained by bypassing the $Ca^{2+}$-precipitation step after removal of hyalin, and, instead, quantitatively precipitating the fertilization product with 30–50% saturating ammonium sulfate (Deits *et al.*, 1984; see Section VI,B). In either case, the resulting precipitate is collected and dissolved in ''chromatography buffer'' (20 m*M* MES, pH 6.0, 1 m*M* EGTA, and 10% glycerol) containing 200 m*M* NaCl and 50 m*M* glycine ethyl ester (Weidman *et al.*, 1985), at a protein concentration of 3–6 mg/ml. After removing any insoluble material by centrifugation at 10,000 *g* for 30 minutes, the soluble protein is applied to a DEAE–Sepharose column equilibrated with the same buffer, allowing 3–5 mg protein/ml sepharose. We have found that the inclusion of 10% glycerol in the chromatography buffers, and the use of DEAE–Sepharose (instead of DEAE–Sephacel or Sephadex), result in less nonspecific protein binding and increased recovery of fractionated fertilization product proteins. Glycine ethyl ester stabilizes the $Ca^{2+}$-dependent, ovoperoxidase-binding protein, proteoliaisin, and 1 m*M* EGTA disrupts the ovoperoxidase–proteoliaisin interaction. Under these conditions, proteoliaisin binds to the column but ovoperoxidase does not. Thus, ovoperoxidase is obtained by simply washing the column with one or two column volumes of a buffer containing 250 m*M* NaCl and 50 m*M* glycine ethyl ester (Weidman *et al.*, 1985). At this stage, it is important that the column is not overloaded and that there is no benzamidine in the sample. Both conditions lead to the contamination of the DEAE-nonbinding fraction with fertilization product proteins that interfere in the subsequent purification of ovoperoxidase. Benzamidine prevents the binding of fertilization product proteins to DEAE and can be difficult to remove by dialysis (Weidman *et al.*, 1985). Fortunately, it is easily removed by precipitation of the fertilization product with either $Ca^{2+}$ or ammonium sulfate, procedures that are normally used to concentrate fertilization product as described above. Low concentrations of benzamidine can be detected in fertilization product by its very strong UV absorbance as compared to the absorbance of most protein solutions ($\lambda_{max} = 229$ nm, $\epsilon = 9.1 \times 10^3$; $\lambda_2 = 268$ nm, $\epsilon = 8.1 \times 10^2$).

The ovoperoxidase-enriched fractions obtained from this procedure are next pooled, concentrated, and dialyzed against "chromatography buffer." Since ovoperoxidase is the only protein in fertilization product that can bind to a cation exchange column, the ovoperoxidase-enriched fractions are next applied to a column of carboxymethyl–Sephadex equilibrated with "chromatography buffer," at 20–30 mg protein/ml of Sephadex. Ovoperoxidase is eluted with 3 column volumes of a linear 0–200 m*M* NaCl gradient in "chromatography buffer." Peak fractions having an $A_{415}$ (heme absorbance) to $A_{280}$ (protein absorbance) ratio greater than 1.3 are pooled and dialyzed against buffer containing 20 m*M* HEPES, pH 7.5, 200 m*M* NaCl, 10% glycerol. This purified ovoperoxidase is generally greater than 98% homogeneous and can be stored frozen (−70°C) for a year or longer. The purified enzyme cannot withstand lyophilization without loss of activity and solubility. See Deits *et al.* (1984) for a summary of the properties of ovoperoxidase.

## E. Purification of Proteoliaisin

The $Ca^{2+}$-dependent ovoperoxidase-binding protein, proteoliaisin, is retained by the DEAE column described in Section VII,D. To recover this protein, the column is eluted with a 200–500 m*M* NaCl gradient in three columns volumes of "chromatography buffer" containing 50 m*M* glycine ethyl ester. Column fractions are collected in tubes containing the protease inhibitor mixture described in Section VI,B (1 μg of each inhibitor per milliliter eluate). Three protein peaks are usually eluted and proteoliaisin resides in the final (highest ionic strength) peak. See Weidman *et al.*, 1985, for details of the *in vitro* assay for this protein. This peak is incompletely resolved from an adjacent peak eluting at slightly lower ionic strength; in some cases the peaks may nearly coincide, giving the appearance of one peak instead of two. Depending on the extent of overlap between these peaks, the proteoliaisin obtained will be 50–80% homogeneous. It is possible to purify this proteoliaisin to homogeneity by either repeating the DEAE-chromatography step using a more shallow gradient or by affinity "chromatography" using purified ovoperoxidase immobilized on nitrocellulose (Weidman *et al.*, 1985). Neither approach is entirely satisfactory for purifying large quantities of proteoliaisin. Our attempts to covalently attach ovoperoxidase to an inert matrix for conventional affinity chromatography have resulted in nearly complete inactivation of the enzyme and loss of proteoliaisin-binding capability. We have recently developed a procedure for purifying proteoliaisin to homogeneity using hydrophobic chromatography (Weidman, 1985). In this method, proteoliaisin-containing fractions from the DEAE column are applied to a column of phenyl–Sepharose in ''chromatography buffer'' containing 500 m*M* NaCl and 50 m*M* glycine ethyl ester at a ratio of 3–5 mg protein/ml Sepharose. When the column is eluted with a linear, decreasing NaCl gradient in three column volumes of the same buffer (500 to 0 m*M* NaCl), a protein peak containing homogeneous proteoliaisin is obtained. Purified proteoliaisin is stable for weeks when stored at 4°C, and can be stored frozen (−70°C) for months without loss of ovoperoxidase-binding activity. Upon thawing, however, a 50% decrease in binding activity occurs over a period of 7–10 days, after which the activity will remain stable for weeks (Weidman *et al.*, 1985).

### F. Purification of Other Fertilization Product Proteins

As described above, elution of fertilization product from DEAE–sepharose yields two protein fractions in addition to those containing ovoperoxidase and proteoliaisin. The first peak eluted by the gradient contains most of the fertilization product proteins having molecular weights < 120,000 (excluding ovoperoxidase); while the second peak, which usually overlaps the proteoliaisin peak, contains a peptide population exhibiting some microheterogeneity with an average molecular weight of ca. 200,000 (see Fig. 4A). Although the function of these components remains to be elucidated, they are apparently involved in fertilization membrane assembly, since fertilization membranes contain similar components (Fig. 4B).

## VIII. Techniques for the Analysis of Fertilization Membrane Proteins

In this section we have summarized additional methods that we have found useful in the isolation and analysis of fertilization membrane proteins. The first issue concerns the migration of fertilization membrane proteins during SDS polyacrylamide gel electrophoresis. For reasons not yet fully understood, the high-molecular-weight fertilization membrane proteins (>120,000 $M_r$) resolve very poorly after electrophoresis under reducing conditions (i.e., after treatment with β-mercaptoethanol or dithiothreitol) and are unsatisfactorily visualized by Coomassie Blue and silver protein staining. In contrast, unreduced samples yield distinct, reproducible banding patterns (Fig. 4A and B, compare lanes 1 and 2). SDS electrophoresis of nonreduced samples in the second dimension under reducing conditions results in reasonably good resolution, allowing molecular weights to be assigned to the peptides. Since proteoliaisin is among these poorly

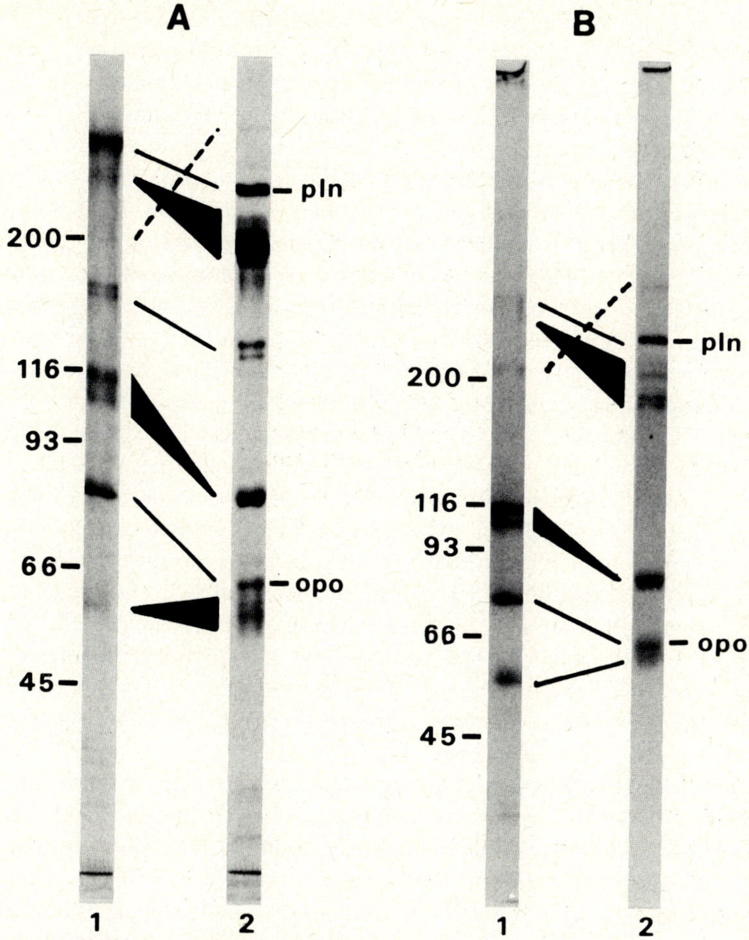


FIG. 4. A comparison of reducing and nonreducing SDS polyacrylamide gel electrophoresis. Samples were either reduced with 1% 2-mercaptoethanol (lanes 1) or not reduced (lanes 2). Electrophoresis was performed using 6–12% polyacrylamide gradient gels as described by Laemmli (1970). Gels were stained with Coomassie Blue in 50% methanol, 10% acetic acid. Values to the left of each figure are molecular weights $\times 10^3$. Lines between lanes indicate the relationship between the reduced and nonreduced peptides, as determined by two-dimensional SDS electrophoresis. pln, Proteoliaisin; opo, ovoperoxidase. (A) 30 μg of fertilization product prepared as described in Section V,A,2, concentrated by ammonium sulfate precipitation at 15% saturation (Section VI,B). Stacking gel not shown. (B) 30 μg of soft fertilization membranes prepared as described in Section IV,B, solubilized in divalent cation-free buffer (Section V,E).

resolved proteins, we routinely use nonreducing gels to monitor the purification of this protein and to detect its presence in fertilization membrane precursor compartments and fertilization membrane intermediates (Weidman, 1985).

Nondenaturing polyacrylamide gel electrophoresis at neutral pH is another useful technique for analyzing fertilization membrane proteins and their interactions. (See Weidman *et al.*, 1985, for a detailed description of this procedure.) Activity stains for ovoperoxidase and proteoliaisin can be used to identify these specific components in solubilized soft fertilization membranes and fertilization product following electrophoresis. Ovoperoxidase is detected directly in the gel using a guaiacol assay cocktail containing $H_2O_2$ and sodium nitroferricyanide (Deits *et al.*, 1984). Although there is a large variety of peroxidative substrates that produce a colored reaction product, ovoperoxidase has very poor catalytic activity toward many of them after electrophoresis; guaiacol seems to be the exception. Unfortunately, the reaction product of guaiacol oxidation is unstable and fades rather quickly. Consequently the staining pattern is difficult to photograph, even when sodium nitroferricyanide is present as a stabilizer. We have observed that oxidized guaiacol in an activity-stained gel can be electrophoretically transferred, with very little loss in stain intensity, to nitrocellulose where the stain is more easily photographed. The enzyme itself is virtually inactive after transfer to nitrocellulose. The transferred guaiacol stain disappears after an hour or so and the nitrocellulose blot can then be stained for proteoliaisin. This staining procedure capitalizes on proteoliaisin's ability to bind specifically to exogenously added ovoperoxidase in the presence of $Ca^{2+}$ (Weidman *et al.*, 1985). Unlike the endogenous enzyme which loses its activity during the various electrophoretic steps, this bound, exogenous enzyme readily oxidizes a number of different substrates. In particular, oxidation of 4-chloro-naphthol produces excellent stains which fade only very slowly (months). Theoretically, this technique could be used to detect the binding of any purified component, that can be labeled for example with $^{125}I$ or visualized enzymatically, to electrophoretically fractionated fertilization membrane and fertilization product proteins. This type of analysis is a powerful tool for dissecting the specific protein interactions that give rise to fertilization membrane assembly.

Another useful technique which we have only just begun to explore is ovoperoxidase-catalyzed iodination of proteins. *In vivo*, ovoperoxidase catalyzes the dityrosine crosslinking of fertilization membrane proteins; in the presence of $I^-$, ovoperoxidase iodinates these proteins instead (Kay, unpublished observations). Endogenous ovoperoxidase iodinates itself and the other proteins of unfractionated, soluble fertilization product under conditions analogous to those used for lactoperoxidase iodination (Morrison, 1980). Purified ovoperoxidase also can be used to iodinate purified proteins to high specific activity. Although $Cl^-$ is not readily oxidized by ovoperoxidase, high concentrations of this anion relative to $I^-$ seem to interfere in the labeling reaction, so methane sulfonate or acetate should be substituted for $Cl^-$ for best results. We have successfully used this procedure to label the fertilization product proteins proteoliaisin and ovoperoxidase (by autoiodination) with $^{125}I$ to 0.5–2.0 mol $I^-$/mol protein, without altering the activity of either protein (Weidman, unpublished observations). Ovoperoxidase immobilized in the insoluble matrix of hard fertilization membranes provides an alternate method for iodinating soluble proteins. This procedure offers the advantage over the use of a soluble enzyme because the iodinating activity is readily removed by gentle centrifugation. This eliminates both the need for elaborate purification schemes to remove soluble, contaminating peroxidase and the difficulties associated with covalently attaching a soluble peroxidase catalyst to an insoluble matrix. Since many fertilization membrane proteins have no known enzymatic activity, ovoperoxidase-catalyzed iodination provides a useful method for detecting, quantifying, and analyzing specific interactions between fertilization membrane components.

## ACKNOWLEDGMENTS

We would like to thank Dr. B. M. Shapiro for his encouragement and assistance in the preparation of this manuscript, and to acknowledge our support by NIH Grant GM 23910.

## REFERENCES

Alliegro, M. C., and Schuel, H. (1983). *J. Cell Biol.* **97,** 180a.
Baginski, R. M., McBlaine, P. J., and Carroll, E. J., Jr. (1982). *Gamete Res.* **6,** 39–52.
Bryan, J. (1970a). *J. Cell Biol.* **44,** 635–644.
Bryan, J. (1970b). *J. Cell Biol.* **45,** 606–614.
Carroll, E. J., Jr. (1976). *In* "Methods in Enzymology" (L. Lorand, ed.), Vol. 45, pp. 343–353. Academic Press, New York.
Carroll, E. J., Jr., and Baginski, R. M. (1978). *Biochemistry* **17,** 2605–2612.
Carroll, E. J., Jr., and Endress, A. G. (1982). *Dev. Biol.* **94,** 252–258.
Carroll, E. J., Jr., and Epel, D. (1975). *Dev. Biol.* **44,** 22–32.
Carroll, E. J., Jr., Lois, A. F., and Lackey, D. A. (1982). *J. Cell Biol.* **95,** 154a.
Decker, G. L., and Lennarz, W. J. (1979). *J. Cell Biol.* **81,** 92–103.
Decker, S. J., and Kinsey, W. H. (1983). *Dev. Biol.* **96,** 37–45.
Deits, T., Farrance, M., Kay, E. S., Medill, L., Turner, E. E., Weidman, P. J., and Shapiro, B. M. (1984). *J. Biol. Chem.* **259,** 13525–13533.
Derbès, M. (1847). *Ann. Sci. Natl. Zool.* **8,** 80–98.
Detering, N. K., Decker, G. L., Schmell, E. P., and Lennarz, W. J. (1977). *J. Cell Biol.* **75,** 899–914.
Epel, D. (1970). *Exp. Cell Res.* **61,** 69–70.
Epel, D., Weaver, A. M., Muchmore, A. V., and Schimke, R. T. (1969). *Science* **163,** 294–296.
Epel, D., Weaver, A. M., and Mazia, D. (1970). *Exp. Cell Res.* **61,** 64–68.
Fodor, E. J. B., Ako, H., and Walsh, K. A. (1975). *Biochemistry* **14,** 4923–4927.
Foerder, C. A., and Shapiro, B. M. (1977). *Proc. Natl. Acad. Sci. U.S.A.* **74,** 4214–4218.
Glabe, C. G., and Vacquier, V. D. (1977). *J. Cell Biol.* **75,** 410–421.
Haggerty, J. G., and Jackson, R. C. (1983). *J. Biol. Chem.* **258,** 1819–1825.

Hall, H. G. (1978). *Cell* **15,** 343–355.
Hylander, B. L., and Summers, R. G. (1981). *Dev. Biol.* **86,** 1–11.
Ishihara, K. (1964). *Exp. Cell Res.* **36,** 354–367.
Justice, R. W., and Carroll, E. J., Jr. (1984). *J. Cell Biol.* **99,** 124a.
Kane, R. E. (1970). *J. Cell Biol.* **45,** 615–622.
Kane, R. E. (1973). *Exp. Cell Res.* **81,** 301–311.
Kay, E. S. (1985). In preparation.
Kay, E. S., and Shapiro, B. M. (1985). *In* "The Biology of Fertilization" (A. Monroy, ed.), Vol. 3, pp. 45–80. Academic Press, New York.
Kay, E., Eddy, E. M., and Shapiro, B. M. (1982). *Cell* **29,** 867–875.
Kinsey, W. H., Decker, G. L., and Lennarz, W. J. (1980). *J. Cell Biol.* **87,** 248–254.
Kopf, G. S., Moy, G. W., and Vacquier, V. D. (1982). *J. Cell Biol.* **95,** 924–932.
Laemmli, U. K. (1970). *Nature (London)* **227,** 680–685.
Loeb, J. (1913). *In* "Artificial Parthenogenesis and Fertilization," pp. 65–73. Univ. of Chicago Press, Chicago.
Lois, A. F., Valdez, E. L., and Carroll, E. J., Jr. (1981). *J. Cell Biol.* **91,** 176a.
Morrison, M. (1980). *In* "Methods in Enzymology," Vol. 17, pp. 214–220. Academic Press, New York.
Moy, G. W., Kopf, G. S., Gauch, G., and Vacquier, V. D. (1983). *Dev. Biol.* **100,** 267–274.
Muchmore, A. V., Epel, D., Weaver, A. M., and Schimke, R. T. (1969). *Biochim. Biophys. Acta* **178,** 551–560.
Niman, H. L., Hough-Evans, B. R., Britten, R. J., Lerner, R. A., and Davidson, E. H. (1984). *Dev. Biol.* **102,** 390–401.
Rossignol, D. P., Earles, B. J., Decker, G. L., and Lennarz, W. J. (1984). *Dev. Biol.* **104,** 308–321.
Sasaki, H., and Epel, D. (1983). *Dev. Biol.* **98,** 327–337.
Schmell, E., Earles, B. J., Breaux, C., and Lennarz, W. J. (1977). *J. Cell Biol.* **72,** 35–46.
Steinhardt, R. A., and Epel, D. (1974). *Proc. Natl. Acad. Sci. U.S.A.* **71,** 1915–1919.
Steinhardt, R. A., Epel, D., Carroll, E. J., Jr., and Yanagimachi, R. (1974). *Nature (London)* **252,** 41–43.
Talbot, C. F., and Vacquier, V. D. (1982). *J. Biol. Chem.* **257,** 742–746.
Vacquier, V. D. (1975). *Dev. Biol.* **43,** 62–74.
Vacquier, V. D. (1976). *J. Supramol. Struct.* **5,** 27–35.
Vacquier, V. D., Epel, D., and Douglas, L. A. (1972). *Nature (London)* **237,** 34–36.
Weidman, P. J. (1985). In preparation.
Weidman, P. J., Kay, E. S., and Shapiro, B. M. (1985). *J. Cell Biol.* **100,** 938–946.

# Chapter 8

# *Purification and Properties of the Egg Plasma Membrane*

WILLIAM H. KINSEY

*Department of Anatomy and Cell Biology*
*University of Miami School of Medicine*
*Miami, Florida*



## I. Introduction

The plasma membrane of eukaryotic cells is now recognized as an important organelle that functions not only to control the intracellular environment but also to receive and transduce outside signals, mediate intercellular recognition and adhesion, and to participate in secretion of material via exocytosis. During fertilization, the egg plasma membrane is involved in binding to and fusing with the sperm, preventing polyspermy (Lynn and Chambers, 1984), exocytosis of the cortical vesicles, "activation" of the egg cytoplasm (Uehara and Yanagimachi, 1976), modification of cytoplasmic pH (Johnson *et al.*, 1976), and transport of nutrients into the cell. While the importance of these functions to fertilization has been appreciated for some time, relatively little is known about the molecular mechanisms by which they are carried out and the control mechanisms by which they are regulated. Subcellular fractionation provides an opportunity to study many plasma membrane functions under defined conditions and thus could be used to work out molecular mechanisms and regulatory parameters. Due to the requirement for large amounts of material during subcellular fractionation, most of the work on egg plasma membranes has been done with echinoderms. There are now several published methods for preparing egg plasma membranes as ghosts (Barber and Mead, 1975), as complexes with other cortical organelles (Detering *et al.*, 1977), or as immobilized membrane sheets (Vacquier, 1975). My own interest has been to prepare plasma membranes from unfertilized and fertilized eggs that are low in contamination by other organelles and that, in vesicle form, are amenable to biochemical analysis. My objective in this article is to describe the details of these methods (Kinsey *et al.*, 1980; Ribot *et al.*, 1983) and to show some of the ways that these membrane preparations can be used.

## II. Plasma Membranes from Unfertilized Sea Urchin Eggs

The preparation of plasma membranes from unfertilized eggs is greatly facilitated by the fact that large sheets of plasma membranes with associated cortical vesicles (cell surface complex) can be isolated quickly and in good yield (Detering *et al.*, 1977). The cortical vesicles can then be released from the plasma membrane sheets (Fig. 1), which are then purified by differential and density gradient centrifugation (Kinsey *et al.*, 1980). This method yields sheets of plasma membrane that retain many vitelline layer components including the sperm receptors.

### A. The Cell Surface Complex

Cell surface complexes are prepared by a slight modification of the original method published in 1977 (Detering *et al.*, 1977). Unfertilized eggs are collected by intracoelomic injection of 0.5 *M* KCl and filtered through coarse Nitex cloth to remove any debris. The egg suspension is then inspected under the light microscope for the presence of oocytes, degenerating eggs, and other cell types. The presence of even a small percentage of these contaminants can have a disastrous effect in that they do not rupture during the initial, gentle homogenization and thus spill contaminating organelles throughout all stages of the purification. They also can cause aggregation of the membrane sheets and greatly reduced recovery. Oocytes and other small cell types can sometimes be removed by allowing the eggs to settle from a 1% vol/vol suspension in filtered seawater

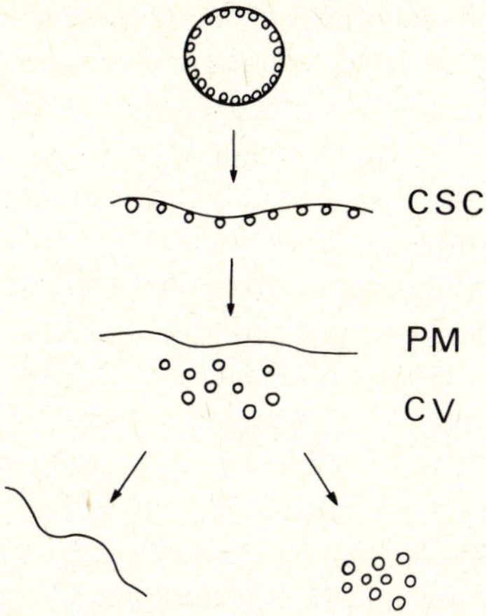


FIG. 1. Schematic representation of the preparation of plasma membranes from unfertilized eggs. The first step is to purify the cell surface complexes (CSC), which consist of sheets of plasma membrane (PM) with the cortical vesicles (CV) attached to the cytoplasmic face of the membrane. The cortical vesicles are then dissociated from the plasma membrane by sucrose treatment and the plasma membranes are purified by differential and density gradient centrifugation.

and aspirating the supernatant which contains the slowly settling oocytes. An egg suspension that contains 1% oocytes or any degenerating eggs will generally yield very poor results.

The following steps should be performed as quickly as possible and without interruption. Since the procedures are lengthy, we do not work in a cold room but use numerous ice buckets, including several rectangular ones containing test tube racks. A team of two people is best, especially for the homogenization and centrifugation steps. The eggs are suspended in filtered seawater (1% vol/vol) then dejellied by adjusting the suspension to pH 5.5 with 1 *N* HCl and allowing the eggs to settle by gravity. The acidic supernatant is aspirated and the eggs are then washed twice with 10 vol of filtered (Whatman #2) seawater. The washing can be done by gravity or with a hand centrifuge and the volume of the packed eggs should be noted. Eight milliliters of packed eggs is then washed twice in 250 ml of "seawater C," then suspended in 80 ml of ice-cold seawater C containing Aprotinin (1 μg/ml, Sigma) as a protease inhibitor. Seawater C is prepared by dissolving NaCl (29.3 g), KCl (0.75 g), $NaHCO_3$ (0.21 g), EGTA (9.5 g), and NaOH (2.5 g) in 900 ml of water, then adjusting the pH to 8.0 with HCl and the final volume to 1 liter. All subsequent steps contain Aprotinin and are at 0°C. This suspension is then homogenized in prechilled size C Potter–Elvehjem homogenizers (Arthur H. Thomas Co., Philadelphia, PA). After each stroke an aliquot of the homogenate is quickly inspected under the microscope to determine whether all of the eggs have broken. The aim is to break open all of the cells and produce large sheets of cell surface complexes which can be seen under phase contrast at 100×. A little too much homogenization or any delays can result in fragmentation of the cell surface complexes into small pieces that will be lost during the later low-speed centrifugation steps. The homogenate is then diluted to 400 ml with ice-cold seawater C containing Aprotinin and aliquoted into 40 12-ml conical centrifuge tubes (on ice) that are centrifuged at 1000–2000 rpm in a clinical or (preferably) a refrigerated bench top centrifuge for 60 seconds. The supernatants are aspirated and each of the pellets is gently resuspended in 1 ml of seawater C by pipetting in and out of a 9-in. Pasteur pipet. Care must be taken not to fragment the large membrane sheets by pipetting more than two or three times or by including bubbles. The resulting suspensions are combined into 20 12-ml tubes which are filled with seawater C, mixed by inversion, and centrifuged as above. Again, the supernatants are aspirated and discarded, and the pellets are carefully resuspended in seawater C. This suspension is divided among four 12-ml tubes which are filled with seawater C, mixed by inversion, and centrifuged again. The supernatant is aspirated and the tightly packed cell surface complexes are used immediately for preparation of plasma membranes.

There are several criteria for evaluating the quality of a preparation of cell surface complexes. Observation by phase-contrast microscopy should reveal large sheets of membrane studded with cortical vesicles without contamination by unruptured eggs or cell fragments. If these larger contaminants occur, they usually settle to the very bottom of the centrifuge tube, and it is often possible to remove the cell surface complexes with a pipet and leave the intact cells behind. The individual sheets of cell surface complexes should be easily dispersed in seawater C. If not, it is usually because fragments of oocytes or degenerating eggs are causing the membranes to clump. Finally, the supernatant of the third centrifugation should be water clear or nearly so. Cloudiness at this stage usually means that the cell surface complexes are fragmenting or that intact eggs remain and are slowly lysing.

## B. Preparation of Plasma Membranes

Having obtained the cell surface complex, the object now is to release the cortical vesicles by replacing all of the ions with sucrose, and then separate the plasma membrane sheets by differential and density gradient centrifugation. It was discovered by chance by Glen Decker that replacing salts with sucrose causes the cortical vesicles to detach. While the mechanism is not known, it is likely that reducing the ionic strength (increasing the dielectric constant) of the solution would reduce the strength of hydrophobic interactions normally favored by a polar, aqueous environment.

Isotonic sucrose (1 *M* and containing 1 μg/ml of Aprotinin) is used to wash down the walls of the four centrifuge tubes containing tightly packed cell surface complex and is then pipetted off without disturbing the pellets. The pellets of cell surface complex are then suspended in 1 ml of sucrose with a Pasteur pipet. The tubes are then filled with sucrose, mixed by inversion, and centrifuged in a refrigerated centrifuge at 1000–2000 rpm for 20 minutes to pellet the cell surface complexes. The supernatants are aspirated and the pellets are suspended in 100 ml of sucrose with gentle homogenization and incubated at 0°C until the cortical vesicles are detached. Again the homogenate should be checked with a microscope to visualize the cortical vesicles and to be sure that the membrane sheets are not fragmenting or forming small vesicles. Normally, large clear membrane sheets with only a few remaining cortical vesicles are obtained. The membranes are then pelleted by centrifugation in a Sorvall SS-34 fixed-angle rotor at 5000 rpm for 20 minutes. The supernatant containing cortical vesicles and small plasma membrane fragments are discarded. The pellets are resuspended in sucrose and either used as is or further purified on a sucrose density gradient. The membranes should be loaded on a single gradient consisting of 12 ml of 60% (wt/vol), 8 ml of 70%, and 8 ml of 78% sucrose, then centrifuged for at least 8 hours in a Beckman SW-28 rotor at 27,000 rpm. The plasma membranes form a band at the 70/78% interface, which can be visualized against a black background, and are collected with a pipet. The membranes are then washed free of sucrose by diluting to 35 ml with seawater C and centrifuging in the SW-28 rotor at 27,000 rpm for 1 hour.

The above method was originally applied to eggs from *Strongylocentrotus purpuratus,* and I have also used it with eggs from *Lytechinus variegatus* although in the latter species the removal of cortical vesicles is less efficient and subsequently the recovery of plasma membranes from the 70/78% interface of the sucrose gradient is lower. The major problem encountered in this method occurs when the cell surface complexes are suspended in sucrose to release the cortical vesicles. Too much agitation can cause the membrane sheets to vesiculate and these small vesicles will be discarded in the supernatant after the centrifugation at 5000 rpm. Occasionally, the sucrose density gradient will contain a

TABLE I

RECOVERY OF PLASMA MEMBRANES FROM *L. variegatus* EGGS

| | Protein (mg) | $Na^+,K^+$-ATPase[a] |
|---|---|---|
| Homogenate | 727 | 100% |
| Unfertilized plasma membranes | 0.7 | 8% |
| Fertilized plasma membranes | 6.2 | 29% |

[a] Ouabain-inhibitable $Na^+,K^+$-ATPase is expressed relative to the activity in the homogenates from which the membranes were prepared.

second, lighter band at the 60/70% interface. This material also consists of plasma membranes and probably represents membranes from which the vitelline layer has been partially removed.

## III. Plasma Membranes from Fertilized Eggs

Fertilization results in extensive modification of the egg surface and underlying cortical cytoplasm (Vacquier, 1981). For this reason, the method developed for preparing plasma membranes from unfertilized eggs is totally ineffective for fertilized eggs. In working with the fertilized egg, I have tried to use the same general approach of preparing large plasma membrane "ghosts" which can be separated from most contaminants by low-speed centrifugation. Fairly stable ghosts can be formed if the egg is homogenized in the presence of $Ca^{2+}$ and $Mg^{2+}$ to maintain the hyaline layer, an extracellular matrix that has some structural integrity and seems to stabilize the plasma membrane as a sheet. Unfortunately, the hyaline layer also traps large quantities of contaminating organelles. To get around this problem, the plasma membranes are first purified in the presence of divalent cations by differential and density gradient centrifugation, then the divalent cations are removed to solubilize the hyaline material and the membranes are purified on a second sucrose gradient containing a calcium chelator (Fig. 2).

Eggs are collected and treated with acidic seawater to remove the jelly coat as described above. Eight milliliters of packed eggs are then washed with 250 ml of a solution of dithiothreitol (5 m*M*) adjusted to pH 9.1 with 1 *N* NaOH. This solution is not stable and should be prepared fresh. The dithiothreitol treatment prevents elevation of the fertilization envelope (Epel *et al.*, 1970), which interferes with the procedure. After about 5 minutes, the eggs are allowed to settle, the supernatant is aspirated as carefully as possible, and the eggs are washed with five 250-ml changes of seawater. Sperm are collected "dry" and stored on ice until use. A fairly high concentration of sperm (about 50 μl sperm/250 ml suspension of eggs) is necessary because the dithiothreitol treatment reduces the fertilizability of the eggs. The sperm are added with stirring sufficient to keep the eggs suspended and then the eggs are allowed to settle and are washed three times with 250 ml of seawater to remove the excess sperm. It is important to remove an aliquot of the fertilized eggs so that cleavage can be monitored, since this is the most reliable indicator that 100% fertilization has been achieved. The above treatments can take 1 or 2 hours, so it is important to start this preparation early in the morning so that the final sucrose gradient can run overnight.

At 20 minutes after fertilization, the settled eggs are suspended in 80 ml of ice-cold (all subsequent steps are on ice) homogenization buffer and homogenized by

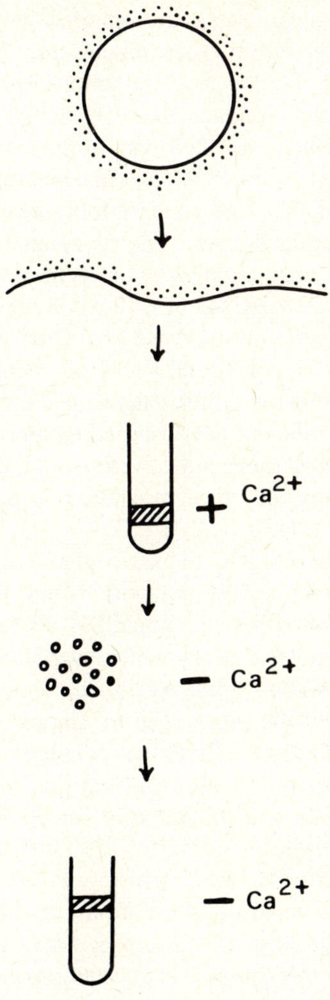


FIG. 2. Schematic representation of the preparation of plasma membranes from fertilized eggs. Fertilized eggs with associated hyaline layer (stippled) are homogenized in the presence of calcium to produce large membrane sheets that are purified on a sucrose gradient containing calcium. The membranes are then homogenized in the presence of a calcium chelator that causes them to vesiculate. The vesicle are then purified on a second sucrose gradient containing calcium chelators.

hand with a size C Potter–Elvehjem homogenizer. Homogenization buffer is made by diluting artificial seawater (see below) with an equal volume of water and adding Aprotinin to a final concentration of 1 μg/ml. The homogenate is diluted to 400 ml with homogenization buffer and aliquoted into 12-ml conical centrifuge tubes, then centrifuged in a refrigerated bench-top centrifuge at 2000 rpm for 90 seconds. The supernatants are aspirated and the pellets suspended in 0.5 ml of homogenization buffer, then pooled and homogenized again. The fertilized ghosts are much sturdier than the unfertilized cell surface complexes and can withstand repeated homogenization by hand. The suspension is then aliquoted into 20 tubes which are filled with homogenization buffer, mixed by inversion, and centrifuged again. The supernatants are discarded, the pellets pooled, homogenized, and aliquoted into four tubes which are filled with homogenization buffer and centrifuged again. The supernatants, which now should be clear, are discarded. The final, rubbery white pellets are suspended in 6 ml of homogenization buffer, homogenized in a 10-ml Potter–Elvehjem homogenizer and loaded onto two discontinuous sucrose gradients. After the third low-speed centrifugation, the ghosts become quite sticky and adhere to the Pasteur pipet, homogenizer, etc. It is useful to minimize the wetted area (fill the pipet only one-third full) and to rinse as much of the adsorbed membranes onto the gradient as possible. I use a clean 9-in. Pasteur pipet to rinse out the membranes stuck to the dirty pipette. The gradients are made by adding sucrose to artificial seawater (NaCl, 24.7 g/liter, $CaCl_2 \cdot 2H_2O$, 1.32 g/liter, $Na_2SO_4$, 3.7 g/liter, HEPES, 1.3 g/liter, $NaHCO_3$, 0.2 g/liter, $MgCl_2 \cdot 6H_2O$, 10.0 g/liter, pH 8.0). Two gradients are poured as follows: 14 ml of 60% (wt/vol), 14 ml of 45%, and 6 ml of 20% sucrose. They are centrifuged in a Beckman SW-28 rotor at 25,000 rpm for 4 hours. The membranes band as a sticky white layer at the 45–60% interface and can be collected with a Pasteur pipet. Again care should be taken to minimize loss of material sticking to the pipet. The membranes from each gradient are placed directly in a size C Potter–Elvehjem homogenizer and homogenized in seawater C containing 1 μg/ml Aprotinin (75 ml final volume), then centrifuged in the Beckman SW-28 rotor at 25,000 rpm for 90 minutes. The pellets are resuspended and homogenized in seawater C (10 ml), then loaded on a second pair of sucrose gradients consisting of sucrose (wt/vol) dissolved in seawater C. These $Ca^{2+}$-free gradients consist of 4 ml of 60%, 5 ml of 45%, overlaid with a 22-ml linear gradient from 20 tp 45% sucrose. These gradients are centrifuged overnight (12 hours) in an SW-28 rotor at 25,000 rpm and 4°C. After centrifugation, the gradients are drained from the top with a Auto-Densiflow sampler (Buchler Instruments, Inc., Fort Lee, NJ) and 1-ml fractions are collected and stored on ice. The plasma membrane containing fractions are present as a single band apparent to the eye, but it is useful to measure the absorbance (light scattering) of each fraction at 310 nm as this provides a permanent record and a reproducible way of collecting the membranes. In my laboratory, we typically pool four fractions on either side of the peak tube (Fig. 3) wash the membranes by diluting the material from each gradient at 35 ml with seawater C and centrifuging in the SW-28 rotor at 27,000 rpm for 90 minutes. The resulting pellet is clear to milky in appearance and can be resuspended in 0.5 ml of seawater C by

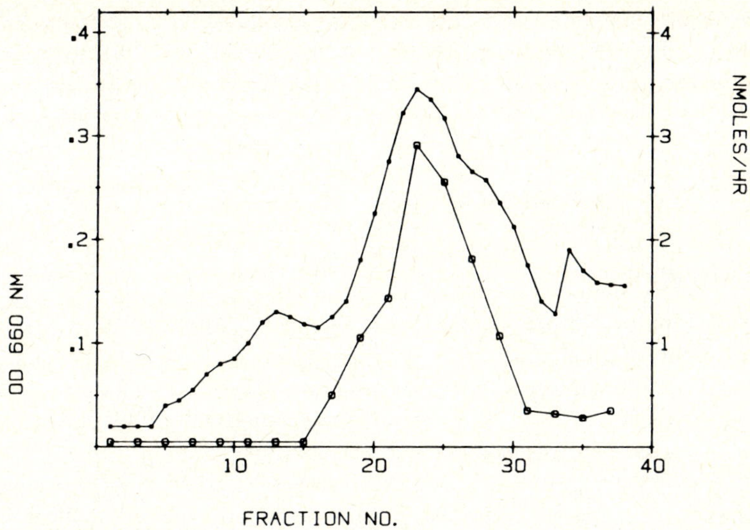


FIG. 3. Purification of plasma membranes from fertilized eggs on a sucrose gradient containing EGTA. Fractions were collected from the second (EGTA-containing) sucrose gradient and aliquots were assayed for protein (Lowry *et al.*, 1951) and for ouabain-inhibitable $Na^+,K^+$-ATPase (Kinsey *et al.*, 1980). Protein content (OD 660 nm) is represented by solid circles (●), and ATPase activity in 0.1 ml is presented as open circles (○).

repeatedly directing the stream from a 100-μl automatic pipetter with a small plastic tip against the bottom of the centrifuge tube. The milky suspension of plasma membranes is removed, and the process is repeated until all of the material has been dislodged from the bottom of the tube. It takes quite a while to get all of the membranes off the bottom of the tube, but more classical methods such as a rubber policeman will result in loss of material.

In general, the methods outlined above are more difficult than many methods used on mammalian tissue. It is crucial that large membrane sheets be maintained through the early phases of the preparation and phase-contrast microscopy is a good way to monitor this. Careful attention during the homogenization and low-speed centrifugation steps are important since overzealous pipetting during resuspension of the low-speed pellets can fragment these membrane sheets. Once the purification is well along, the major problem becomes loss of the membranes due to adsorption to the glassware. Siliconized glassware or plasticware have not helped this problem in my experience. It is more important to reduce the area to which the membranes are exposed and to make a real effort to recover the large clumps that tend to stick to pipets, etc. The yields of plasma membranes can vary quite a lot depending on the experience of the personnel. The first attempt usually results in no plasma membranes. The second and third attempts will go a lot faster and the yield will improve but it usually takes five or six preparations by the same two individuals before a consistent yield is obtained.

## IV. Characterization of Egg Plasma Membranes

### A. Morphology

The morphology of plasma membranes isolated from unfertilized eggs is that of large sheets of membrane easily visible by phase-contrast microscopy (Kinsey *et al.*, 1980). Ultrastructurally, these membrane sheets are seen to be associated with an extensive, "fluffy" layer presumably composed of protein or glycoproteins. The occasional presence of cortical vesicles adhering to the side of the membrane opposite this layer suggests that the "fluffy" layer represents the remnants of the vitelline layer and possibly some adsorbed proteins from ruptured coritcal vesicles. Much of the "fluffy" layer can be removed by homogenizing the membranes in a chaotropic salt such as 0.5 *M* KI or by extensive homogenization in seawater C (Fig. 4A). These procedures also fragment the membrane sheets, producing a population of sealed vesicles which would be useful for ion transport studies, etc. The plasma membranes from fertilized eggs (Fig. 4B) are obtained in the form of small vesicles that appear to be well sealed and exhibit sodium gradient-dependent transport of various metabolites (E. G. Schneider, personal communication).

### B. Enzymatic Activity of the Egg Plasma Membrane

One of the major goals in preparing plasma membranes from sea urchin eggs is often to study a plasma membrane-associated enzyme. The sea urchin is one of the few systems where milligram quantities of egg plasma membrane can be prepared and is therefore a useful material for fertilization biologists interested in cell surface enzymes. The sea urchin egg plasma membrane is similar to many mammalian cells in that it is enriched in the ouabain-inhibitable, $Na^+$,$K^+$-ATPase and alkaline phosphatase (Table II). A tyrosine-specific protein kinase is

---

FIG. 4. Ultrastructure of plasma membrane vesicles from unfertilized and fertilized eggs. Plasma membranes prepared from unfertilized *S. purpuratus eggs* (A) were homogenized by hand in seawater C to produce membrane vesicles that were then pelleted by centrifugation in a Beckman SW-50.1 rotor at 45,000 *g* for 1 hour. Plasma membranes from fertilized *L. variegatus* (B) eggs were already in vesicle form and were prepared immediately for electron microscopy. The samples were fixed 1% glutaraldehyde in seawater C and postfixed in 1% $OsO_4$. Sections were stained with uranyl acetate and lead citrate. Magnifications approximately 10,000×.

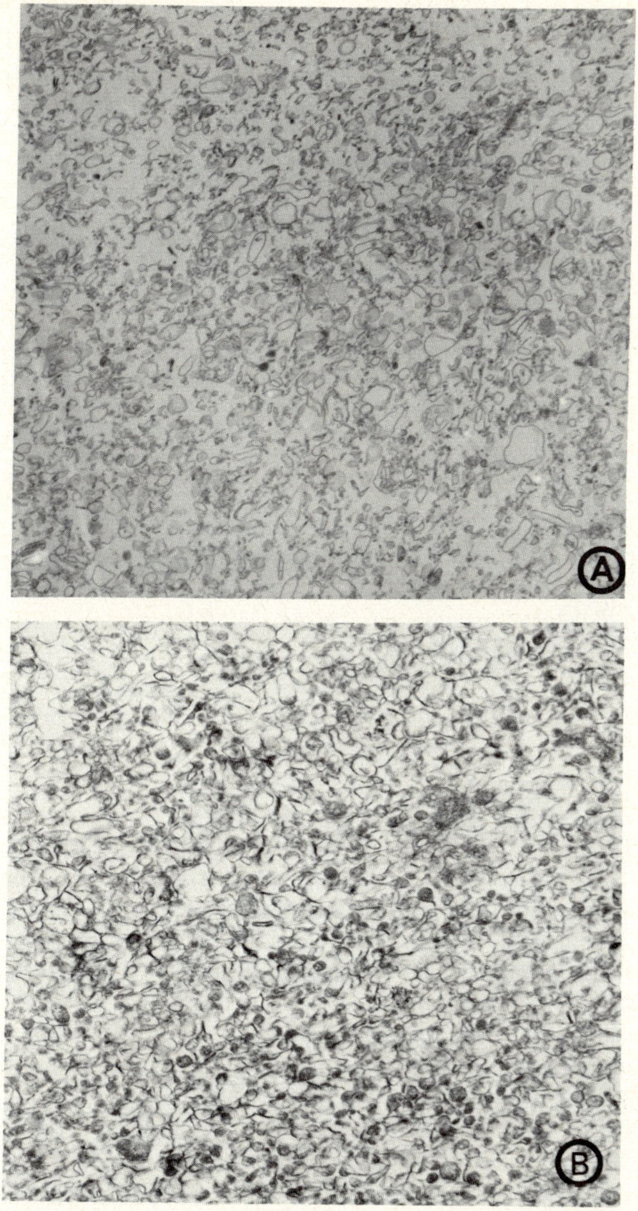
A
B

TABLE II

ENZYMATIC ACTIVITY OF PLASMA MEMBRANES FROM *L. variegatus* EGGS[a]

| Enzyme | Unfertilized | Fertilized |
|---|---|---|
| $Na^+,K^+$-ATPase | 516 | 130 |
| Alkaline phosphatase | — | 4.7 |
| Tyrosine kinase | 2.3 | 8.7 |

[a] Enzyme activity is presented as specific activity per milligram protein. The ouabain-inhibitable $Na^+,K^+$-ATPase is expressed in nanomoles of inorganic phosphate produced per milligram protein per hour. Alkaline phosphatase is expressed in nanomoles of para-nitrophenyl phosphate hydrolyzed per milligram protein per hour. Tyrosine protein kinase activity is expressed as picomoles of phosphate transferred to a synthetic peptide substrate per milligram protein per minute.

also enriched in the plasma membrane although this class of enzymes has not been characterized well enough to be considered a marker enzyme. We have been unable to detect 5′-nucleotidase in the sea urchin egg and galactosyltransferase exists at such low levels (Schneider and Lennarz, 1976) that we were unable to use it as a marker enzyme. The details of these enzyme assays have been published elsewhere (Kinsey *et al.*, 1980; Ribot *et al.*, 1983).

## C. Protein Composition

The protein content of these membrane preparations is best measured by colorimetric assay (Lowry *et al.*, 1951), however, the sample should be dissolved in 1 ml of 1% wt/vol SDS instead of distilled water. The standard curve should also be prepared in SDS since it does reduce the sensitivity of the assay.

The plasma membranes from unfertilized eggs contain relatively more protein than membraness from fertilized eggs as shown by the buoyant density (1.23 vs 1.14 g/ml) and the protein-to-phospholipid ratio (2.6 vs 1.8 wt%). Analysis of plasma membranes from fertilized and unfertilized egg by SDS–polyacrylamide gel electrophoresis (Ribot *et al.*, 1983) has shown that the protein composition is basically similar. However, proteins of approximately 280, 160, 120, and 25 kDa are present in greatly reduced amounts in plasma membranes from fertilized eggs. Proteins of similar size have been shown to be released from the egg surface at fertilization (Carroll and Epel, 1981), probably the result of limited proteolysis following the cortical reaction (Shapiro, 1975). Analysis by two-dimensional gel electrophoresis (Fig. 5) reveals that actin is a major component associated with the egg plasma membrane and is enriched in plasma membrane

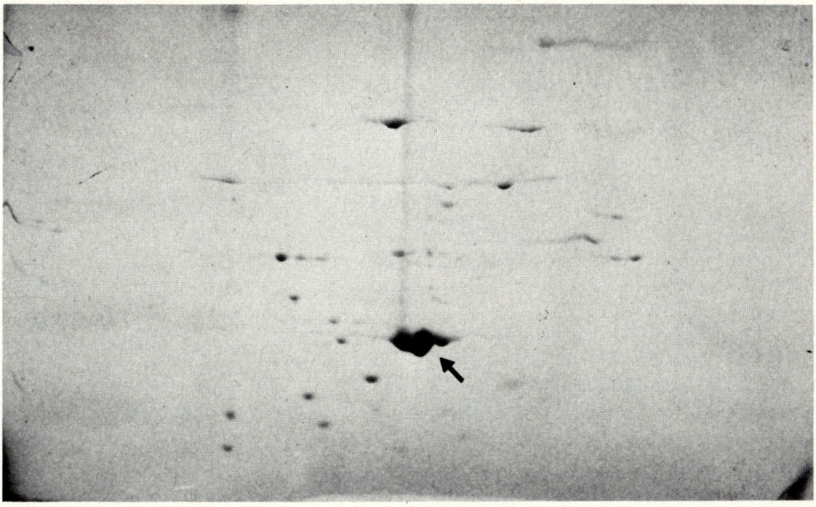

FIG. 5. Analysis of plasma membrane proteins by two-dimensional SDS–polyacrylamide gel electrophoresis. Plasma membranes prepared from fertilized *L. variegatus* eggs were separated by two-dimensional polyacrylamide gel electrophoresis (Rubin and Leonardi, 1983). The pH gradient is from pH 8.5 (left) to pH 4.3 (right). The second dimension (10% SDS-gel) was run from top to bottom and then stained with Coomassie Blue. Actin is the major membrane-associated protein and is indicated by the arrow.

from fertilized eggs (Ribot *et al.*, 1983). Actin is known to become associated with the egg cortex following fertilization (Burgess and Schroeder, 1977; Begg *et al.*, 1978; Spudich and Spudich, 1979), and this association is apparently strong enough to withstand the disruptive forces encountered during subcellular fractionation.

A technical problem that we have encountered in the analysis of the egg plasma membrane by SDS–gel electrophoresis is that the membrane lipids often cause streaking and poor resolution. To overcome this we routinely extract the lipids from a 0.1-ml suspension of membranes with 4 ml of chloroform:methanol (2:1 vol/vol) and pellet the denatured protein in a clinical centrifuge. The chloroform is carefully removed and the protein pellet is dried under nitrogen before being solubilized for electrophoresis. If very small quantities of membrane proteins are involved, add 0.1 ml of methanol to the sample first to precipitate the protein, then add 4 ml of the chloroform–methanol mixture (2:1).

In summary, the methods presented in this article allow the preparation of useful quantities of plasma membranes from the sea urchin egg. These membranes exhibit enzyme activities and at least some of the transport properties that are characteristic of other eukaryotic plasma membranes. These preparations are routinely used in my laboratory to study fertilization-dependent changes in tyrosine protein kinase activity and they should be useful for other enzymes as well. Additionally, the methods may enable the study of changes in actin–membrane interactions and in lipid or glycoprotein composition at fertilization and possibly through the early cleavage stages.

## Acknowledgment

Supported by NIH-HD14846 and Research Career Development Award 00620.

## References

Barber, M., and Mead, J. (1975). *Wilhelm Roux Arch.* **177,** 19–27.
Begg, D. A., and Rebhun, L. (1979). *J. Cell Biol.* **83,** 241–248.
Burgess, D. R., and Schroeder, T. E. (1977). *J. Cell Biol.* **74,** 1032–1037.
Carroll, E. J., and Epel, D. (1981). *Dev. Biol.* **87,** 374–378.
Detering, N. K., Decker, G. L., Schmell, E. D., and Lennarz, W. J. (1977). *J. Cell Biol.* **75,** 899–914.
Epel, D., Weaver, A. M., and Mazia, D. (1970). *Exp. Cell Res.* **61,** 64–68.
Johnson, J. D., Epel, D., and Paul, M. (1976). *Nature (London)* **262,** 661–664.
Kinsey, W. H., Decker, G. L., and Lennarz, W. J. (1980). *J. Cell Biol.* **87,** 248–254.
Lowry, O. H., Rosebrough, N. J., Farr, A. L., and Randall, R. J. (1951). *J. Biol. Chem.* **193,** 265–275.
Lynn, J. W., and Chambers, E. L. (1984). *Dev. Biol.* **102,** 98–109.
Ribot, H., Decker, S. J., and Kinsey, W. H. (1983). *Dev. Biol.* **97,** 494–499.
Ribot, H. D., Eisenman, E. A., and Kinsey, W. H. (1984). *J. Biol. Chem.* **259,** 5333–5338.
Rubin, R. W., and Leonardi, C. L. (1983). *In* "Methods in Enzymology" (S. Fleischer and B. Fleischer, eds.), Vol. 96, pp. 184–191. Academic Press, New York.
Schneider, E. G., and Lennarz, W. J. (1976). *Dev. Biol.* **102,** 10–20.
Shapiro, B. M. (1975). *Dev. Biol.* **46,** 88–102.
Spudich, A., and Spudich, J. A. (1979). *J. Cell Biol.* **82,** 212–226.
Uehara, T., and Yanagimachi, R. (1976). *Biol. Reprod.* **15,** 467–470.
Vacquier, V. D. (1975). *Dev. Biol.* **43,** 62–75.
Vacquier, V. D. (1981). *Dev. Biol.* **84,** 1–26.

# Chapter 9

# *Nuclei and Chromosomal Proteins*

DOMINIC L. POCCIA AND GEORGE R. GREEN

*Department of Biology*
*Amherst College*
*Amherst, Massachusetts*



## I. Introduction

Our aim here is to explain in detail some of the procedures which may be employed in the isolation of nuclei and nuclear proteins from the sea urchin embryo. We will emphasize procedures we have used or developed. Most of our work has focused on pronuclear proteins, especially those of the male pronuclei.

We will emphasize techniques developed especially for their study. Most of these techniques can be adapted for, or applied directly to, nuclei from embryos and other sources.

In studying male pronuclei, two problems common to all nuclear isolation procedures occur in exaggerated form: those of purity and yield. Purity and yield tend to be low because of the low nuclear/cytoplasmic protein mass ratio characteristic of fertilized eggs. We have solved these problems in the sea urchin by using moderately polyspermic eggs, thereby increasing the male pronucleus/cytoplasm ratio. Polyspermy also renders the contribution of the female pronucleus to the isolate quantitatively insignificant.

Nuclei of early embryos tend to be relatively fragile, and endogenous DNases can degrade the chromatin during isolation. We routinely use aqueous ethanol without divalent cations to minimize breakage and degradation. The procedure even allows isolation of female pronuclei from unfertilized eggs in sufficient quantity and purity to characterize their histones (Green and Poccia, 1985). Nuclei isolated by this procedure have been found suitable for structural studies (Savić *et al.*, 1981; Poccia *et al.*, 1984), analyses of histone variants and their posttranslational modifications (Green and Poccia, 1985; Poccia *et al.*, 1981), and enzymatic assays (Poccia *et al.*, 1978b).

Following nuclear isolation further purification is often necessary to separate individual molecular species and to characterize the small quantities of protein obtainable from embryonic sources. For these purposes we have developed a two-dimensional gel electrophoretic system that optimally separates histones from contaminants and from each other and have designed an apparatus that allows quantitative recovery of proteins from fixed and stained gel spots. The proteins may then be subjected to various forms of microanalysis.

All of the procedures described below work well on *Strongylocentrotus purpuratus*. We will point out any differences observed for other sea urchin species with which we have worked: *Lytechnius pictus*, *L. variegatus*, and *Arbacia punctulata*. All of the procedures given in detail are ones we have tested in our own laboratory, but we will refer where appropriate to other potentially useful methods.

## II. Inducing Polyspermy

To achieve high yields of clean male pronuclei, eggs are polyspermically fertilized using a modification of a procedure developed by Mazia (1974). By interfering with the fast block to polyspermy (Whitaker and Steinhardt, 1982; Nuccitelli and Grey, 1984), the procedure allows many sperm to penetrate the egg before the fertilization membrane becomes an effective barrier.

## A. Determination of Egg and Sperm Concentrations

The number of eggs in a suspension may be counted directly after Amido Black staining by the method of Fry and Gross (1970). The number of eggs can also be estimated as approximately $2 \times 10^6$/ml (Hinegardner, 1967). Sperm suspended in 1% glutaraldehyde in seawater can be counted in a Petroff–Hauser counter. It is most convenient to do this once for a set of serially diluted samples whose turbidities are then measured ($A_{350}$). Turbidity can then be used as a quick measure of sperm concentration in subsequent experiments. In most cases, however, it is not necessary to know accurately the concentration of sperm required to achieve polyspermy.

## B. Pretreatment of Eggs and Fertilization

The following procedure is used to achieve high and moderate degrees of polyspermy. All steps are carried out at the temperature used for rearing embryos (15°C for *S. purpuratus*).

1. After injection of 1 ml of 0.5 *M* KCl into its coelomic cavity, collect eggs from a female into a beaker of Millipore-filtered seawater.
2. Let the eggs settle and pour off the supernatant.
3. Resuspend 2–4 ml of settled eggs in 100 ml of seawater. Pour the suspension through four layers of surgical gauze and one layer of 53-μm Nitex to remove jelly and debris (e.g., spines). Let the eggs settle.
4. Resuspend eggs in 100 ml 30 m*M* $NH_4Cl$ in seawater adjusted to pH 8.0 with NaOH and stir gently for 1 minute. Let settle (about 10–12 minutes).
5. Suspend sperm (routinely two drops of thick semen in 25 ml of seawater). If pronuclei are to be isolated subsequently, 3 m*M* aminotriazole is included in the sperm suspension to inhibit hardening of the fertilization membranes (Showman and Foerder, 1979).
6. Immediately suspend the eggs from the $NH_4Cl$-seawater in the sperm suspension and stir gently for 1 minute. Let the eggs settle.
7. Resuspend eggs in fresh seawater. Culture by shaking in a 50-ml polypropylene culture tube on a rotary table (if < 25 ml) or in a beaker using a plastic paddle driven by a 60-rpm clock motor.

Notes: Paddles can be made by cutting squares from plastic lids of coffee cans, making two small slits near the center, and slipping the plastic square over a Lucite shaft. The ammonium chloride pretreatment allows very high degrees of polyspermy, as many as several hundred male nuclei/egg in *L. pictus* (Fig. 1; Poccia *et al.*, 1978a). If desired, nucleated and enucleated egg halves can be prepared before polyspermic fertilization (Nishioka and Mazia, 1977; Krystal and Poccia, 1979). Other methods that achieve low degrees of polyspermy use

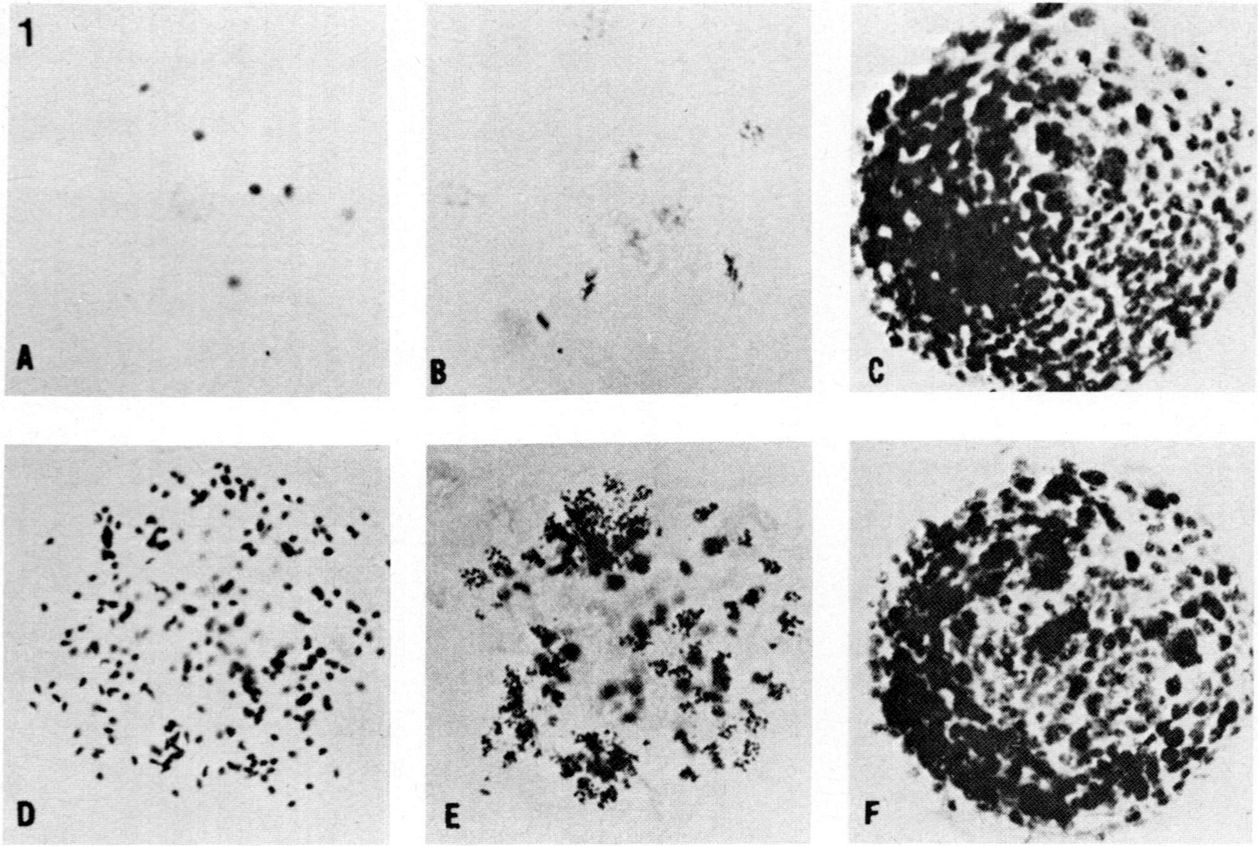
1
A
B
C
D
E
F

urethane (Longo and Anderson, 1970), nicotine (Clark, 1936; Longo and Anderson, 1969), or procaine (Belisle and Byrd, 1981; Vacquier and Brandriff, 1975) treatments, or simply no pretreatment but very high sperm concentrations. It is always advisable as a control to prepare a small culture of monospermically fertilized eggs to check for normal development (Hinegardner, 1967).

## C. Determining the Degree of Polyspermy by Nuclear Staining

It is useful to know the average number of male pronuclei/egg or degree of polyspermy ($\bar{n}$) in individual experiments. By counting the male nuclei in 100 fertilized eggs one can obtain an accurate number. Depending upon the species, male nuclei can be visualized by light microscopy at 200–400× magnification by differential interference microscopy in unfixed cells, or by phase-contrast microscopy in acetic acid/ethanol-fixed cells, or by bright field microscopy after staining with either Feulgen (Humason, 1967; Poccia *et al.*, 1978a) or aceto-orcein (see below). Feulgen-stained preparations are particularly good for photomicrography (Fig. 1).

We find the simplest procedure is one adapted from Mazia (1974). A small aliquot of the culture is removed and hand centrifuged to pellet the cells. Cells are resuspended in a few drops of seawater and several milliliters of ethanol/acetic acid (3:1) are added. After at least 10 minutes, a drop of the suspension is placed on a slide, the ethanol is allowed to evaporate, and a drop of 2% orcein in 75% acetic acid is added, then a cover slip. After 10 minutes, excess stain is removed from the edges of the cover slip with bibulous paper, and the preparation is examined. Nuclei are dark red. As shown in Fig. 2, the procedure also works for embryonic nuclei up to at least late larval stages (Poccia and Hinegardner, 1975).

Occasionally filtering the stain and allowing the color to develop in stained slides for several minutes are important. Sampling the culture at 15 minutes postfertilization is optimal for determining the degree of polyspermy; male pronuclei have not had enough time to fuse extensively (which would reduce the apparent degree of polyspermy), yet they will have undergone sufficient decondensation to allow them to be distinguished from unincorporated surface sperm, which remain triangular in shape and highly refractile by phase-contrast optics.

The degree of polyspermy can be varied in a given batch of eggs by fertilizing with different concentrations of sperm. In any experiment, the degree of polyspermy seems to depend on the particular batch of gametes and therefore cannot be predicted accurately from the sperm/egg ratio at fertilization.

---

FIG. 1. Feulgen-stained nuclei and chromosomes of polyspermic *L. pictus* eggs. (A–C) $\bar{N} = 6$. (D–F) $\bar{N} = 197$. Time postfertilization: (A) 30 minutes; (B) 3 hours; (C) 22 hours; (D) 15 minutes; (E) 5 hours; (F) 23 hours. (From Poccia *et al.*, 1978a.)

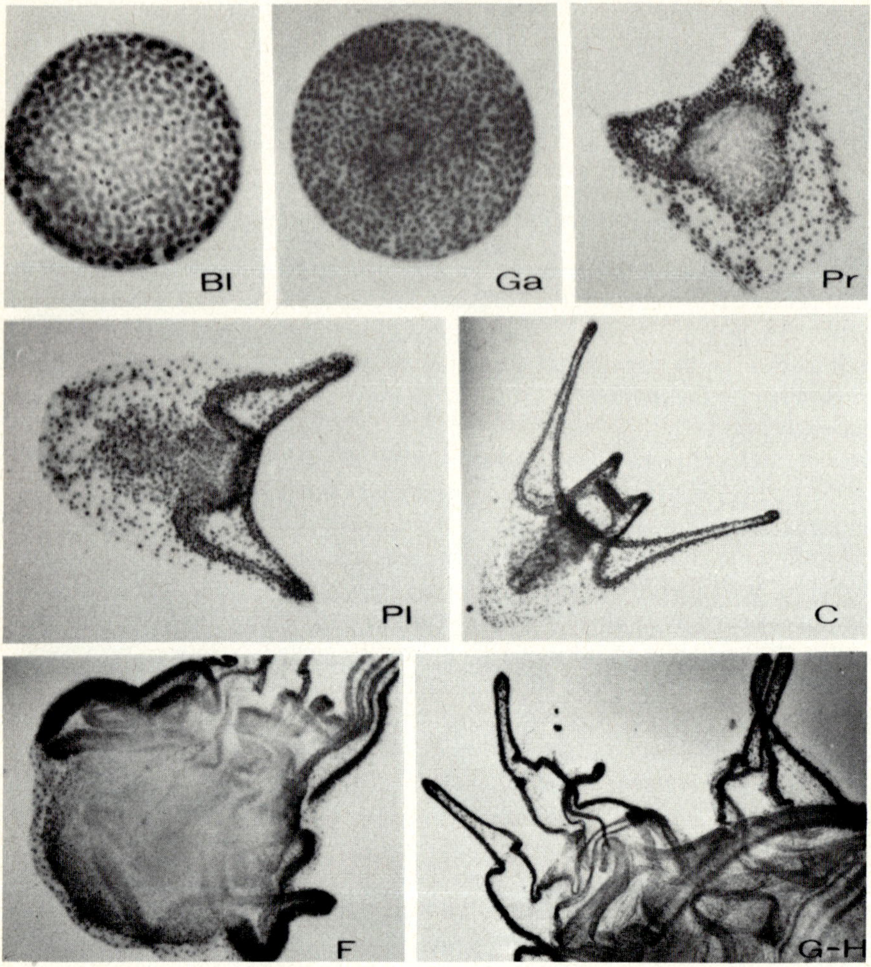


FIG. 2. Orcein-stained embryos of *L. pictus*. Specimens were fixed in ethanol–acetic acid (3:1) and stained and cleared in 2% orcein–75% acetic acid. First four stages are blastula (Bl), gastrula (Ga), prism (Pr), and pluteus (Pl). Stages C, F, and G–H refer to young feeding larva, half-grown larva, and almost mature larva. (From Poccia and Hinegardner, 1975.)

## III. Cell Cycle Parameters

Polyspermically fertilized sea urchin eggs represent a highly synchronized population of cells whose cell cycle parameters are easily evaluated. The two most useful measures are the time of initiation of DNA synthesis and the time of first chromosome condensation (Fig. 3 and Table I). In addition, since the

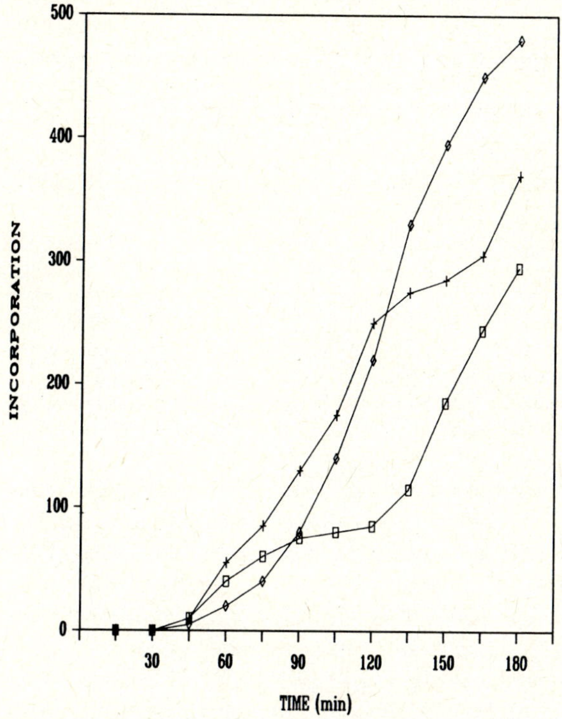


FIG. 3. Cumulative acid-insoluble incorporation of [$^3$H]thymidine in polyspermic eggs. Incorporation (cpm × $10^{-3}$). □ = 24, + = 49, ◇ = 95 male pronuclei/egg. (From Poccia *et al.*, 1978a.)

$NH_4Cl$ pretreatment activates the egg cell cycle without raising a barrier to fertilization, these eggs may be fertilized at any time within the activated cell cycle (Mazia, 1974; Krystal and Poccia, 1979), thus allowing introduction of male pronuclei during different phases. This makes polyspermy analogous to nuclear transplantation or to the production of heterophasic homokaryons by cell fusion (Harris, 1974; Gurdon and Woodland, 1968; Johnson and Rao, 1971). For example, one may obtain premature chromosome condensation merely by following the procedures in Section II,B but not adding sperm until some time after the 10-minute $NH_4Cl$ treatment (Krystal and Poccia, 1979).

## A. Phases

### 1. DNA SYNTHESIS

DNA synthesis after fertilization is easily monitored by including 10 μCi/ml [$^3$H]thymidine in the culture or by preloading a 10% (v/v) suspension of unfer-

TABLE I

TIME AFTER ACTIVATION OF MAXIMUM CHROMOSOME CONDENSATION IN POLYSPERMICALLY FERTILIZED *L. pictus* EGGS

| Degree of polyspermy (number of male nuclei/egg ± SD) | Time until male chromosome condensation (hr) | Eggs with condensed chromosomes (%) |
|---|---|---|
| 6.5 ± 3 | 1.5 | 98 |
| 28 ± 8 | 2.5 | 95 |
| 197 ± 51 | 5.0 | 79 |

[a] Eggs were treated with $NH_4Cl$, pH 8, for 15 minutes, inseminated with increasing concentrations of sperm, and incubated at 15°C. Aliquots were fixed for cytology every 10 minutes. One hundred eggs were counted at 15 minutes postfertilization to determine the degree of polyspermy and 100 eggs were counted for determination of percentage at time of maximum condensation.

tilized eggs in the presence of 10 μCi/ml [$^3$H]thymidine for 3 hours (Krystal and Poccia, 1979; Hinegardner *et al.*, 1964). Incorporation into 5% trichloroacetic acid-insoluble material is measured in several aliquots taken from the culture at different times (Poccia *et al.*, 1984). If desired, acid-soluble pools (mostly thymidine triphosphate) can be measured after washing the eggs free of exogenous label (Nishioka *et al.*, 1984). Similar procedures can be employed to measure RNA synthesis (Poccia *et al.*, 1985). It is important not to include aminotriazole in extensively washed cultures since the denuded eggs become very sticky and tend to lyse. The time of initiation of DNA synthesis is relatively insensitive to the degree of polyspermy, occurring at about 30 minutes postactivation (Poccia *et al.*, 1978a). The length of the DNA synthesis phase (S phase), however, increases at very high degrees of polyspermy (Fig. 3; Poccia *et al.*, 1978a). Labeled eggs may also be examined autoradiographically, demonstrating that all male nuclei replicate after 30 minutes (Poccia *et al.*, 1978a).

DNA synthesis is effectively inhibited by addition of 5 μg/ml of Aphidicolin to the culture (Ikegami *et al.*, 1978; Poccia *et al.*, 1984; Nishioka *et al.*, 1984). Aphidicolin is added from a 5 mg/ml stock solution in dimethyl sulfoxide, and shuts off synthesis to >95% within approximately 1 minute. Inhibition is partially reversible by washing the eggs in fresh seawater (Poccia *et al.*, 1984, 1985). Aphidicolin can be prepared in the laboratory (Ikegami *et al.*, 1978; Poccia *et al.*, 1984) or purchased (Sigma Chemical Co., St. Louis, MO).

### 2. MITOSIS

The time of mitosis is usually about 90 minutes postactivation in *S. purpuratus*. It is evaluated by the staining procedures described in Section II,C described above. Metaphase chromosomes are easily distinguished from other mitotic stages (Mazia, 1974; Krystal and Poccia, 1979). At high degrees of polyspermy, the time to reach first mitosis may extend to several hours (Table I; Poccia *et al.*, 1978a).

## B. Premature Chromosome Condensation

Prematurely condensed chromosomes can be observed by the same methods used for mitotic chromosomes. If sperm nuclei are introduced before the S phase of the egg, male chromosomes have a double thickness; if sperm nuclei are introduced after the S phase of the egg, male chromosomes have a single thickness, indicating that they have not replicated and have condensed prematurely (Mazia, 1974; Fig. 4).

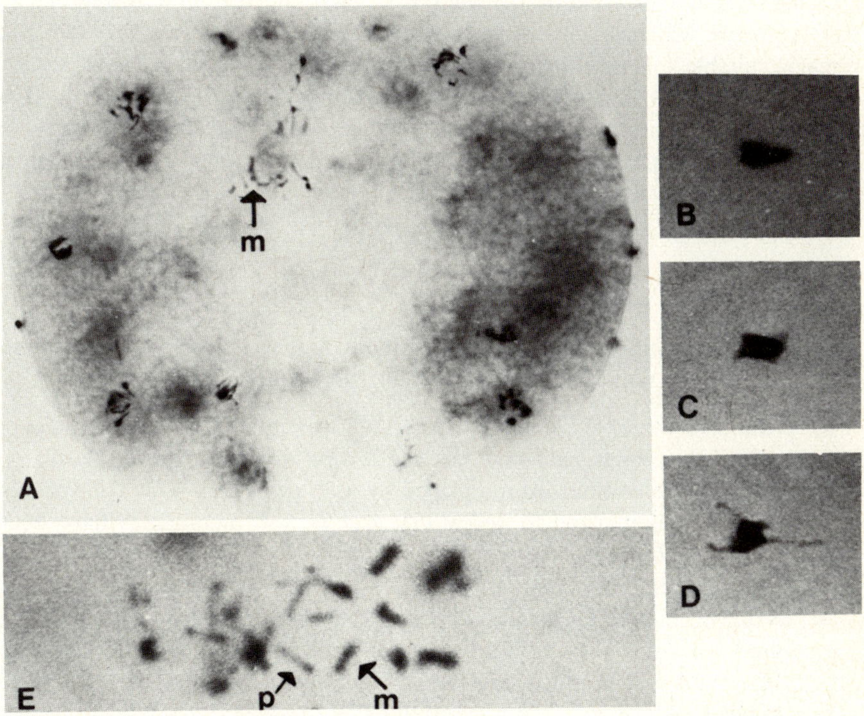


FIG. 4. Morphology of prematurely condensed chromosomes. (A) Premature chromosome condensation 30 minutes after polyspermic fertilization of an ammonia-activated prometaphase egg; (B–D) male pronuclei at 7, 14, and 20 minutes following fertilization; (E) prematurely condensed male chromosomes intermingled with maternal chromosomes at 33 minutes postfertilization. p, Prematurely condensed male chromosomes; m, maternal chromosomes. (From Krystal and Poccia, 1979.)

## IV. Isolation of Nuclei

Several procedures are available for isolating nuclei from embryonic sea urchins. Nuclei from cleavage stage embryos and pronuclei tend to be relatively fragile, and endogenous DNases often degrade the DNA during isolation. We have avoided these problems by adapting a procedure originally developed for isolation of the mitotic apparatus (Mazia *et al.*, 1972). The isolation is done in the cold and in the absence of ions, particularly $Mg^{2+}$ and $Ca^{2+}$, in order to reduce nucleolytic and proteolytic activity. The embryos are immediately exposed to ice-cold 30% ethanol containing 0.2% Triton X-100, which dissolves the cell membrane while coagulating cytoplasm and leaching out soluble components. The cytoplasm is subsequently dissolved in 10% ethanol–0.2% Triton X-100, from which demembranated nuclei are collected by centrifugation. Ethanol stabilizes the nuclei to centrifugation and does not seem to be deleterious to the chromatin, which retains active enzyme activities (Poccia *et al.*, 1978b) and remains suitable for digestion studies with micrococcal nuclease (Savic *et al.*, 1981) and DNase I. Degradation of proteins during isolation is minimal, as is adsorption of proteins to the chromatin (Poccia *et al.*, 1981). Proteins whose *in vivo* association with the nuclei depends on the structural integrity of the nuclear envelope may not be retained by these techniques.

### A. Pronuclei

The following procedure is a modification of that given by Poccia *et al.* (1981). It is more rapid and results in cleaner nuclei.

1. Add 250 ml of ice-cold seawater to 50 ml of a 2% (v/v) culture of polyspermic eggs, and pour the suspension through a 30-ml plastic syringe whose end has been replaced by a 53-μm Nitex screen. Inspect in the light microscope to verify that fertilization membranes have been removed. Repeat if necessary.
2. Let eggs settle at 4°C.
3. Remove as much seawater as possible and resuspend in 10 ml of 1 *M* glucose. Transfer to a 50-ml polypropylene culture tube. Underlay with 10–15 ml of 1 *M* glucose using a 25-ml pipet. Centrifuge at 300 *g* for 5 minutes.
4. Aspirate the supernatant and resuspend the pellet uniformly in a minimal volume of 1 *M* glucose. Resuspend in 40 ml 30% ethanol–0.2% Triton X-100–1 m*M* phenylmethylsulfonylfluoride (PMSF). Incubate for 5 minutes at 4°C. Centrifuge at 800 *g* for 5 minutes.
5. Resuspend the pellet in a minimal volume of the 30% ethanol solution, and add 40 ml 10% ethanol–0.2% Triton X-100–1 m*M* PMSF. Incubate

for 5 minutes to disperse the cytoplasm and centrifuge at 1800 *g* for 10 minutes. The pellet contains purified pronuclei lacking membranes.

Notes: If aminotriazole was not included at fertilization, fertilization membranes will interfere with proper lysis, resulting in contamination of the nuclear pellet. If the eggs are not kept cold, cytoplasmic dispersal may be impaired. Appropriate controls for protein degradation or modification during isolation, and for evaluation of cosedimenting contamination, are described in Poccia *et al.* (1981).

Female pronuclei can be isolated by the above procedure (Green and Poccia, 1985). They have also been isolated by Hinegardner (1962) using lysis in 2 m*M* $MgCl_2$ followed by centrifugation through a sucrose step gradient, and by Albanese *et al.* (1980) using lysis in buffered lithium chloride followed by centrifugation through 0.5 *M* sucrose.

## B. Sperm Nuclei

Sperm cells are washed in the 10% ethanol–Triton X-100 solution given above to remove seminal fluid, or used directly for nuclear isolation. Shearing the cells in a Sorvall Omnimixer (DuPont) for 2 minutes at full speed (Salik *et al.*, 1981) or sonicating will efficiently remove tails and membranes, including those of the midpiece. The nuclei can be pelleted at 2500 *g* for 5 minutes.

Sperm nuclei can also be prepared by nonethanolic procedures. Semen (100 μl) is suspended in 1 ml of 0.15 *M* NaCl–1% Triton X-100–1 mg/ml dithiothreitol–1m*M* $CaCl_2$–10 m*M* Tris-HCl, pH 8, in a 1.5-ml microfuge tube and the suspension is sonicated in a bath sonicator for 1 minute. Nuclei are collected at 2500 *g* for 1 minute, resuspended and resonicated in the same buffer.

An alternative method is to shear a sperm suspension in 2 *M* sucrose–10 m*M* $CaCl_2$ for 2 minutes on ice in an Omnimixer (Maller *et al.*, 1976). This breaks off tails but leaves the midpiece attached (Fig. 5A). Inclusion of 1% Triton X-100 in the solution solubilizes all membranous components and results in a nuclear preparation equivalent to the other procedures. Further purification can be achieved by modest dilution of the sample and centrifugation through 1.7 *M* sucrose–10 m*M* $CaCl_2$ at 10,000 *g* for 20 minutes. Centrioles may remain attached (Fig. 5B; Maller *et al.*, 1976).

Other procedures are given by Mizuno *et al.* (1973), as modified by Spadafora *et al.* (1976). See also Azorin *et al.* (1983) for sea cucumber sperm, Zalenskaya *et al.* (1981) for starfish sperm, and Marushige and Marushige (1978) for general procedures.

## C. Condensed Chromosomes

Isolation of prematurely condensed and mitotic chromosomes in a form suitable for histone analysis can be accomplished with the ethanol–Triton X-100

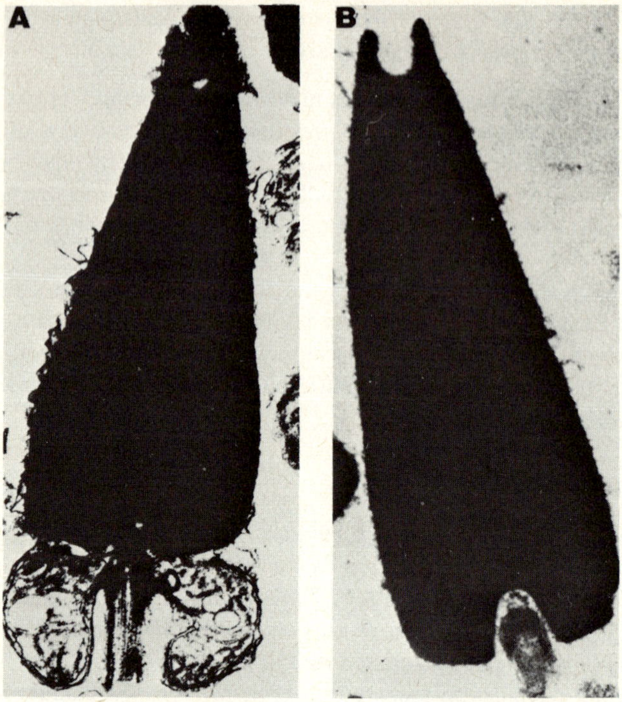


FIG. 5. Isolated sperm nuclei. Sperm nuclei prepared (A) without and (B) with Triton X-100 in 2 *M* sucrose–10 m*M* $CaCl_2$.

procedures used for interphase nuclei above. Alternatively, one may use the technique of Salmon and Jenkins for isolation of the mitotic apparatus in which the lysis buffer containing EGTA–$MgCl_2$–PIPES–DTT–Nonidet P-40–PMSF may be used (Salmon and Jenkins, 1977; Krystal and Poccia, 1981).

## D. Embryonic Nuclei

The ethanol–Triton X-100 procedure given above can be used for embryonic nuclei, but several other methods are available. These include lysis in 0.15 *M* NaCl–0.15 *M* sodium citrate followed by centrifugation through 1 *M* sucrose (Loeb *et al.*, 1969, for *S. purpuratus* and *S. franciscanus,* modified by Ord and Stocken, 1978); lysis in 20 m*M* NaCl–10 m*M* $MgCl_2$–20 m*M* Tris-HCl, pH 7.5–1% Triton X-100 (Keichline and Wasserman, 1979, for *S. purpuratus,* modified by Katula, 1983); lysis in 2 m*M* $MgCl_2$ followed by a sucrose step gradient (Hinegardner, 1962, for *S. purpuratus,* modified by Carroll and Ozaki, 1979); lysis in 1 *M* sucrose–2% Triton X-100–2 m*M* $MgCl_2$ followed by a sucrose step gradient (Thaler *et al.*, 1969, for *A. punctulata,* modified by Seale and Aronson, 1973); lysis in 0.5 *M* sucrose–0.05 *M* Tris–0.25% Triton X-100 (Hogan and Gross, 1972, for *A. punctulata, S. purpuratus,* and *L. pictus,* modified by Arceci and Gross, 1980); and lysis in 0.1 *M* LiCl–0.1 *M* sodium phosphate–0.5 m*M* PMSF–0.024 *M* EDTA followed by centrifugation through 0.5 *M* sucrose (Albanese *et al.*, 1980, for *P. lividus,* modified by Shaw *et al.*, 1981, for *S. purpuratus,* and Chambers and Shaw, 1984, for *A. punctulata*). For general procedures and considerations in designing nuclear isolation procedures, see Smuckler *et al.* (1976), Busch and Daskal (1977), Birnie (1978a), and Hoffmann and Chalkley (1976).

## E. Chromatin

Chromatin is isolated from nuclei by a modification of the procedure of Marushige and Ozaki (1967) involving two washes in 10 vol of cold 0.075 *M* NaCl–0.025 *M* $Na_2$EDTA–0.1% Triton X-100–0.01 *M* Tris-HCl, pH 8–0.1 m*M* PMSF–1% thiodiglycol at 1500 *g* for 10 minutes followed by three washes in 0.01 *M* Tris-HCl, pH 8–1% thiodiglycol at 10,000 *g* for 15 minutes. If desired, the chromatin gel can be pelleted through 1.7 *M* sucrose–0.01 *M* Tris-HCl, pH 8.0 at 10,000 *g*. An independent purification step is equilibrium buoyant density gradient ultracentrifugation in 40% w/v Metrizamide (Sigma) (Poccia *et al.*, 1978b; Birnie, 1978b). For histone analysis, one wash in saline–EDTA and one wash in Tris-HCl is usually sufficient.

# V. Protein Analysis

The use of powerful one- and two-dimensional polyacrylamide gel electrophoretic methods has revolutionized the analysis of chromosomal proteins. The sensitivity of Coomassie Blue and in particular of silver staining techniques allows detection of submicrogram quantities of proteins. Western protein blotting is useful for identification of individual histones if specific antibodies are available (Towbin *et al.*, 1979; Salik *et al.*, 1981). Elution of spots from gels followed by chemical microanalysis makes possible detailed characterizations of the individual proteins.

We describe below some of the methods we have applied to histones. Many of these procedures can be adapted to study nonhistone chromosomal proteins and indeed any proteins that can be separated on polyacrylamide gels. A wide variety of microtechniques for protein analysis (Hunkapillar *et al.*, 1984) should be applicable to the purified eluted proteins. Controls for modification, degradation, and contamination of male pronuclear proteins have been described (Poccia *et al.*, 1981).

## A. Protein Separation

### 1. Histone Extraction

Histones can be extracted from nuclei or chromatin with 0.4 *N* $H_2SO_4$ or with protamine sulfate. In the first case, an equal volume of cold 0.8 *N* (0.4 *M*) $H_2SO_4$ is added to the suspension of nuclei (<0.5 mg/ml DNA) while vortexing. The mixture is stirred for 8–16 hours at 4°C. The precipitate is collected at 10,000 *g* for 10 minutes. The supernatant is removed and the pellet is washed with a small volume of 0.4 *N* $H_2SO_4$. The combined supernatants are made 20% in trichloroacetic acid (TCA) by addition of one-quarter volume of cold 100% (w/v) TCA. After 30 minutes the precipitated histones are collected and washed once with cold acetone–0.1% $H_2SO_4$ and twice with acetone. The pellet is dried under vacuum.

Acid-extracted histones may be used without precipitation for electrophoresis in acidic gel systems (Bonner *et al.*, 1980). To inhibit oxidation and proteolysis, 2-mercaptoethanol and PMSF may be added. When small quantities of protein are to be analyzed electrophorectically, extraction of nuclei or chromatin with buffers containing protamine sulfate, using a procedure adapted from Richards and Shaw (1982), is often preferable. At room temperature, the homogeneously suspended sample is mixed with 3 vol of protamine buffer [10.7 *M* urea–6.7% acetic acid–6.7% 2-mercaptoethanol–3.3% thiodiglycol–1.3 m*M* PMSF–1.3% protamine sulfate (Sigma, grade X)–0.04% Crystal Violet] and sonicated for 1 minute in a bath sonicator. After 1 hour incubation at room temperature, the sample is clarified at 10,000 *g* for 2 minutes. It may be stored at −20°C indefinitely. For further histone isolation and purification procedures see *Methods Cell Biol.* **16** (1968).

### 2. Gel Electrophoresis

Histones are resolved with high resolution by electrophoresis in several gel systems. Most useful are the acetic acid–urea system (Panyim and Chalkley, 1969), the acid–urea–Triton X-100 system (Zweidler, 1978), the acid–urea–cetyltrimethylammonium bromide (CTAB) system (Bonner *et al.*, 1980), and the sodium dodecyl sulfate (SDS) system (Laemmli, 1970). In acetic acid–urea gels, relative mobility is influenced by the concentration of urea that unfolds the proteins to different degrees. In acid–urea–Triton X-100 gels, effects of hydrophobicity resulting in differential binding of detergent are imposed on the urea effects. The ionic detergents CTAB and SDS denature the histones and minimize inherent charge differences, so that mobility is proportional to molecular size. Used in various two-dimensional combinations, these systems can resolve histone variants and their modified forms (Fig. 6; Savić and Poccia, 1978; Bonner *et al.*, 1980; Davie, 1982). Additional systems are described by Hardison and Chalkley (1978).

We use 12 cm tall, 1.5 mm thick slab gels for the first dimension and 2.0 mm thick slab gels for the second dimension. Standard acetic acid–urea gels contain 2.5 *M* urea–5% acetic acid–12% acrylamide–0.08% bisacrylamide. Polymerization is initiated with 1 mg/ml ammonium persulfate and 5 μl/ml tetramethylethylenediamine (TEMED). Acid–urea–Triton X-100 gels are the same but in addition contain 6 *M* urea and 6 m*M* Triton X-100. Gels are overlaid to 1 cm with 2.5 *M* urea–5% acetic acid or 6 *M* urea–6 m*M* Triton X-100–5% acetic acid, respectively, and preelectrophoresed for 6–12 hours. We use 2 cm stacking gels of 5.4 *M* urea–9 m*M* potassium acetate–6% polyacrylamide–0.04% bisacrylamide polymerized under bright fluorescent light with 0.0003% riboflavin from a freshly prepared 200× stock solution and 5 μl/ml TEMED. The stacking gel is preelectrophoresed with 6.0 *M* urea–0.01 *M* potassium acetate–60 mg/ml cysteamine at 150 V for 30–60 minutes. Gels are run at 150 V. The electrode buffer is 5% acetic acid. CTAB gels are the same as acetic acid–urea gels but with 0.15% CTAB in the upper reservoir during the run. SDS gels are composed of 15% polyacrylamide–0.2% bisacrylamide–0.37 *M* Tris-HCl,

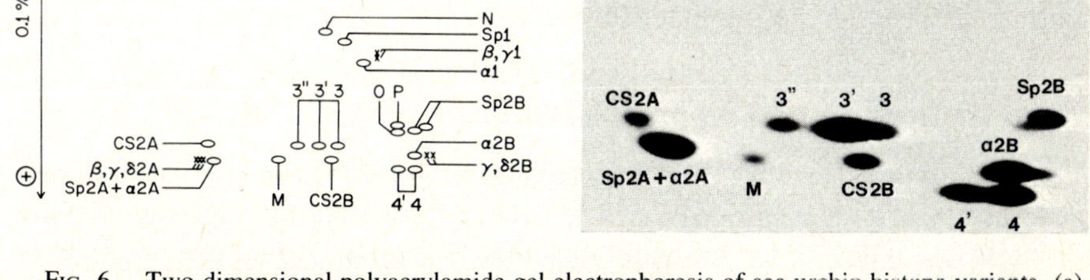


FIG. 6. Two-dimensional polyacrylamide gel electrophoresis of sea urchin histone variants. (a) Diagrammatic representation of the relative positions of histone variants separated on a 6 *M* urea–6 m*M* Triton X-100–0.9 *N* acetic acid, pH 2.8 gel in the first dimension and a 0.1% SDS–0.37 *M* Tris–HCl, pH 8.7 gel in the second dimension. (b) Portion of a gel separating a mixture of standard histones from sperm and actinomycin D-treated blastulae. The β, γ, δ, N, O, and P proteins are absent from this gel. (From Poccia *et al.*, 1981.)

pH 8.8–0.1% SDS. The 2 cm high stacking gel is 5% polyacrylamide–0.13% bisacrylamide–0.125 *M* Tris-HCl, pH 6.8–0.1% SDS. Polymerization is with 0.5 mg/ml ammonium persulfate and 0.5 μl/ml TEMED. The electrode buffer is 0.025 *M* Tris–glycine, pH 8.3–0.1% SDS.

In order to excise lanes accurately, first dimension gels are usually stained. If the second dimension contains an ionic detergent, the gel may be fixed and stained in 40% ethanol–5% acetic acid–1% cysteamine–0.05% Coomassie Blue. Lanes are cut out, and washed three times over a period of 1 hour with 5% acetic acid or 0.06 *M* Tris-HCl, pH 6.8, for gels containing CTAB or SDS, respectively. Detergents are omitted from these washes since they can extract proteins from the gels, leading to quantitative errors. Alternatively, first dimension acetic acid–urea or acid–urea–Triton X-100 gels may be surface stained for 5 minutes in 0.005% Coomassie Blue in water. Excised lanes may be applied directly to any of the four gel types for the second dimension.

Gel slices are slipped between the glass plates of the second gel apparatus and secured to the stacking gel with 1% agarose melted in 0.1% SDS–1% 2-mercaptoethanol–0.06 *M* Tris-HCl, pH 6.8 for SDS gels. For acid gels, agar is melted in water, cooled to 60°C and adjusted to 5% acetic acid–10 mg/ml cysteamine. Preelectrophoresis is not required if cysteamine is included in the agar.

Staining of the second dimension is performed with 0.05% Coomassie Blue dissolved in 40% ethanol–5% acetic acid. For greater sensitivity, silver staining can be used (Burk *et al.*, 1983). If protein or dye recovery are required for subsequent analysis (see below), gels are stained for 48 hours in 500 ml of the Coomassie Blue stain containing 1% cysteamine at room temperature with continuous shaking. The gel and container are rinsed with water and destained in 500 ml of 20% ethanol–5% acetic acid–1% cysteamine for 24 hours, leaving a light blue background.

Nonhistone chromosomal proteins may be separated by isoelectric focusing or SDS gels, either separately or in combination (O'Farrell and O'Farrell, 1977). Nonequilibrium pH gradient electrophoresis allows separation of both histones and nonhistones on the same gel (O'Farrell *et al.*, 1977).

## B. Purification and Characterization

Analytical gels can be used as preparative gels if combined with microanalysis. As a general procedure, it is necessary to elute the proteins. We describe below a device for the rapid quantitative elution of multiple spots taken from polyacrylamide gels.

### 1. The Eluter

The device we have developed for multiple elution also can be used for quantification and concentration of individual proteins and for passive or elec-

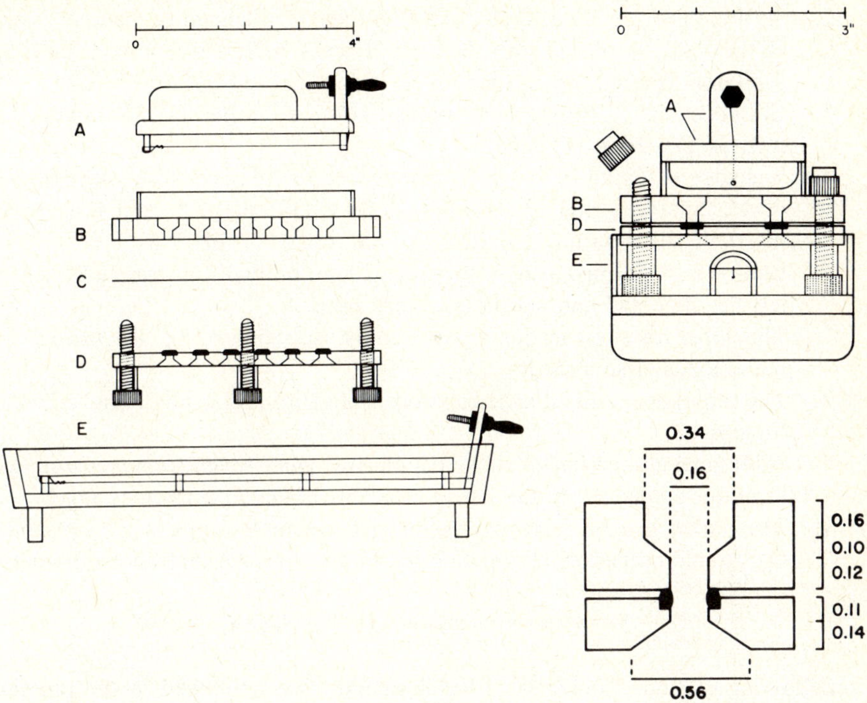


FIG. 7. The eluter. Side view, end view, and cross-section of a well. (A) Upper reservoir electrode and cover; (B) upper reservoir and sample wells; (C) dialysis membrane; (D) membrane clamping plate; (E) lower reservoir and electrode. All dimensions in inches. (From Green *et al.*, 1982.)

trophoretic dialysis (Green *et al.*, 1982). It is shown in Fig. 7. It is easy and inexpensive to construct and can be used on fixed and stained gels, allowing simple positive identification of starting material. Eluates are collected in <30-μl volumes. Buffer, dye, and SDS can be removed from the proteins by dialysis or ion-pair extraction before further characterization of the proteins (Green *et al.*, 1982; Henderson *et al.*, 1979).

### 2. Purification and Concentration of Individual Proteins

Assembly of the eluter is described in Green *et al.* (1982). The following elution procedure has been modified from the original to minimize protein loss due to convection in the sample wells, which sometimes occurs due to heating in the recovery zone.

1. The recovery zone is filled with 100 μl of 0.01 *M* Tris-HCl, pH 6.8–0.2% SDS.

2. A filter paper disk, cut slightly smaller than the diameter of the upper part of the well, is placed into the well. Excess buffer is removed with an aspirator, leaving the surface of the paper exposed but moist.
3. Gel slices cut from fixed and stained gels are washed for 30 minutes in 0.01 *M* Tris-HCl, pH 6.8.
4. The slice is placed in the well and attached to the paper disk with 100 μl of 1% agarose–0.01 *M* Tris-HCl–0.2% SDS, pH 6.8, and the upper reservoir filled with 0.2% SDS–0.03 *M* Tris-HCl, pH 8.8.
5. Elution is performed at 50 V for about 1 hour until all Coomassie Blue has left the slice and entered the recovery zone.
6. The upper reservoir buffer is removed by aspiration so that the surface of the embedded slice is dry.
7. The filter paper with attached agarose and gel slice is gently removed with forceps.
8. Additional elution buffer above the blue recovery zone is removed with an aspirator connected to the barrel of a 25-μl Hamilton microsyringe.
9. The eluted protein is recovered from the sample wells with a plastic-tipped micropipettor, taking care not to puncture the dialysis membrane.

### 3. Quantification of Gel Spots

Following elution, Coomassie Blue is quantitatively retained in the protein recovery zone and may be measured spectrophotometrically to estimate the amount of protein in the gel slice. If the amount of dye is sufficient, the sample may be read in a standard spectrophotometer after its removal with a micropipet and dilution (Fig. 8). Linearity of response is excellent up to at least 10 μg protein/band. Accuracy may be improved for $< 1$ μg by photographing the undiluted samples in the cassette on a light box and densitometrically scanning the negative.

For absolute measurements, intensities are compared with standard proteins of known amounts run in parallel, with a background measurement taken from a region of the gel free from protein. For accurate estimates, precautions must be taken to avoid loss of protein between first and second dimension gels (see Section V,A,2). Gels should be free of oxidation, degradation, and electrophoretic artifacts that could decrease the amount of protein in any spot.

### 4. Peptide Mapping

Eluted proteins can be chemically or enzymatically cleaved. We have used cyanogen bromide, *N*-bromosuccinimide, trypsin, chymotrypsin, submaxillary protease, and *S. aureus* protease on eluted histones.

Chemical cleavage is performed directly on fixed and stained gel slices wash-

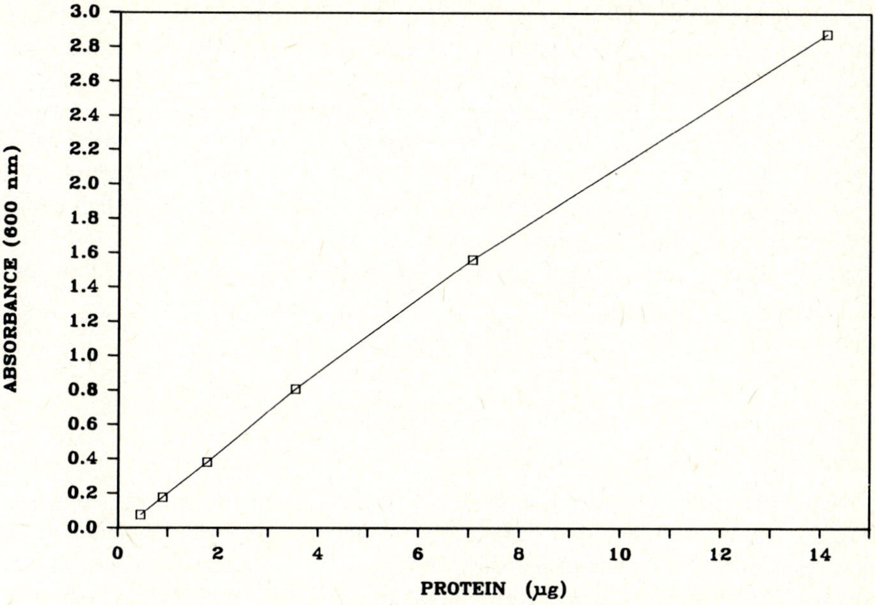


FIG. 8. Quantification of Coomassie Blue-stained gel bands. Sea urchin total sperm histones of *S. purpuratus* were run in twofold serial dilutions on a 0.1% SDS gel. Stained bands of Sp H2B histones were excised and eluted. Absorbance of bound dye after dilution to 0.5 ml was read in a Perkin–Elmer Lambda 1 UV/VIS spectrophotometer. Values were corrected for background ($A$ = 0.020).

ed in distilled water for 30 minutes to remove acetic acid and ethanol, using a modification (Green and Poccia, 1985) of the method of Nikodem and Fresco (1979). The slice is incubated for 3 hours at room temperature with shaking in 5 g cyanogen bromide dissolved in 20 ml acetonitrile–40 ml 0.125 *M* Tris-HCl, pH 6.8–40 ml 0.6 *N* HCl (or 100 ml of 10 mg/ml *N*-bromosuccinimide in 5% acetic acid). Cyanogen bromide cleaves at methionine and *N*-bromosuccinimide cleaves at tyrosine residues. Fragments are analyzed on a 15% SDS gel after equilibration of the slice with SDS buffer as in the two-dimensional electrophoretic procedure given above.

Tryptic mapping is done on *in vitro* labeled proteins to increase the detection sensitivity of peptides. Chromatin proteins are labeled with tritium by reductive methylation using the method of Kuhn and Wilt (1980). After electrophoresis and elution, 50 µg of sperm histone is added to each sample as carrier. Coomassie Blue and SDS are removed by ion pair extraction. Each precipitated protein sample is dissolved in 1% ammonium bicarbonate, pH 8.0 and digested with 1 µg of TPCK–trypsin (Sigma) for 3 hours at 37°C. Trypsin (1 µg) is added and digestion continued for an additional 3 hours. Solvent is evaporated in a Savant Speed-Vac concentrator. The residue is dissolved in loading buffer and electrophoresed on a 50% polyacrylamide gel (West and Bonner, 1980). Peptides are visualized by fluorography (Bonner and Laskey, 1974) after soaking the gel in 10% glutaraldehyde for 30 minutes.

In combination with *in vivo* labeling, the mapping procedures can be used to analyze sites of posttranslational modification (Green and Poccia, 1985). General methods for studying posttranslational modifications are given in *Methods Cell Biol.* **19** (Stein *et al.*, ed., 1978) and in Vols. 106 and 107 of ''Methods in Enzymology'' (Wold and Moldave, eds., 1984). General techniques and strategies for peptide mapping are given by DeLange (1978), and Hsiang and Cole (1978).

## VI. Conclusions

The isolation procedures described here work well on several species of sea urchin. They result in rapid isolation of nuclei with minimal degradation of protein and DNA. The nuclei are suitable for structural studies as well as protein analyses. The microanalytical procedures are especially useful when quantities of material are limited. We hope that these may be of use in the study of nuclear proteins from many organisms, most particularly of male pronuclear proteins from other echinoderms and marine organisms, since comparative data on nuclear protein transitions in the first cell cycle after fertilization are lacking.

### Acknowledgment

This work was supported by a grant from the National Institute of Health (HD 12982).

### References

Albanese, I., Di Liegro, I., and Cognetti, G. (1980). *Cell Biol. Int. Rep.* **4,** 201–210.

Arceci, R. J., and Gross, P. R. (1980). *Dev. Biol.* **80,** 186–209.

Azorin, F., Olivares, C., Jordan, A., Perez-Gau, L., Cornudella, L., and Subirana, J. A. (1983). *Exp. Cell Res.* **148,** 331–344.

Belisle, B. W., and Byrd, W. (1981). *J. Cell Biol.* **91,** 11a.

Birnie, G. D. (1978a). *Methods Cell Biol.* **17,** 13–26.

Birnie, G. D. (1978b). *Methods Cell Biol.* **18,** 23–40.

Bonner, W. P., and Laskey, R. A. (1974). *Eur. J. Biochem.* **46,** 83–88.

Bonner, W. P., West, M. H. P., and Stedman, J. D. (1980). *Eur. J. Biochem.* **109,** 17–23.

Burk, R. K., Eschenbruch, M., Leuthard, P., and Steck, G. (1983). *In* ''Methods in Enzymology'' (C. H. W. Hirs and S. N. Timasheff, eds.), Vol. 91, pp. 247–253. Academic Press, New York.

Busch, H., and Daskal, Y. (1977). *Methods Cell Biol.* **16,** 1–44.

Carroll, A. G., and Ozaki, H. (1979). *Exp. Cell Res.* **119,** 307–315.

Chambers, S. A. M., and Shaw, B. R. (1984). *J. Biol. Chem.* **259,** 13458–13463.

Clark, J. M. (1936). *Biol. Bull.* **70,** 361–384.
Davie, J. R. (1982). *Anal. Bioch.* **120,** 276–281.
DeLange, R. J. (1978). *Methods Cell Biol.* **18,** .69–188.
Fry, B. J., and Gross, P. R. (1970). *Dev. Biol.* **21,** 125–146.
Green, G. R., and Poccia, D. (1985). *Dev. Biol.,* **208,** 235–245.
Green, G. R., Poccia, D., and Herlands, L. (1982). *Anal. Biochem.* **123,** 66–73.
Gurdon, J. B., and Woodland, H. R. (1968). *Biol. Rev.* **43,** 233–267.
Hardison, R., and Chalkley, R. (1978). *Methods Cell Biol.* **17,** 235–252.
Harris, H. (1974). "Nucleus and Cytoplasm." Clarendon, Oxford.
Henderson, L. E., Oroszlan, S., and Konigsberg, W. (1979). *Anal. Biochem.* **93,** 153–157.
Hinegardner, R. T. (1962). *J. Cell Biol.* **15,** 503–508.
Hinegardner, R. T. (1967). *In* "Methods in Developmental Biology" (F. H. Wilt and N. K. Wessells, eds.), pp. 139–155. Cromwell-Collier, New York.
Hinegardner, R. T., Rao, B., and Feldman, D. E. (1964). *Exp. Cell Res.* **36,** 53–61.
Hoffmann, P. J., and Chalkley, R. (1976). *Anal. Biochem.* **76,** 539–546.
Hogan, B., and Gross, P. R. (1972). *Exp. Cell Res.* **72,** 101–114.
Hsiang, M. W., and Cole, R. D. (1978). *Methods Cell Biol.* **18,** 189–228.
Humason, G. L. (1967). "Animal Tissue Techniques." Freeman, San Francisco.
Hunkapiller, M. W., Strickler, J. E., and Wilson, K. J. (1984). *Science* **226,** 304–311.
Ikegami, S., Taguchi, T., and Ohashi, M. (1978). *Nature (London)* **275,** 458–460.
Johnson, R. T., and Rao, P. T. (1971). *Biol. Rev.* **46,** 97–155.
Katula, K. S. (1983). *Dev. Biol.* **98,** 15–27.
Keichline, L. D., and Wasserman, P. M. (1979). *Biochemistry* **18,** 214–219.
Krystal, G. W., and Poccia, D. (1979). *Exp. Cell Res.* **123,** 207–219.
Krystal, G. W., and Poccia, D. (1981). *Exp. Cell Res.* **134,** 41–48.
Kuhn, O., and Wilt, F. H. (1980). *Anal. Biochem.* **105,** 274–280.
Kuhn, O., and Wilt, F. H. (1980). *Anal. Biochem.* **105,** 274–280.
Laemlli, U. (1970). *Nature (London)* **227,** 680–685.
Loeb, L. A., Fansler, B., Williams, R., and Mazia, D. (1969). *Exp. Cell Res.* **57,** 298–304.
Longo, F. J., and Anderson, E. (1969). *J. Cell Biol.* **46,** 308–325.
Longo, F. J., and Anderson, E. (1970). *J. Cell Biol.* **47,** 646–665.
Maller, J., Poccia, D., Nishioka, D., Kidd, P., Gerhart, J., and Hartman, H. (1976). *Exp. Cell Res.* **99,** 285–294.
Marushige, K., and Ozaki, H. (1967). *Dev. Biol.* **16,** 474–488.
Marushige, Y., and Marushige, K. (1978). *Methods Cell Biol.* **17,** 59–74.
Mazia, D. (1974). *Proc. Natl. Acad. Sci. U.S.A.* **71,** 690–693.
Mazia, D., Petzelt, C. H. R., Williams, R. O., and Meza, I. A. (1972). *Exp. Cell Res.* **70,** 325–332.
Mizuno, S., Whiteley, H. R., and Whiteley, A. H. (1973). *Differentiation* **1,** 339–347.
Nikodem, V., and Fresco, J. R. (1979). *Anal. Biochem.* **103,** 382–386.
Nishioka, D., and Mazia, D. (1977). *Cell Biol. Int. Rep.* **1,** 23–30.
Nishioka, D., Killian, C. E., Chacon, C. T., and Sgagias, M. K. (1984). *J. Cell. Physiol.* **118,** 27–33.
Nuccitelli, R., and Grey, R. D. (1984). *Dev. Biol.* **103,** 1–17.
O'Farrell, P. Z., and O'Farrell, P. H. (1977). *Methods Cell Biol.* **16,** 407–420.
O'Farrell, P. Z., Goodman, H. M., and O'Farrell, P. H. (1977). *Cell* **12,** 1133–1142.
Ord, M. G., and Stocken, L. A. (1978). *Cell Differ.* **7,** 271–281.
Panyim, S., and Chalkley, R. (1969). *Biochemistry* **10,** 3972–3979.
Poccia, D., Krystal, G., Nishioka, D., and Salik, J. (1978a). *ICN-UCLA Symp. Mol. Cell. Biol.:Cell Reprod.* pp. 197–206.
Poccia, D., LeVine, D., and Wang, J. C. (1978b). *Dev. Biol.* **64,** 273–283.

Poccia, D., Salik, J., and Krystal, G. (1981). *Dev. Biol.* **82,** 287–296.
Poccia, D. L., and Hinegardner, R. T. (1975). *Dev. Biol.* **45,** 81–89.
Poccia, D. L., Greenough, T., Green, G. R., Nash, E., Erickson, J., and Gibbs, M. (1984). *Dev. Biol.* **104,** 274–286.
Poccia, D., Wolff, R., Kragh, S., and Williamson, P. (1985). *Biochim. Biophys. Acta* **824,** 349–356.
Richards, R. G., and Shaw, B. R. (1982). *Anal. Biochem.* **121,** 69–82.
Salik, J., Herlands, L., Hoffmann, H. P., and Poccia, D. (1981). *J. Cell Biol.* **90,** 385–395.
Salmon, E. D., and Jenkins, R. J. (1977). *J. Cell Biol.* **75,** 295a.
Savić, A., and Poccia, D. (1978). *Anal. Biochem.* **88,** 573–579.
Savić, A., Richman, P., Williamson, P., and Poccia, D. (1981). *Proc. Natl. Acad. Sci. U.S.A.* **78,** 3706–3710.
Seale, R. L., and Aronson, A. I. (1973). *J. Mol. Biol.* **75,** 633–645.
Shaw, B. R., Cognetti, G., Sholes, W. M., and Richards, R. G. (1981). *Biochemistry* **20,** 4971–4978.
Showman, R. M., and Foerder, C. A. (1979). *Exp. Cell Res.* **120,** 253–256.
Smuckler, E. A., Koplitz, M., and Smuckler, D. E. (1976). *In,* ''Subnuclear Components: Preparation and Fractionation'' (G. D. Birnie, ed.), pp. 1–58. Butterworths, London.
Spadafora, C., Bellard, M., Compton, J. L., and Chambon, P. (1976). *FEBS Lett.* **69,** 281–285.
Thaler, M. M., Cox, M. C. L., and Villee, C. A. (1969). *J. Cell Biol.* **45,** 846–850.
Towbin, H., Staehelin, T., and Gordon, J. (1979). *Proc. Natl. Acad. Sci. U.S.A.* **76,** 4350–4354.
Vacquier, V. D., and Brandriff, B. (1975). *Dev. Biol.* **47,** 12–31.
West, M. H. P., and Bonner, W. M. (1980). *Biochemistry* **19,** 3238–3245.
Whitaker, M. J., and Steinhardt, R. A. (1982). *Q. Rev. Biophys.* **15,** 593–666.
Zalenskaya, I. A., Pospelov, V. A., Zalensky, A. O., and Vorob'ev, V. I. (1981). *Nucleic Acids Res.* **9,** 473–486.
Zweidler, A. (1978). *Methods Cell Biol.* **17,** 223–233.

# Chapter 10

## *The Protein Synthetic Machinery: Ribosomes and Cell-Free Systems*

MERRILL B. HILLE

*Department of Zoology*
*University of Washington*
*Seattle, Washington*

MICHAEL V. DANILCHIK

*Department of Molecular Biology*
*University of California*
*Berkeley, California*



## I. Introduction

Echinoid eggs and asteroid oocytes have very low rates of protein synthesis although both mRNA and ribosomes are stored in these eggs and oocytes (echinoids: Spieth and Whiteley, 1981; Epel, 1978; Humphreys, 1971; Infante and

Nemer, 1967; Goustin and Wilt, 1981; asteroids: Houk and Epel, 1974; Johnson and Epel, 1982). After fertilization of sea urchin eggs the numbers of ribosomes in polyribosomes increase 25-fold before the first cleavage without the synthesis of new mRNAs (Humphreys, 1971; Infante and Nemer, 1967; Goustin and Wilt, 1981). Protein synthesis also increases when anucleate halves of sea urchin eggs are parthenogenetically activated (Denny and Tyler, 1964). Today there is still a controversy over whether translational control in eggs is due to masking of mRNAs, inhibition of ribosomes, or a deficiency in discriminatory and/or initiation factors.

Monroy and co-workers were the first to suggest that ribosomes regulate the rates of protein synthesis in sea urchin eggs (Monroy *et al.*, 1965). In fact, monoribosomes from eggs were found to contain tightly bound inhibitory proteins (Metafora *et al.*, 1971; Gambino *et al.*, 1973; Hille, 1974), and subsequently were found to be less active than ribosomes derived from the polyribosomes of embryos (Danilchik and Hille, 1981). Others have observed, however, that ribosomes from unfertilized eggs and embryos do not differ significantly in their relative translational activities (Stavy and Gross, 1967, 1969; Clegg and Denny, 1974; Ilan and Ilan, 1978). They interpreted their results as evidence that the block to translation could not be at the level of ribosomes. Their conclusion was based solely on the comparison of monoribosomes from unfertilized eggs with monoribosomes from embryos. One might expect, however, that if translational rates were regulated by ribosomes, the most active ribosomes would be found in the polyribosomes. Thus, the question of whether ribosomes regulate protein synthesis in eggs is currently unresolved.

There are several approaches that can be used in analyzing protein synthesis machinery. *In vivo,* the kinetics of reactions can be monitored; the polyribosome sizes and distribution can be measured; and the cells can be perturbed by changing internal ion concentration with external treatments such as ammonia or acetate that change internal pH, or calcium ionophores that change internal calcium concentrations. These methods have given insight into the ionic events responsible for the activation of protein synthesis in fertilized eggs (Steinhardt and Epel, 1974; Johnson *et al.*, 1976; Grainger *et al.*, 1979; Brandis and Raff, 1979; Winkler *et al.*, 1980). In addition, specific steps in protein synthesis can be blocked *in vivo* by treatment of the cells with membrane-soluble inhibitors, and components such as mRNA can be injected into eggs and zygotes (Colin, this volume). These latter methods have given insights into possible rate-limiting steps in eggs (Hille *et al.*, 1981, 1985). The relationships between the ionic events that stimulate protein synthesis, and the changes in the protein synthetic machinery that allow an increased rate of protein synthesis, however, are not known and must be determined by analyzing the separate and isolated components of the synthetic machinery.

Advantages of using *in vitro* systems are that individual components can be tested, and specific responses to a large variety of inhibitors of protein synthesis that are not membrane soluble can be measured (Vazquez, 1979). Some disadvantages of *in vitro* systems are that the high concentrations in the cytosol are not mimicked, proteases and nucleases are released, and important factors may be lost. Nonetheless, the only way to sort out the functions of the various particles and factors is to test them individually. In heterologous *in vitro* systems isolated components can be evaluated without the complications of other parts of the endogenous cytosol. For instance, sea urchin ribosomes can be tested *in vitro* in reconstituted lysates from rabbit reticulocytes that are depleted of rabbit ribosomes. The total amounts of protein synthesis can be measured in such a system, as well as the dissociation of ribosomes into subunits and their movement onto polyribosomes. Such a system has the potential of revealing any aberrant steps in the activities of the ribosomes.

*In vitro* methods can also be used to analyze other components of the protein synthesis machinery. The roles of ribosome-associated initiation factors in regulating the rates of protein synthesis in echinoid eggs and embryos is being actively studied by Lopo (Lopo *et al.*, 1985a) in homologous systems from sea urchin eggs and embryos.

A goal of *in vitro* studies is to mimic both the repression of protein synthesis in eggs, and the derepression of the inhibited components by using only the discrete molecules and ions involved in the repression and activation processes.

## II. Strategy

Listed below are some of the practical matters to be taken into account in designing *in vitro* cell-free systems. In particular, the rationale for the isolation and testing of ribosomes is described. Similar ideas could be applied to the isolation of other components of the protein synthesis machinery.

1. In any *in vitro* study one of the major concerns is to isolate molecules and particles in their native states. Conditions should be designed to mimic the ionic concentrations, the reduced state, and the osmolality of the cytoplasm. The ionic concentrations of sea urchin eggs have been carefully measured by Rothschild and Barnes (1953).

When cells are lysed, organelles which ordinarily restrict the activity of degradative enzymes are often lysed, even under the gentlest conditions. These organelles include, besides lysosomes and nuclei, the cortical granules of unfertilized eggs. Hence, homogenization conditions should be designed to limit lysis of organelles and should include, when required, protease and nuclease inhibitors.

2. Monoribosomes constitute the bulk (at least 99%) of the ribosome population in unfertilized eggs (Humphreys, 1971); however, polyribosomes constitute only from 20% (two-cell embryos) to 60% (hatched blastula embryos) of the ribosome population of early embryos of sea urchins. Since polyribosomes

should contain the most active population of ribosomes, if the ribosomes regulate translational rates, any comparative study of ribosomes should include the careful isolation and comparison of the two population of ribosomes (Section III).

3. In the past the ribosomes in polyribosomes have not been extensively studied. This oversight is in part due to the inequity of comparing a population of free monoribosomes with a population of mRNA-bound ribosomes. Moreover, it has been difficult, until recently, to release ribosomes from polyribosomes in their native state. We describe a physiological method for releasing and separating ribosomes from polyribosomes which maintains ribosome activity (Section III,C). Any treatment of polyribosomes must be conducted in parallel with monoribosomes.

4. *In vitro* systems used to test isolated ribosomes must be free of endogenous ribosomes and ribosomal subunits because active ribosomes from the *in vitro* test system could easily out-compete test ribosomes. Ideally, they should also be free of any materials that will activate the ribosomes. An *in vitro* system prepared from reticulocyte lysates is described (Section IV). A complex particle such as a ribosome that is found to have a reduced activity in a translational system could be further tested in partially purified lysates, or with the addition of specific inhibitors of initiation or elongation to determine which of its functions is repressed.

5. The final description of the repression of translation would include the changes that occur in regulatory molecules after fertilization, such as chemical modification and/or their location in the cytoplasm. The isolation of a protein fraction from monoribosomes that is inhibitory to protein synthesis is described (Section V), although nothing is yet known of its mechanism of repression or how the inhibitory protein(s) might be modified after fertilization. Methods for localizing RNAs or factors that might be physically segregated in the cytoplasm are not discussed here, but are available (Capco and Jeffery, 1981, 1982; Jeffery, 1982; Moon *et al.*, 1983; Jeffery *et al.*, 1983; Angerer *et al.*, 1984; Cox *et al.*, 1984; Yablonka-Reuveni and Hille, 1983; Steinert *et al.*, 1984).

6. Rather than search for a repressed component, another approach is to supplement the egg cytoplasm with various active components. Cell-free lysates from sea urchin eggs and embryos can be supplemented with factors from reticulocyte lysates. The preparation of cell-free lysates is briefly reviewed in Section VI. The injection of material directly into the cytoplasm of eggs is described in this volume by Colin and by Kishimoto.

## III. The Isolation of Ribosomes and Polyribosomes

### A. Solutions

1. Precautions: Ionic solutions are filtered through nitrocellulose disks with 0.45-μm pores that have been prerinsed with glass-distilled water to remove

detergents. Glassware is baked at 160°C for 5 hours to destroy residual ribonuclease. Plasticware, if not new and sterile, may need to be washed in 0.1% SDS at 60°C, and rinsed in glass-distilled water.

2. Seawater: All seawater is filtered through nitrocellulose disks with 0.45-μm pores. Streptomycin sulfate (100 mg/ml) is added and the seawater is aerated by rapid pouring.

3. Stop buffer: 100 m*M* potassium acetate, 80 m*M* KCl, 9.9 m*M* $K_2HPO_4$, 101. m*M* $KH_2PO_4$, 5 m*M* magnesium acetate, pH 6.8, has a final concentration of 210 m*M* $K^+$. 2× stop buffer: stop buffer at double strength.

4. Homogenizing medium: Stop buffer plus 1 mg/ml soybean trypsin inhibitor and 1 m*M* dithiothreitol.

*Special notes:* Dithiothreitol (1 mg/ml) is required to preserve the native state of the monoribosomes. Some researchers use reduced glutathione (0.5 mg/ml) instead (Winkler and Steinhardt, 1981).

Two effective protease inhibitors are 1 m*M* EGTA and 1 mg/ml soybean trypsin inhibitor (Type I-S, Sigma, No. T-9003) (Yablonka-Reuveni and Hille, 1983). A combination of several protease inhibitors can be used: Aprotinin (Sigma), leupeptin, benzamidine, and *p*-tosyl-L-arginine methyl ester (Lopo *et al.*, 1985b). EGTA may be effective in inhibiting nucleases. Bentonite is not recommended as a nuclease inhibitor since it absorbs ribosomes in proportion to its concentration (Hille, unpublished). RNasin ribonuclease inhibitor (Promega Biotec) is effective but has not been used in the isolation of ribosomes. Heparin is toxic to protein synthesis but can be used to prepare mRNAs that are precipitated free of heparin before translation.

5. Sucrose gradient solutions: Sucrose solutions are prepared according to the International Critical Tables as percentage sucrose by weight to total solution weight (w/w). Fifty milliliters of 2× stop buffer and the requisite amount of sucrose are mixed, then the solution is brought to 100 ml. For 20, 25, and 30% sucrose (w/w), 21.7, 27.6, and 33.9 g sucrose are added, respectively. Just before preparing the gradients, dithiothreitol and soybean trypsin inhibitors are added to the cold sucrose solution to final concentrations of 1 m*M* and 1 mg/ml, respectively.

## B. Preparation of Monoribosomes and Polyribosomes from Eggs and Embryos

Since ribosomes are prepared in parallel from the eggs and embryos, the methods are similar and interwoven in our discussion.

### 1. Preparation of Metabolically Active Eggs and Embryos

Gametes are obtained by intracoelomic injection of 0.55 *M* KCl, or by electrical shock (30 to 50 V) with electrodes placed in the seawater on opposite sides of a small shallow dish containing the adults. If embryos are to be prepared, one-half of the eggs from each spawned sea urchin are fertilized, and incubated as a 1% culture at 12 to 15°C until the desired stage, and then the embryos are rapidly concentrated for the subsequent procedure. Continuous flow centrifugation is used for the rapid concentration of cells.

The unfertilized eggs and embryos are prepared for homogenization in a manner designed for isolation of the polyribosomes at their maximum *in vivo* size. Embryos that are concentrated by settling or by centrifugation at the end of a growth period, must be reincubated to aerate the cells in order to allow reformation of polyribosomes. If the same batch of eggs is used for unfertilized egg and embryo preparations, the eggs may need to be treated before the embryos to avoid prolonged storage of the eggs. The eggs and embryos of the desire stage, which have been concentrated, are aerated by tumbling as 5–10% suspensions in Erlenmeyer flasks that are large enough to provide generous surface-to-volume ratios; or by any other vigorous means that allows adequate aeration. [Bubbling with $O_2$ is not recommended since it drives off $CO_2$ and activates eggs by raising the pH of the cytoplasm (MacKintosh and Bell, 1969)]. After 30 to 60 minutes incubation with adequate aeration, the polyribosomes are reformed to a maximum extent.

### 2. Homogenization of Eggs and Embryos

The egg and embryo cultures are arrested in 15 to 20 vol of stop buffer at −2°C. The solution should not go above −1°C during the procedure. Rapid cooling of cells arrests the polyribosomes of embryos at maximum sizes. Slow cooling results in smaller polyribosomes, presumably because the initiation of protein synthesis is more sensitive to the metabolic depletion of GTP than elongation. Eggs and embryos are collected from the stop buffer by settling at 0°C or by centrifugation at 400 $g_{max}$ and 0°C. Some eggs are fragile after long periods in stop buffer and thus should be rapidly collected and homogenized, or collected in stop buffer containing 0.5 *M* dextrose. The cells are washed in a small volume of stop buffer and homogenized in 1–5 cell volumes of homogenizing medium by about three forced passes through a 20-gauge needle. [Homogenization with a syringe decreases breakage of nuclei (Fromson and Nemer, 1970; Dworkin and Infante, 1976).] The homogenate is centrifuged twice for 15 minutes at 14,500 $g_{max}$ to remove membranes yolk and mitochondria. The upper three-quarters of the supernatant is saved.

### 3. Preparation of Monoribosomes and Polyribosomes

Ribosomes are prepared basically as described by Danilchik and Hille (1981). Monoribosomes are conveniently collected as a discrete 80 S fraction separate from polyribosomes on preparative step-gradients. Under these conditions ribosomes can be collected as a pellet.

Ten-milliliter aliquots of the postmitochondrial supernatants are layered on step gradients designed for the Beckman SW 27 rotor, containing 8 ml of 30% sucrose (w/w), 6 ml of 25% sucrose, and 5 ml of 20% sucrose. The homogenate is overlayered with 8.5 ml of the homogenizing medium and centrifuged for 4 hours at 92,000 $g_{max}$ to float the monoribosomes at the 30/25% interface and to pellet the polyribosomes. The centrifugation conditions can be varied for convenience if the product of the force and time is kept constant. Very high centrifugal forces may, however, dissociate protein from the ribosomes (Baierlein and Infante, 1974). The final position of the monoribosomes can be seen as a cloudy layer and withdrawn by a syringe or monitored as absorbance at 260 nm. Floated monoribosomes are pelleted by centrifugation for 12 hours at 54,000 $g_{max}$ (or 20,000 $g_{max}$ for 20 hours) after dilution of the sucrose with homogenization medium to about 10% sucrose. Polyribosomes and monoribosome pellets are dissolved in homogenizing medium at a concentration of 50–200 $A_{260nm}$/ml. (One $A_{260nm}$/ml is defined as that material which gives 1 $A_{260nm}$ in a 1-cm path length.) The dissolved pellets are cleared by centrifugation at 9000 $g_{max}$ for 10 minutes and stored in liquid nitrogen or at −80°C.

Monoribosomes from eggs, which are used for preparation of ribosome-associated proteins (Section V), can be prepared in quantity by pelleting them from the postribosomal supernatant. The supernatant is layered over 2.5 ml of 15% sucrose and 15 ml of 10% sucrose in homogenizing buffer in 38 ml SW 27 Beckman tubes and centrifuged overnight at 100,000 $g_{max}$.

## C. Isolation of Ribosomes from Polyribosomes

To compare the activities and structures of ribosomes from polyribosomes with monoribosomes it is first necessary to release the active ribosomes from the polysomal mRNA by nondenaturing means. This release is accomplished by incubating the polyribosomes in a ribosome-free lysate that has no chain initiation activity. This method is preferred over EDTA or puromycin dissociation of ribosomes, which either decrease the size of ribosomes or introduce inhibitors. The factors for peptide elongation and release of the ribosomes are supplied by a postribosomal supernatant of rabbit reticulocytes (Section IV,A). Initiation of new peptide synthesis is retarded by excluding hemin from the lysing medium. The absence of ribosomes from the reticulocyte lysate is checked by analytical sucrose gradient centrifugation. Conditions during incubation for "runoff" of ribosomes from polysomes are 20% by volume ribosome-free lysate, 170 m*M* KCl, 1.5 m*M* magnesium acetate, 15 m*M* creatine phosphate, 1 m*M* ATP, 0.3 m*M* GTP, 22 IU/ml creatine phosphokinase, 10 μ*M* amino acids, 10 m*M* Tris-HCl, pH 6.8 to 7.2. Both blastula polysome and egg monoribosomes are incubated in the initiation-deficient lysate at a concentration of 10–50 $A_{260nm}$ units/ml for 1 hour at 26°C for 1 hour. The mixture is cooled to 0°C and 10-ml aliquots are layered on sucrose step gradients and centrifuged to float the monoribosomes as described above. The 80 S fractions are concentrated by centrifugation, suspended in homogenizing medium, and frozen as in Section III,B.

Complete disintegration of the polyribosomes can be tested by incubation of the ribosomes in the heterologous lysate (Section IV,B,1) in the presence of the initiation inhibitor aurintricarboxylic acid. The resulting runoff ribosomes from morula and blastula embryos, prepared in the above manner, are found to require new initiation for translation, indicating that they are essentially free of polyribosomal mRNA (Danilchik and Hille, 1981).

## IV. A Ribosome-Free Translation System for Testing Echinoid Ribosomes

### A. Preparation of Ribosome-Free Lysates

Postribosomal supernatants for rabbit reticulocyte lysates that are supplemented with initiation factors and mRNA are effectively used to test ribosomal activities. Reticulocytes are obtained from anemic New Zealand White rabbits according to the methods of Gilbert and Anderson (1970) and Crystal *et al.* (1974) with the following modifications. The reticulocytes are washed the last time in 135 m$M$ KCl, 7.5 m$M$ magnesium acetate, pH 7.0, at 0°C to remove $Na^{+}$, which is inhibitory; and then lysed in 1.5 weight equivalents of reticulocyte lysing buffer. Lysing buffer consists of 1.67 m$M$ dithiothreitol (added just before use) and 33 $\mu M$ hemin. For preparing hemin stock see Jackson and Hunt (1983).

Ribosome-free reticulocyte lysates are prepared by centrifugation of the lysates at 80,000 $g_{max}$ for 4 hours at 0°C. The upper three-fourths of the ribosome-free supernatant is stored in aliquots in liquid nitrogen or at −80°C. This ribosome-free lysate is also depleted in initiation factors and mRNAs. Therefore, the ribosomal pellet is washed with 0.5 $M$ KCl to obtain crude initiation factors as described by Crystal *et al.* (1974). The factor preparation is partially purified by ammonium sulfate precipitation and dialysis as described by Schreier and Staehelin (1973). In addition, globin mRNA is obtained from ethanol precipitations of the lower quarter of the ribosome-free lysate, or from ribosomal pellets by extraction with phenol–chloroform using the method of Palmiter (1974). The ethanol-precipitated lysates give the greatest yields if they are centrifuged from the ethanol, dried, dissolved in pH 7.4 buffer and 0.7% SDS, then digested at 37°C for 30 minutes with 0.6 mg/ml Pronase (Calbiochem) or with proteinase K before phenol extraction. The Pronase must be previously self-digested at a concentration of 10 mg/ml for 1 hour at 35°C to destroy ribonucleases. Poly(A)-containing mRNA is purified from the total reticulocyte RNA by oligo(dT)-cellulose chromatography using the method of Aviv and Leder (1972). Other methods of mRNA purification are discussed by Lizardi (1983).

## B. The Heterologous Reticulocyte-Sea Urchin Cell-Free Translation System

### 1. Kinetic Analysis of Ribosomal Activity

The activity of sea urchin ribosomes can be measured in ribosome-free reticulocyte lysates that are supplemented with poly($A^+$)-mRNA and with initiation factor-rich KCl washes from reticulocyte ribosomes. In this system only the ribosomes are of sea urchin origin, thus any differences in activities can be ascribed to differences in the ribosomes themselves or to factors that are isolated with the ribosomes.

Typically a 30- to 50-μl reaction mixture containing sea urchin ribosomes, reticulocyte cytosol minus ribosomes, and an ATP/GTP regenerating system is incubated at 15–26°C. Five-microliter samples are removed at 2- to 5-minute intervals. Final concentrations in the incubation mixture are 80 m*M* KCl, 100 m*M* potassium gluconate, 15 m*M* creatine phosphate, 1 m*M* ATP, 0.3 m*M* GTP, 22 IU/ml creatine phosphokinase, 10 μ*M* amino acids minus the labeled amino acid, L-[$^3$H]leucine, L-[$^3$H]valine, or an L-[$^3$H]amino acid mixture (ICN, Biomedicals, Inc., Irvine, CA), 10 m*M* Tris-HCl, 0.5–1.5 m*M* magnesium acetate, 3 mg/ml initiation factor preparation, and 40 μg/ml globin mRNA, 20% by volume ribosome-free reticulocyte lysate, and sea urchin ribosomes. Addition is in the order indicated. All components are kept on ice. Final salt concentrations must take into account any contribution from the addition of initiation factors and ribosomes. Optimal conditions in our hands are 60–80 m*M* $Cl^-$, 150–180 m*M* $K^+$, and pH 7.2 and 7.4. The pH of the reaction is measured at the end of the each incubation with an MI-410 Microcombination pH probe (Microelectrodes, Inc., Londonderry, NH). This electrode accurately measures 2- to 5-μl samples. Mock incubation mixtures are typically prepared before the assays to determine the amount of KOH required to give the desired pH. The $Mg^{2+}$ optimum depends on the particular lysate used and on the amounts of mRNA and ribosomes added and must, therefore, be determined empirically.

At specific intervals during the incubation of the samples, 5-μl aliquots are withdrawn, spotted on one end of a strip of filter paper cut as 7 × 30 mm rectangles from sheets of Whatman 3MM paper. The paper is allowed to dry, then placed into ice-cold 10% (w/w) trichloroacetic acid. Samples are then rapidly and simultaneously plunged into 7.5% trichloroacetic acid at 85–90°C for 3 minutes in a porous basket made from the bottom of a polyethylene bottle. Although the hot trichloroacetic acid treatment is intended to hydrolyze [$^3$H]aminoacyl-tRNAs, small peptides can also be hydrolyzed with extended treatment. Thus, samples must be simultaneously and identically treated. After the treatment with hot trichloroacetic acid the samples are washed successively in 7.5% trichloroacetic acid, 95% ethanol (twice), 50% ethanol:50% ether, and 100% ether. The dried samples are solubilized from the filter papers in 4.0 ml Bio-Vials (Beckman) with 50 μl of water and 350 μl of NCS tissue solubilizer (Amersham) before counting in 2.5 ml of toluene-based PPO/POPOP scintillant.

### 2. Sucrose Gradient Analysis of Ribosomal Subunit and Polyribosome Formation

Analysis of the formation of ribosomal subunits and of polyribosomes gives additional information of the activity deficiencies of ribosomes. Such information is readily obtained during the kinetic analysis of ribosome activity described above (Section IV,B,1). Aliquots of the reaction mixture, generally 50 μl, are rapidly chilled at specific intervals by pipetting into 6 vol of −2°C buffer (100 m*M* KCl, 5 m*M* magnesium acetate, 20 m*M* Tris-HCl, pH 7) and 100 μ*M* emetine. The samples are then layered on 4.9-ml linear sucrose gradients [15–50% (w/w)] made up in the above buffer. The samples are overlayered with 250 μl of the buffer, then centrifuged on a Beckman SW 50.1 rotor. Centrifugation at 233,000 $g_{av}$ at 0°C for 2.5 hours is appropriate for polyribosome analysis; for analysis of ribosomal subunits, the gradients are spun longer. Profiles of the subunits and polyribosomes are obtained by pumping from the bottom with 60% sucrose and monitoring for RNA absorbance at 260 nm with a Gilford spectrophotometer. The baseline absorption is determined with gradients overlayered with 600 μl of the above buffer. The areas under the various peaks can be integrated.

### 3. Determination of Elongation Rates

The *in vitro* elongation rates of the ribosomes can be measured by a modification of the methods of Fan and Penman (1970). Fractions from the gradients described above (Section IV,B,2) are precipitated with trichloroacetic acid, trapped on glass fiber filters (Gelman), and analyzed for tritium to determine the incorporation of the [$^3$H]amino acids into nascent globin chains in the polyribosome region and into released globin chains, which remain in the supernatant at the top of the gradient. The transit times are calculated according to the method of Fan and Penman (1970).

## V. Isolation of Inhibitory Proteins

Monoribosomes from sea urchin eggs contain a tightly bound protein fraction which is inhibitory to peptide synthesis (Metafora *et al.*, 1971; Gambino *et al.*, 1973; Hille, 1974). In contrast, polyribosome populations do not contain extractable inhibitory proteins. These inhibitors are not removed with 0.7 *M* chloride salts, which are sufficient to remove most initiation factors and other loosely bound proteins; rather they elute with 1.4 *M* chloride salts.

To remove loosely associated proteins, monoribosomes (about 100 $A_{260\ nm}$) are suspended in 1–2 ml of a salt solution with a final concentration of 0.7 *M* KCl, 18 m*M* magnesium acetate, 40 m*M* buffer, pH 7.0–7.5, and 1 m*M* dithiothreitol. The suspension is mixed gently for 4 hours at 0°C, and then pelleted by centrifugation at 100,000 $g_{max}$ and 2°C for 5 hours. To elute the inhibitory proteins, the pellet monoribosomes are resuspended in 1–2 ml of 1.0 *M* $NH_4Cl$, 0.4 *M* KCl, 18 m*M* magnesium acetate, 50 m*M* buffer, pH 7.0–7.5, and 1 m*M* dithiothreitol. These ribosomes are mixed as above and centrifuged for 5 hours to overnight. The resultant upper two-thirds of the postribosomal supernatant, containing the inhibitory activity, is dialyzed against 0.1 m*M* EDTA, 50 m*M* Tris-HCl, pH 7.0, 1 m*M* dithiothreitol. It may also require concentration by negative-pressure dialysis (e.g., in a Micro-ProDiCon apparatus, Bio-Molecular Dynamics, Beaverton, OR) or by any other convenient method (Metafora *et al.*, 1971). The dialyzed supernatant is cleared of the insoluble precipitated protein by low-speed centrifugation (15 minutes at 10,000 *g*). The final clear supernatant contains the inhibitory protein(s) and may be frozen, preferably in liquid nitrogen.

The inhibitory activity of these low salt-soluble proteins can be tested in any complete *in vitro* protein synthesis system. The rat liver system is described in Hille (1974). Recently we have used whole lysates from rabbit reticulocytes (Section IV,A). Incubation conditions are similar to those described in Section IV,B except that a 40% by volume of whole lysate substitutes for initiation factor, mRNA, and ribosome-free lysate additions. The volume of lysate added to the assay is adjusted to give about 7.5 $A_{260\ nm}$ of ribosomes per milliliter in the final assay solution. An equivalent of 1.4 *M*-eluted proteins from three times the assay concentration of ribosomes is nearly 100% inhibitory. That is if there are 0.23 $A_{260\ nm}$ of ribosomes in a 50-μl assay, then the proteins eluted from 0.7 $A_{260\ nm}$ of ribosomes, as described in this section, are strongly inhibitory.

## VI. Homologous Cell-Free Echinoid Translational Systems

Homologous cell-free systems have the advantage that deficiencies in protein synthesis can be analyzed by supplementing the cell-free lysates with factors derived from other cells (Lopo *et al.*, 1985c). Lesions in the translational machinery can also be tested for by assaying for specific translation steps (Winkler *et al.*, 1985). In addition, a spectrum of factors can also be isolated from the embryo lysates and their effects on the egg lysates can be monitored.

A superior sea urchin translational system was developed in 1981 by Winkler using *Lytechinus pictus* eggs (Winkler and Steinhardt, 1981). We will not reproduce his methods here, but we will comment on points we have found critical for the preparation of active lysates. The homogenization medium should contain inhibitors of proteases and a reducing agent such as glutathione or dithiothreitol. Soybean trypsin inhibitor and several other protease inhibitors are satisfactory. The homogenization medium should contain, in addition, 0.3 *M* glycine, 150–180 m*M* $K^+$ (or a combination of $K^+$ and $Na^+$), and no more than 80 m*M* $Cl^-$. EGTA-buffered calcium is not required, but EGTA is. Glycerol is not essential. Dialysis is not necessary since the pH of the lysate can be adjusted with KOH and measured with a microelectrode (see Section IV,B). For accurate comparisons of two lysates, the specific activity of the radiolabeled amino acid should be known; hence, the concentrations of the appropriate endogenous amino acid should be determined for each lysate. A simple method for determining endogenous amino acid concentrations is to find the concentration of cold amino acids required to reduce the rate of incorporation into proteins by half. The concentration of added amino acids should at this point equal that of the endogenous amino acids plus that contributed by the radioactive label. A final note on testing lysates is that several proteins such as globin and histones contain very few methionine residues; therefore, [$^3$H]valine, [$^3$H]leucine, or a mixture of [$^3$H]amino acids is preferred for analysis of the lysates.

Lopo *et al.* (1985b,c) have recently developed a lysate system from *Strongylocentrotus purpuratus* eggs and embryos. This lysate requires, in addition to several protease inhibitors, high concentrations of RNasin, a ribonuclease inhibitor (Promega Biotec, Madison, WI). Lopo *et al.* (1985a) discuss the methods for isolating echinoid initiation factors from echinoids as compared to methods for mammalian or bacterial factors.

## VII. Concluding Remarks

The block in the translation of mRNA in sea urchin eggs remains a mystery. We have presented several techniques that will hopefully be useful in solving the problem. The emphasis of the techniques is on methods of isolating and analyzing ribosomes and inhibitory proteins associated with monoribosomes. Methods of isolating and analyzing initiation factors are described by Lopo *et al.* (1985a). Several pieces of evidence suggest that mRNA is not rate limiting in eggs (Moon *et al.*, 1982; Hille *et al.*, 1985; Colin and Hille, 1986) or zygotes (Hille *et al.*, 1985; Colin and Hille, 1986). A combination of studies including the use of heterologous cell-free systems with purified sea urchin components, the use of homologous sea urchin cell-free systems, and the purification of molecules and components should yield useful information. Finally, comparative studies with oocytes and eggs from other echinoderms will tell us whether the regulatory mechanisms evolved by sea urchins exist in other organisms.

## Acknowledgments

This research was generously supported by the National Institutes of Health. A general medical sciences grant, HD 11070, to M.B.H. and training grant support, GM 07270, to M.V.D., are gratefully acknowledged. We wish to thank Anita Colin, Allen Lewis, Kathy Moore, and Peter Mullen who have contributed to the improvements of these methods.

## References

Angerer, L. M., DeLeon, D. V., Angerer, R. C., Showman, R. M., Wells, D. E., and Raff, R. A. (1984). *Dev. Biol.* **101,** 477–484.

Aviv, H., and Leder, P. (1972). *Proc. Natl. Acad. Sci. U.S.A.* **69,** 1408–1412.

Baierlein, R., and Infante, A. A. (1974). *In* "Methods in Enzymology" (K. Moldave and L. Grossman, eds.), Vol. 30, part F, pp. 328–345. Academic Press, New York.

Brandis, J. W., and Raff, R. A. (1979). *Nature (London)* **278,** 467–468.

Capco, D. C., and Jeffery, W. R. (1981). *Nature (London)* **294,** 255–257.

Capco, D. C., and Jeffery, W. R. (1982). *Dev. Biol.* **89,** 1–12.

Clegg, K. B., and Denny, P. C. (1974). *Dev. Biol.* **37,** 263–272.

Colin, A. M., and Hille, M. B. (1986). *Dev. Biol.*, in press.

Cox, K. H., DeLeon, D. V., Angerer, L. M., and Angerer, R. C. (1984). *Dev. Biol.* **101,** 485–502.

Crystal, R. G., Elson, N. A., and Anderson, W. F. (1974). *In* "Methods in Enzymology" (K. Moldave and L. Grossman, eds.), Vol. 30, Part F, pp. 101–127. Academic Press, New York.

Danilchik, M. V., and Hille, M. B. (1981). *Dev. Biol.* **84,** 291–298.

Denny, P. C., and Tyler, A. (1964). *Biochem. Biophys. Res. Commun.* **14,** 245–249.

Dworkin, M. B., and Infante, A. A. (1976). *Dev. Biol.* **53,** 73–90.

Epel, D. (1978). *Curr. Top. Dev. Biol.* **12,** 185–246.

Fan, H., and Penman, S. (1970). *J. Mol. Biol.* **50,** 655–670.

Fromson, D., and Nemer, M. (1970). *Science* **168,** 266–267.

Gambino, R., Metafora, S., Felicetti, L., and Raisman, J. (1973). *Biochim. Biophys. Acta* **312,** 377–391.

Gilbert, J. M., and Anderson, W. F. (1970). *J. Biol. Chem.* **245,** 2342–2349.

Goustin, A. S., and Wilt, F. A. (1981). *Dev. Biol.* **82,** 32–40.

Grainger, J. L., Winkler, M. M., Shen, S. S., and Steinhardt, R. A., (1979). *Dev. Biol.* **68,** 396–406.

Hille, M. B. (1974). *Nature (London)* **278,** 469–471.

Hille, M. B., Hall, D. C., Yablonka-Reuveni, Z., Danilchik, M. V., and Moon, R. T. (1981). *Dev. Biol.* **86,** 241–249.

Hille, M. B., Danilchik, M. V., Colin, A., and Moon, R. T. (1985). *In* "The Cellular and Molecular Biology of Invertebrate Development" Bell W. Baruch Series in Marine Biology (R. H. Sawyer and R. M. Showman, eds.), Vol. 15, in press. University of South Carolina Press, Columbia, South Carolina.

Houk, M., and Epel, D. (1974). *Dev. Biol.* **40,** 298–309.

Humphreys, T. (1971). *Dev. Biol.* **26,** 201–208.

Jackson, R. J., and Hunt, T. (1983). *In* "Methods in Enzymology" (S. Fleischer and B. Fleischer, eds.), Vol. 96, pp. 51–75. Academic Press, New York.

Ilan, J., and Ilan, J. (1978). *Dev. Biol.* **66,** 375–385.

Infante, A. A., and Nemer, M. (1967). *Proc. Natl. Acad. Sci. U.S.A.* **58,** 681–688.

Jackson, R. J., and Hunt, T. (1974). *In* "Methods in Enzymology," Vol. 96, pp. 50–74. Academic Press, New York.

Jeffery, W. R. (1982). *J. Cell Biol.* **95,** 1–7.

Jeffery, W. R., Tomlinson, C. R., and Brodeur, R. D. (1983). *Dev. Biol.* **99,** 408–417.

Johnson, C. H., and Epel, D. (1982). *Dev. Biol.* **92,** 461–469.

Johnson, J. D., Epel, D., and Paul, M. (1976). *Nature (London)* **262,** 661–664.

Lizardi, P. M. (1983). *In* "Methods in Enzymology" (S. Fleischer and B. Fleischer, eds.), Vol. 96, Part J, pp. 24–38. Academic Press, New York.

Lopo, A. C., Lashbrook, C. C., Infante, A. A., Infante, D., and Hershey, J. W. B. (1985a). *Arch. Biochem. Biophys.*, submitted.

Lopo, A., Lashbrook, C. C., and Hershey, J. W. B. (1985b). *Dev. Biol.*, submitted.

Lopo, A., MacMillan, S., and Hershey, J. W. B. (1985c). *Dev. Biol.*, submitted.

MacKintosh, F. R., and Bell, E. (1969). *J. Mol. Biol.* **41,** 365–380.

Metafora, S., Felicetti, L., and Gambino, R. (1971). *Proc. Natl. Acad. Sci. U.S.A.* **68,** 600–604.

Monroy, A., Maggio, R., and Rinaldi, A. (1965). *Proc. Natl. Acad. Sci. U.S.A.* **54,** 107–111.

Moon, R. T., Danilchik, M. V., and Hille, M. B. (1982). *Dev. Biol.* **93,** 389–403.

Moon, R. T., Nicosia, R. F., Olsen, C., Hille, M. B., and Jeffery, W. R. (1983). *Dev. Biol.* **95,** 447–458.

Palmiter, R. D. (1974). *Biochemistry* **13,** 3606–3615.

Rothschild, L., and Barnes, H. (1953). *J. Exp. Biol.* **30,** 534–544.

Schreier, M. H., and Staehelin, T. (1973). *J. Mol. Biol.* **73,** 329–349.

Spieth, J., and Whiteley, A. H. (1981). *Wilhelm Roux's Arch.* **190,** 111–117.

Stavy, L., and Gross, P. R. (1967). *Proc. Natl. Acad. Sci. U.S.A.* **57,** 735–742.

Stavy, L., and Gross, P. R. (1969). *Biochim. Biophys. Acta* **182,** 193–202.

Steinert, G., Felsani, A., Kettmann, R., and Brachet, J. (1984). *Exp. Cell Res.* **154,** 203–212.

Steinhardt, R., and Epel, D. (1974). *Proc. Natl. Acad. Sci. U.S.A.* **71,** 1915–1919.

Vazquez, D. (1979). "Inhibitors of Protein Synthesis." Springer-Verlag, Berlin.

Winkler, M. M., Steinhardt, R. A., Grainger, J. L., and Manning, L. (1980). *Nature (London)* **287,** 558–560.

Winkler, M. M., and Steinhardt, R. A. (1981). *Dev. Biol.* **84,** 432–439.

Winkler, M. M., Nelson, E. M., Lashbrook, C., and Hershey, J. W. B. (1985). *Dev. Biol.* **107,** 290–300.

Yablonka-Reuveni, Z., and Hille, M. B. (1983). *Biochemistry* **22,** 5205–5212.

# Chapter 11

## *Tubulin-Containing Structures*

KATHY A. SUPRENANT

*Department of Physiology and Cell Biology*
*University of Kansas*
*Lawrence, Kansas*



## I. Introduction

Echinoderm eggs and embryos are appropriate material for detailed studies of microtubule function during mitosis and meiosis in single cells (e.g., eggs and oocytes), as well as microtubule chemistry in different tubulin-containing organelles (e.g., cilia, flagella, and mitotic and meiotic spindles). Fertilization triggers the synchronous development of large mitotic spindles (20 μm in length) that are easily visualized in the light microscope (for reviews see Inoue, 1981; McIntosh, 1981) and isolated in mass quantities. Sea urchin eggs and embryos have been used extensively for classical as well as current studies of cell division (see Harvey, 1956; Inoue, 1981). In addition, microtubules are necessary for the motility of sperm flagella (see Gibbons, 1977, 1981), the cytoplasmic movements of sperm and egg pronuclei during syngamy (Longo and Anderson, 1968; Bestor and Schatten, 1981; Schatten *et al.*, 1982), ciliogenesis at the hatching bastula stage (Auclair and Siegel, 1966; Burns, 1973; Stephens, 1977; Bibring and Baxandall, 1981), as well as the maintenance of cell shape and morphogenetic movements during gastrulation (Gibbins *et al.* 1969; Tilney and Gibbins, 1969). With the ability to isolate and reassemble most of these tubulin-containing structures *in vitro,* as well as the ability to identify the alpha- and beta-tubulin genes (Alexandraki and Ruderman, 1981), it is possible to examine microtubule biology in echinoderm embryos from the genomic to phenotypic level.

The purpose of this article is to describe methods for the purification of echinoderm egg and spindle tubulins, as well as microtubule-associated proteins, mitotic organizing centers, and, briefly, cilia and flagella. I have not attempted to write a comprehensive review of the methodologies and their significance, or to give a complete history of microtubule-related techniques. I present my personal preferences and methods with which I am most familiar. In addition to a detailed description of the methods, I have included a discussion of some of the relevant aspects of the biochemistry of echinoderm tubulin and microtubule reassembly *in vitro*. Hopefully our understanding of mitosis and microtubule function will benefit from promising new research and an ability to isolate and characterize these tubulin-containing structures *in vitro.*[1]

## II. Unfertilized Egg and Oocyte Tubulin

Tubulin is an abundant protein in echinoderm gametes and embryos (Borisy and Taylor, 1967; Bryan, 1972; Raff *et al.*, 1971; Bibring and Baxandall, 1977), comprising as much as 1–5% of the total soluble protein in unfertilized eggs

[1]Since the initial purification (Weisenberg *et al.*, 1968; Weisenberg and Timasheff, 1970) and *in vitro* assembly of vertebrate brain tubulin (Weisenberg, 1972a; Borisy and Olmsted, 1972), significant progress has been made toward understanding the chemistry, structure, and function of the brain tubulin molecule *in vitro*. Several books, review articles, and a staggering number of manuscripts have been written about microtubules (for recent reviews see Raff, 1979, 1984; Cleveland, 1983; Roberts and Hyams, 1979; Dustin, 1978).

(Raff and Kaumeyer, 1973; Pfeffer *et al.*, 1976). This soluble pool of tubulin can be purified from echinoderm eggs or embryos by fractionation on (1) DEAE–Sephadex in phosphate buffers (Kuriyama, 1977; Suprenant and Rebhun 1983, 1984a) or (2) DEAE–Sephacel in PIPES buffers (Detrich and Wilson, 1983). Egg tubulin, in a soluble egg extract, is adsorbed to the DEAE–resin, removed by stepwise salt elution, concentrated, and assembled into microtubules. The assembly properties and tubulin yields are approximately the same with each method. The only notable differences among the preparations are (1) the cost—phosphate buffers and Sephadex resins are much less expensive than chromatographic procedures using PIPES buffers and DEAE–Sephacel, and (2) the more recent preparations (Suprenant and Rebhun, 1983, 1984a; Detrich and Wilson, 1983) reassemble egg microtubules in the absence of glycerol, a solvent known to bind tightly to the tubulin molecule (Detrich *et al.*, 1976).

Using these methods, microtubules have been assembled from tubulin purified from the sea urchins *Hemicentrotus pulcherrimus, Anthocidaris crassispina* (Kuriyama, 1977), *Strongylocentrotus purpuratus* (Suprenant and Rebhun, 1983; Detrich and Wilson, 1983), the starfish, *Asterias amurensis* (Kuriyama, 1977), *Asterias vulgaris* (Figs. 6 and 7, this article), and the sand dollar, *Echinarachnius parma* (Figs. 6 and 7, this article). Assembly competent tubulin has also been obtained in this manner from oocytes of the mollusc, *Spisula solidissima* (Suprenant and Rebhun, 1984a).

A DEAE–Sephadex batch-elution method, modified from the chromatographic procedures of Suprenant and Rebhun (1983, 1984a), is described below. The appropriate ratios of eggs, DEAE–resin, and buffers have been provided in detail, such that large quantities of eggs (50–200 ml packed) can be processed efficiently. For smaller volumes of eggs (5–25 ml), where the tubulin concentration will be limiting, a single-column fractionation, described elsewhere, is recommended (Suprenant and Rebhun, 1983, 1984a).

## A. General Methods

### 1. Tubulin

Tubulin is a labile molecule that is easily denatured in solution. The rapid isolation of tubulin from cell homogenates is essential because the assembly activity of microtubule protein decays with a half-life of 3.5 hours in a brain extract (Olmsted and Borisy, 1975) and as rapidly as 10 minutes in a yeast tubulin isolate (Kilmartin, 1981, 1982). Microtubules are assembled *in vitro* by warming a concentrated tubulin sample to 35–37°C, in a high ionic strength buffer (pH 6.4–6.9) that contains $Mg^{2+}$, GTP, and is depleted of calcium (Weisenberg, 1972b; Olmsted and Borisy, 1975). The reassembly of microtubules can be monitored by turbidity (Gaskin *et al.*, 1974), viscosity (Olmsted and Borisy, 1975), or sedimentation (Johnson and Borisy, 1975), and can be visualized by dark-field light microscopy (Kuriyama and Miki-Noumura, 1975; Suprenant and Dentler, 1982), or in thin sections or by negative staining for electron microscopy. The purification of tubulin can be analyzed by polyacrylamide gel electrophoresis. Before attempting to purify and reassemble echinoderm spindle and egg tubulins, it is helpful to isolate and monitor the assembly of microtubules from a more abundant source of tubulin, such as cow, pig, or chick brain. Several detailed reviews on brain microtubule purification methods have recently been published (Murphy, 1982; Lee, 1982).

#### 2. Gametes

The method, used to obtain gametes, does not matter for the DEAE-purification methods as long as the egg jelly is solubilized or stripped, and the oocytes can be defolliculated. Eggs can be obtained by either intracoelomic injection of isotonic KCl or by macerating the ovaries. The eggs or oocytes should be washed free of dermal secretions and coelomic fluids with filtered (0.45-μm filter) seawater (natural, artificial, or $Ca^{2+}$ free, where appropriate) and then washed free of sea salts, with, for example, an isotonic glycerol medium. It is important to check an aliquot of eggs for fertilization because batches of deteriorating eggs, often obtained late in the breeding season or from an overripe animal, do not fertilize well nor do they make good sources of assembly competent tubulin.

Most of the methods in this article are described for the purple sea urchin, *Strongylocentrotus purpuratus* (Pacific Biomarine, Venice, CA). Eggs are shed by intracoelomic injection of isotonic KCl (0.52 *M*) and washed twice with artificial seawater (Cavanaugh, 1956) by sedimenting in a hand-cranked centrifuge. The egg jelly is removed by 2 washes with isotonic buffer containing 19 parts isotonic sodium chloride to 1 part potassium chloride (''isotonic 19:1'': 500 m*M* NaCl, 27 m*M* KCl, 2 m*M* EDTA, pH 7.80, at 23°C). Dejellied eggs are gently washed once with a buffered glycerol medium (0.9 *M* glycerol, 10 m*M* $MgCl_2$, 10 m*M* sodium phosphate, pH 6.80) and pelleted in a hand-cranked centrifuge.

## B. DEAE–Sephadex Batch Elution

#### 1. Homogenization

The packed eggs are resuspended in an equal volume of homogenization buffer (20 m*M* sodium phosphate, pH 6.80, 1 m*M* $MgCl_2$, 5 m*M* EGTA, 1 *M* glycerol, 0.1 m*M* GTP, sodium salt) at 4°C, and are homogenized by three to five passes of a glass Dounce homogenizer (pestle A). All subsequent steps are at 4°C unless noted otherwise. The extract is centrifuged at 100,000 *g* for 60 minutes and the clear supernatant fluid, containing the soluble tubulin, is carefully removed by inserting a Pasteur pipet just below the opaque lipid layer at the top. The lipid must be avoided because it will stick to the DEAE–Sephadex, reducing the flow rate and the available binding sites.

### 2. Batch Elution

Two volumes of clear supernatant fluid are mixed with 1 vol of gravity-packed DEAE–Sephadex, equilibrated the previous day in 20 m*M* sodium phosphate, pH 6.80, 1 m*M* $MgCl_2$, 1 m*M* EGTA, and 0.1 m*M* GTP. The DEAE slurry is gently mixed for 20 minutes on ice and is then sedimented by centrifugation at 2000 *g* for 2 minutes in five to six conical centrifuge tubes (50 ml) and the supernatant fluid is aspirated and discarded. The nontubulin egg proteins are eluted from the DEAE first, by three 10-minute washes with 0.35 *M* NaCl in 20 m*M* sodium phosphate, pH 6.80, 1 m*M* $MgCl_2$, 1 m*M* EGTA, and 0.1 m*M* GTP (sodium salt) (30 ml buffer/10 ml Sephadex/tube). After each wash the Sephadex beads are sedimented at 2000 *g* and the supernatant fluid is aspirated and discarded. Egg tubulin is eluted from the DEAE–Sephadex by two washes, 10 minutes each, with 0.65 *M* NaCl in the same buffer (20 ml of buffer/10 ml Sephadex/tube). The supernatant fluids (tubulin-containing fraction) are removed and saved. Residual Sephadex beads in the 0.65 *M* NaCl fraction are removed by passing the solution through a plug of glass wool in a funnel.

### 3. Ammonium Sulfate Fractionation

The dilute tubulin-containing fractions are concentrated by ammonium sulfate precipitation by dissolving 313 g ammonium sulfate (ultrapure) per liter of the 0.65 *M* NaCl fraction (50% saturation). This is carried out on ice with a pH electrode immersed in the solution which is gently stirred with a magnetic stirring bar. The pH is continually adjusted to pH 6.80 by the dropwise addition of 1 *N* KOH as the ammonium sulfate is dissolved and the proteins are precipitated. This should not take longer than 10–15 minutes. The precipitated protein is sedimented at 27,000 *g* for 20 minutes and is resuspended in assembly buffer (100 m*M* PIPES/$Na^+$, pH 6.80 at 23°C, 1 m*M* $MgCl_2$, 1 m*M* EGTA, 1 m*M* GTP) by gentle homogenization in a Potter–Elvehjem tissue grinder with a Teflon pestle. The volume of assembly buffer added should not exceed 10–15% of the original egg volume, or else the protein will be too dilute to reassemble. The resuspended egg tubulin solution will be turbid and is centrifuged again at 27,000 *g* for 2 minutes to remove denatured and aggregated protein.

### 4. Desalting

The ammonium sulfate is removed from the tubulin sample by passage over Sephadex G-25 (medium). Plastic disposable columns, each containing 5–10 ml of Sephadex G-25 equilibrated in assembly buffer, are placed on top of conical centrifuge tubes (15 ml) and centrifuged at 2500 *g* for 5 minutes in a low-speed swinging bucket rotor to remove excess buffer. The tubulin fraction is layered on top of the "dried" Sephadex and the column is transferred to a clean centrifuge tube. It is important that the volume of the sample to be desalted does not exceed 15% of the volume of the packed column. The unit is then centrifuged at 2500 *g* for 5 minutes, resulting in the rapid desalting of the tubulin samples, with minimum dilution.

## C. *In Vitro* Microtubule Assembly

The concentrated tubulin is assembled into intact microtubules by incubation at 37°C for 30–40 minutes. The rate and extent of microtubule polymerization can be monitored by an increase in turbidity at 350 nm (Fig. 1). The assembly of small samples (1–2 μl) can be monitored directly by dark-field light microscopy (Fig. 2) (Suprenant and Rebhun, 1984a; Suprenant and Dentler, 1982). The presence of microtubules must be confirmed by either negative staining or pelleting, fixing (0.5% glutaraldehyde, 0.5% tannic acid in assembly buffer), embedding, and thin sectioning for transmission electron microscopy (Fig. 3). The assembled microtubules are sedimented at 40,000 *g* at 30°C for 30 minutes and

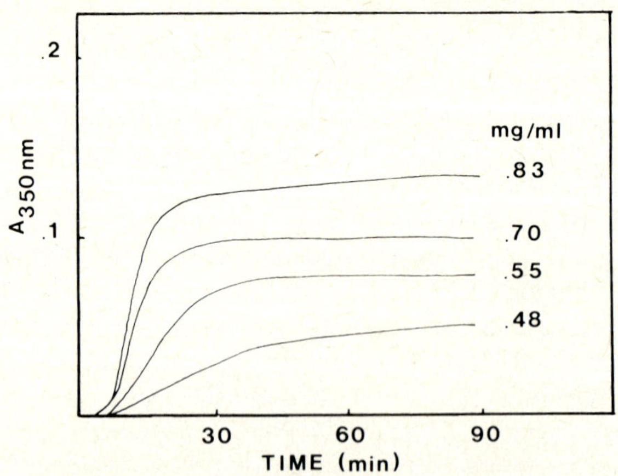


FIG. 1. *In vitro* assembly of DEAE-purified *S. purpuratus* egg tubulin at 37°C. Turbidometric scans (absorbance) at 350 nm at various tubulin concentrations. Reproduced with permission from Suprenant and Rebhun (1983).

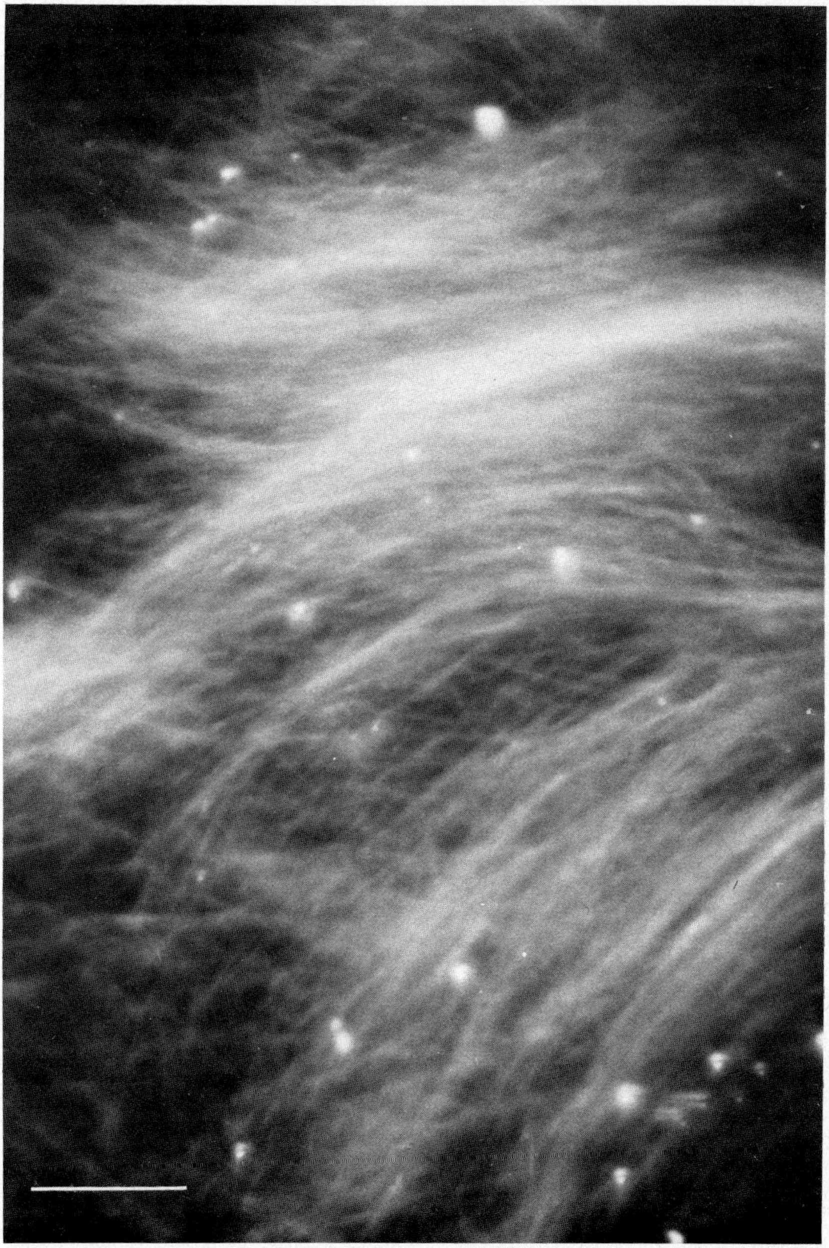

FIG. 2. Dark-field light micrograph of DEAE-purified *Spisula* oocyte tubulin reassembled *in vitro* (bar = 10 μm). Reproduced with permission from Suprenant and Rebhun (1984a).

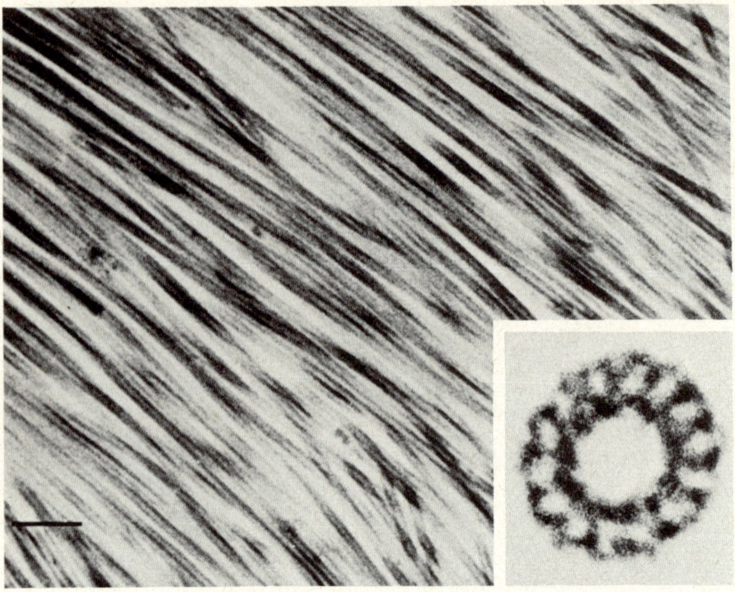

FIG. 3. Thin section of third-cycle microtubule pellet reassembled from DEAE-purified *S. purpuratus* egg tubulin (bar = 0.18 μm). Inset: cross-section of 14 protofilament microtubule. Reproduced with permission from Suprenant and Rebhun (1983).

resuspended in ice-cold assembly buffer by gentle homogenization with a Potter–Elvehjem tissue grinder. The volume of assembly buffer added depends upon the yield of polymer; but, as a general rule, the volume of buffer added is decreased by 50% for each successive cycle of assembly. The homogenate is left on ice for 30 minutes to cold depolymerize the microtubules and the cold-depolymerized microtubule protein is centrifuged at 100,000 *g* for 30 minutes at 4°C to remove aggregated as well as denatured protein. The cold supernatant fluid, containing active tubulin subunits, is warmed to 37°C to reassemble microtubules and the above process is repeated a second and third time. The yield of egg tubulin after three cycles of assembly is approximately 100–200 μg/ml of packed eggs. Egg tubulin can be stored at −70°C in solution prior to microtubule assembly by dropwise freezing in liquid nitrogen (Detrich and Williams, 1978) or as a pellet of microtubules. Samples frozen at −70°C retain their assembly competence for at least 6 months (longest period tested).

If microtubule assembly does not occur, the protein concentration may be below the critical protein concentration necessary for microtubule polymerization. For *S. purpuratus* egg and spindle tubulins, the critical concentration at 37°C is known to be 0.15–0.20 mg/ml (Keller and Rebhun, 1982; Suprenant and

Rebhun, 1983; Detrich and Wilson, 1983); however, for other echinoderm tubulins it has not been determined.

## III. Mitotic Spindle Tubulin

The physical nature and composition of the mitotic apparatus and the force-producing mechanisms during anaphase that pull the chromosomes to the poles and push the two poles further apart are still enigmas. It has been over 30 years since Mazia and Dan (1952) first isolated mitotic spindles from sea urchin zygotes, yet there are still no mitotic apparatus preparations that adequately model chromosome movements *in vitro*. Nevertheless, progress has been made toward isolating spindles that maintain some *in vivo* properties.

Current models of mitosis that consider treadmilling of tubulin subunits (Margolis *et al.*, 1978; Margolis and Wilson, 1981), sliding of antiparallel microtubules (McIntosh *et al.*, 1969; McIntosh, 1981), zippering of adjacent microtubules (Bajer and Mole-Bajer, 1975), and equilibrium of microtubule subunits (Inoue and Sato, 1967; Inoue and Ritter, 1975) are all based upon properties of microtubules. Spindle function is, therefore, at least partially dependent upon the chemistry of spindle tubulin, the major component of the spindle microtubules. The biochemical characterization of native spindle tubulin is not only important for understanding the chemistry, motility, and stability of the spindle fibers, but also for describing the interaction of the microtubules with other spindle structures such as the mitotic organizing centers, kinetochores, and membranes, as well as with potential regulatory molecules such as microtubule-associated proteins, ATPases, and calcium-binding proteins.[2]

The successful isolation of native spindle tubulin depends upon the reversible stabilization of the spindle microtubules. In other words, the labile microtubules of the mitotic apparatus *in vivo* must be stabilized during isolation *in vitro* and then gently solubilized under conditions where the tubulin is still competent to reassemble. One approach for the isolation of mitotic spindles has been to lyse embryos directly into buffers that were originally developed for the cold-reversible polymerization of brain microtubules (Weisenberg, 1972b; Borisy and Olmsted, 1972). Spindles were first isolated in this manner from the surf clam *Spisula solidissima* (Rebhun *et al.*, 1974). Surf clam eggs were lysed at room temperature in a microtubule-reassembly buffer that contained 0.1 *M* MES or

[2]Recently several excellent reviews have been written on mitotic apparatus function (Inoue, 1981), ultrastructure (Fuge, 1977a,b), methods for purification (Forer, 1969; McIntosh, 1977; Zimmerman *et al.*, 1977; Turner and McIntosh, 1977; Sakai, 1978; Zimmerman and Forer, 1981; Salmon, 1982), calcium regulation (Rebhun, 1976, 1977; Harris, 1978, 1981; Petzelt, 1979; Nagle and Egrie, 1981), membranes (Hepler and Wolniak, 1984), and ATPases (Pratt, 1984).

PIPES buffers, pH 6.8–6.9, 0.5–1 m*M* $MgCl_2$ 1–5 m*M* EGTA, a nonionic detergent, Triton X-100, and a proteolytic inhibitor, TAME. The isolated spindles were stable at room temperature and were, at least initially, cold labile. Moreover, the isolated spindles that maintained native cold lability were also able to incorporate chick brain tubulin in the presence of 1 m*M* GTP, resulting in an increase in spindle length and birefringence. Similar results have been reported with oocytes of the marine worm *Chaetopterus* (Inoue *et al.*, 1974).

Recently, Rebhun and co-workers (Rebhun *et al.*, 1975, 1982; Keller and Rebhun, 1979, 1982; Suprenant and Rebhun, 1984a) have isolated and reassembled mitotic and meiotic spindle tubulin *in vitro.* Their procedure for the isolation of first cleavage spindles from the sea urchin *S. purpuratus,* and the *in vitro* reassembly of spindle microtubules, is described below.

## A. General Considerations

The conditions for fertilization, such as temperature, composition of the "seawater," and the methods for removing the jelly layer, tough fertilization membrane, and extracellular hyaline layer, are crucial for synchronous development (see Salmon, 1982). The isolation of mitotic spindles is also dependent upon (1) lysing the cell, usually by detergent solubilization, osmotic shock, or mechanical shear, (2) stabilizing the spindle outside the cell without promoting nonspecific associations, by controlling the pH, ionic strength, and solvent conditions used in the isolation media, and (3) separating the spindles from cortices and other cytoplasmic constituents by centrifugation or filtration. Conditions that stabilize the mitotic spindle microtubules can make it difficult to lyse the cell, and vice versa; thus, spindle isolation methods can be species specific and dependent upon the "history" of the preparation.

## B. Fertilization and Embryo Development

For the isolation of *S. purpuratus* first-cleavage spindles, 10–15 ml of dejellied eggs is diluted in 150–200 ml of artificial seawater (Cavanaugh, 1956) in a 1-liter beaker and is gently stirred in a water bath at 16–18°C. The eggs are fertilized by the addition of 1 ml of a dilute sperm suspension (two drops of "dry" sperm in 10 ml of seawater), and after 1 minute, the elevation of the fertilization membrane is monitored by phase-contrast microscopy. The eggs are diluted to 800–1000 ml with artificial seawater and the fertilization membranes are stripped by aspirating the egg suspension, under a slight vacuum, through bolting mesh (Tekto, Inc., Elmsford, NY). The fertilization membranes will harden rapidly and must be removed within 5 minutes of fertilization. The embryos are allowed to settle to the bottom of a large finger bowl and the fertilization membranes and other debris that do not settle are aspirated away. The developing embryos are washed two to three times with 500 ml of artificial seawater and are allowed to settle. The development of the embryos is monitored by polarization or bright-field optics. The time to reach first mitotic metaphase is 80–85 minutes at 18°C and 90–100 minutes at 16°C.

## C. Mass Isolation of Spindles

At the onset of metaphase, the embryos are pelleted by centrifugation in a hand-cranked centrifuge in four conical tubes (15 ml), gently washed free of sea salts with "isotonic 19:1" (see Section II,A,2), and resuspended in 10 vol of isolation medium (100 m*M* PIPES/$K^+$, pH 6.50, at 23°C, 1 m*M* $MgCl_2$, 5 m*M* EGTA, and 0.1% Triton X-100). *S. purpuratus* spindles are isolated into this lower pH microtubule-reassembly buffer because they are not stable at higher pH values (pH 6.80) in the absence of glycerol. The embryos are lysed at room temperature by vortexing or shaking, and the release of the spindles is monitored by polarization microscopy. The spindles are pelleted by centrifugation at 1000 *g* for 4 minutes in a clinical centrifuge and are washed once with pH 6.5 isolation medium, twice with isolation medium without detergent, and twice with 100 m*M* PIPES, pH 6.80, 1 m*M* $MgCl_2$, and 1 m*M* EGTA at room temperature (Fig. 4).

## D. Reassembly of Spindle Microtubules

The solubilization of spindle tubulin by cold depolymerization of the spindle microtubules is monitored by the loss of spindle birefringence. Isolated spindles are resuspended in 0.3–0.5 ml of 100 m*M* PIPES/$K^+$, pH 6.80, at 23°C, 1 m*M* EGTA, 1 m*M* $MgCl_2$, and 0.1 m*M* GTP, and placed on ice for 30–45 minutes. The loss of spindle birefringence is facilitated by repeated passes of the spindle preparation through a 23-gauge needle, with caution not to denature the sample. Spindle remnants with no detectable birefringence are centrifuged at 100,000 *g* for 30 minutes and the soluble spindle tubulin in the 100,000 *g* supernatant is removed, made 1 m*M* in GTP, and polymerized into microtubules at 37°C. The reassembled spindle microtubules are sedimented at 40,000 *g* for 30 minutes. Spindle tubulin can be purified further by cycles of temperature-dependent assembly and disassembly as described for egg tubulin (Section II,C). Similar methods can be used to isolate assembly competent tubulin from the surf clam, *Spisula solidissima* (Keller and Rebhun, 1979; Murphy, 1980; Suprenant and Rebhun, 1984a) and from the sea urchin, *Arbacia punctulata* (Rebhun *et al.*, 1982).

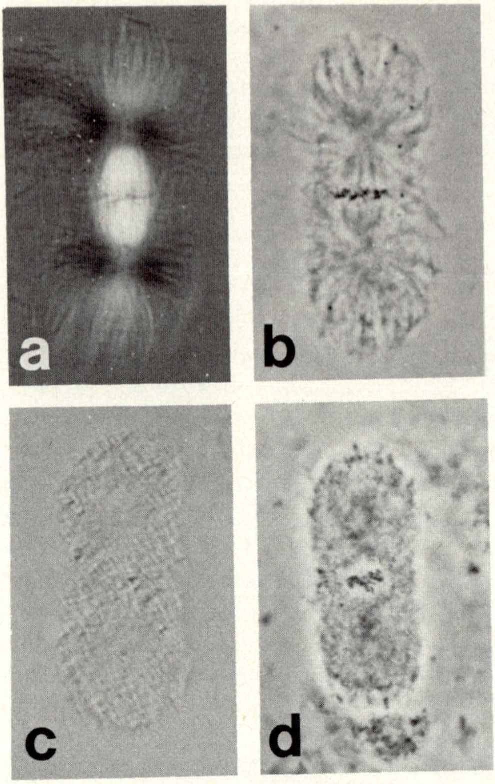


FIG. 4. Mitotic spindles isolated from the sea urchin, *S. purpuratus*. As isolated, mitotic spindles are brightly birefringent [(a), polarization optics] and chromosomes are easily visualized at the metaphase plate [(b), phase-contrast optics]. The remnant of the spindle left after extraction of the tubulin is not birefringent [(c), polarization], but maintains its overall organization [(d), phase contrast]. Reproduced from Keller and Rebhun (1982) with permission.

## IV. Cilia and Flagella

The unique organization of the outer doublet microtubules of cilia and flagella (Warner, 1972) and the ability to selectively solubilize chemically distinct tubulins from the central pair, A- and B-subfiber microtubules (Gibbons, 1965; Behnke and Forer, 1967; Witman *et al.*, 1972; Stephens, 1975; Linck and Langevin, 1981), have provided opportunities for examining the potential relationship between tubulin chemistry and microtubule function (Bibring *et al.*, 1976; Stephens, 1978). The isolation of cilia and flagella from echinoderms is reviewed by R. E. Stephens (this volume). In addition, Farrell (1982) has written an excellent methodological review on the purification and reassembly of tubulin from outer doublet microtubules.

## V. Properties of Microtubules Reassembled from Egg, Spindle, and Flagellar Tubulins

### A. Composition

Egg microtubules, reassembled *in vitro* after DEAE-chromatography, are composed of greater than 98% tubulin (Kuriyama, 1977; Suprenant and Rebhun, 1983; 1984a; Detrich and Wilson, 1983) that migrates as equimolar quantities of alpha- and beta-subunits on SDS–PAGE. There are no high-molecular-weight proteins similar to the assembly promoting, microtubule-associated proteins of vertebrate brain microtubule preparations; however, third cycle *S. purpuratus* egg microtubule preparations contain two prominent low-molecular-weight proteins, $M_r$ 39,000 and 36,000, which comprise 2% by mass of the total microtubule protein. The function and prevalence of the low-molecular-weight proteins are unknown at this time. Similarly, microtubules assembled from flagellar outer doublet tubulin, derived either by sonication (Kuriyama, 1976; Binder and Rosenbaum, 1978; Farrel and Wilson, 1978; Farrell *et al.*, 1979) or thermal fractionation (Linck and Langevin, 1981), are composed of greater than 95–98% tubulin.

Spindle microtubules reassembled *in vitro* are composed of tubulin and an $M_r$ 80,000 protein (Fig. 5). The $M_r$ 80,000 protein is present in third-cycle microtubule preparations approximately 0.5 to 5.0% by mass in the case of *S. purpuratus* (Keller and Rebhun, 1982) and as much as 20% by mass in *Arbacia punctulata* preparations (Rebhun *et al.*, 1982). The $M_r$ 80,000 protein is not required for the efficient polymerization of spindle tubulin (Fig. 5A) (Keller and Rebhun, 1982) but may be involved in the cross-bridging of spindle microtubules observed *in vitro* (Rebhun *et al.*, 1982).

### B. Morphology

#### 1. Microtubules

In general, purified echinoderm egg and flagellar tubulins reassemble into smooth-walled, singlet microtubules with an outer diameter of 25–30 nm (Figs. 3 and 7) (Kuriyama, 1976, 1977; Binder and Rosenbaum, 1978; Farrell and Wilson, 1978; Suprenant and Rebhun, 1983; Detrich and Wilson, 1983). The

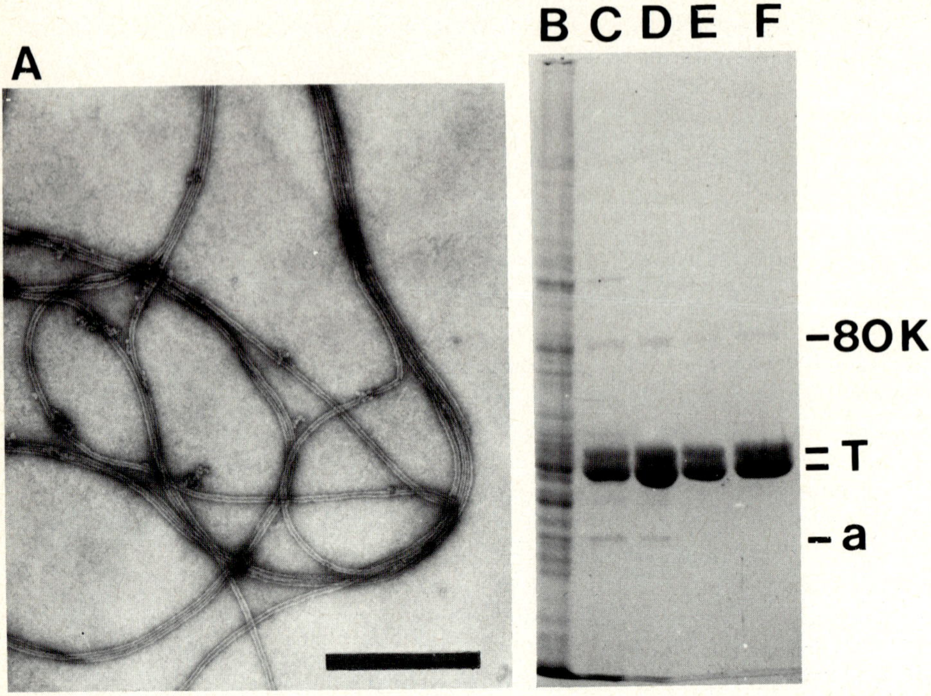


FIG. 5. Purification and assembly of tubulin from nonglycerol isolated spindles. Isolated spindles are exposed to 0°C for 45 minutes to extract the tubulin. The extracted tubulin reassembles into a microtubule when warmed to 37°C and can be further purified by two additional cycles of temperature-dependent polymerization. Spindle tubulin reassembles at a critical concentration of 0.15–0.20 mg/ml (37°C). SDS–polyacrylamide gel electrophoresis on a 7.5% gel confirms that most of the tubulin is extracted from the spindle remnant (lane B), and that little else besides tubulin was in the extracted 0°C supernatant (lane C). After assembly and pelleting, the microtubules are composed of nearly pure tubulin (T), a small amount of actin (a) and a protein of approximate $M_r$ 80,000 (lane D). After two additional cycles of purification (lanes E and F), only the 80,000 $M_r$ protein remains with the tubulin. The 80,000 $M_r$ protein can be removed by chromatography of the third cycle spindle microtubules. The critical concentration for the purified tubulin is also 0.15–0.20 mg/ml. In addition, the microtubules assembled from the DEAE–cellulose-purified spindle tubulin are intact and smooth walled when examined by negative staining for electron microscopy (panel A) (bar = 0.5 μm). Reproduced from Keller and Rebhun (1982) with permission.

lack of filamentous projections is indicative of microtubules assembled in the absence of high-molecular-weight microtubule-associated proteins (Murphy and Borisy, 1975; Sloboda *et al.*, 1975; Dentler *et al.*, 1975; Vallee *et al.*, 1984). However, bovine brain microtubule-associated proteins will stimulate the rate and final extent of assembly of *S. purpuratus* egg and flagellar tubulins as well as decorate the microtubules with filamentous projections, indicating that a microtubule-associated protein-binding site is present on the egg and flagellar tubulin molecule (Binder and Rosenbaum, 1978; Farrell and Wilson, 1978; Suprenant and Rebhun, 1983). Microtubules assembled *in vitro* from *S. purpuratus* spindle tubulin are also smooth walled (Keller and Rebhun, 1982), but those from *Arbacia punctulata* appear to be laterally cross-bridged (Rebhun *et al.*, 1982). The cross-bridging is believed to be a function of the $M_r$ 80,000 protein in these *Arbacia* microtubule preparations.

Similar to other microtubules reassembled *in vitro* (Pierson *et al.*, 1978), egg, spindle, and flagellar microtubules are composed of 14 and 15 protofilaments in cross-section (Figs. 3 and 7A). In contrast, most cytoplasmic microtubules *in vivo* are composed of 13 protofilaments (Tilney *et al.*, 1973). The significance of the different numbers of protofilaments is not understood, although it has been suggested that the number of protofilaments *in vivo* is determined by nucleation sites for microtubule assembly (Scheele *et al.*, 1982) and alternatively by the microtubule-associated proteins (Bohm *et al.*, 1984).

### 2. Macrotubules

Echinoderm tubulin is also capable of reassembling into *macro*tubules, rather than *micro*tubules, under identical *in vitro* assembly conditions. Unlike tubulin purified from sea urchin and sand dollar eggs, oocyte tubulin from the starfish, *Asterias vulgaris* (Figs. 6 and 7) and surf clam, *Spisula solidissima* (Suprenant and Rebhun, 1984a) reassembles into 50- to 70-nm macrotubules during the first assembly cycle following DEAE-chromatography. Unlike microtubules in which the protofilaments are parallel to the long axis of the polymer, macrotubules are assembled from ribbons of protofilaments coiled at an oblique angle. Surf clam macrotubules are composed of sheets of 26–27 protofilaments coiled at a 36° angle to the long axis (Suprenant and Rebhun, 1984a). In the starfish polymer, protofilaments are coiled at a 40° angle, giving rise to a slightly larger outer diameter macrotubule (Fig. 7). Using the method of Warfield and Bouck (1975), approximately 29–36 protofilaments are calculated to compose the starfish macrotubule sheets.

Although the conditions that lead to macrotubule versus microtubule formation *in vitro* are not resolved, it is clear that comparative studies with these marine invertebrate tubulins may suggest answers to how protofilament number is regulated *in vivo* and may reveal potential differences in meiotic and mitotic tubulins that are isolated from germinal vesicle-stage oocytes and pronuclear eggs, respectively.

## C. Assembly

In general, the polymerization of egg, spindle, and flagellar tubulins is similar to vertebrate brain tubulin. Sea urchin tubulin requires magnesium (0.5–1.0 m*M*), high ionic strength (0.10–0.15 m*M* KCl), and GTP (0.5–1.0 m*M*) for

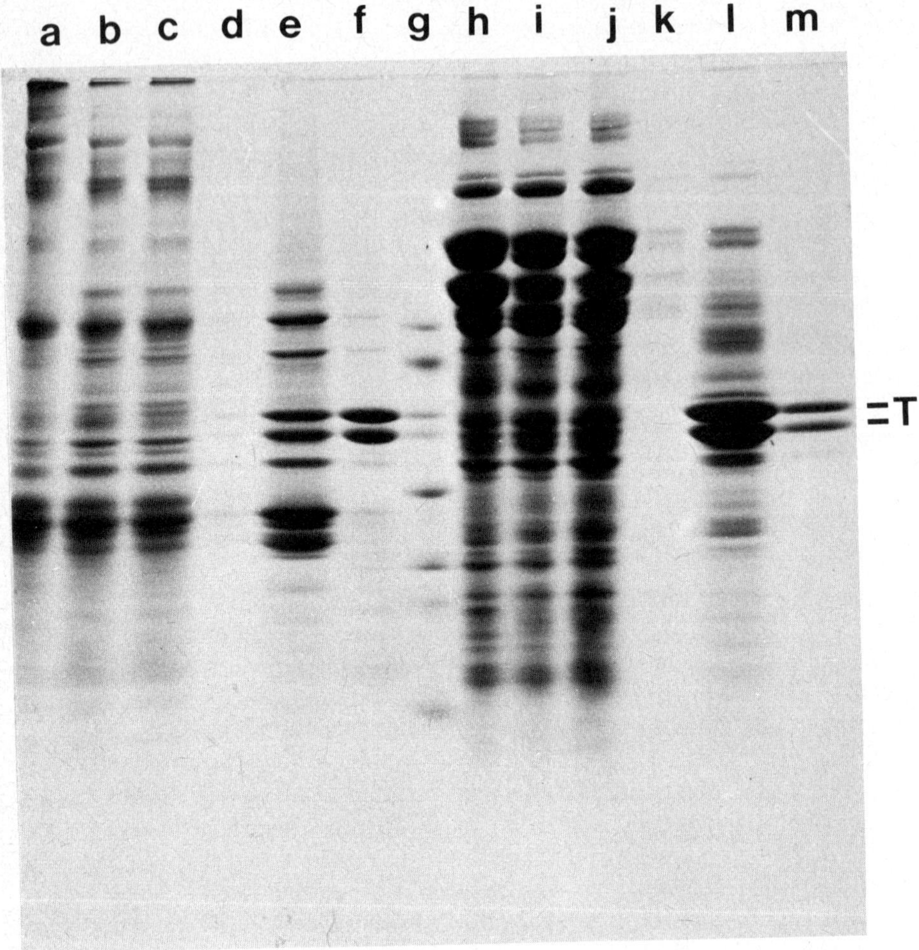


FIG. 6. Purification of starfish (*Asterias vulgaris*) oocyte tubulin and sand dollar (*Echinarachnius parma*) egg tubulin by DEAE–Sephadex column chromatography and one cycle of *in vitro* assembly (procedures of Suprenant and Rebhun, 1983, 1984a). Proteins are resolved on an SDS–polyacrylamide gel with a 4–16% acrylamide gradient. Lanes a–f, starfish preparation. Lanes h–m, sand dollar preparation. Lane g, molecular weight standards, phosphorylase *b* (94,000), bovine serum albumin (67,000), alpha-tubulin (58,000), beta-tubulin (55,000), ovalbumin (43,000), carbonic anhydrase (30,000), soybean trypsin inhibitor (20,000), and alpha-lactalbumin (14,400). Approximately 15 ml of starfish oocytes and 4 ml of sand dollar eggs were homogenized and centrifuged. The total soluble proteins in the 100,000 *g* supernatant fluids (lanes a and h) were concentrated by ammonium sulfate precipitation (lanes b and i) and loaded on a DEAE–Sephadex column. Most of the proteins were either not absorbed to the column (lanes c and j) or were eluted from the column in 0.35 *M* NaCl elution buffer (Lanes d and k). The tubulin-containing fractions were eluted with 0.65 *M* NaCl (lanes e and l), concentrated, and reassembled at 22°C. Greater than 95% of the protein in the first cycle starfish and sand dollar polymers is tubulin (lanes f and m). Unpublished data from K. A. Suprenant and L. I. Rebhun.

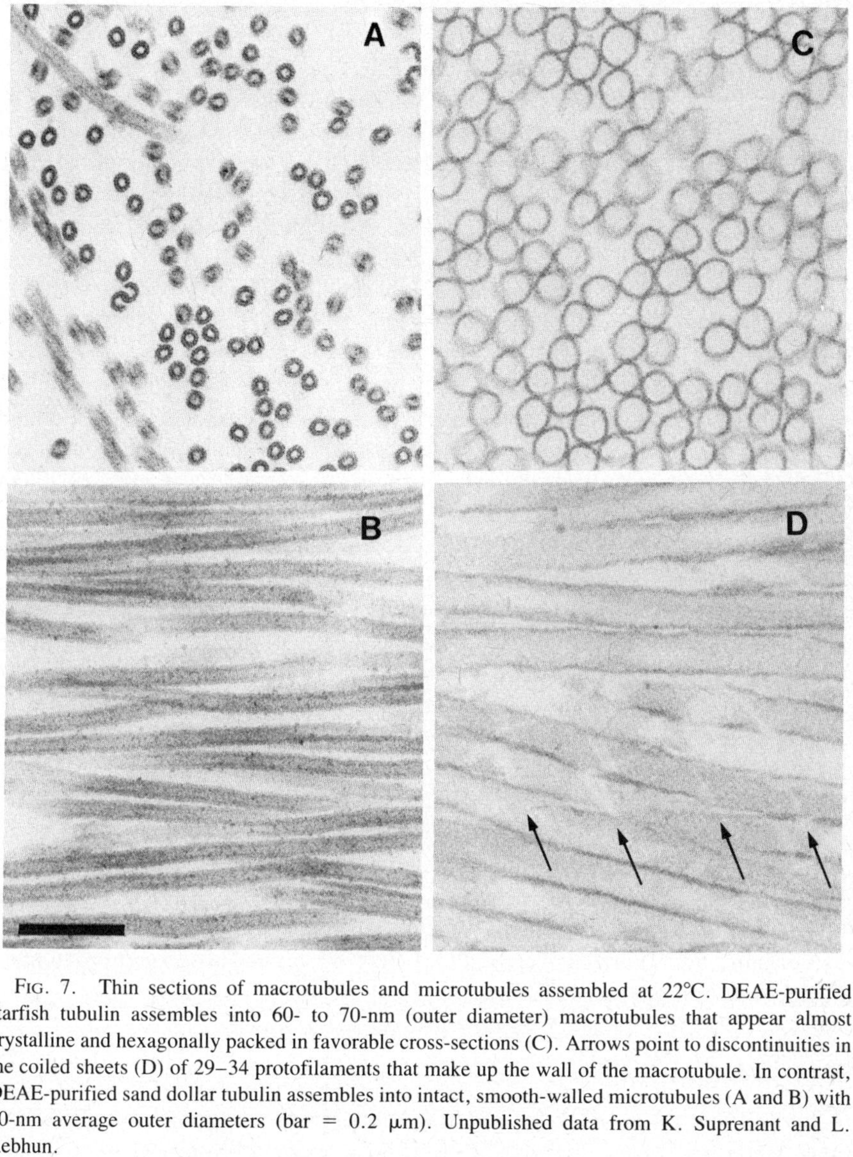


FIG. 7. Thin sections of macrotubules and microtubules assembled at 22°C. DEAE-purified starfish tubulin assembles into 60- to 70-nm (outer diameter) macrotubules that appear almost crystalline and hexagonally packed in favorable cross-sections (C). Arrows point to discontinuities in the coiled sheets (D) of 29–34 protofilaments that make up the wall of the macrotubule. In contrast, DEAE-purified sand dollar tubulin assembles into intact, smooth-walled microtubules (A and B) with 30-nm average outer diameters (bar = 0.2 μm). Unpublished data from K. Suprenant and L. Rebhun.

assembly, although in some cases, egg tubulin will polymerize in ATP (Kuriyama, 1977). Unfertilized and fertilized egg tubulin binds 0.97 and 0.80 mol of [$^3$H]GTP, respectively, per 100,000 g of protein at the E-site (Kuriyama, 1977) and microtubule assembly is prevented by low temperature, calcium ions, colchicine, and sulfhydryl reagents (Kuriyama, 1976, 1977). Similar to brain microtubules, egg microtubule assembly occurs at a specific critical protein concentration in a process similar to the nucleated condensation–polymerization process described by Oosawa and Kasai (1962) for actin polymerization. Below the critical concentration no microtubule assembly takes place and above it the mass of microtubules formed is proportional to the initial tubulin concentration.

Differences are apparent in the extent to which sea urchin and vertebrate brain tubulins assemble at their respective physiological temperatures. At 37°C, sea urchin egg, spindle, and flagellar tubulins assemble in the absence of microtubule-associated proteins at protein concentrations at least 10-fold lower than mammalian brain tubulin that is free of microtubule-associated proteins (Binder and Rosenbaum, 1978; Farrell and Wilson, 1978; Keller and Rebhun, 1982; Suprenant and Rebhun, 1983; Detrich and Wilson, 1983). Moreover, at urchin physiological temperatures of 15–18°C, egg and spindle tubulins assemble into microtubules at protein concentrations as low as 0.7 mg/ml (Keller and Rebhun, 1982; Suprenant and Rebhun, 1983). Interestingly, isolated flagellar outer doublet tubulin from the same urchin is not able to assemble at these lower temperatures and protein concentrations. This may reflect differences in the assembly properties of flagellar tubulin or may be due to the harsh conditions necessary to solubilize outer doublet tubulins.

The differences in the assembly properties of cytoplasmic and flagellar tubulin, as well as between sea urchin and vertebrate brain tubulin, are supported by recent experiments that demonstrate that the tubulin populations are chemically distinct. Sea urchin tubulin is much less complex than brain tubulin when examined by isoelectric focusing; only 2 alpha- and 2 beta-tubulins have been found in sea urchins (Bibring *et al.*, 1976; Suprenant and Rebhun, 1983) whereas a minimum of 15 variants of brain tubulin have been resolved (George *et al.*, 1981; Field *et al.* 1984). In addition, heterogeneity has been observed among sea urchin cytoplasmic and flagellar microtubules examined by gel electrophoresis (Bibring *et al.*, 1976), peptide mapping (Stephens, 1978), or colchicine-binding (Wilson *et al.*, 1984). Together, these data suggest that the different functions of cytoplasmic and flagellar microtubules may be regulated through chemically distinct sites on the tubulin molecule.

## D. Calcium

Intracellular calcium levels have long been thought to regulate spindle formation (Heilbrun, 1921), and the *in vivo* and *in vitro* assembly of microtubules (Weisenberg, 1972a); and, although there may be other calcium targets in the spindle, the calcium sensitivity of the microtubule component is likely to be integral to spindle function. Sea urchin spindle microtubules *in vivo* (Kiehart, 1981) and *in vitro* (Salmon and Segall, 1980), as well as egg and spindle tubulins isolated and reassembled *in vitro,* are very sensitive to depolymerization by micromolar calcium, especially at low temperatures and protein concentrations (Nishida and Kumagai, 1980; Keller *et al.*, 1982; Suprenant and Rebhun, 1984b). Additional factors, presumed to be important for calcium regulation of microtubule assembly, such as calmodulin and microtubule-associated proteins (see Lee and Wolff, 1982, 1984), are present in unfertilized eggs (Head *et al.*, 1979; Burgess, 1981; Vallee and Bloom, 1983; Scholey *et al.*, 1984), and in the mitotic spindle (Vallee and Bloom, 1983; Hamaguchi and Iwasa, 1980). It is reasonable to consider that the disassembly of microtubules which moves the chromosomes to the poles during anaphase (Inoue and Sato, 1967), as well as the lack of assembled microtubules in the unfertilized egg, are due in part to the intrinsic calcium sensitivity of the tubulin and associated regulatory molecules. The intracellular calcium levels are postulated to be controlled by fluctuations in a $Ca^{2+}$-ATPase (see Petzelt, 1979; Pratt, 1984; Hepler and Wolniak, 1984), presumed to be associated with a membrane component of the mitotic apparatus and able to actively sequester calcium in a manner analogous to the sarcoplasmic reticulum of muscle (Harris, 1975, 1981; Silver *et al.*, 1980, 1983; Nagle and Egrie, 1981; Hepler, 1977; Hepler and Wolniak, 1984).

## VI. Microtubule-Associated Proteins

The chromatographic procedures used to purify egg tubulin (Section II) remove most if not all of the microtubule-associated proteins that are probably important for the regulation of microtubule function *in vivo* (for review see Vallee *et al.*, 1984). Attempts to polymerize microtubules directly from unfertilized egg and oocyte homogenates under conditions that support the reassembly of vertebrate brain microtubules have failed (Kane, 1975; Weisenberg, 1973; Weisenberg and Rosenfeld, 1975), although a limited amount of polymerization is obtained if short pieces of rat brain microtubules are added as "seeds" (Burns and Starling, 1974). Several laboratories have suggested that RNA or other polyanions (Bryan, 1975; Bryan *et al.*, 1975), polysaccharides (Naruse and Sakai, 1981), or proteases (Asnes and Wilson, 1981) prevent microtubule assembly in egg extracts, although the effectiveness of these inhibitors *in vivo* is not known.

Microtubules can be induced to assemble in egg extracts by the assembly promoting drug taxol (Schiff *et al.*, 1979; Vallee, 1980). Generally 10–20 $\mu M$ taxol[3] is added to a high-speed extract obtained from egg homogenates. The

[3]Taxol is not commercially available, but can be obtained (in milligram quantities) by writing to the Natural Products Branch of the National Cancer Institute. Taxol is generally made as a concentrated stock (5 m$M$) in DMSO and is stored frozen at −20°C.

extract is warmed to 30–37°C for 5–15 minutes to assemble microtubules and the polymerized microtubules are centrifuged through a cushion of sucrose (10–20% w/v) in a taxol-containing microtubule reassembly buffer. These reassembled microtubules are observed in the electron microscope to have filamentous projections and cross-bridges (Vallee and Bloom, 1983; Scholey *et al.*, 1984), similar to the projections produced by the high-molecular-weight microtubule-associated proteins of mammalian brain. Prominent among the sea urchin microtubule-associated proteins are (1) a group of high $M_r$ proteins > 250,000 with a dynein-like $Mg^{2+}$-ATPase activity (Scholley et al., 1984), similar to cytoplasmic dynein (Pratt, 1980; Pratt *et al.*, 1980; Hisanaga and Sakai, 1983), and (2) a protein, presumed to be identical to the $M_r$ 80,000 protein that copolymerizes with spindle microtubules *in vitro* (Keller and Rebhun, 1982). Antibodies to several of the putative microtubule-associated proteins (e.g., $M_r$ 235,000, 205,000, 150,000, and 37,000) stain the mitotic spindle of *L. variegatus,* suggesting that these proteins may be structural or functional components of the spindle (Vallee and Bloom, 1983).

## VII. Mitotic Organizing Centers

Microtubules within the mitotic spindle originate from an amorphous spherical mass at the spindle pole called the centrosome, or in more general terms, the mitotic organizing center (Pickett-Heaps, 1969). Mitotic organizing centers are important for the initiation and directional assembly of ordered arrays of microtubules within the spindle and in the cytoplasm of most cells, yet little is known about their composition or activity (see Raff, 1979). Mitotic organizing centers can be isolated and analyzed in at least two ways in developing echinoderm embryos.

First, Weisenberg has shown that homogenates of activated surf clam oocytes contain a sedimentable material that will induce the formation of aster-like structures *in vitro* (Weisenberg, 1973; Weisenberg and Rosenfeld, 1975). *In vitro* mitotic organizing activity from surf clams is obtained by (1) homogenization (4°C) of activated oocytes in a high ionic strength buffer (0.5 *M* MES, 1 m*M* EGTA, pH 6.5–6.6), (2) low-speed centrifugation (4°C) of the extract (5 minutes at 2000 *g*), and (3) incubation (23°C) of the sedimentable fraction from activated eggs with the supernatant fluids of unfertilized *or* activated eggs. The assembly of sea urchin asters *in vitro* is additionally dependent upon a high concentration of tubulin, low free calcium (5 m*M* EGTA), and the presence of glycerol (15% minimum) (Weisenberg, 1978).

A second method involves turning sea urchin eggs into factories for the production of mitotic organizing centers *in vivo*. Several two-step procedures for artificial parthenogenesis (Loeb, 1913) will induce the formation of as many as 40 cytasters with an egg, each with numerous centrioles at the centers (Dirksen, 1961; Sachs and Anderson, 1970; Mazia, 1977; Kato and Sugiyama, 1971; Miki-Noumura, 1977; Kuriyama and Borisy, 1983). The first step, which generally leads to monaster formation, includes treatment of the eggs with acidic (Loeb, 1913), basic (ammonia) (Paweletz and Mazia, 1979), thymol (von Ledebur-Villiger, 1972), or procaine-containing seawater (Moy *et al.*, 1977). Numerous cytasters are induced by a second incubation step in either hypertonic (Loeb, 1913), $D_2O$ (Mazia, 1977), or ethanol-containing seawater. Recently, the various combinations of the first and second steps were reexamined for cytaster formation (Kuriyama and Borisy, 1983). They found that the only first and second combinations that do not produce cytasters in either *S. purpuratus* or *L. pictus* eggs are acid–ethanol, thymol–ethanol, acid–$D_2O$, and thymol–$D_2O$. All other combinations induce cytaster formation. Cytasters can be isolated in mass, using the same methods used for mitotic spindle isolation (Moy *et al.*, 1977; Mazia, 1978; Kuriyama and Borisy, 1983). The ability to produce and isolate mass quantities of asters both *in vivo* and *in vitro* should be useful for biochemical analysis of mitotic organizing activity.

## VIII. Particulate Tubulin

It is important to note that not all the tubulin in the egg and embryo is soluble. A substantial portion of tubulin in eggs and oocytes (Weisenberg, 1972b, 1973; R. Raff, 1975) as well as in other cell types appears to be particulate and sedimentable (see E. C. Raff, 1979, for review). The role and even actual existence of particulate tubulin in the living cell are unknown at this time. It is intriguing to speculate that particulate tubulin may be a modified storage form of tubulin found in the oocyte and egg—the cell's way of compartmentalizing the tubulin, thereby decreasing the pool of assembly competent tubulin.

## IX. Concluding Remarks

"An egg is an egg and its eggness lies in the storage of macromolecules, messengers and spatial predetermination of the future embryo; storage of what is expressed in later development."

D. Mazia (1978)

Many of the components presumed necessary for the assembly and organization of mitotic spindle microtubules are present in the unfertilized sea urchin egg, including assembly competent tubulin, microtubule-associated proteins, and latent mitotic organizing centers. Despite the storage of microtubule components in the unfertilized egg, there are few, if any, assembled microtubules (Harris *et al.*, 1980a,b; Bestor and Schatten, 1981). Fertilization triggers the assembly of microtubules, as well as many other events such as the alkalinization of the cytoplasm from pH 6.80 to 7.30 (Steinhart, 1978). This increase in intracellular pH has been shown to promote the formation of actin filaments in the sea urchin cortex (Begg and Rebhun, 1979), and it would be significant if similar pH effects on microtubule assembly were also observed. On the contrary, alkaline pH disassembles brain microtubules (Regula *et al.*, 1981), and although similar effects have not been demonstrated for sea urchin egg tubulin, it appears that the alkalinization of the cytoplasm alone does not promote assembly. The increase in intracellular pH could, however, release an inhibitor molecule or protein from egg tubulin, allowing assembly to occur in the presence of microtubule-associated proteins and mitotic organizing centers. Historically, those who study microtubules have focused on some of those factors that promote microtubule assembly. Inhibitors of assembly have, with the exception of calcium ions and antimitotic drugs like colchicine, only been considered theoretically, albeit in great detail (Hill and Kirschner, 1983). Yet small proteins, similar in action to the actin-binding protein profilin (see Korn, 1982), could complex with tubulin in the unfertilized egg, keeping the pool of assembly competent tubulin subunits below the *in vivo* critical concentration necessary for assembly. Sea urchin eggs may prove to be extremely valuable in the search for these putative regulatory factors.

While an egg may be an egg, an egg is not an oocyte. It is probably more than just a coincidence that tubulin purified from starfish and surf clam *oocytes* assembles into *macro*tubules (see Section VI,B, and Suprenant and Rebhun, 1984a), while tubulin purified from sea urchin and sand dollar *eggs* assembles into *micro*tubules. Sea urchin and sand dollar eggs have undergone meiotic maturation divisions during oogenesis prior to spawning and fertilization. By contrast, starfish and surf clam oocytes are spawned in a state of meiotic arrest. In addition, there are major metabolic differences between sea urchin and starfish eggs (see Meijer and Guerrier, 1984; Johnson and Epel, 1982) and differences in microtubule organization (Otto and Schroeder, 1984a,b). Just as it is short sighted to base all our knowledge of microtubule function on brain tubulin, it is equally short sighted to assume that sea urchin eggs are paradigms of microtubule function.

Finally, our present understanding of the physiological regulators of mitosis and microtubule function will benefit tremendously from the future isolation of mitotic spindles that can be reactivated to move chromosomes *in vitro*. The ongoing characterization of tubulin-containing structures and regulatory factors from echinoderm embryos will be invaluable for a careful dissection of the control elements active in these mitotic models.

## Acknowledgments

I thank Dr. L. I. Rebhun and Dr. W. L. Dentler for their support, Drs. S. Frost and P. R. Burton for a critical reading of the article, and Dr. T. C. S. Keller for generously providing Figs. 4 and 5. Parts of this work were supported by grants to L.I.R. (NIH GM 26784), W.L.D. (NIH GM 32556), and K.A.S. (ACS PF-2415).

## References

Alexandraki, D., and Ruderman, J. V. (1981). *Mol. Cell Biol.* **1,** 1125–1137.

Asnes, C. F., and Wilson, L. (1981). *Arch. Biochem. Biophys.* **207,** 75–80.

Auclair, W., and Siegel, B. S. (1966). *Science* **154,** 913–915.

Bajer, A. S., and Mole-Bajer, J. (1975). *In* "Molecules and Cell Movement" (S. Inoue and R. E. Stephens, eds.), pp. 77–96. Raven, New York.

Begg, D. A., and Rebhun, L. I. (1979). *J. Cell Biol.* **83,** 241–248.

Behnke, O., and Forer, A. (1967). *J. Cell Sci.* **2,** 169–192.

Bestor, T. H., and Schatten, G. (1981). *Dev. Biol.* **88,** 80–91.

Bibring, T., and Baxandall, J. (1977). *Dev. Biol.* **55,** 191–195.

Bibring, T., and Baxandall, J. (1981). *Dev. Biol.* **83,** 122–126.

Bibring, T., Baxandall, J., Denslow, S., and Walker, B. (1976). *J. Cell Biol.* **69,** 301–312.

Binder, L., and Rosenbaum, J. L. (1978). *J. Cell Biol.* **79,** 500–515.

Bohm, K. T., Vater, W., Fenske, H., and Unger, E. (1984). *Biochim. Biophys. Acta* 800, 119–126.

Borisy, G. G., and Olmsted, J. B. (1972). *Science* **177,** 1196–1197.

Borisy, G. G., and Taylor, E. W. (1967). *J. Cell Biol.* **34,** 525–533.

Bryan, J. (1972). J. Mol. Biol. **66,** 157–168.

Bryan, J. (1975). *Am. Zool.* **15,** 649–660.

Bryan, J., Nagle, B. W., and Doenges, K. H. (1975). *Proc. Natl. Acad. Sci. U.S.A.* **72,** 3570–3574.

Burgess, W. H. (1981). *J. Biol. Chem.* **257,** 1800–1804.

Burns, R. G. (1973). *J. Cell Sci.* **13,** 55-67.

Burns, R. G., and Starling, R. (1974). *J. Cell Sci.* **14,** 411–419.

Cavanaugh, G. M., ed. (1956). "Formulae and Methods IV of the Marine Biological Laboratory Chemical Room." Marine Biological Laboratory, Woods Hole, Massachusetts.

Cleveland, D. W. (1983). *Cell* **34,** 330–332.

Dentler, W. L., Granett, S., and Rosenbaum, J. L. (1975). *J. Cell Biol.* **65,** 237–241.

Detrich, H. W., III, and Wilson, L. (1983). *Biochemistry* 22, 2453–2462.

Detrich, H. W., III, and Williams, R. C., Jr. (1978). *Biochemistry* **17,** 3900–3907.

Detrich, H. W., III, Berkowitc, S. A., Kim, H., and Williams, R. C., Jr. (1976). *Biochem. Biophys. Res. Commun.* **68,** 961–968.

Dirksen, E. R. (1961). *J. Biophys. Biochem. Cytol.* **11,** 244–247.

Dustin, P. (1978). "Microtubules." Springer-Verlag. Berlin and New York.

Farrell, K. W. (1982). *Methods Cell Biol.* **24,** 61–78.

Farrell, K. W., and Wilson, L. (1978). *J. Mol. Biol.* **121,** 393–410.

Farrell, K. W., Morse, A., and Wilson, L. (1979). *Biochemistry* **18,** 950–911.

Field, D. J., Collins, R. A., and Lee, J. C. (1984). *Proc. Natl. Acad. Sci. U.S.A.* **81,** 4041–4045.

Forer, A. (1969). *In* "Handbook of Molecular Cytology" (A. Lima-de-Faria, ed.), pp. 553–601. North-Holland Publ., Amsterdam.

Fuge, H. (1977a). *In* "Mitosis; Facts and Questions" (M. Little, N. Paweletz, C. Petzelt, H. Ponstingl, D. Schroeter, and H.-P. Zimmermann, eds.), pp. 51–68. Springer-Verlag, Berlin and New York.

Fuge, H. (1977b). *Int. Rev. Cytol. Suppl.* **6,** 1–58.

Gaskin, F., Cantor, C. R., and Shelanski, M. L. (1974). *J. Mol. Biol.* **89,** 737–758.

George, H. J., Misra, L., Field, D. J., and Lee, J. C. (1981). *Biochemistry* **20,** 2402–2409.

Gibbins, J. R., Tilney, L. G., and Porter, K. R. (1969). *J. Cell Biol.* **41,** 201–226.

Gibbons, I. R. (1965). *Arch. Biol. (Liege)* **76,** 317–374.

Gibbons, I. R. (1977). *In* "International Cell Biology, 1976–1977" (B. R. Brinkley and K. R. Porter, eds.), pp. 348–347. Rockefeller Univ. Press, New York.

Gibbons, I. R. (1981). *J. Cell Biol.* **91,** 107s–124s.

Hamaguchi, Y., and Iwasa, F. (1980). *Biomed Res.* **1,** 502–509.

Harris, P. J. (1975). *Exp. Cell Res.* **94,** 409–425.

Harris, P. J. (1978). *In* "Cell Cycle Regulation" (J. R. Jeter, Jr., I. L. Cameron, G. M. Padilla, and A. M. Zimmerman, eds.), pp. 75–104. Academic Press, New York.

Harris, P. J. (1981). *In* "Mitosis–Cytokinesis" (A. M. Zimmerman and A. Forer, eds.), pp. 29–57. Academic Press, New York.

Harris, P., Osborn, M., and Weber, K. (1980a). *Exp. Cell Res.* **126,** 227–236.

Harris, P. J., Osborn, M., and Weber, K. (1980b). *J. Cell Biol.* **84,** 668–679.

Harvey, E. B. (1956). "*The American Arbacia.*" Princeton Univ. Press, Princeton, New Jersey.

Head, J. F., Mader, S., and Kaminar, B. (1979). *J. Cell Biol.* **80,** 211–218.

Heilbrun, L. V. (1921). *J. Exp. Zool.* **30,** 211–237.

Hepler, P. K. (1977). *In* "Mechanisms and Control of Cell Division" (T. L. Post and E. M. Gifford, eds.), pp. 212–223. Dowden, Hutchinsen & Ross, Stroudsburg, Pennsylvania.

Hepler, P. K., and Wolniak, S. M. (1984). *Int. Rev. Cytol.* **90,** 169–238.

Hill, T. L., and Kirschner, M. W. (1983). *Int. Rev. Cytol.* **84,** 185–234.

Hisanaga, S., and Sakai, H. (1983). *J. Biochem. (Tokyo)* **93,** 87–98.

Inoue, S. (1981). *J. Cell Biol.* **91,** 131s–147s.

Inoue, S., and Ritter, H., Jr. (1975). *In* "Molecules and Cell Movement" (S. Inoue and R. E. Stephens, eds.), pp. 3–30. Raven, New York.

Inoue, S., and Sato, H. (1967). *J. Gen. Physiol.* **50,** 259–292.

Inoue, S., Borisy, G. G., and Kiehart, D. P. (1974). *J. Cell Biol.* **62,** 175–184.

Johnson, C. H., and Epel, D. (1982). *Dev. Biol.* **92,** 461–469.

Johnson, K. A., and Borisy, G. G. (1975). *In* "Molecules and Cell Movement" (S. Inoue and R. E. Stephens, eds.), pp. 119–139. Raven, New York.

Kane, R. E. (1975). *J. Cell Biol.* **66,** 305–315.

Kato, K. H., and Sugiyama, M. (1971). *Dev. Growth Differ.* **13,** 359–366.

Keller, T. C. S. III, and Rebhun, L. I. (1979). *Biol. Bull. (Woods Hole)* **157,** 374–375.

Keller, T. C. S. III, and Rebhun, L. I. (1982). *J. Cell Biol.* **93,** 788–796.

Keller, T. C. S. III, Jemiolo, D. K., Burgess, W. H., and Rebhun, L. I. (1982). *J. Cell Biol.* **93,** 797–803.

Kiehart, D. P. (1981). *J. Cell Biol.* **88,** 604–617.

Kilmartin, J. V. (1981). *Biochemistry* **20,** 3629–3633.

Kilmartin, J. V. (1982). *In* "Microtubules in Microorganisms" (N. Morris and P. Cappuccinelli, eds.), pp. 171–178. Dekker, New York.

Korn, E. D. (1982). *Physiol. Rev.* **62,** 672–737.

Kuriyama, R. (1976). *J. Biochem. (Tokyo)* **80,** 153–165.

Kuriyama, R. (1977). J. Biochem. (Tokyo) **81,** 1115–1125.
Kuriyama, R., and Borisy, G. G. (1983). *J. Cell Sci.* **61,** 175–189.
Kuriyama, R., and Miki-Noumura, T. (1975). *J. Cell Sci.* **19,** 607–620.
Lee, J. C. (1982). *Methods Cell Biol.* **24,** 9–30.
Lee, Y. C., and Wolff, J. (1982). *J. Biol. Chem.* **257,** 6306–6310.
Lee, Y. C., and Wolff, J. (1984). *J. Biol. Chem.* **259,** 1226–1230.
Linck, R. W., and Langevin, G. L. (1981). *J. Cell Biol.* **89,** 323–337.
Loeb, J. (1913). "Artificial Parthenogensis and Fertilization" Univ. of Chicago Press, Chicago.
Longo, F., and Anderson, E. (1968). *J. Cell Biol.* **39,** 339–368.
McIntosh, J. R. (1977). *In* "Mitosis; Facts and Questions" (M. Little, N. Paweletz, C. Petzelt, H. Ponstingl, D. Schoeter, and H.-P. Zimmermann, eds.), pp. 167–184. Springer-Verlag, Berlin and New York.
McIntosh, J. R. (1981). *In* "International Cell Biology, 1980–1981" (H. G. Schweiger, ed.), pp. 359–368. Springer-Verlag, Berlin and New York.
McIntosh, J. R., Hepler, P. K., and VanWie, D. G. (1969). *Nature (London)* **224,** 659–663.
Margolis, R. L., and Wilson, L. (1981). *Nature (London)* **298,** 705–711.
Margolis, R. L., Wilson, L., and Kiefer, B. (1978). *Nature (London)* **272,** 450–452.
Mazia, D. (1977). *In* "Mitosis; Facts and Questions" (M. Little, N. Paweletz, C. Petzelt, H. Ponstingl, D. Schroeter, and H.-P. Zimmermann, eds.), pp. 196–219. Springer-Verlag, Berlin and New York.
Mazia, D. (1978). *In* "Cell Reproduction" (E. R. Dirksen, D. M. Prescott, and C. F. Fox, eds.), pp. 1–13. Academic Press, New York.
Mazia, D., and Dan, K. (1952). *Proc. Natl. Acad. Sci. U.S.A.* **38,** 826–838.
Meijer, L., and Guerrier, L. (1984). *Int. Rev. Cytol.* **86,** 129–198.
Miki-Noumura, T. (1977). *J. Cell Sci.* **24,** 203–216.
Moy, G., Brandriff, B., and Vacquier, V. (1977). *J. Cell Biol.* **73,** 788–793.
Murphy, D. B. (1980). *J. Cell Biol.* **84,** 235–245.
Murphy, D. B. (1982). *Methods Cell Biol.* **24,** 31–49.
Murphy, D. B., and Borisy, G. G. (1975). *Proc. Natl. Acad. Sci. U.S.A.* **72,** 2696–2700.
Nagle, B. W., and Egrie, J. C. (1981). *In* "Mitosis/Cytokinesis" (A. M. Zimmerman and A. Forer, eds.), pp. 337–361. Academic Press, New York.
Naruse, H., and Sakai, H. (1981). *J. Biochem. (Tokyo)* **90,** 581–587.
Nishida, E., and Kumagai, H. (1980). *J. Biochem. (Tokyo)* **87,** 143–151.
Olmsted, J. B., and Borisy, G. G. (1975). *Biochemistry* **14,** 2996–3005.
Oosawa, R., and Kasai, M. (1962). *J. Mol. Biol.* **4,** 10–21.
Otto, J. J., and Schroeder, T. E. (1984a). *Dev. Biol.* **101,** 274–281.
Otto, J. J., and Schroeder, T. E. (1984b). *Dev. Biol.* **103,** 493–503.
Paweletz, N., and Mazia, D. (1979). *Eur. J. Cell Biol.* **20,** 37–44.
Petzelt, C. (1979). *Int. Rev. Cytol.* **60,** 53–92.
Pfeffer, T. A., Asnes, C. F., and Wilson, L. (1976). *J. Cell Biol.* **69,** 599–607.
Pickett-Heaps, J. D. (1969). *Cytobios* 3, 257–280.
Pierson, G. B., Burton, P. R., and Himes, R. H. (1978). *J. Cell Biol.* **76,** 223–228.
Pratt, M. M. (1980). *Dev. Biol.* **74,** 354–378.
Pratt, M. M. (1984). *Int. Rev. Cytol.* **87,** 83–105.
Pratt, M., Otter, T., and Salmon, E. D. (1980). *J. Cell Biol.* **86,** 738–745.
Raff, E. C. (1979). *Int. Rev. Cytol.* **59,** 1–96.
Raff, E. C. (1984). *J. Cell Biol.* **99,** 1–10.
Raff, R. (1975). *Am. Zool.* **15,** 661–678.
Raff, R. A., and Kaumeyer, J. F. (1973). *Dev. Biol.* **32,** 309–320.
Raff, R. A., Greenhouse, G., Gross, K. W., and Gross, P. R. (1971). **50,** 516–527.

Rebhun, L. I. (1976). *Am. Zool.* **16,** 469–482.
Rebhun, L. I. (1977). *Int. Rev. Cytol.* **49,** 1–54.
Rebhun, L. I., Rosenbaum, J. L., Lefebvre, P., and Smith, G. (1974). *Nature (London)* **249,** 113–115.
Rebhun, L. I., Jemiolo, D. K., Ivy, N., Mellon, M., and Nath, J. P. (1975). *Ann. N.Y. Acad. Sci.* **253,** 362–377.
Rebhun, L. I., Suprenant, K. A., Keller, T. C. S., and Folley, L. (1982). *Cell Differ.* **11,** 367–371.
Regula, C. S., Pfeiffer, J. R., and Berlin, R. D. (1981). *89,* 45–53.
Roberts, K., and Hyams, J. S., eds. (1979). *"Microtubules."* Academic Press, New York.
Sachs, M. I., and Anderson, E. (1970). *J. Cell Biol.* **47,** 140–158.
Sakai, H. (1978). *Int. Rev. Cytol.* **55,** 23–48.
Salmon, E. D. (1982). *Methods Cell Biol.* **25,** 69–105.
Salmon, E. D., and Segall, R. R. (1980). *J. Cell Biol.* **86,** 355–365.
Schatten, G., Schatten, H., Bestor, T. H., and Balczon, R. (1982). *J. Cell Biol.* **94,** 455–465.
Scheele, R. B., Bergen, L. G., and Borisy, G. G. (1982). *J. Mol. Biol.* **154,** 485–500.
Schiff, P. B., Fant, J., and Horwitz, S. B. (1979). *Nature (London)* **277,** 665–667.
Scholey, J. M., Neighbors, B., McIntosh, J. R., and Salmon, E. D. (1984). *J. Biol. Chem.* **259,** 6516–6525.
Silver, R. B., Cole, R. D., and Cande, W. Z. (1980). *Cell* **19,** 505–516.
Silver, R. B., Saft, M. S., Taylor, A. R., and Cole, R. D. (1983). *J. Biol. Chem.* **258,** 13287–13291.
Sloboda, R. D., Rudolph, S. A., Rosenbaum, J. L., and Greengard, P. (1975). *Proc. Natl. Acad. Sci. U.S.A.* **72,** 177–181.
Steinhardt, R. A. (1978). *Nature (London)* **272,** 253–254.
Stephens, R. E. (1975). *In* "Molecules and Cell Movements" (S. Inoue and R. E. Stephens, eds.), pp. 181–206. Raven, New York.
Stephens, R. E. (1977). *Dev. Biol.* **61,** 311–329.
Stephens, R. E. (1978). *Biochemistry* **17,** 2882–2891.
Suprenant, K. A., and Dentler, W. L. (1982). *J. Cell Biol.* **93,** 164–174.
Suprenant, K. A., and Rebhun, L. I. (1983). *J. Biol. Chem.* **258,** 4518–4525.
Suprenant, K. A., and Rebhun, L. I. (1984a). *J. Cell Biol.* **98,** 253–266.
Suprenant, K. A., and Rebhun, K. A. (1984b). *Cell Motil.* **4,** 333–350.
Tilney, L. G., and Gibbins, J. R. (1969). *J. Cell Biol.* **41,** 227–250.
Tilney, L. G., Bryan, J., Bush, D. J., Fujiwara, K., Mooseker, M. S., Murphy, D. B., and Snyder, D. A. (1973). *J. Cell Biol.* **59,** 267–275.
Turner, J. L., and McIntosh, J. R. (1977). *Methods Cell Biol.* **16,** 373–379.
Vallee, R. B. (1980). *J. Cell Biol.* **92,** 435–442.
Vallee, R. B., and Bloom, G. S. (1983). *Proc. Natl. Acad. Sci. U.S.A.* **80,** 6259–6263.
Vallee, R. B., Bloom, G. S., and Theurkauf, W. E. (1984). *J. Cell Biol.* **99,** 38s–44s.
Von Ledebur-Villiger, M. (1972). *Exp. Cell Res.* **72,** 285–308.
Warfield, R. K. N., and Bouck, G. B. (1975). J. Mol. Biol. **93,** 117–120.
Warner, F. R. (1972). *Adv. Cell Mol. Biol.* **2,** 193–235.
Weisenberg, R. C. (1972a). *Science* **177,** 1104–1105.
Weisenberg, R. C. (1972b). *J. Cell Biol.* **54,** 266–278.
Weisenberg, R. C. (1973). *Am. Zool.* **13,** 981–987.
Weisenberg, R. C. (1978). *In* "Cell Reproduction" (E. R. Dirksen, D. M. Prescott, and C. F. Fox, eds.), pp. 359–389. Academic Press, New York.
Weisenberg, R. C., and Rosenfeld, A. (1975). *J. Cell Biol.* **64,** 146–158.
Weisenberg, R. C., and Timasheff, S. N. (1970). *Biochemistry.* **9,** 4110–4116.
Weisenberg, R. C., Borisy, G. G., and Taylor, E. W. (1968). *Biochemistry* **7,** 4466–4479.

Wilson, L., Miller, H. P., Pfeffer, T. A., Sullivan, K. F., and Detrich, H. W. III (1984). *J. Cell Biol.* **99,** 37–41.

Witman, G. B., Carlson, K., and Rosenbaum, J. L. (1972). *J. Cell Biol.* **54,** 540–555.

Zimmerman, A. M., and Forer, A. (1981). *In* "Mitosis/Cytokinesis" (A. M. Zimmerman and A. Forer, eds.), pp. 327–336. Academic Press, New York.

Zimmerman, A. M., Zimmerman, S., and Forer, A. (1977). *Methods Cell Biol.* **16,** 361–372.

# Chapter 12

# *Isolation of Embryonic Cilia and Sperm Flagella*

R. E. STEPHENS

*Marine Biological Laboratory*
*Woods Hole, Massachusetts*



## I. Introduction

With the exception of the assembly of the mitotic apparatus at each division, the first major tubulin-containing structures generated by the developing sea urchin embryo are the cilia that arise at the blastula stage. Studies with actinomycin D (e.g., Gross *et al.*, 1964) indicate that no new mRNA synthesis is required, suggesting that ciliary proteins are coded on preexisting maternal messages. Studies with protein synthesis inhibitors indicate that the cilia are assembled mainly from preexisting protein pools (Auclair and Siegel, 1966). Even the regeneration of cilia can take place quite normally in the presence of inhibitors of protein synthesis (Burns, 1973). Just as most of the tubulin for the first cleavage mitotic apparatus is derived from the unfertilized egg (Bibring and Baxandall, 1977), most of the tubulin that forms the ciliary outer doublets comes from a preexisting pool (Bibring and Baxandall, 1981). The bulk of the tubulin in the unfertilized egg differs at a primary structural level from that found in the cilia, indicating that most tubulin is destined for mitotic apparatus formation (Stephens, 1978), with the ciliary tubulin pool consequently representing only a small fraction of the total. Much of the tubulin incorporated into regenerating cilia is also derived from preexisting pools (Burns, 1973; Stephens, 1977), but multiple deciliation increases the rate of both tubulin and tubulin mRNA synthesis (Stephens, 1977; Merlino *et al.*, 1978).

Although these various inhibition and regeneration studies suggest that the ingredients needed to manufacture cilia are already provided in the unfertilized egg, the story is not that simple. Early studies of specific protein synthesis demonstrate no striking qualitative changes throughout early development, even when analyzed by high-resolution two-dimensional methods (cf. Brandhorst, 1976, and references therein). However, much can go unnoticed in a sea of protein turnover. When only the proteins incorporated into cilia are analyzed by isolating the cilia after appropriate isotope pulse–chase experiments, some striking, developmentally relevant differences are seen (Stephens, 1977). Such an approach confirms the contention that a large pool of tubulin and most other ciliary proteins preexists in the unfertilized egg. However, immediately before ciliogenesis, and also at each regeneration, one particular protein is made *de novo* in a "quantal" amount, sufficient for only one round of ciliogenesis. This protein is a structural component of the outer doublet wall or junction (cf. Linck, 1976), prompting speculation that its limited amount may regulate the length of the cilia that are able to form from the abundant pool of other precursor proteins (Stephens, 1977). Using a more effective inhibitor of protein synthesis than previous workers, Burns (1979) has been able to inhibit the regrowth of cilia in animalized embryos, arguing "that a residual pool of limiting precursor was required to maintain the assembly–disassembly equilibrium in favour of the assembled organelle."

Isolated cilia are useful for developmental studies concerning gene expression, organelle morphogenesis, and kinetic/length determination. Given an adequate supply of healthy sea urchin embryos and the most basic technical skills, the isolation of embryonic cilia, particularly using successive regenerations, is relatively easy. For biochemical comparison of microtubule-based motile systems, sperm flagella offer an abundant source of material and involve equally simple isolation techniques.

## II. Isolation Methods

### A. Embryonic Cilia

#### 1. Background

All current isolation methods are based on the hypertonic salt technique devised by Auclair and Siegel (1966). This approach involves brief exposure of embryos to seawater having approximately twice the tonicity as normal seawater, then returning the embryos to normal seawater, whereupon cilia will regenerate. Other methods that have proved useful for deciliation of protozoa have not received wide acceptance with sea urchin embryos. To strip cilia after regeneration from hypertonically deciliated embryos, Auclair and Siegel (1966) used the calcium–ethanol procedure commonly used for *Tetrahymena* cilia. Although it is quite thorough and yields very clean cilia after gradient centrifugation, this treatment does not normally permit regeneration due to shock to the embryos. The use of dibucaine, also very popular for *Tetrahymena* deciliation, appears not to be effective in marine organisms.

### 2. Method

The following outline presupposes that the worker is capable of fertilizing eggs without producing polyspermy and is able to grow the embryos through hatching without significant cell death or loss of synchrony.

*a. Culture.* Ideally, embryos at a concentration of 1% (v/v) should be grown as a monolayer in large culture dishes, employing Millipore-filtered seawater containing 25 μg/ml each of penicillin and streptomycin, buffered with 10 m*M* Tris-HCl, pH 8.3, and containing 0.1 m*M* EDTA. For maximum oxygen exchange, the seawater should be no more than 1 cm deep. The temperature and hatching time depend, of course, upon the species used. Once morphogenetic "landmarks" are known for several temperatures, the time course of development is easily predicted (e.g., Stephens, 1972), allowing temperature to be used as a tool to produce embryos at convenient times and stages for cilia isolation (Fig. 1).

For scaling up the procedure, large volumes containing higher egg concentrations may be mechanically stirred in a variety of culture flasks. I have used a 3-liter Erlenmeyer flask containing 1 liter of 5% egg suspension, magnetically stirred with an egg-shaped stir bar on the bottom, for both *Strongylocentrotus droebachiensis* (a delicate, cold-water form) and *Tripneustes gratilla* (a sturdy, tropical urchin). This simple set-up should be used with caution since an improperly chosen stir bar or too high a speed will grind up the embryos. A more expensive but foolproof approach is to use tissue culture flasks having suspended magnetic stir bars or paddles.

*b. Harvesting Clean Embryos.* First, place the hatched culture in a beaker (or several 250-ml beakers for large cultures) and then decant the swimming embryos from any unhealthy embryos that settle. This may be repeated several times to ensure that the culture consists only of swimming embryos. The embryos should be washed at least once with a minimum of 10 vol of Millipore-filtered, Tris-buffered, EDTA–seawater (see above) to rid the culture of any fine cell debris. This and the following steps are best accomplished through the use of a hand-driven centrifuge equipped with two 50-ml conical tube buckets (e.g., a

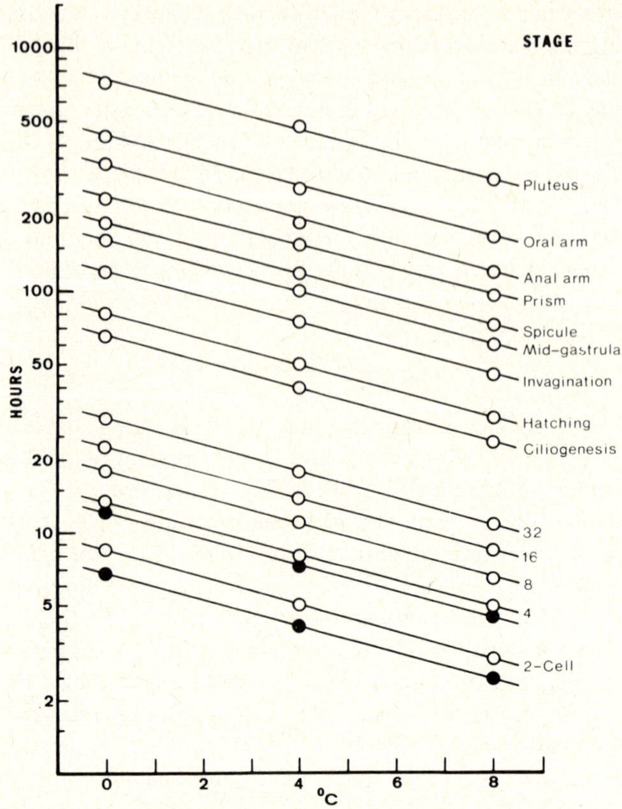


FIG. 1. Time of developmental stage versus culture temperature for the sea urchin *Strongylocentrotus droebachiensis*. Filled circles (●) denote the time of metaphase. These data encompass the full physiological temperature range for this organism (cf. Stephens, 1972).

Boekel four-place unit with two buckets removed to allow slower stoppage). Most embryos can be sedimented without injury at <1000 rpm, corresponding to one handle rotation per second (1:16 gear ratio) for 10 seconds. They may be recovered without injury by quickly drawing (or pouring) off the supernatant, gently adding fresh medium, and then resuspending the pellet by wafting the medium against the packed pellet with a Pasteur pipet whose tip is broken off to an inside diameter of about 3 mm. Once the wash medium is free of cell debris, as judged by phase-contrast microscopy, sediment the embryos and draw off the wash medium quickly and completely.

*c. Deciliation.* To deciliate, add 10 vol of Tris-buffered, EDTA–seawater that contains an additional 0.5 mol/liter (29.2 g) of sodium chloride, at the normal growth temperature of the embryos. Suspend the pellet as above and stir intermittently, and gently, with a glass rod to prevent the embryos from settling. Check the progress of deciliation with a phase-contrast microscope, leaving the embryos in the hypertonic medium no longer than necessary for complete deciliation, typically about 2 minutes.

*d. Separation.* Sediment the deciliated embryos as prescribed above and carefully pipet the supernatant into a chilled centrifuge tube, taking care not to include any embryos. If embryos or embryo fragments are accidently included, they may be removed by recentrifuging the supernatant in a fresh tube containing a small ball of cotton (<1 cm) at about 2000 rpm (two turns per second) for 10–15 seconds. This will trap the debris and not allow it to resuspend as the centrifuge comes to a stop.

*e. Regeneration.* The pellet of packed embryos may be suspended in sterile, Tris-buffered EDTA–seawater and allowed to regenerate cilia, using the original growth conditions (Section II,A,2,a above). They can then be used for a second harvest of cilia, typically cleaner than the virgin cilia, by repeating the steps outlined in Sections II,A,2,b–d.

*f. Cilia Recovery.* Recover the cilia from the chilled supernatant from the separation step (Section II,A,2,d) by centrifugation at 10,000 *g* for 10 minutes at 4°C in a standard refrigerated centrifuge. To prevent proteolysis, phenyl methylsulfonyl fluoride (PMSF) may be added to the supernatant immediately after removal from the deciliated embryos. This is done conveniently by adding 0.2 ml of a 1% solution of PMSF in 2-propanol to 40 ml of ciliary supernatant, with thorough mixing.

### 3. Variations

Hatched embryos may be treated with trypsin for prolonged periods of time, resulting in extensive animalization, characterized by growth of all of the embryonic cilia to two to three times their normal length (Riederer-Henderson and Rosenbaum, 1979; Burns, 1979). This may be done by treating newly hatched embryos with 50 μg/ml of trypsin (bovine Type II: Sigma Chemical Co., St. Louis, MO) in sterile, Tris-buffered, EDTA–seawater for 12–20 hours. When the desired degree of animalization is achieved, the embryos are thoroughly washed and then deciliated as described above. Although this method has proven successful with *Arbacia punctulata,* its application to other species may require adjustment of conditions.

### 4. Cautions

It is important to prevent bacterial growth, particularly prior to hatching, since overgrowth with bacteria can destroy the embryos and may produce heavy bacterial and blastomere contamination of the isolated cilia fraction. This is less of a

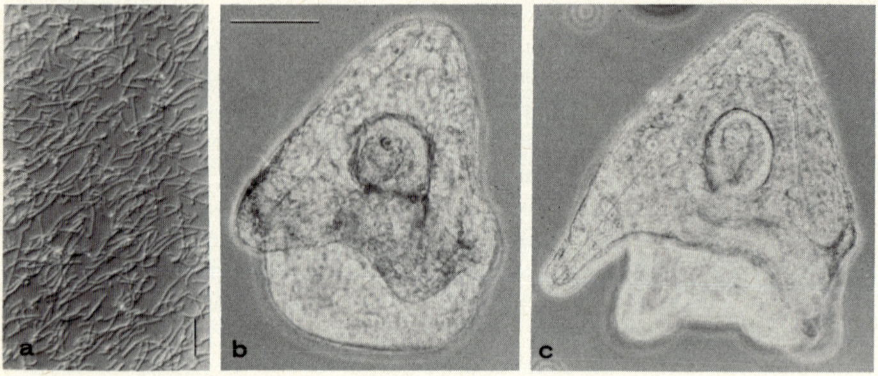


FIG. 2. Isolated blastula cilia (a) by differential interference-contrast microscopy, a pluteus larva of *S. droebachiensis* that was deciliated 10 times (b), and an untreated larva from the same culture (c), both by phase-contrast microscopy. The scale marker is 10 μm in (a) and 100 μm in (b).

problem during regeneration since the embryos are washed thoroughly as a consequence of the initial deciliation, and little or no cell dissociation occurs at later stages. When this procedure is done properly, very pure cilia will result and they generally need no further purification (Fig. 2a). The use of a hand centrifuge is not required but it is highly recommended so that the steps may be performed quickly and gently, minimizing pellet dispersal that generally accompanies rotor deceleration in a conventional (e.g., clinical) centrifuge. Excessive packing of the embryo pellet requires excessive force to disperse the embryos, resulting in cell dissociation and subsequent loss. I have used this procedure for 10 successive deciliations in large-scale preparations of the 2 species noted above, with losses typically no more than 30% of the original packed embryo volume. In spite of such repeated abuse, the embryos develop surprisingly normally (Fig. 2b–c).

## B. Sperm Flagella

### 1. BACKGROUND

Most sea urchin sperm tail isolation procedures currently in use are based upon a simple decapitation–differential centrifugation method originally devised to produce more amounts of axonemal tubulin for biochemical comparison (Stephens, 1970). Less convenient variations involve decapitation by forcing sperm through a syringe needle and separating heads from tails on a sucrose gradient, or by suspension in detergent solution. Both approaches result in membrane removal and possible loss of protein components during isolation.

### 2. Method

This outline presupposes that the worker is able to obtain sperm by KCl injection, electrical stimulation, or excision of gonads.

*a. Sperm Recovery.* Shed urchins into, or mince gonads in, Tris-buffered EDTA–seawater (above) and filter the resulting sperm suspension through four layers of cheesecloth or a single layer of Miracloth (Calbiochem-Behring, La Jolla, CA) to remove gonadal debris. Chill to 4°C. Centrifuge the sperm suspension at 5000 *g* for 5 minutes to produce pellets that may be stored (dry) on ice for several days. It is most convenient to adjust the original dilution such that the final pellet volume is about 2 ml.

*b. Decapitation.* With a stirring rod, suspend 2 ml of "dry" sperm in 40 ml of Tris-buffered EDTA–seawater and homogenize by about 10 strokes in a glass–Teflon or Dounce homogenizer until at least 90% of the sperm are decapitated. Five seconds in a Sorvall Omnimixer with a 50-ml vessel also works well. Check the progress with a phase-contrast microscope to avoid excessive homogenization.

*c. Head/Tail Separation.* Centrifuge at 1000 *g* for 5 minutes to sediment to sperm heads and intact sperm. Carefully decant the supernatant of sperm tails into a fresh tube, discarding the pellet and any interface material.

*d. Tail Recovery.* Centrifuge the supernatant at 10,000 *g* for 5 minutes to sediment the sperm tails. Discard the resulting supernatant and resuspend the pellet in fresh medium. Check for the presence of contaminating sperm heads and motile sperm; some will usually be present.

*e. Clean-Up.* Repeat the separation and tail recovery step (Section II,B,2,c and d) until the resuspended sperm tails are totally free of sperm heads. Generally one additional cycle will produce a pure preparation (Fig. 3), given that the decantation in the initial separation step is done carefully. Finally, spin down the clean tails for further use.

### 3. Complications

This method works successfully with at least six echinoid species. Two problems can arise, however, that make the differential centrifugation less efficient. Excessive homogenization will break the tails into small pieces, lowering their recovery unless longer centrifugation or higher speeds are used. This in turn will result in the contamination of the tail fragments with mitochondria and membranous debris that normally would remain in the high-speed supernatant. Secondly, the presence of mucus, found in certain species or at certain times of the breeding season, will cause loss of tails due to the sedimentation of tail–mucus aggregates into the low-speed pellet. This often can be minimized by gently washing the intact sperm several times with Tris-buffered EDTA–seawater be-

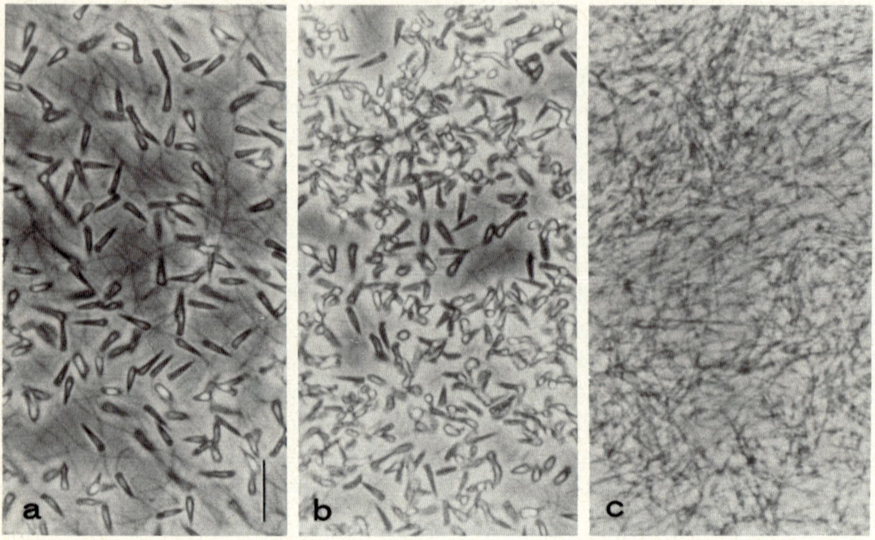


FIG. 3. Whole sperm (a), sperm heads (b), and sperm flagella (c) from *S. droebachiensis*. The scale marker is 10 μm. Phase-contrast microscopy.

fore homogenization, taking care not to decapitate the sperm during resuspension. Avoid pipetting; a stirring rod and natural swimming action will readily resuspend a pellet resulting from centrifugation at 1000 *g* for 5 minutes.

## III. Fractionation Methods

### A. Demembranation

Resuspend pellets of cilia or flagella in at least 10 vol of a solution containing 1% Triton X-100 (or NP-40) in 3 m*M* $MgCl_2$ and 30 m*M* Tris-HCl, pH 8, followed by centrifugation for 10 minutes at 10,000 *g* to totally solubilize the membrane (Stephens, 1970, 1977, 1978). This results in the production of intact 9 + 2 axonemes for further fractionation and also serves to solubilize any cellular contaminants. An additional wash with 50 vol of 10 m*M* Tris-HCl, pH 8, will dilute the detergent and magnesium sufficiently for further fractionation.

### B. Outer Doublet Microtubules

Resuspend sperm flagellar axonemes in 10 vol of 0.6 *M* KCl, 10 m*M* Tris, pH 8, and 0.1 m*M* dithiothreitol for 30 minutes at 4°C to solubilize both central pair members and most of the dynein ATPase, producing nearly pure outer doublet microtubules. Recover the outer doublets by centrifuging at 10,000 *g* for 5 minutes. As a cleaner but longer alternative, resuspend sperm flagellar axonemes in 10 vol of 1 m*M* Tris-HCl, pH 8, and 0.1 m*M* EDTA and dialyze at 0–4°C against the same medium for 18–36 hours. To recover the outer doublet fraction from the dialysis procedure as a tight pellet, increase the ionic strength by making the solution 10–20 m*M* in Tris-HCl, pH 8, prior to centrifugation at 35,000 *g* for 10 minutes.

In contrast to the relatively simple fractionation of sperm flagella, the identical low ionic strength dialysis (or high-salt extraction) of ciliary axonemes results in the solubilization of one-half of the dynein ATPase (outer arm only?), one central pair member, and all of the B-subfibers, yielding an "outer fiber" fraction that is in reality a ring of A-tubules (Stephens, 1977, 1978).

The B-subfibers can be "melted" away from outer doublets of sperm flagella (Stephens, 1970). Resuspend the outer doublets produced by low ionic strength dialysis, in about 10 vol of 1 m*M* Tris, pH 8, containing 1 m*M* dithiothreitol. Heat the suspension to 40°C for 2 minutes (for cold water *S. droebachiensis*) or 5–10 minutes at ~50°C (for temperate or tropical sea urchins), chill quickly, add Tris to 10 m*M*, and centrifuge at 50,000 *g* for 30 minutes to recover the intact A-tubules as a pellet and the solubilized B-tubulin and minor structural proteins as the supernatant.

## C. Comparative Biochemistry

These isolation and fractionation procedures for sea urchin cilia and flagella (Fig. 4), coupled with the isolation of total cytoplasmic tubulin from the unfertilized egg by vinblastine precipitation, yield sufficient material to allow an analysis of tubulin from the respective organelles (Stephens, 1978). As shown by two-dimensional tryptic peptide mapping, both the alpha and beta chains of cytoplasmic tubulin, mainly mitotic in fate, are significantly different than the equivalent chains from either cilia or flagella. The outer doublet tubulins from sperm flagella are clearly distinguishable from those of the A-subfiber of embryonic cilia, although these resemble each other more that they do the equivalent chains of cytoplasmic tubulins. Intermediate are the flagellar central pair tubulins, distinguishable from outer doublet tubulins but also resembling cytoplasmic tubulins. Most of the primary structural differences are difficult to reconcile on the basis of simple posttranslational modification, suggesting that separate gene products are involved in specifying cytoplasmic, ciliary, and flagellar tubulins. This idea is consistent with the unusually large number of tubulin genes found in the sea urchin (Alexandraki and Ruderman, 1981).

Considering the marked fractionation differences seen in embryonic cilia versus sperm tail flagella, it will be of interest to determine whether other structural

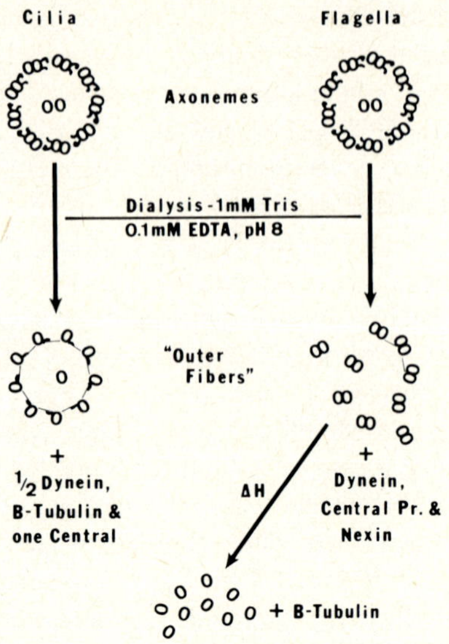


FIG. 4. Fractionation scheme contrasting the behavior of ciliary and flagellar 9 + 2 axonemes. In the former, a ring of nine singlet A-tubules and one central pair microtubules remain stable while, in the latter, true outer doublets are stable and both central pair members dissolve. Mild heat treatment, in turn, solubilizes the B-subfiber, leaving flagellar A-tubules.

differences exist in the accessory proteins responsible for production of 9 + 2 symmetry. Dynein also extracts quite differently in these two morphologically identical organelles and a comparative analysis may prove to be a valuable future endeavor.

## ACKNOWLEDGMENTS

This work was supported by USPHS Grant GM 20,644 from the National Institute for General Medical Sciences.

## REFERENCES

Alexandraki, D., and Ruderman, J. V. (1981). *Mol. Cell. Biol.* **1,** 1125–1137.
Auclair, W., and Siegel, B. W. (1966). *Science* **154,** 913–915.
Bibring, T., and Baxandall, J. (1977). *Dev. Biol.* **55,** 191–195.
Bibring, T., and Baxandall, J. (1981). *Dev. Biol.* **83,** 122–126.
Brandhorst, B. P. (1976). *Dev. Biol.* **52,** 310–317.
Burns, R. G. (1973). *J. Cell Sci.* **13,** 55–67.
Burns, R. G. (1979). *J. Cell Sci.* **37,** 205–215.
Gross, P. R., Malkin, L. I., and Moyer, W. D. (1964). *Proc. Natl. Acad. Sci. U.S.A.* **51,** 407–414.

Linck, R. W. (1976). *J. Cell Sci.* **20,** 405–439.
Merlino, G. T., Chamberlain, J. P., and Kleinsmith, L. J. (1978). *J. Biol. Chem.* **253,** 7078–7085.
Riederer-Henderson, M. A., and Rosenbaum, J. L. (1979). *Dev. Biol.* **70,** 500–509.
Stephens, R. E. (1970). *J. Mol. Biol.* **47,** 353–363.
Stephens, R. E. (1972). *Biol. Bull.* **142,** 132–144.
Stephens, R. E. (1977). *Dev. Biol.* **61,** 311–329.
Stephens, R. E. (1978). *Biochemistry* **17,** 2882–2891.

# Chapter 13

## *Components of the Actin-Based Cytoskeleton*

**R. E. KANE**

*Pacific Biomedical Research Center*
*Kewalo Marine Laboratory*
*University of Hawaii*
*Honolulu, Hawaii*



## I. Introduction

Sea urchin eggs offer many well-known advantages as experimental material for investigations in cell biology: they are relatively large cells, are easily obtained in quantity and, most important, undergo synchronous cell divisions and early embryonic development. Mitosis and cytokinesis have undoubtedly been investigated more extensively in these cells than in any other and their synchronous divisions have made them the primary source material for isolation of the mitotic apparatus and, more recently, for investigations of the role of cortical actin and myosin in furrowing. Although these features are advantageous in the isolation and investigation of the proteins of the actin-based cytoskeleton, these cells also have a significant disadvantage in the presence of large amounts of storage protein present in the yolk granules. This distorts the relative amounts of cytoskeletal protein in comparison to that measured in other cell types: actin and myosin play a major role in cell division and early embryonic events, yet these proteins make up a misleadingly small fraction of the *total* cell protein. More importantly, this large amount of storage protein presents a serious experimental problem when extracting the soluble egg proteins, as its presence can reduce the relative amounts of actin and the actin-associated proteins. This requires that procedures and solutions be chosen that avoid or minimize the breakdown of these granules and the release of protein. It is important to select the tonicity of the media used to avoid swelling of the granules and control the free calcium, which triggers breakdown (Hultin, 1950). By observing these precautions the soluble cytoplasmic proteins can be obtained from unfertilized and fertilized eggs. Actin-based cytoskeletal proteins make up a significant fraction of these preparations. The behavior of these cytoplasmic extracts has provided a useful model for the investigation of the role of these proteins in the events of division and early development.

This article will be concerned with the methods we have developed for the extraction and study of proteins of the actin-based cytoskeleton, using as material primarily the gametes of the Hawaiian sea urchin, *Tripneustes gratilla*. In addition to being one of the more common local species, this sea urchin also offers the advantage of large size, so that 10–20 ml of gametes can be obtained from a ripe animal. Furthermore, its breeding period extends over more than 10 months of the year. Most of these methods have been used satisfactorily with several other sea urchin species, and it is assumed that they are transferable with appropriate modifications to most. One condition which should be kept in mind in this regard as a possible source of divergence is temperature. The normal sea temperature to which these Hawaiian animals are exposed is 24–25°C, with little seasonal variation. This is very favorable from the standpoint of the investigator, as gametes develop at room temperature, with no arrangements for temperature control being necessary. The behavior of extracts is undoubtedly temperature dependent, so appropriate adjustments of temperature when working with other species will have to be made.

The focus will be on the cytoplasmic extract as a model system for studying cytoplasmic behavior. The detailed procedures to be described include the preparation of extracts, the recovery of actin, myosin, and several other actin-associated proteins through the use of procedures based on the specific properties of these extracts, methods for the study of the interaction of these components *in vitro*, and the relation of these reactions to cellular events. Other methods for the isolation of actin (Mabuchi and Spudich, 1980; Miki-Noumura and Oosawa, 1969; Miki-Noumura, 1969), myosin (Mabuchi, 1976), tropomyosin (Ishimoda-

Tagaki, 1978) and several actin-modulating proteins (Hosoya *et al.*, 1982; Hosoya and Mabuchi, 1984; Wang and Spudich, 1984) will not be reviewed.

## II. Preparation of Extracts

### A. Obtaining and Handling Eggs

Sexes cannot be determined externally in *T. gratilla,* but fairly vigorous shaking of the animals (aboral side down) causes ripe animals to shed a small quantity of gametes and allows the sexes to be separated. The animals can be shed out by injection of 0.56 *M* KCl (isotonic with our seawater); electrical stimulation is unsuccessful. The animals shed slowly, and repeated injections of small quantities of KCl over 15–30 minute are the most effective. Eggs are separated from pigment cells and other contaminants by several washes with seawater and are viable for several hours if maintained at room temperature with slow stirring by means of a suspended magnetic stirrer. Just before fertilization or extraction eggs are dejellied by titrating to pH 5.0 on a pH meter, using dilute HCl in seawater, and then rapidly washed several times in fresh seawater by hand centrifugation. We have observed that a pause of ~5 minutes after dejellying improves fertilizability, but it is important that the eggs be gently agitated during this time and fertilized promptly after this, as the close packing that occurs when dejellied eggs are allowed to settle reduces viability, the first effect seen being the formation of "tight" membranes upon fertilization.

Many of our procedures involve measurements based on a known volume of eggs. All egg transfers in the procedures to be described are carried out using 50-ml conical centrifuge tubes in a hand centrifuge for speed and convenience; for reproducible egg volume measurements the eggs are transferred to a 15-ml tube and centrifuged in an electric centrifuge at 220 *g* for 30 seconds. This provides a useful comparative measure and has no ill effects on the cells. Higher speeds and longer times will pack the eggs to smaller volume but reduce viability.

### B. Fertilization and Membrane Removal

If extracts are to be made of fertilized eggs, the fertilization membrane and hyaline layer must be removed. Treatment with isotonic glycine is effective with the eggs of *T. gratilla* (Kane, 1973). This procedure allows several milliliters of eggs to be treated quickly, but has the disadvantage that one step of the process must be accomplished within an extremely short time period. 0.75 to 1 ml of dejellied eggs can be treated in each 50-ml centrifuge tube; two tubes can be handled easily and four with experience. A small volume of a dense sperm suspension is added to 0.75 to 1 ml of eggs suspended in 30–40 ml of seawater, stirred with a glass rod, and timing started. Hand centrifugation is begun at 20 seconds, and the supernatant poured off and replaced with glycine at 40–50 seconds (maximum 1 minute). The solution used is 1 *M* glycine, adjusted to pH 8.4 with NaOH. The fertilization membrane lifts and is solubilized, and the products of cortical granule release remain in solution. We have used this method successfully with other Hawaiian species (*Colobocentrotus atratus, Pseudoboletia indiana*) and find that adjustment for the time of membrane separation is critical as the cells must be in the glycine solution at the time the membrane forms. We use glycine solution at room temperature; appropriate modification will be necessary for colder water species. At approximately 5 minutes after fertilization the blastomeres are gently sedimented, and the glycine is replaced with the medium used to culture the embryos; this solution is usually changed again 5 minutes later to remove any debris from broken cells. Any cloudiness of the supernatants indicates that cell breakdown is occurring due to overly vigorous centrifugation or stirring.

Such membraneless eggs are "sticky" and are usually grown in calcium-free seawater to prevent clumping. The blastomeres of *T. gratilla* secrete additional hyalin protein at later stages (Kane, 1973) and will adhere to form embryos if grown in normal seawater after glycine treatment. These cells can be cultured successfully in calcium-free seawater after membrane removal and will continue to divide, separating at each division. To insure that the cells are hyalin free for some experiments we have utilized calcium-free media with reduced magnesium. A magnesium concentration of 0.1 *M* will gel hyalin and seawater contains approximately one-half this amount. A culture medium without divalent ions would be ideal, such as the so-called "19:1" medium of Chambers (1940), but blastomeres of *T. gratilla* will not complete first cleavage in this medium. We have determined that the minimum concentration of magnesium to support continued cleavage is about 2–3 m*M* and the addition of 5 m*M* $MgSO_4$ to a 19:1 mixture of isotonic (0.56 *M*) NaCl and KCl will support continued cleavage in this species, the cells separating completely at each division with no evidence of membrane or hyalin. Measured seawater pH is 8.3 and cleavage rate is slowed at pH 8.0; we routinely buffer all media used for growing embryos with 10 m*M* Tris, pH 8.4–8.5.

## C. Extraction of Soluble Cytoplasmic Protein

To prevent granule lysis during extraction the medium must be isotonic or hypertonic with sufficient EGTA to control free calcium, which causes breakdown of yolk (Hultin, 1950) and cortical (Vacquier, 1975) granules. Our standard medium contains 0.9 *M* glycerol, 100 m*M* PIPES buffer, 5 m*M* EGTA, pH 6.8–7.0. This concentration of buffer is more than sufficient to control pH during cell breakdown and the EGTA concentration is twice the observed minimum of 2–3 m*M*. The final osmolarity of this medium is 1.3 times that of seawater; hypertonicity helps to prevent granule swelling and breakdown.

The eggs, unfertilized or fertilized, are collected in 50-ml centrifuge tubes, and their volume is estimated by centrifugation in these tubes at 220 *g* for 1 minute. The eggs are then washed in 10–15 times their volume with a 19:1 mixture of NaCl and KCl containing 2 m*M* EDTA to remove divalent salts and washed once more with 2–3 times their volume of cold extraction medium. Further steps are carried out at 0–4°C. The final egg pellet is suspended in 1.5 times its volume of fresh extraction medium, transferred to a large Dounce homogenizer and homogenized until few whole eggs remain. This ratio of eggs to medium was chosen to give a very concentrated extract; higher ratios become too viscous to centrifuge efficiently, while more dilute extracts do not display the same gelation behavior. We find that approximately 20 passes in the homogenizer are required; it is preferable to have some unbroken eggs than attempt to eliminate them, as this effort usually leads to some granule breakdown. A simple indicator of yolk granule breakdown during extraction is the presence of significant amounts of a major yolk protein at ~180 kDa; we also consider the absence of precipitable hyalin (Kane, 1973) in the extract to be an indicator of the absence of cortical granule breakdown in extracts of unfertilized eggs.

Total volume of extract to be prepared is limited by the capacity of the centrifuge rotor. A swinging bucket rotor *must* be used as the granule pellet is not tightly packed and will redistribute in an angle rotor. We run either a total of 30 ml in a Beckman SW 50.1 rotor at 40,000 rpm for 1 hour or 60 ml in a SW 41 titanium rotor at 37,000 rpm for 1.5 hour at 4°C. The centrifuged homogenate of *T. gratilla* consists of a small pigment pellet at the bottom, orange–yellow granules (contributing approximately one-half the total volume), a clear extract, and then a thin, floating lipid layer. The latter can be removed by quick aspiration, and the clear layer can be removed by a syringe attached to polyethylene catheter tubing. Approximately one-half of the original extract volume is recovered as clear supernatant. The pellets are loose and the supernatant must be removed slowly to avoid remixing; it is often possible to return to a tube a second time and recover slightly more material. The clear extract is dialyzed overnight at 0–4°C against a low ionic strength medium (5 m*M* PIPES, 0.1 m*M* EGTA, 0.1 m*M* ATP, pH 6.8–7.0) at a ratio of 30 ml of supernate to 1 liter of dialysis medium. A small precipitate forms on dialysis and is removed by centrifugation at 25,000 *g* for 20 minutes at 4°C. 1 m*M* EGTA and 1 m*M* sodium azide are added to the final extract, which has a protein concentration of 10–12 mg/ml. Actin is a major component of these extracts and there are no major differences in the composition of extracts of unfertilized and fertilized eggs (Fig. 1). Extracts are stable on ice for 1 day and can be frozen in liquid nitrogen and stored at −80°C.

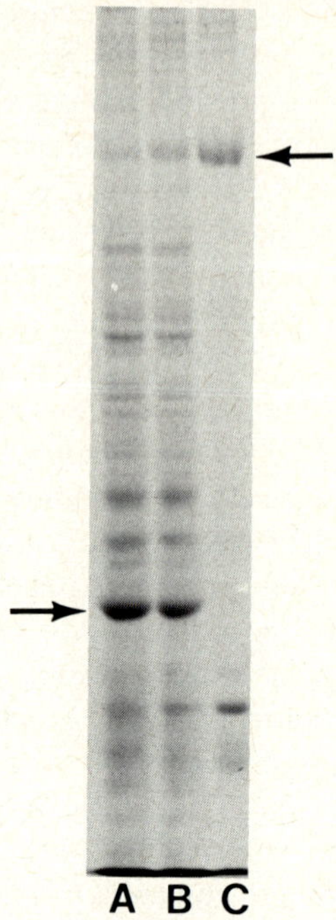


FIG. 1. SDS–polyacrylamide gel of an electrophoresed extract from unfertilized eggs (A) and eggs at 1 hour after fertilization (B). Left arrow indicates actin band. Components of yolk granules of unfertilized eggs in (C); right arrow indicates ~180K yolk polypeptide used as indicator of yolk granule breakdown during extraction.

We have tested the effect of the inclusion of several proteolysis inhibitors (phenylmethylsulfonyl fluoride, leupeptin, α-macroglobulin, tosylarginine methyl ester) and have detected no apparent change in the behavior of the resulting extracts in the procedures to be described below. Inhibitors are not routinely included; the control of granule lysis by hypertonicity and EGTA may prevent the release of proteolytic agents.

## III. Behavior of Extracts and Preparation of Cytoskeletal Proteins

### A. Myosin

Myosin is present in soluble form after extracts have been dialyzed to low ionic strength. It was first seen as part of a precipitate which forms on extended storage of extract at 0°C. It was later determined that precipitation of myosin could be induced in extract immediately after dialysis by the addition of 8 m*M* ATP; SDS–polyacrylamide gel electrophoresis showed the resulting precipitate to consist almost entirely of myosin (Kane, 1980). Myosin is routinely prepared by adding 8 m*M* ATP to extract, allowing it to incubate for 3 hours on ice and collecting the precipitate, which appears as short-beaded chains by phase-contrast microscopy, by centrifugation at 4000 *g* for 10 minutes. The precipitate is washed in 10 m*M* PIPES buffer, pH 7.4, and dissolved in 1 *M* KCl containing the same buffer. We conventionally use a volume of KCl that is one-fortieth the volume of extract processed; this yields a final myosin concentration of 2.5–3.0 mg/ml. For storage it is diluted with an equal volume of glycerol and held at −20°C. The yield is typically 0.075 mg/ml of extract, or 0.75% of the protein of an extract with a concentration of 10 mg/ml.

Myosin precipitation occurs with ATP, GTP, ITP, and ATP-γ-S, but not with nonmetabolizable ATP analogs nor with ADP or AMP. Light chain phosphorylation is apparently not involved in this ATP-induced precipitation of myosin at low ionic strength, although it does influence the activity of myosin in contractile gels (Burgess *et al.*, 1984). The possibility of heavy chain phosphorylation in the precipitation of myosin has not been investigated. ATP-induced precipitation provides a simple one-step procedure for separating myosin from extract that can be carried out at low temperatures, where other reactions such as actin polymerization are minimized, and yields a relatively pure product.

## B. Actin and Extract Gelation

Actin makes up a significant fraction of soluble cytoplasmic protein recovered in the extract. This actin does not polymerize in extracts of *T. gratilla* eggs at the physiological temperature (24–25°C) of these animals, although purified sea urchin egg actin has been shown to have a critical concentration similar to other actins (Mabuchi and Spudich, 1980). Polymerization is induced by warming extracts to 35–40°C in the presence of low concentrations of ATP and KCl. Minimum ATP concentration is 0.5 m*M;* we use 1 m*M* in all cases. The KCl concentration required varies somewhat among extracts and test samples containing 1 m*M* ATP and KCl in the range of 10–50 m*M* are warmed to 40°C for 30 minutes. In most cases 15–25 m*M* KCl yields a gel composed of a network of delicate filaments visible by phase-contrast microscopy (Kane, 1975); at lower KCl concentrations the yield is reduced, while at higher concentrations the gel becomes amorphous with little visible filamentous structure. A KCl concentration that yields gel with a clear filamentous structure in good quantity is chosen and added to the total extract with 1 m*M* ATP. The mixture is warmed to 40°C for a minimum of 30 minutes and a maximum of 60 minutes.

The adaptation of this procedure to extracts of other sea urchin species will require systematic testing of the temperature necessary to induce actin polymerization. An extract of the eggs of *T. gratilla* containing the appropriate concentration of ATP and KCl and warmed to the normal physiological temperature of these animals exhibits very limited actin polymerization after 6 hours, whereas polymerization is complete in less than 1 hour at 40°C. As other species generally have lower environmental temperatures, lower polymerization temperatures will presumably be required and must be determined by trial and error.

When gelatin is complete, the tubes are removed from the water bath and centrifuged for 20 minutes at 25,000 *g* at room temperature to sediment the gel. The supernatant from this centrifugation is discarded and the pellet is stirred to a slurry in a wash solution containing 50 m*M* KCl, 1 m*M* EGTA, and 10 m*M* PIPES, pH 6.9. All further steps are carried out at 0–4°C. The gel is repelleted at 25,000 *g* for 20 minutes in a weighed tube and the supernatant is discarded. The pellet weight is determined and an equal volume of 2 *M* KCl, 10 m*M* PIPES, pH 6.9, is added, followed by 2 times the pellet weight of 1 *M* KCl similarly buffered. The gel is dispersed in this solution by brief homogenization and after 15–30 minutes on ice it is recentrifuged at 25,000 *g* for 20 minutes to remove any insoluble material. Approximately 1 mg of protein is obtained from 1 ml of the usual extract having a protein concentration of 10 mg/ml.

The resulting solution contains all of the gel components; the bulk of the F-actin present can be separated by an ATP-induced aggregation to bundles, a process apparently unique to sea urchin actin (Bryan and Kane, 1982). This process requires a high actin concentration (the critical actin concentration is of the order of 0.5 mg/ml), high KCl (1 *M*) and 1–2 m*M* ATP. Bundle formation is complete overnight at 0°C. The resulting aggregates are easily visible in the phase microscope (Kane, 1975) and are collected by centrifugation at 25,000 *g*, 20 minutes in a weighted tube. The supernatant is removed and saved, and the actin pellet is drained, weighed, and dispersed in 9 times the pellet weight of 50 m*M* KCl, 10 m*M* PIPES, pH 6.9, with gentle homogenization. The final actin solution is clarified at 25,000 *g* for 20 minutes and stored with the addition of 1 m*M* sodium azide; the protein concentration is 4–6 mg/ml. Polyacrylamide gel electrophoresis shows actin to be the only significant component (Kane, 1975); the bundling process can be repeated if further purification is desired.

The saved supernatant solution contains the other gel components plus a small amount of actin. This actin is removed by dialyzing the solution against 50 m*M* KCl, 0.1 m*M* EGTA, 10 m*M* PIPES, pH 6.9, to reduce the salt concentration, causing the residual actin present to combine with the other components to form a small amount of gel on overnight dialysis. Optionally, the supernatant solution can be centrifuged at 100,000 *g* for 3 hours at 4°C to remove the remaining F-actin before dialysis, thereby eliminating most of the loss on the dialysis step. The gel pellet resulting from dialysis is removed by centrifugation at 25,000 *g* for 20 minutes and discarded and 1 m*M* EGTA and sodium azide are added to the supernatant. The components of this solution are those which were originally combined with the actin removed at the earlier step. The major component is a 58-kDa protein called fascin with smaller and variable amounts of a 220-kDa polypeptide. The final protein concentration is 0.5–1 mg/ml. Composition of the myosin, actin, and supernatant solutions by SDS–polyacrylamide gel electrophoresis is shown in Fig. 2. Recombination of actin and supernatant solutions in approximately equal protein amounts reconstitutes gel with the same unique ultrastructural banding pattern as seen in the gel formed in the extract (Kane, 1976). A similar pattern was later seen (Burgess and Schroeder, 1977) and analyzed (Spudich and Amos, 1979) in the actin cores of egg microvilli and fascin was shown to localize in the microvilli after fertilization (Otto *et al.*, 1980). This supernatant solution has been used without further purification in most experiments but, if desired, the fascin and 220-kDa components can be separated by fractionating gel on agarose and DEAE–cellulose columns (see Bryan and Kane, 1978, for details). By this method it was demonstrated that the banding pattern seen in gel was a property of the actin–fascin combination only; the structural basis of this pattern is described in DeRosier *et al.* (1977).

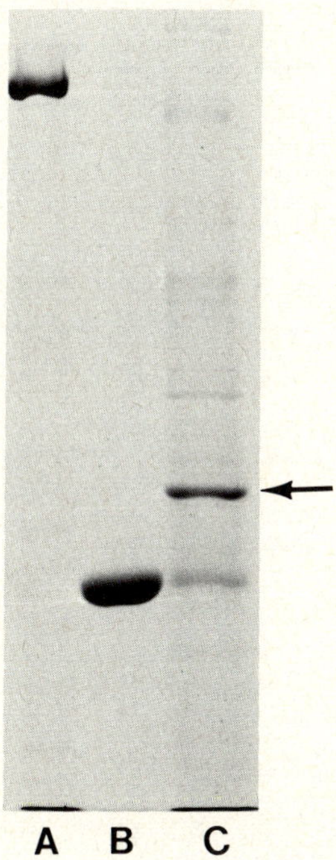


FIG. 2. SDS–polyacrylamide gel of myosin (A), actin (B), and supernatant solution (C) prepared from extract. Right arrow indicates fascin; 220K content was low in this sample.

## C. Contraction of Extracts

Actin polymerization is most rapid at 40°C and the conditions described above maximize the yield of gel. It was later determined that this time–temperature cycle inactivates the myosin present in the extract, explaining why these gels are stable and do not contract (Kane, 1980). A reduced temperature and shorter exposure retains myosin activity while allowing some actin polymerization. With extracts of *T. gratilla* we have found that a temperature of 35°C and incubation time of 10 minutes results in the formation of a more fragile gel that contracts to a small volume within 1 hour at room temperature. An increase in KCl to 50 m*M* and ATP to 4 m*M* is required for this rapid contraction; the presence of 1 m*M* $MgCl_2$ also usually accelerates contraction. Protein yield is reduced to 10% of that recovered at 40°C, or approximately 0.1 mg/ml of extract. These contractile gels have a more complex composition, containing myosin and a 250-kDa polypeptide in addition to the components of the so-called ''structural'' gel (Kane, 1980). The components of structural and contractile gels as seen in SDS–polyacrylamide gels are compared in Fig. 3. The amount of fascin is reduced in these

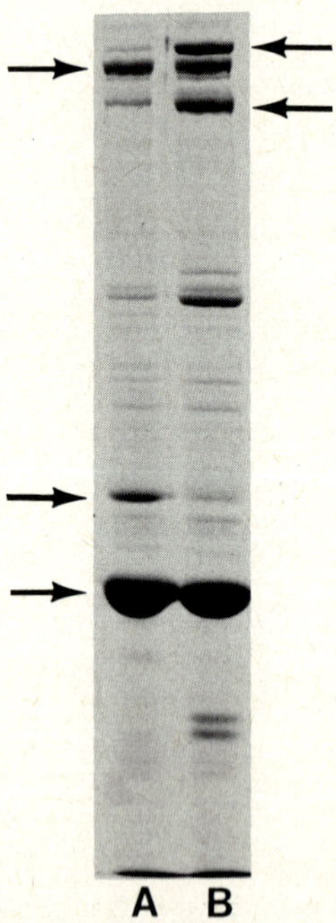


FIG. 3. SDS–polyacrylamide gel of a structural gel formed at 40°C (A) and a contractile gel formed at 35°C (B). Left arrows indicate (from bottom) actin, fascin, 220K. Right arrows indicate myosin and 250K. From Bryan and Kane (1982).

gels and this loss can be shown to be associated with contraction. Gels formed at the reduced incubation time and temperature, but in the presence of the lower ATP and KCl concentrations used for structural gels, do not contract and retain the usual amount of fascin. If these gels are then suspended in a medium with the ATP and KCl concentrations required for contraction, they will contract and their fascin content is then reduced (Kane, 1982).

## D. Reassembly of Contractile Gels

The components separated by procedures described in preceding sections can be recombined under appropriate conditions to form a contractile gel (Kane, 1983). We have found that the following conditions are important in this procedure:

1. A sufficiently high concentration of ATP and KCl must be maintained during assembly to prevent actomyosin aggregation. We begin with a high ATP–KCl solution and dilute it by successive additions of the protein components.
2. The final ATP concentration in the mixture must be at least 0.1 m*M* for complete contraction.
3. The final KCl concentration controls contraction rate and must be in the range of 60–100 m*M*; 80 m*M* is a useful compromise.
4. A weight ratio of actin:myosin:gel proteins of at least 1.0:0.3:0.6 is necessary.

To satisfy these conditions simultaneously the following solutions are combined in the order given, using the protein solutions prepared as previously described:

1 *M* KCl, 2 m*M* ATP
Actin in 50 m*M* KCl, 10 m*M* PIPES
Myosin in 0.5 *M* KCl/50% glycerol
Supernatant solution in 50 m*M* KCl, 10 m*M* PIPES, 1 m*M* EGTA
10 m*M* PIPES, pH 6.9
80 m*M* KCl, 10 m*M* PIPES

The relative amounts of the protein solutions to be added to achieve the above protein ratios will depend on their concentration, which varies in each preparation; these volumes will in turn affect the KCl concentration at each step. To determine the volume of each solution to be combined in the order given, simultaneous equations are solved by means of a computer program which is supplied with the ratios of actin, myosin, and gel proteins desired, the protein concentrations of each of the stock solutions, and the final volume and KCl concentration required. Examples of such calculations are given in Kane (1983).

For test purposes, a volume of 100 μl is assembled and observed in a 6 × 50 mm tube with a 6× binocular microscope. The solution is clear at 0°C; within a few minutes of transfer to room temperature it becomes turbid due to gel formation, and this gel then separates from the walls of the container and contracts to a small volume. For details of the effects of the various parameters on contraction events see Kane (1983).

The rate of contraction of these reassembled gels is regulated by the phosphorylation of the myosin light chain. Although myosin can be precipitated from extract by a number of nucleotides, ITP-precipitated myosin is much less active in inducing contraction in these gels than myosin precipitated by ATP or ATP-γ-S. ITP is not a substrate for myosin light chain kinase and myosin precipitated by this nucleotide is not phosphorylated; subsequent incubation of this myosin in extract containing ATP increases its phosphorylation level and effectiveness in inducing contraction (Burgess *et al.*, 1984).

## IV. Problems and Prospects

Several useful procedures have been developed for the separation and manipulation of these cytoskeletal components, based on the observed properties of cytoplasmic extracts, but their mechanisms and biological significance remain obscure. Among the questions for future investigation are the following:

The precipitation of myosin from extract by ATP is simple, rapid, and produces a relatively pure product, although the role of the nucleotide in this process is unknown. It proceeds with equal effectiveness with nucleotides known not to be substrates for myosin light chain kinase, and later experiments have shown that light chain phosphorylation is not required for precipitation (Burgess *et al.*, 1984). Its mechanism is obviously of interest and possibly of functional significance in regulating the state of myosin in the egg, where it acts in limited areas for limited times, notably in the cleavage furrow (Mabuchi and Okuno, 1977).

Actin remains in the unpolymerized form well above its critical concentration in egg extracts at the physiological temperature; how warming overcomes this inhibition and how this might related to the normal regulation of actin polymerization is unknown. This question is of particular interest with regard to the use of this method with extracts of the eggs of sea urchins existing at lower temperatures. Will there be some relation between the normal growth temperature and that required to initiate actin polymerization? The extreme case of sea urchins that develop close to 0°C, such as *S. droebachiensis,* would be of interest in this regard.

The "structural gel" formed either in extract at 40°C or assembled from components contains fascin in a fixed relation to actin and has a characteristic fine structure, whereas contracted gels contain little or no fascin (Kane, 1980). If gelation and contraction are separated by the method outlined above (Kane, 1982), a fascin-containing gel is formed, but after the completion of contraction almost no fascin remains. Does myosin displace fascin from actin and is this related to the control of the structural and contractile roles of actin in the cell? Is there a homolog of fascin in non-sea urchin cells?

What is the role of the 220K protein? It is a major component of both structural and contractile gels and is clearly distinct from the 250K protein present in contractile gels, which resembles filamin or actin-binding protein of other cells. The 220K localizes with actin in the starfish egg (Otto and Schroeder, 1984), but its binding to actin, fascin, or both is undetermined.

Our knowledge of the localization and roles of these proteins during the early cleavages that follow fertilization is very limited; information on the possible roles of these proteins in the cell movements and shape changes of sea urchin development is virtually nonexistent. However, even our present, very limited evidence indicates that the interactions of these proteins can provide a very flexible and responsive cytoskeleton for these embryonic processes.

## Acknowledgment

This work is supported by NIH Grant GM 14363.

## References

Bryan, J., and Kane, R. E. (1978). *J. Mol. Biol.* **125,** 176–179.
Bryan, J., and Kane, R. E. (1982). *Methods Cell Biol.* **25,** 176–199.
Burgess, D. R., and Schroeder, T. E. (1977). *J. Cell Biol.* **74,** 1032–1037.
Burgess, D. R., Kane, R. E., and Yabkowitz, R. (1984). *J. Cell Biol.* **99,** 34a.
Chambers, R. (1940). *Cold Spring Harbor Symp.* **8,** 144–153.
DeRosier, D. J., Mandelkow, E., Silliman, A., Tilney, L., and Kane, R. E. (1977). *J. Mol. Biol.* **113,** 679–695.
Hosoya, H., and Mabuchi, I. (1984). *J. Cell Biol.* **99,** 994–1001.
Hosoya, H., Mabuchi, I., and Sakai, H. (1982). *J. Biochem.* **92,** 1853–1862.
Ishimoda-Tagaki, T. (1978). *J. Biochem.* **83,** 1757–1762.
Hultin, T. (1950). *Exp. Cell Res.* **1,** 272–283.
Kane, R. E. (1973). *Exp. Cell Res.* **81,** 301–311.
Kane, R. E. (1975). *J. Cell Biol.* **66,** 305–315.
Kane, R. E. (1976). *J. Cell Biol.* **71,** 704–714.
Kane, R. E. (1980). *J. Cell Biol.* **86,** 803–809.
Kane, R. E. (1982). *Cell Differ.* **11,** 285–287.
Kane, R. E. (1983). *J. Cell Biol.* **97,** 1745–1752.
Mabuchi, I. (1976). *J. Mol. Biol.* **100,** 569–582.
Mabuchi, I., and Okuno, M. (1977). *J. Cell Biol.* **74,** 251–263.
Mabuchi, I., and Spudich, J. A. (1980). *J. Biochem.* **87,** 785–802.
Miki-Noumura, T. (1969). *Dev. Growth Differ.* **11,** 219–231.
Miki-Noumura, T., and Oosawa, F. (1969). *Exp. Cell Res.* **56,** 224–232.
Otto, J. J., and Schroeder, T. E. (1984). *Dev. Biol.* **101,** 263–273.

Otto, J. J., Kane, R. E., and Bryan, J. (1980). *J. Cell Motil.* **1,** 31–40.
Spudich, J. A., and Amos, L. A. (1979). *J. Mol. Biol.* **129,** 319–331.
Vacquier, V. D. (1975). *Dev. Biol.* **43,** 62–74.
Wang, L.-L., and Spudich, J. A. (1984). *J. Cell Biol.* **99,** 844–851.

# Chapter 14

# *Cytology and Immunocytochemistry*

**PATRICIA J. HARRIS**

*Department of Biology*
*University of Oregon*
*Eugene, Oregon*



## I. Introduction

Important transition points in the cell cycle, or certain embryonic stages of development, are very often characterized by drastic changes of intracellular architecture. Cells entering mitosis break down an elaborate interphase microtubule system and build a mitotic apparatus in its place; filamentous actin becomes an important feature of the invaginating cells of the sea urchin gastrula. An important tool for studying these changes has been the use of specific fluorescent probes, particularly indirect immunofluorescence microscopy with antibodies made to an ever-widening array of cytoskeletal proteins. These probes can be injected into individual cells or embryos, for example fluorescein-labeled actin (Wang and Taylor, 1979; Wang *et al.*, 1982) and tubulin (Wadsworth and Sloboda, 1983; Salmon *et al.*, 1984), or they can be used for staining of large synchronous populations of fixed material (Harris *et al.*, 1980; Balczon and Schatten, 1983). The obvious advantage of fluorescence microscopy is that it provides a bridge between ordinary light microscopy and electron microscopy. Specific structural systems can be chemically identified and visualized intact within a whole cell and do not require three-dimensional reconstructions from serial sections.

Although fluorescent probes have been particularly useful in the study of relatively flat vertebrate tissue culture cells, application of fluorescence microscopy to large spherical invertebrate eggs has presented some particular difficulties. Some of these difficulties arise from the cell thickness and subsequent loss of resolution, high background, and loss of adherence to the substrate during processing. Seasonal availability, egg size and transparency, temperature requirements, and abnormalities to be expected at the very beginning and toward the end of the spawning season are other factors that must be taken into account (see Hinegardner, 1975). Furthermore, antibodies made against vertebrate antigens may not react very well in such phylogenetically distant organisms as echinoderms (see Osborn and Weber, 1982). It is also useful to complement fluorescence studies by taking parallel samples for whole mounts of aldehyde-fixed cells for phase microscopy, or osmium-fixed cells for observing yolk and membrane distributions.

The techniques described here have been worked out primarily for use with *Strongylocentrotus purpuratus* and *Lytechinus pictus,* but with some possible exceptions they should be generally applicable. Fluorescence methods are evolving rapidly, and very likely those described here will be superceded. However, the basic considerations are the same and should be useful with later developments.

## II. Whole Mounts of Fixed and Embedded Material

### A. Sampling and Fixation

Whole mounts are useful for establishing a developmental timetable, where paired samples are taken for other studies. They do not provide good resolution, but they are adequate to establish the timing of such developmental markers as pronuclear fusion, streak stage, nuclear membrane breakdown, furrowing, and in later stages, cell number, cilia formation, and gastrulation. Large numbers of cells can thus be examined to determine the distribution of stages within each sample. The stage found in greatest number can then be considered representative of that time point. Such paired sampling can be very useful, even when the corresponding sample is to be used for cytological studies. For example, in indirect immunofluorescence microscopy there can be a considerable loss of cells during the processing, and it is important to know which of the survivors are typical for that sample.

Stirred cultures of eggs or embryos in 1-liter beakers are kept in a water bath at the appropriate temperature for the species being used. The stirrers consist of timing motors with a 1-second revolution mounted on a Lucite cover fitted to the beaker. A plastic rod with a three-quarter inch wide paddle at the end is mounted on the motor shaft. The temperatures generally used have been 18°C for *Lytechinus pictus* and 15°C for *Strongylocentrotus purpuratus,* although there are indications that 15°C may be close to or exceeding the upper limit for normal development (Hinegardner, 1975; Schroeder and Battaglia, 1985). Samples are taken for whole mounts by pipetting aliquots of the suspended cells into 15-ml glass centrifuge tubes and spinning down gently with a hand centrifuge. The volume taken should be adjusted to yield a sample of about 0.1 ml of lightly packed eggs. Seawater is drawn off and the embryos resuspended in the fixative.

The fixative found most useful is 2% glutaraldehyde in 0.4 *M* sodium acetate at pH 6.0. The low pH causes an aggregation of microtubules as well as the adherence of numerous small, dense particles along their lengths, which makes microtubule-based structures such as asters and spindle quite visible by phase-contrast microscopy. Figure 1a shows an *S. purpuratus* zygote at first division, with reforming nuclei and extensive late telophase asters. The temperature of the fixative is maintained at the same temperature as the cell culture for the first 10 or 15 minutes and then allowed to come to room temperature. Length of fixation does not seem to be critical and the entire set of samples can be set aside until the primary samples for chemistry or other purposes have been processed.

## B. Mounting

After 1 to several hours, or even overnight, the samples can be gently collected, using a hand centrifuge, the fixative aspirated off, and the eggs or zygotes resuspended in 15 ml of ice-cold 0.4 *M* sodium acetate buffer to wash. After a brief wash (10–15 minutes), they are then dehydrated in the centrifuge tubes through an ethanol series, beginning with cold 50 and 70%, two changes of 95% at room temperature, and finally three changes of 100%. The samples can be moved quite rapidly through the lower steps, but at least a half hour should be allowed for each change of 100%. Following the ethanol, they are passed through a 1:1 mixture of ethanol–propylene oxide, then into propylene oxide, and finally into a 1:1 mixture of propylene oxide and Araldite embedding medium containing two times the normal concentration of catalyst. The samples can be parked in this mixture overnight for infiltration.

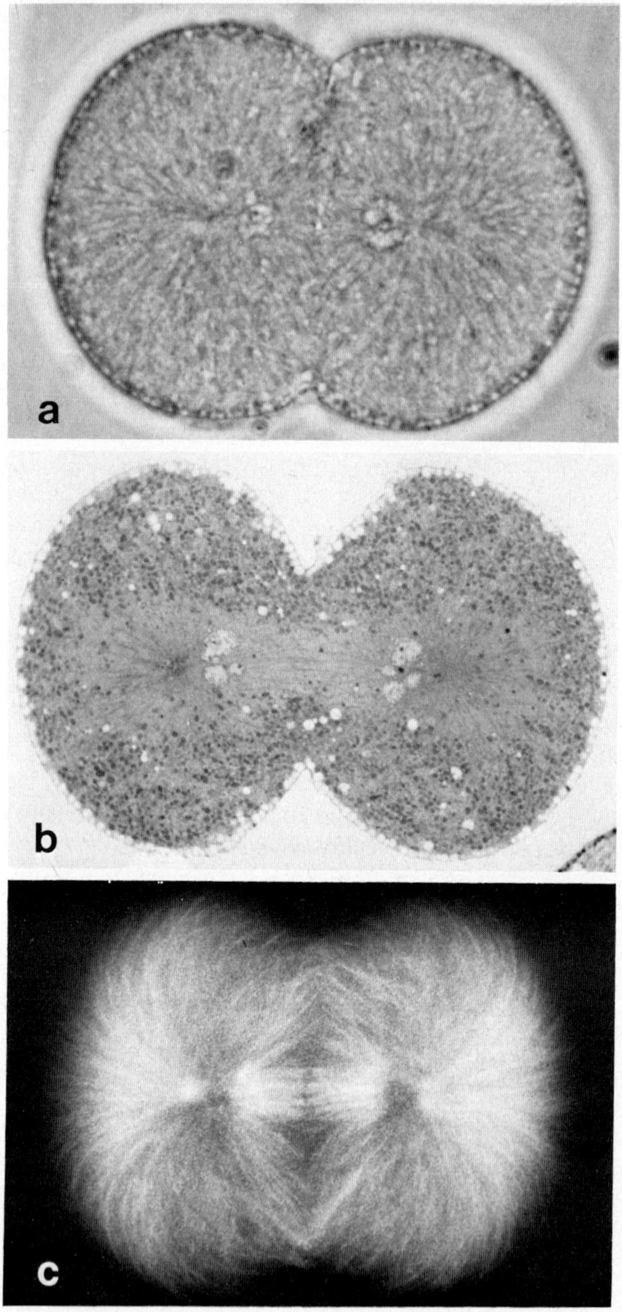
a
b
c

The Araldite mixture used is based on Luft's (1961) formula, and contains 27 ml Araldite 502, 23 ml hardener (DDSA), and 0.75–1 ml catalyst (DMP-30). Although Araldite resin was used here simply because it is the standard all-purpose embedding medium in our laboratory, there are numerous other formulations with less viscosity that can be used to greater advantage for aldehyde-fixed eggs (see Glauert, 1975, p. 132). The choice, however, often depends on the cutting qualities with either glass or diamond knives and is not entirely a matter of viscosity alone.

Glutaraldehyde-fixed eggs without osmium postfixation settle rather slowly in this somewhat viscous mixture, but can be spun down with a hand centrifuge. If the eggs have settled to the bottom after standing overnight, it is best not to resuspend them, as the increased viscosity makes them very difficult to concentrate. The settled eggs are then drawn up into a Pasteur pipet and transferred to a 10-ml beaker, taking care to transfer as little of the clear solution as possible. The eggs will probably spread around the bottom of the beaker and it is a good idea to let them sit a few minutes to allow evaporation of remaining propylene oxide. Then 1–2 ml of final embedding mix is added and the eggs are very gently, but thoroughly, stirred into the epoxy. Drops of the egg-containing mixture can then be placed on microscope slides and covered with a cover slip. It is important to have the eggs as concentrated as possible and the drop of egg mixture small so that the resulting sandwich is thin enough for use with oil immersion objectives. If there is too much epoxy under the cover slip, warming will help to thin it, but it does not help to apply pressure, as the eggs are very brittle and tend to shatter before very much epoxy can be forced out. Polymerization is carried out in an embedding oven at 60°C. On warming, however, any residual propylene oxide tends to form numerous bubbles under the cover slip.

## III. Sectioned and Stained Material for Bright-Field Light Microscopy

### A. Fixation

A more time-consuming method for monitoring stages of development, but one which provides considerably more accuracy, utilizes stained sections of

---

FIG. 1. First cleavage of *Strongylocentrotus purpuratus* zygotes, comparing three methods for examining cell structure. (a) Whole amount, fixed in glutaraldehyde and embedded in epoxy resin Reconstituting nuclei and late telophase asters are clearly visible. (b) Stained section of a zygote fixed initially with glutaraldehyde and postfixed with osmium. Yolk granules can be seen outlining the yolk-free regions of the asters and spindle. (c) Indirect immunofluorescence microscopy of a zygote fixed in cold methanol and stained with an antibody to yeast α-tubulin, showing the extensive array of microtubules in the telophase asters.

epoxy-embedded material. Yolk granules which were deliberately made invisible in the whole-mount preparations are densely stained and clearly delineate the clear areas containing aggregates of membrane-bounded vesicles. The degree to which microtubular structures can be seen within these clear areas, for example, in the asters and spindles of mitotic cells, depends on the fixation.

In our hands, glutaraldehyde fixatives that generally provide excellent fixation for microtubules in vertebrate tissues have not been satisfactory for fixing microtubules in sea urchin eggs and early embryos. At neutral pH, microtubules appear as linear arrays of small, dense granules, with few if any intact tubules remaining. On the other hand, 1% $OsO_4$ in 0.4 *M* sodium acetate at pH 6.0 preserves microtubules and membrane-containing structures quite well, but leaves the remainder of the cytoplasm less well preserved. An extensive series of experiments was carried out to examine shrinkage, swelling, and quality of preservation with relation to pH, composition, and concentration of the buffer and fixatives (Harris, 1978). From this work and more recent investigations, we have found the following fixatives to be the most useful for the purposes outlined here.

### 1. 1% $OsO_4$ in 0.4 *M* Sodium Acetate, pH 6.0

This is a very simple fixative to use. Osmium tetroxide is made up to 2% in glass-distilled water. Sodium acetate at 0.8 *M* is adjusted to pH 6.0, and the two solutions are mixed 1:1. The fixative can be made up immediately before use, or can be prepared in larger quantities and stored in small aliquots (3–5 ml) in small, tightly capped vials at −20°C. The vials should be further sealed with Parafilm and then placed in a sealed container. Osmium tetroxide is extremely volatile and once thawed should not be put into the refrigerator for storage, as only extreme precautions will prevent everything in the refrigerator from becoming fixed by osmium vapors.

Fixation is adequate after $\frac{1}{2}$ to 1 hour at room temperature and should not be extended much beyond an hour, and certainly not overnight. If samples need to be parked before further processing, it is best to leave them in the 95 or 100% ethanol, following the dehydration protocol described for the whole-mount specimens.

This fixative is good for preserving microtubules for electron microscopy. However, osmium even at low pH does not aggregate microtubules, and they are thus not as visible for light microscopy as in glutaraldehyde-fixed cells.

### 2. 2% Glutaraldehyde in 0.4 *M* Sodium Acetate, pH 6.5, followed by 1% $OsO_4$ in 0.4 *M* Sodium Acetate, pH 6.0

This two-step procedure is the fixative of choice if observations are to be limited to light microscopy. It combines the visibility of microtubules (with accompanying granular aggregates), with the better preservation of lipid-containing membranes, yolk granules, and oil droplets provided by osmium tetroxide (Fig. 1b). Initial fixation in glutaraldehyde is the same that described for the whole-mount preparations, except that the pH is slightly higher. After at least 1 hour in the fixative at room temperature, the samples are washed two times for 15 minutes each in cold 0.4 *M* sodium acetate at pH 6.0. Postfixation should not exceed 1 hour to prevent extraction, and is followed by a brief wash in cold buffer before dehydration.

## B. Embedding and Sectioning

Samples are dehydrated as described for whole mounts and taken through propylene oxide to the 1:1 mixture of propylene oxide and catalyzed Araldite for infiltration. There is less difficulty with the heavy osmium-fixed eggs to concentrate them, as they settle very rapidly, but for this reason they should be resuspended occasionally with a light spin on a vortex mixer or by bubbling air through them with a Pasteur pipet. After several hours or overnight, eggs are allowed to settle or are sedimented by hand centrifugation, and transferred with a Pasteur pipet to a polyethylene bottle stopper. These stoppers (Kimble or Nalgene) are available in various standard cork sizes, with the range from 0 (small) to 8 (about 3 cm diameter) being the most useful. The size used will depend on the concentration of eggs in the final embedding plastic, as the ultimate objective is to obtain a monolayer of eggs as closely packed as possible.

The 1:1 propylene oxide–Araldite mixture carried over with the sample is removed from the polyethylene stopper as completely as possible, and the sample is allowed to stand for a few minutes to evaporate off remaining propylene oxide. Final Araldite embedding mix is then added and the sample is carefully, but thoroughly, stirred into the plastic with a small spatula. The amount of plastic added to each stopper should not exceed a depth of 2–3 mm, so that all eggs are able to sink to the bottom of the mold before polymerization is complete. The polyethylene stoppers can withstand embedding oven temperatures up to 60°C, but at higher temperatures will begin to melt or distort. After polymerization for 12–24 hours, the hardened disk with egg monolayer can be flipped out simply by pressing on the bottom of the flexible mold.

The disk can then be examined under a dissecting microscope and the desired region marked with a sharp needle. If the surface is not smooth enough for good visibility, a drop of water or glycerine spread on the surface will make the eggs clearly visible. The marked region can then be sawed out with a jeweler's saw and mounted on a precast epoxy mounting pin with a rapid-setting two-component epoxy glue.

One-micrometer sections from trimmed blocks containing up to 200 eggs can be cut easily with a glass knife and floated off onto the water surface of the trough. Sections are cut one at a time and transferred with a sharpened needle to a drop of water on a microscope slide, where the section will float on the surface. Sharpened tungsten needles are prepared from approximately 3-in. lengths of tungsten wire. The wire is held firmly with pliers, heated to incandescence in a flame, and then plunged into a small beaker containing sodium nitrite. This initiates an exothermic reaction which essentially dissolves the tungsten. By moving the wire back and forth, a very fine point can be formed. Protecting the hands and eyes is advisable, as the molten reactants often splatter. The sharpened needle can then be mounted in a pin vise for use.

It is convenient to place the drops of water on the microscope slide in uniform arrays for mounting sequentially individual serial sections, or serial groups of 5–10 sections. The slides are then placed on a slide warmer or warm hot plate to allow sections to flatten and dry down. Usually they adhere tightly during subsequent staining if the slides are clean and free of grease.

### C. Staining and Mounting

There are numerous stains that can be used for thick epoxy sections. The one most commonly used for electron microscopy survey purposes is a solution containing 1% Methylene Blue and 1% Azure II in 1% borax (Richardson *et al.*, 1960). The sections are covered with several drops of staining solution and warmed gently for about a minute. Excess stain is drained off and the sections are washed thoroughly in distilled water. Slides are dried by gentle warming. Cover slips can be mounted using a catalyzed epoxy mixture or simply with a drop of immersion oil.

For determining the number or percentages of cells in a particular stage, a section parallel to the face of the block and passing through the middle of the monolayer containing 100–200 cells should be adequate. If the scoring depends on a particular orientation or eccentric structure, probably more than one block will be necessary. For scoring stages, a large-format Polaroid print or a camera lucida drawing to identify individual cells is very useful, especially if it is necessary to examine earlier or later serial sections of the same cell.

## IV. Indirect Immunofluorescence Microscopy

### A. Choice of Antibodies

The success of immunofluorescence microscopy will depend largely on the quality of the antibodies that are used, i.e., the specificity and affinity of the antibody for the antigen, the titer of the elicited antibody, nonspecific staining, etc. These characteristics will be influenced by the phylogenetic relationship of the antigen source and the animal eliciting the antibody, purification of the antibody by antigen-affinity chromatography, and other aspects of antibody preparation. Unless one carries out all the procedures, including purification of the antigen and preparation of the antibodies, many of the factors influencing the quality of the antibody will be beyond the control of the investigator.

Detailed methods for preparing conventional polyclonal antibodies, as well as monoclonal antibodies derived from hybridoma techniques, have been published elsewhere (for example, see Johnstone and Thorpe, 1982). General methods for immunofluorescence and immunocytochemistry have been described (Osborn and Weber, 1982) and include detailed protocols for single and double labeling which are applicable to sea urchin material and will not be repeated here.

The type of antibody one uses will depend on the question being asked. For example, if one wishes to visualize the entire array of microtubules in a cell, polyclonal antibodies may be preferred, while monoclonals can be useful to distinguish between subpopulations of microtubules within the same cell, for example, the tyrosinated vs the nontyrosinated α-tubulin (Gundersen *et al.*, 1984). Antibodies elicited to such highly conserved proteins as tubulin and actin show a wide range of cross-species or cross-phyla reactivity, and it is possible to stain sea urchin and even plant microtubules with antibodies to mammalian brain tubulin. As a rule, it is preferable to use antibodies to proteins from the organism being stained—in this case, the sea urchin. Methods for purification of tubulin from sea urchin flagella (Bibring *et al.*, 1976; Farrell, 1982) and from unfertilized eggs, using taxol (Vallee and Bloom, 1983), and preparation of antibodies have been described. Tube feet have proved a good source of actin (Kabat-Zinn and Singer, 1981). With less conserved proteins and greater evolutionary divergence between mammalian and invertebrate antigens, the chances for good cross-reactivity become much less. The best advice is to try different combinations of antibodies, as well as variations in fixation, which may affect the immunological reactivity of the antigen.

## B. Handling of Eggs and Embryos for Indirect Immunofluorescence Microscopy

### 1. Removal of External Coats

Sea urchin eggs, unlike vertebrate tissue culture cells, are invested with relatively impermeable outer layers that prevent entrance into the cell of large antibody molecules. A number of methods have been developed for the removal of these extraneous coats. Several reviews discuss these in some detail (Berg, 1967; Hinegardner, 1975).

For unfertilized eggs, the jelly coat and, more importantly, the vitelline layer, must first be removed or made permeable. The jelly coat is easily removed by passing the eggs through a fine mesh Nitex cloth, the mesh size depending on the diameter of the egg, or by a brief treatment with acidified seawater, pH 5.5 to 5.8 (Hagström, 1959).

Commonly used methods for removal of the vitelline layer utilize proteolytic enzymes (Epel, 1970) or thiol reagents (Epel *et al.*, 1970). Pretreatment with nonelectrolytes such as urea (Moore, 1930), glycerol, or glycine (Kane, 1970, 1973) at a concentration isotonic to seawater and pH 7 to 8 are also effective in preventing formation of the fertilization membrane.

After fertilization, there are numerous ways that the fertilization membrane can be removed. In many species, the forming membranes can be stripped within the first 1–2 minutes after they begin to rise, but care must be taken to remove calcium to prevent the demembranated eggs from sticking. Nonelectrolytes such as 1.0 *M* urea (see, e.g., Kane, 1962; Schroeder, 1979) and proteolytic enzymes (Berg, 1967; Vacquier and Mazia, 1968) have also been used effectively if applied within the first 30–45 seconds after sperm addition.

Sulfhydryl-reducing agents can also be used to prevent the hardening of the fertilization membranes, after which they can then be stripped mechanically. A method employing mercaptoethylgluconamide (MEGA) has been our method of choice since it was first introduced by Mazia *et al.* (1961). The rationale for its use was that it prevented the formation of disulfide bonds which are responsible for the hardening process. Furthermore, it is a large molecule that does not enter the cell. The method is very simple and effective, and as far as development to the feeding pluteus stage it apparently has no detrimental effects on development. Mercaptoethylgluconamide is currently available from Vega Chemicals, Tucson, Arizona.

Mercaptoethylgluconamide solution is made by dissolving 0.1 g MEGA in about 20 ml seawater. The resulting solution is somewhat acidic and is brought to pH 8 with NaOH. About 100 ml of concentrated egg suspension (0.5 ml packed eggs/10 ml) is fertilized by the addition of several drops of freshly prepared diluted sperm. The eggs are kept agitated by pouring from one beaker to another to insure rapid and synchronous fertilization. A sample is then removed and examined under the microscope. As soon as fertilization membranes have just begun to lift (about 30 seconds with *Strongylocentrotus purpuratus*), the MEGA solution is added to the egg suspension and the mixture is again stirred thoroughly. The eggs are then transferred to a 1000-ml beaker and incubated at 15°C (or preferably lower) with gentle stirring.

The resulting membranes are exceptionally high and thin, and can be stripped by pouring the egg suspension through a silk or Nitex cloth with a mesh size depending on the diameter of the particular species of egg. A sieve is easily made by cutting the bottom from a 50-ml plastic centrifuge tube and fastening a piece of mesh over the uncut end with a rubber band.

If stripping is delayed about 15 minutes until the hyaline layer is formed, the demembranated eggs do not clump. They are allowed to settle, the seawater with stripped membranes, excess sperm, and MEGA is drawn off, and the eggs are resuspended in normal seawater for development. During the settling period there may be some danger of anoxia, resulting in less synchronous development. If synchrony is of primary importance, a single wash is sufficient, as residual MEGA seems to have no effect on development. If later procedures involve fluorescent staining of chromosomes, residual sperm in the culture will appear as brightly lighted spots and often make it difficult to get good pictures. A second wash by settling, decanting, and resuspending in fresh seawater removes practically all the MEGA and excess sperm.

With the above treatment the hyaline layers remain intact, thus allowing normal development beyond the first division. Hyaline layers can be removed by introducing EGTA to the egg suspension to a final concentration of 10 m*M* at the time MEGA is added. After 10–15 minutes the eggs can be stripped and washed in calcium-free seawater, and finally resuspended in either calcium-free or normal seawater for further development.

A similar method, which also prevents the hardening of the newly forming fertilization membranes, utilizes aminotriazole, an inhibitor of the ovoperoxidase released during the cortical reaction (Showman and Foerder, 1979). As with the MEGA, the softened membranes can be stripped mechanically and the hyaline layers remain, unless further treated with calcium-free seawater.

### 2. Sample Preparation on Cover Slips

The methods used for indirect immunofluorescence microscopy of sea urchin eggs and embryos follow closely the protocol described by Osborn and Weber (1982), with some modifications to accommodate special problems of large eggs. One serious problem of handling cells that do not normally attach to glass surfaces is the loss of cells during the prolonged processing. The following procedure seems to work quite well with minimum loss.

1. Coat cover slips with polylysine (Mazia *et al.*, 1975) by immersing in a 1 mg/ml solution of polylysine for 5–10 minutes. Drain and allow to air dry without rinsing. There is better retention of cells if the cover slips are coated just before use. Fresh preparation of the polylysine also seems to be more effective, although a stock solution made up and frozen in small aliquots and thawed just before use also work quite well.

2. Prepare a humid chamber by placing several layers of filter paper in a 9-cm plastic Petri dish. The filter paper should be moist, with excess water blotted off. Place three or four coated cover slips on the wet filter paper, and then add several drops of egg suspension from the culture to each. The concentration of egg suspension and amount placed on each cover slip should be adjusted so that after all eggs settle to the glass surface there will be only a single layer of cells, and

where none overlaps or touches another. This will allow each egg to make good contact with the surface and allow some room for flattening.

3. Cover the Petri dish and allow the eggs to settle to the cover slip, about 2–5 minutes. During this time it is essential that the temperature be kept the same as that of the remaining culture. The dish can be placed in a cold room at the desired temperature, or if a water bath is being used for incubating the culture, the dish can be floated on the water surface. When the eggs have adhered firmly, the water is drained and the cover slip is ready for fixation and staining.

### 3. Fixation

A number of different methods have now been described for processing sea urchin eggs for indirect immunofluorescence microscopy. The simplest of these is cold methanol–EGTA (Harris *et al.*, 1980) (see Fig. 1c).

*a. Cold methanol-EGTA.* 1. Remove each cover slip with settled eggs from the humid chamber, using a pair of curved jeweler's forceps (Dumont #7). Drain most, but not all, of the water from the egg layer by touching the cover slip to the edge of a dry filter paper, and then immerse the cover slip carefully in cold 90% methanol–EGTA in a 5-cm Petri dish. The EGTA is made up as a 0.5 *M* stock at pH 6.0 and added in appropriate dilution to make the final desired concentrations. The methanol–EGTA is kept in a −20°C freezer and poured out immediately before use.

The amount of EGTA added to the methanol depends on the sample that is being processed. As microtubules are sensitive to calcium, it is important that calcium in the residual seawater (10 m*M*) as well as any calcium that might be released within the cell should be removed rapidly. Eggs retaining their hyaline layers also possess an added "sink" of calcium. For eggs with hyaline layers grown in normal seawater, concentrations of EGTA up to 50 m*M* may be necessary for good microtubule preservation. For eggs in calcium-free seawater without hyaline layers, and later embryos, concentrations from 5–10 m*M* appear to be adequate. Higher concentrations of EGTA cause the dissociation of ciliary axonemes. The cover slips in the cold methanol can be returned to the freezer for the fixation period of 6–10 minutes, but in our experience there seems to be no negative effect if the temperature of the fixative warms to room temperature during this time.

2. Remove the cover slips from the methanol and transfer without allowing to dry to another Petri dish containing phosphate-buffered saline (PBS) (see Osborn and Weber, 1982 for composition). A note of caution here: Be sure the cover slip is completely immersed in the PBS before releasing it. Otherwise it takes off in a jet-propelled fashion on the surface, allowing the eggs to air dry before it is recaptured.

After several minutes of rehydration, examine the cover slips under the microscope and pick the one or two best ones for further processing. Generally the appearance depends on the length of time the eggs have settled—long enough to stick well, but not long enough for excessive flattening. "Just the right length of time" must be determined empirically, as well as just how much water is drained from the eggs before immersing in cold methanol. Eggs at the periphery of the drop of eggs on the cover slip tend to flatten more than those in the center, due to surface tension. If flattened too much, they become somewhat distorted. If more than one layer of eggs settle on the cover slip, those lying on top will come off, either in the methanol or in the subsequent rehydration, with the danger that some of the lower layer may also be pulled off. Once again, these details can be appreciated only through actual practice.

*b. 3.7% Formaldehyde in PBS, Cold Methanol.* The formaldehyde is prepared by dissolving 18.5 g of paraformaldehyde powder in 500 ml PBS. The fixation procedure is similar to that described for cold methanol alone, except that the cover slips with attached eggs are first placed in the formaldehyde for 10 minutes, then into cold methanol for 6 minutes, and finally rehydrated in PBS before staining. There is no loss of antigen reactivity, with good staining characteristics for tubulin, actin, and at least one intermediate filament antibody. Arrays of microtubules, however, appear distorted, especially the mitotic asters. This fixative may be more useful for later embryonic stages.

*c. Preextraction, Aldehyde Fixation.* One of the disadvantages of the cold methanol fixation, especially when one is interested primarily in the elements of the cytoskeleton, is that the great bulk of lipid-containing yolk and membranes remain to obscure the image of the stained structures. Several procedures have now been described for preextraction of the eggs prior to fixation either with methanol or aldehydes. Balczon and Schatten (1983) extracted eggs in bulk in a glycerol-containing buffer, after which the eggs are affixed to polylysine-coated cover slips for incubation with antibodies. While background is greatly reduced, distortions were inevitably introduced by handling the loose, fragile eggs. Hollenbeck and Cande (1985) and O'Brien *et al.* (1984) have now described methods (almost identical) for extracting eggs already immobilized on polylysine cover slips. The eggs are immersed for 2–4 minutes in a buffer containing 0.3% Triton X-100, 6 m*M* $MgSO_4$, 10 m*M* EGTA at pH 6.8–6.9, and then fixed in either 1% glutaraldehyde in extraction buffer or 0.1% glutaraldehyde and 1% formaldehyde in extraction buffer. Following fixation and washing, excess aldehydes are reduced with sodium borohydride (4 mg/ml) in PBS for 10 minutes to eliminate autofluorescence. The eggs are then incubated with antibodies, washed, and mounted. The results are a considerable improvement over cold methanol alone, and are less distorted than the bulk-extracted eggs.

Similar extraction procedures should be useful for other cell components, but one needs to be sure that the target protein is not also being eliminated by the extraction, or that in the case of glycerol-containing buffers, microtubule polymerization is not being induced.

## 4. Indirect Immunofluorescence Procedures

The procedures outlined here are essentially the same as those described by Osborn and Weber (1982), which should be consulted for details.

1. Following fixation and washing in PBS, the cover slips with eggs are drained and returned to the Petri dish moist chamber used previously for settling the eggs onto the cover slips. The filter paper in the Petri dish should be marked with pencil or waterproof ink to identify individual cover slips.
2. Five to 10 μl of first antibody is pipetted onto each cover slip and carefully spread over the surface, using the plastic tip of the Eppendorf or similar point. The Petri dish is covered and placed in a high-humidity incubator at 37°C. If the incubator has no humidity control, the Petri dish and cover slips should be placed in an additional humidity chamber, made by putting wet paper towels in the bottom of any plastic box with a tightly fitted cover (shoe box, cake box, etc.) before putting into the incubator.
3. After 1 to 1½ hours the cover slips are washed three times to remove excess first antibody. This can be done by holding the cover slip with jeweler's forceps and dipping several times in each of three 30-ml beakers containing PBS. After each wash, the cover slip is drained by touching its edge to dry filter paper. If there is danger of the eggs or embryos being lost from the cover slip, or otherwise damaged by this procedure, the cover slips can be placed in individual small dishes containing PBS. Washing can be accomplished with slight agitation and the PBS drawn off after each successive wash.
4. After washing, the cover slips are again drained and returned to their appropriate places in the Petri dish moist chamber. Five to 10 μl of second antibody with fluorescent label is added and the cover slips are incubated at 37°C as described previously.
5. After 1 to 1½ hours the cover slips are washed as before and are then ready for mounting.

## 5. Mounting

Simple and commonly used mounting media based on one first described by Coons and Kaplan (1950) consist of various mixtures of glycerol and buffered saline solution, preferably at an alkaline pH (8.5) for maximum fluorescence of fluorescein label. Although simple, there are several disadvantages: (1) a liquid medium is inconvenient and requires sealing of the cover slip to the slide, (2) there is a rapid loss (over several weeks) of fluorescence, even when stored in the dark at 4°C, and (3) there is danger in a liquid glycerol-staining medium of cell constituents translocating over a period of time (Aubin *et al.*, 1979).

Several media which set to a semisolid within 1 to 2 hours have been developed. These incorporate polyvinyl alcohol, which is available under the trade names Gelvatol (Monsanto Co., St. Louis, MO), Mowiol 4-88 (Hoechst, Frankfurt, F.R.G.; Calbiochem-Behring, San Diego, CA) or Elvanol (du Pont Co., Ltd., London). The following is a basic recipe, of which there are many minor variants currently in use, with regard to quantities and sequence of mixing (see Osborn and Weber, 1982; Goldman, 1968).

A 20% w/v solution of polyvinyl alcohol in 0.2 *M* Tris buffer, pH 8.5, is made up by first adding a small amount of buffer to the polyvinyl alcohol powder or flakes and mixing to make a slurry, making sure that all the powder is wetted and in suspension. After adding the remainder of the buffer to completely disperse the polyvinyl alcohol, the mixture is heated to 50–60°C for about 10 minutes with occasional stirring to dissolve the resin. To 80 ml of the cooled Tris-buffered resin, add 40 ml of analytical grade glycerol. The amount can be varied, but if too much glycerol is used, the mixture fails to set; if too little, bubbles tend to form (Rodrigues and Deinhardt, 1960). Mix thoroughly and clarify by centrifuging at 5000 *g* for 15 minutes. The batch can be divided into aliquots in glass vials and stored at −20°C for at least a year without deterioration. The solidified medium prevents redistribution of fluorescent label and apparently greatly extends the lifetime of the fluorescent label over long storage times (years) at 4°C in the dark (Osborn and Weber, 1982).

Aside from gradual bleaching during storage, rapid photobleaching of the fluorescent label under intense illumination in the microscope remains a serious problem. The apparent implication of oxygen or oxygen-induced free radicals in photobleaching has led to a search for suitable antioxidant as additives to the mounting medium. Among these are sodium dithionite (Gill, 1979), *p*-phenylenediamine at a concentration of 1 mg/ml glycerol (Johnson and Nogueira Aranjo, 1981), ascorbic acid at 0.1–0.2 *M* in citrate-buffered glycerol, and *n*-propylgallate at 2–5% w/v in glycerol (Giloh and Sedat, 1982).

While these mixtures reduce photobleaching soon after preparation, up to 15-fold in the case of *n*-propyl gallate, there is sometimes a deterioration of the specimen and degradation of fluorescence after several days of storage (Langanger *et al.*, 1983). Langanger *et al.* (1983) have found that addition of 1,4-diazabicyclo[2.2.2]octane at a concentration of 100 mg/ml to Gelvatol not only reduces photobleaching, but also protects against loss of fluorescence during storage in the dark at 4°C. As the other compounds were examined in buffered glycerol, it is possible that their performance as protectants would be improved in the semisolid polyvinyl alcohol-containing media.

## C. Methods of Reducing Background

Even under the best of circumstances, there may be a relatively high general background arising from nonspecific staining. This can be recognized as induced fluorescence which appears after incubating the cells with conjugated second antibody alone. It is not an immunological reaction, but apparently is due to a charge effect. We have found two procedures relatively successful in bringing down this background in sea urchin eggs.

#### 1. Dilution

If the titer of the antibody is high enough, it is often possible simply to dilute the antibody beyond the range of nonspecific binding. Usually protein concentrations of 0.5 mg/ml will provide sufficiently high specific staining and reasonably low nonspecific binding.

#### 2. Absorption with Sea Urchin Egg Acetone Powder

To make acetone powder, homogenize unfertilized eggs for 3–5 minutes in about 10 vol of cold (−76°C) acetone, using a small power homogenizer. Then centrifuge and remove the acetone. After three such extractions, the pelleted extracted eggs are dried in a vacuum desiccator and finally ground to powder in a mortar. The powder keeps well and can be stored indefinitely at −20°C.

For incubating small amounts of antibody, 2-ml polyethylene microcentrifuge tubes can be used. The amount of powder to make a final concentration of approximately 100 mg/ml is placed in the tube and wetted with PBS. Excess PBS is removed before adding antibody. By prewetting the powder there is less loss of antibody, which can run as high as 20%.

The appropriate amount of antibody is added and the tube shaken to suspend the powder. Two or three successive incubations of the powder–antibody mixture for 1 hour each at 37°C are usually sufficient.

## V. Other Fluorochrome Probes

### A. Bisbenzimide (Hoechst 33258) Stain for DNA

In studies of the cell cycle, and particularly of different stages of mitosis, it is important to stain the chromosomes as well as the microtubules. A convenient stain is bisbenzimide (Hoechst 33258). It can be made up as a stock solution of 1 mg/ml in methanol and used at a dilution of 1:200 directly in the cold methanol

---

FIG. 2. Three stages of the first division cycle of *Strongylocentrotus purpuratus*. Paired micrographs show cells double stained with an antibody to yeast α-tubulin (a,c,e) and with bisbenzimide (Hoechst 33258) for DNA (b,d,f). Streak stage (a,b) early prophase (c,d), and anaphase (e,f).

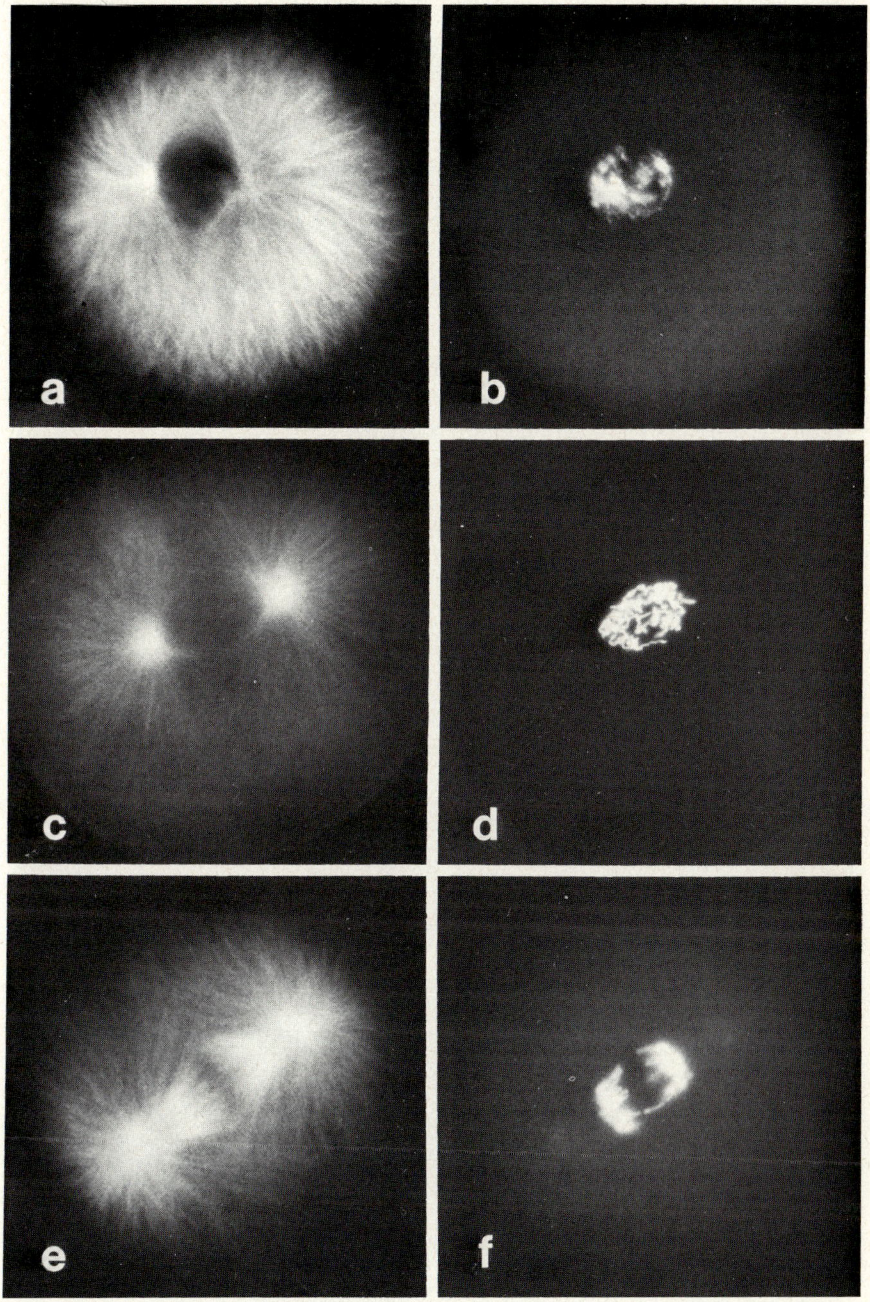
a
b
c
d
e
f

fixative. Following the 6-minute combined fixation–staining, cover slips should be rinsed with methanol before rehydrating and incubating with antibodies.

The excitation wavelength of the Hoechst stain is in the near ultraviolet, thus requiring UV-transmitting objectives such as the Zeiss Neofluars and an appropriate set of excitation and barrier filters. Figure 2 shows paired micrographs of eggs from three stages of mitosis, comparing tubulin FITC fluorescence with Hoechst-stained chromosomes in the same cell.

## B. NBD–Phallacidin and Rhodamine–Phalloidin Stains for Actin

These are fluorescent derivatives of the mushroom toxin, phalloidin, which binds specifically to filamentous F-actin, but not to unpolymerized G-actin (Barak *et al.*, 1980). They are very useful probes for actin when there is only weak cross-reaction of sea urchin actin with antibodies to actin from vertebrate sources. NBD–phallacidin can be detected with the microscope filter combination used for fluorescein; the rhodamine–phalloidin can be used with the rhodamine filters. Both fluorochromes can be obtained from Molecular Probes, Inc., Junction City, Oregon.

NBD–phallacidin can be used diluted 1:5 with 1% bovine serum albumin and applied to cover slips of fixed cells as described previously for antibody incubations. It works very well after formaldehyde fixation, but less well after several minutes of cold methanol. However, cold methanol fixation for 8 seconds followed by 20 minutes in NBD–phallacidin works quite well. Figure 3a shows a

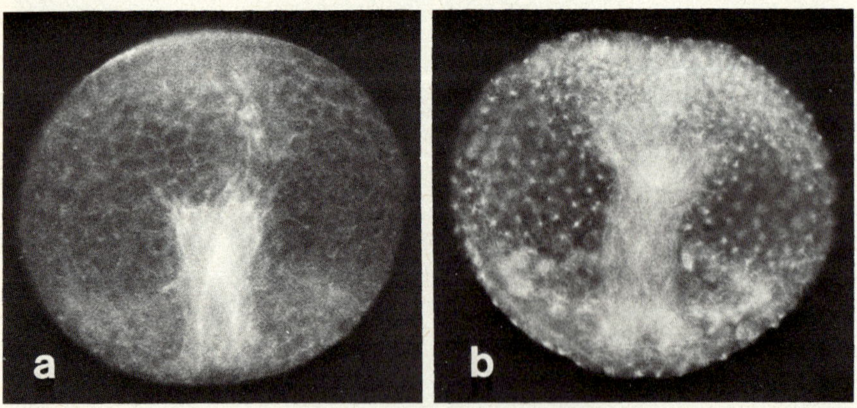


FIG. 3. *Lytechinus pictus* gastrulae. (a) Embryo fixed for 8 seconds in cold methanol and stained with NBD–phallacidin for actin. Note outlines of epithelial cells and large amount of actin in the invaginating cells. (b) Embryo fixed for 6 minutes in cold methanol and stained with an antibody to yeast α-tubulin. Small dots in the center of each epithelial cell are ciliary basal bodies.

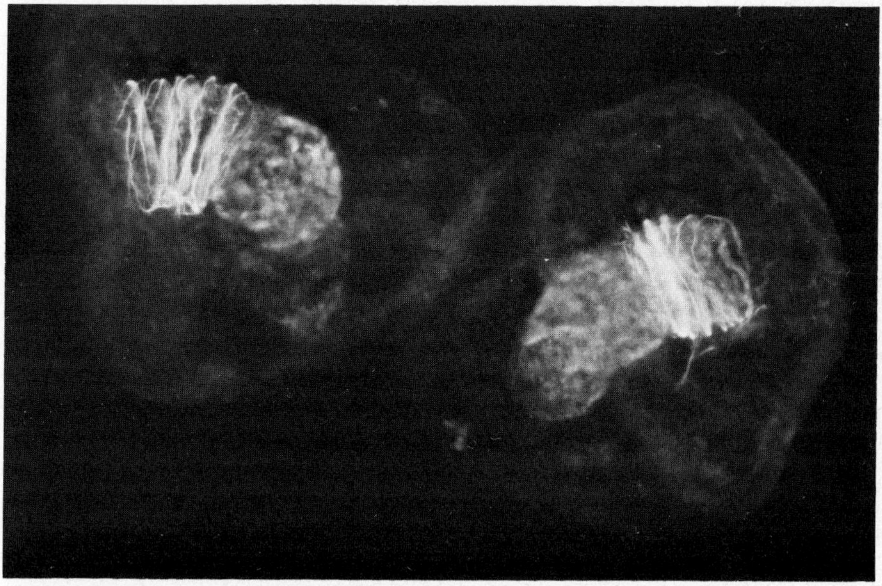

FIG. 4. Early plutei of *Strongylocentrotus purpuratus* stained with NBD–phallacidin after 8 seconds fixation in cold methanol. Muscle bands around the esophagus are highly stained. Identical staining is seen after formaldehyde fixation.

*Lytechinus pictus* gastrula fixed briefly in cold methanol and stained for actin with NBD–phallacidin. Note intense staining of invaginated column of cells, and bands of actin around individual cells of the outer cell layer. Compare with Fig. 3b, which shows a similar embryo stained for tubulin. Figure 4 shows two *Strongylocentrotus purpuratus* early plutei stained with NBD–phallacidin, showing muscular rings around the esophagus. Although these embryos were fixed for 8 seconds in cold methanol, the staining is identical to that seen after formaldehyde fixation.

## ACKNOWLEDGMENTS

I thank Drs. M. Osborn and K. Weber for their hospitality during my stay in their laboratory in Göttingen, F.R.G., where many of the applications of immunofluorescence microscopy to sea urchin material were first developed, and Dr. J. Wehland for kindly supplying tubulin antibodies. This work was supported in part by Grant PCM 84-09573 from the National Science Foundation.

## REFERENCES

Aubin, J. E., Weber, K., and Osborn, M. (1979). *Exp. Cell Res.* **124,** 93–109.
Balczon, R., and Schatten, G. (1983). *J. Cell Motil.* **3,** 213–226.

Barak, L. S., Yocum, R. R., Nothnagel, E. A., and Webb, W. W. (1980). *Proc. Natl. Acad. Sci. U.S.A.* **77,** 980–984.

Berg, W. E. (1967). *In* "Methods in Developmental Biology" (F. H. Wilt and N. K. Wessels, eds.), pp. 767–776. Crowell, New York.

Bibring, T., Baxandall, J., Denslow, S., and Walker, B. (1976). *J. Cell Biol.* **69,** 301–312.

Coons, A. H., and Kaplan, M. H. (1950). *J. Exp. Med.* **91,** 1–13.

Epel, D. (1970). *Exp. Cell Res.* **61,** 69–70.

Epel, D., Weaver, A. M., and Mazia, D. (1970). *Exp. Cell Res.* **61,** 64–68.

Farrell, K. W. (1982). *Methods Cell Biol.* **24,** 61–78.

Gill, D. (1979). *Experientia* **35,** 400–401.

Giloh, H., and Sedat, J. W. (1982). *Science* **217,** 1252–1255.

Glauert, A. M. (1975). "Fixation, Dehydration and Embedding of Biological Specimens." North-Holland Publ., Amsterdam.

Goldman, M. (1968). "Fluorescent Antibody Methods." Academic Press, New York.

Gundersen, G. G., Kalnoski, M. H., and Bulinski, J. C. (1984). *Cell* **38,** 779–789.

Hagström, B. E. (1959). *Exp. Cell Res.* **17,** 256–261.

Harris, P. (1978). *In* "Cell Reproduction: In Honor of Daniel Mazia" (E. R. Dirksen, D. M. Prescott, and C. F. Fox, eds.), pp. 505–514. Academic Press, New York.

Harris, P., Osborn, M., and Weber, K. (1980). *J. Cell Biol.* **84,** 668–679.

Hinegardner, R. (1975). *In* "The Sea Urchin Embryo" (G. Czihak, ed.), pp. 10–25. Springer Verlag, Berlin.

Hollenbeck, P., and Cande, W. Z. (1985). *Eur. J. Cell Biol.* **37,** 140–148.

Johnson, G. D., and de C. Nogueira Araujo, G. M. (1981). *J. Immunol. Methods* **43,** 349–350.

Johnstone, A., and Thorpe, R. (1982). "Immunochemistry in Practice." Blackwell, Oxford.

Kabat-Zinn, J., and Singer, R. H. (1981). *J. Cell Biol.* **89,** 109–114.

Kane, R. E. (1962). *J. Cell Biol.* **12,** 47–55.

Kane, R. E. (1970). *J. Cell Biol.* **45,** 615–622.

Kane, R. E. (1973). *Exp. Cell Res.* **81,** 301–311.

Langanger, G., De Mey, J., and Adam, H. (1983). *Mikroskopie (Vienna)* **40,** 237–341.

Luft, J. H. (1961). *J. Biophys. Biochem. Cytol.* **9,** 409–414.

Mazia, D., Mitchison, J. M., Medina, H., and Harris, P. (1961). *J. Biophys. Biochem. Cytol.* **10,** 467–474.

Mazia, D., Schatten, G., and Sale, W. (1975). *J. Cell Biol.* **66,** 198–200.

Moore, A. R. (1930). *Protoplasma* **9,** 9–17.

O'Brien, E. T., Asai, D. J., Jacobs, R. S., and Wilson, L. (1984). *J. Cell Biol.* **99,** 182a.

Osborn, M., and Weber, K. (1982). *Methods Cell Biol.* **24,** 97–132.

Richardson, K. C., Jarrett, L., and Finke, E. H. (1960). *Stain Technol.* **35,** 313–323.

Rodriquez, J., and Deinhardt, F. (1960). *Virology* **12,** 316–317.

Salmon, E. D., Leslie, R. J., Saxton, W. M., Karow, M. L., and McIntosh, J. R. (1984). *J. Cell Biol.* **99,** 2165–2174.

Schroeder, T. E. (1979). *Dev. Biol.* **70,** 306–326.

Schroeder, T. E., and Battaglia, D. E. (1985). *J. Cell Biol.* **100,** 1056–1062.

Showman, R. M., and Foerder, C. A. (1979). *Exp. Cell Res.* **120,** 253–255.

Vacquier, V. D., and Mazia, D. (1968). *Exp. Cell Res.* **52,** 459–468.

Vallee, R. B., and Bloom, G. S. (1983). *Proc. Natl. Acad. Sci. U.S.A.* **80,** 6259–6263.

Wadsworth, P., and Sloboda, R. D. (1983). *J. Cell Biol.* **97,** 1249–1254.

Wang, Y.-L., and Taylor, D. L. (1979). *J. Cell Biol.* **82,** 672–679.

Wang, Y., Heiple, J. M., and Taylor, D. L. (1982). *Methods Cell Biol.* **25,** 1–11.

# Chapter 15

## *Scanning Electron Microscopy of Embryos*

JOHN B. MORRILL

*Division of Natural Sciences*
*New College of the University of South Florida*
*Sarasota, Florida*

## I. Introduction

Since its commercial introduction in the 1960s, the scanning electron microscope has been used more and more by biologists in general, developmental biologists in particular, and specifically echinoderm embryologists studying microanatomical and macromolecular features associated with morphogenesis and cellular differentiation. Because of the instrument's ability to combine great depth of focus at magnifications from 10 to 100,000× with the method of image formation that makes recessed areas appear dark and elevated areas cast shadows, scanning electron microscopic (SEM) images have an apparent three-dimensional quality (Hayes, 1980) that become true three-dimensional images when stereo-pair SEM photos (Figs. 34–36) are viewed with a stereoscope. The major application of SEM by embryologists has been for morphological form and surface analysis at magnifications less than 10,000× (10 k×) via secondary electron imagery. In addition to secondary electron images other informational signals produced by the primary electrons may be detected and imaged in various ways with appropriate signal detectors—backscatter electron, X ray, and cathodluminescence.

SEM not only bridges the gap between light microscopy and transmission electron microscopy (TEM) but provides new information regarding shapes and geographic distribution of morphological entities from whole embryos (Figs. 1 and 2), intracellular organelles (Figs. 3 and 4), cellular processes (Figs. 7, 9, and 10), and extracellular coats (Figs. 4 and 8). Removal of the external cellular coats of embryos allows an unparalleled view of the arrangements of cells during major morphogenetic events such as gastrulation (Fig. 5). SEM analysis of fractured embryos provides new information on the geography of cellular patterns, cellular shapes (Figs. 6 and 7) and cellular processes (Figs. 9 and 10) heretofore visualized only in sections of embryos.

The current generation of SEM instruments with clean, high-vacuum systems, improved electron emission sources, and accessory SE detectors are capable of 6- to 3-nm resolutions. Even at a beam voltage of 10 kV macromolecular features of the surfaces of filopodia (Fig. 10), the basal lamina (Figs. 10 and 11), and extracellular matrices may be imaged at 30 k×. Thus high-resolution SEM is eminently feasible for observing ultrastructural geographies of cytodifferentiation of cellular organelles and the differentiation of specific macromolecules of cellular surfaces, extracellular coats, and extracellular matrices. When combined with cytochemistry, immunocytochemistry, and selective removal or alteration of macromolecules, SEM adds a new and flexible dimension to studies on supramolecular structures and their roles in embryogenesis. Indeed, nearly all cytological and many cytochemical methods can be adapted to provide additional information beyond cell topography with SEM (Albrecht and Wetzel, 1979).

Numerous publications on SEM techniques for biological specimens have appeared in the last 10 years. I include here introductory texts by Hayat (1978), Postek *et al.* (1980), and Gabriel (1982); a six-volume series on SEM techniques edited by Hayat (1974–1978); and detailed, technique reviews on cells and tissues in culture (Bell and Revel, 1980; Bell, 1984) and biological specimen preparation (Revel *et al.*, 1984; Murphy and Roomans, 1984). In addition the journal *Scanning Electron Microscopy* is a rich source of ancillary SEM techniques and their potential applications for echinoderm embryologists. Throughout this article I will reference selected chapters from the above as well as other sources in the literature that I have found particularly relevant for SEM of echinoderm embryos.

This article provides descriptions of wet chemical, alcohol dehydration and critical point-drying methods we have developed for routine preparation of sea urchin embryos for SEM analysis in the secondary electron mode. The article is directed to those who only occasionally use SEM. Therefore, artifacts of specimen preparation and operational parameters of the SEM scope will be reviewed.

Except for Figs. 3 and 24, the photomicrographs in this article are of the sea

---

FIG. 1. Light micrograph of early pluteus larva.

FIG. 2. SEM of early pluteus; ×250.

FIG. 3. TEM of apical region of cells of blastular wall of *Strongylocentrotus purpuratus* showing large, acidic, cortical vacuoles, an intercellular space, and tight junction between two cells. ×20,000.

FIG. 4. SEM of fractured blastula. Contents of two acidic vacuoles are preserved. The black cavities are vacuoles in which the contents were lost when the embryo was dry fractured. ×20,000.

FIG. 5. SEM of deciliated, dehyalinated, early gastrula viewed from the vegetal pole showing the shapes and arrangements of cells and the depth of field at ×1000 and stage tilt of 30°.

FIG. 6. SEM of dry-fractured, early gastrula at same stage as embryo in Fig. 5 showing shapes of cells in gastrular wall, the primary mesenchyme cells, and the fibrous extracellular matrix after dehydration and critical point drying. ×750.

FIG. 7. Luminal side of early, blastular wall showing the domed-shaped, inner ends of the cells and the fine cellular processes connecting neighboring cells. ×2500.

FIG. 8. Apical surface of ectodermal region of a lithium-induced exogastrula showing a cilium, the stretched hyaline layer, and the depth of field at ×20,000.

FIG. 9. View of a portion of a lateral arm of primary mesenchyme cells of a dry-fractured late gastrula showing the cells anchored to the gastrular wall and basal lamina by filopodia. ×3000.

FIG. 10. Higher magnification (×30,000) of boxed area in Fig. 9 showing the character of the luminal side of the basal lamina and the insertion of primary mesenchyme cell filopodia in the basal lamina.

FIG. 11. View of outer side of an abnormal basal lamina of a dry-fractured lithium-induced exogastrula showing the loose, fibrous nature of the basal lamina and protruding cellular process of a primary mesenchyme cell. ×5000.

FIG. 12. Higher magnification (×30,000) of box in Fig. 11 showing beaded nature of the dried fibers of the basal lamina.

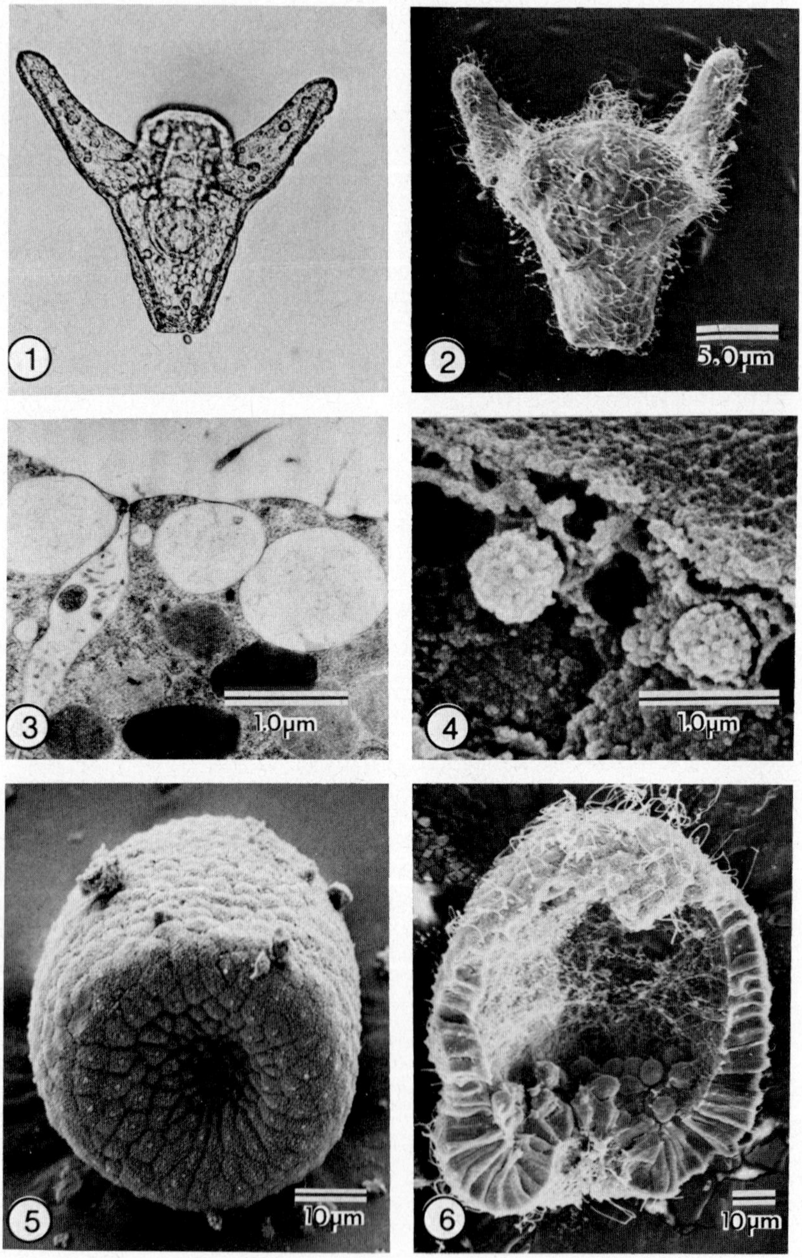


FIGS. 1–6. See legends on p. 265.

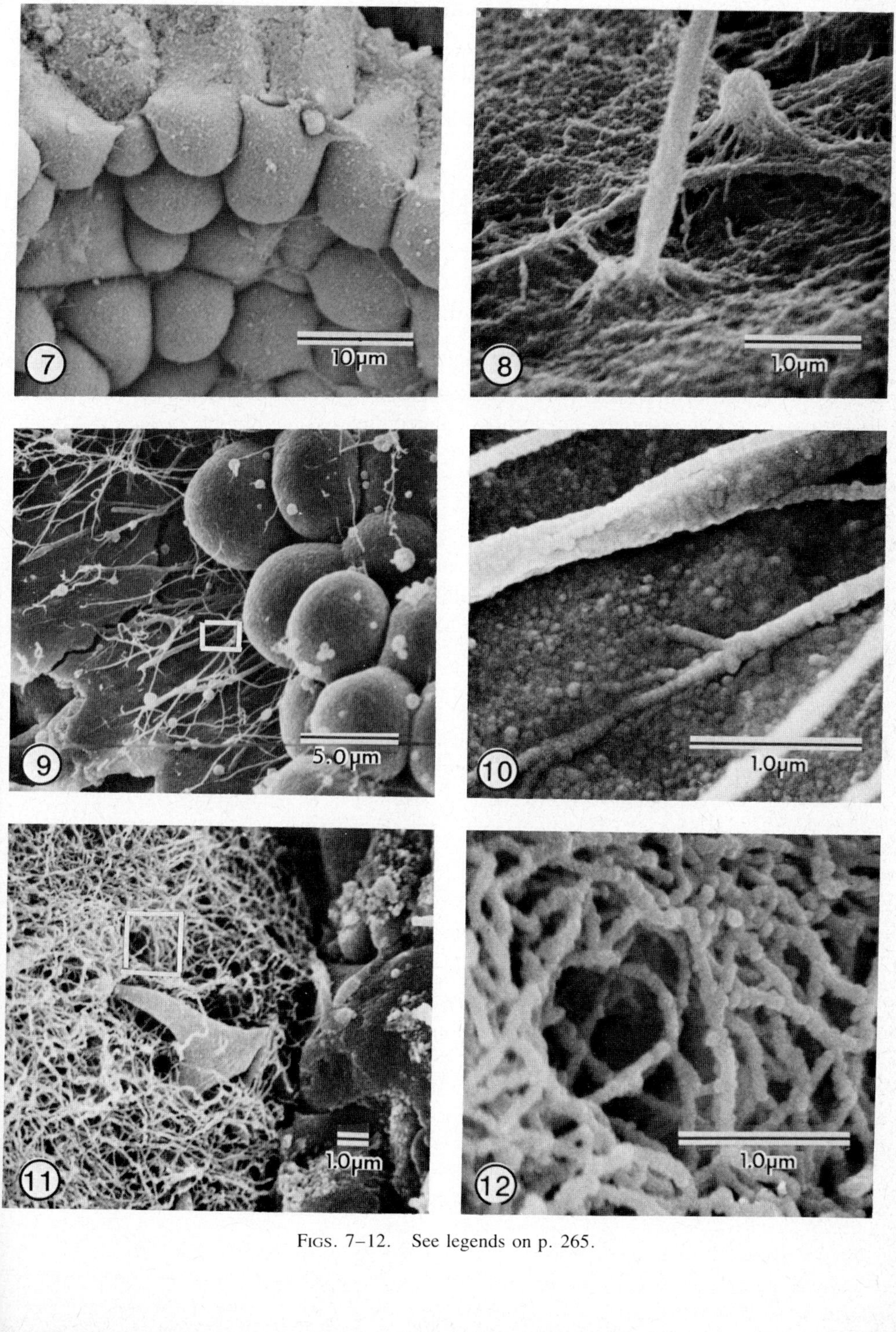


FIGS. 7–12. See legends on p. 265.

urchin *Lytechinus variegatus* embryo. The figures illustrate various SEM techniques rather than specific developmental morphologies. The magnifications included in the figure legends are those of the original photos on an ISI-SS40 SEM microscope.

## II. Specimen Preparation Supplies

Before one begins to prepare embryos for SEM analysis, it is advisable to assemble the necessary glassware, chemicals, fixatives, processing containers, and SEM supplies for mounting, storing, and photographing specimens. In addition one should coordinate their fixation and dehydration schedule with the available use of a critical point drier because specimens should be processed through critical point drying without interruption. To minimize surface contamination of specimens, all glass and plasticware should be thoroughly washed, rinsed, air dried, and stored in dust-free containers. A supply of Freon 12 gas containers and nozzle should be on hand at all times to remove dust and lint from various surfaces. All solutions should be filtered through 20-μm filters; for filtering distilled or deionized water, buffers, and alcohols, a set of 20-μm Nalgene Filter Units is useful. Small volumes of fixative and other solutions may be filtered through syringes with filter adapters.

A supply of the following items are needed for processing of embryos in suspension: 5- and 10-ml glass-stoppered graduated cylinders for preparing the fixative(s) and buffer rinses; disposable, Pasteur pipets and 2-ml rubber bulbs; one-piece, disposable polyethylene pipets whose tips may be drawn out by warming near a Bunsen burner; Wheaton snap cap, glass bottles (4 ml-l dr); cleaned EM embedding capsules (BEEM, size 00) in which the caps and tapered bottoms have been cut off with a razor blade (Fig. 13A and B) and the center of the caps punched out with a hole borer to make cap rims; 40-μm mesh Nitex filter cloth that may be cut into 1.5-cm squares. A square is placed over the end of a cut BEEM capsule, held in place by a cap rim, and the excess filter cloth trimmed off with a razor blade (Fig. 13C and D), thereby producing a specimen basket that is processed through fixation, dehydration, and critical point drying (Fig. 13). Each basket requires two cap rims and pieces of filter cloth—one for the bottom of the basket and processing up to 100% ethanol and the other for the top of the basket before it is placed in the critical point drier.

Chemicals needed for the preparation of specimens include those for removing extracellular coats and prefixation preparations of the embryos, buffer chemicals, fixatives, and fixative additives such as tannic acid, cetylpyridinium chloride, and Ruthenium Red. For buffers we routinely prepare a stock 0.4 *M* solution of cacodylate buffer (pH 7.4) with anhydrous, sodium cacodylate and an *s*-collidine buffer from a sym-collidine buffer kit. Because the number of samples that can be dried at any one time is a limiting factor and the intervals between sample preparations vary, it is judicious and economical to use 10-ml solutions of 10% EM grade glutaraldehyde in sealed ampoules and 2- or 10-ml solutions of 4% osmium tetroxide in sealed ampoules. Opened ampoules with unused fixative may be sealed with parafilm and stored (in wide-mouth screw top jars sealed with plastic tape) in a refrigerator for a few days. All of the above chemicals are available from various EM supply companies. Because the fixatives and buffers are toxic, as many operations as possible through dehydration should be done in a fume hood. Plastic gloves and gas-tight goggles are added protections. All contaminated, disposable materials and solutions should be placed in a sealed container and removed from the laboratory as soon as possible. Reusable glassware can be thoroughly rinsed in the fume hood or rinsed under a water faucet outside the building. It is of no small importance that toxic chemicals and glassware and their use be in an area separated from that used for physiological experiments. For preparing buffered fixatives a pH meter dedicated to toxic solutions is necessary.

For dehydration a supply of nondenatured 95% ethanol is needed for preparing 100 to 200 ml of a filtered series of ethanol solutions (10, 20, 30, 40, 50, 70, 80, 90, and 95%). Dry, absolute 100% ethanol is also required. Since 100% ethanol is extremely hygroscopic, it is best to use ethanol in sealed pint bottles. On first opening a pint bottle, 20–50 ml may be removed, and an equivalent amount of molecular sieve drying agent (Linde pellets, Type 4A) should be added to the remaining ethanol and allowed to settle before using the ethanol. After three-quarters of the ethanol has been used, discard the remaining ethanol, drying agent, and bottle.

Materials for mounting and dry-fracturing critical point-drying embryos include SEM stub mounts for the SEM scope one is likely to use; appropriate stub mount storage boxes, a roll of scotch, 3M double-sided tape with liner for preparing small squares of tape to mount dried embryos on stubs; insect pins (Carolina Biological Supply Co., Burlington, NC) for transferring embryos to the sticky tape; a bottle of colloidal silver paint; 12 mm round and 12 × 12 mm square #2 cover slips (Carolina); #11 pointed, disposable scalpel blades; a bottle of Microstik adhesive and adhesive tabs (Ted Pella, Inc., Tustin, CA); a micromanipulator and femur clamp for holding either cover slip or a #11 scalpel blade; and an SEM specimen mount support (Ladd Research Industries, Inc., Burlington, VT) to hold SEM stubs in a steady upright position during the mounting and dry fracturing of embryos.

Unmounted and mounted CPD specimens should be stored in desiccators. A 2-lb coffee can with a plastic lid and layer of desiccant on the bottom is an inexpensive desiccator that will hold three stub mount storage boxes ($2\frac{5}{8} \times 3$ in.) and snap-cap Wheaton bottles. Seal-N-Save boilable cooking pouches sealed with a Seal-N-Save iron (Sears, Roebuck, Co.) are also inexpensive desiccators that are useful if one needs to carry or ship mounted specimens to an EM facility.

For recording SEM images we take full-frame 4 × 5 in. photos with Polaroid Land P/N Type 55 film because of the simultaneous production of a high-quality negative and a fine-grain print for study purposes. This type of film is particularly useful for stereo pair photos since the prints can be viewed within minutes with an inexpensive, portable Geoscope mirror stereoscope (Pelco) near the SEM scope.

## III. Specimen Preparation

Specimen preparation includes treatment of embryos prior to fixation, fixation, dehydration, critical point drying, mounting, and fracturing specimens, and metal coating. Specimens can be processed from initial fixation through metal coating in 8 to 10 hours. Thus it is no more time consuming to prepare specimens for SEM analysis than to prepare permanent whole mounts for light microscopy.

### A. Prefixation Preparation of Embryos

Normally SEM of cleavage stage embryos requires the removal of the jelly coat and vitelline membrane before fertilization or the removal of the fertilization membrane after fertilization by standard procedures described elsewhere in this volume. The hyaline layer may also be removed by standard procedures to expose the outer surfaces of blastomeres and cells of postcleavage stage embryos. Embryos may be deciliated by a 30–60 second treatment with 2× seawater or 1 *M* sodium acetate. The literature abounds with additional procedures for removing, altering, labeling or selective preserving extra- and intracellular structures and macromolecules prior to fixation.

For small volumes (0.1–0.5 ml) of embryos in suspensions, the embryos are conveniently processed during treatment in glass or plastic tubes (12 mm i.d.) with Nitex mesh attached to the bottoms of the tubes by rubber "O" rings. Of course, both embryos, dissociated cells, and cell reaggregates may be affixed to cover slips or membranes with polylysine or protamine sulfate.

### B. Fixatives and Fixation

The main objectives of fixation of embryos for SEM are threefold: (1) to preserve the forms of cells, cellular processes, and whole embryos with minimal alterations from the living state with regard to volume, morphology, and spatial relationships; (2) to preserve macromolecular entities on surfaces of cells, the apical and basal laminae, and in the extracellular spaces without destroying or altering their chemical reactivities; and (3) to minimize preparative artifacts in order to visualize membrane-associated events at the ultimate resolving powers of modern SEM instruments (Arro *et al.*, 1984). These objectives further require the protection of specimens against dehydration, critical point drying, metal or carbon coating, vacuum, and exposure to the electron beam. For a detailed analysis of fixatives and fixation see Hayat (1981).

Because of the several osmotically active cellular and extracellular compartments in developing embryos, I have not found an optimal fixative that meets all of the above objectives. However, I offer here four fixatives that preserve the morphology of whole embryos, cells, and some extracellular materials.

### 1. Glutaraldehyde/Cacodylate Fixative

This fixative is prepared by mixing 2.5 ml of 10% glutaraldehyde with 7.5 ml of 0.4 *M* cacodylate buffer (pH 7.4) in a 10-ml graduated cylinder. Embryos are fixed for 1 to 2 hours at 20–25°C, rinsed briefly in distilled water, and dehydrated. This fixative preserves overall morphology of even deciliated embryos (Fig. 24).

### 2. Glutaraldehyde/Osmium Tetroxide Cocktail Fixative (D. Galileo, Personal Communication)

This fixative is prepared by mixing ice-cold solutions of 2× artificial seawater (2 ml), 10% glutaraldehyde (1 ml), and 4% osmium tetroxide (2 ml) in a 5-ml graduated cylinder in an ice bath. Embryos are fixed for 1 hour at 4°C, rinsed five times with cold distilled water and dehydrated (10–100% ethanol). To optimize the preservation of anionic extracellular substances 10 mg of 50% Ruthenium Red (Sigma Chemical Co., St. Louis, MO) or 50 mg of cetylpyridinium chloride is added to the fixative. All solutions and glassware must be ice cold to prevent the formation of contaminating precipitates during fixation.

This isotonic fixative (1200 mosmol) minimizes osmotic effects that occur when embryos are fixed in glutaraldehyde and postfixed in osmium tetroxide. Prepluteus embryos fixed with this method exhibit little shrinkage (Fig. 16) compared to embryos fixed only in isotonic glutaraldehyde-*s*-collidine (Fig. 17). The hyaline layer (Fig. 16), basal lamina (Fig. 10), cellular processes (Figs. 7 and 9), and dried extracellular fibers (Figs. 6 and 38) and vesicles (Fig. 9) are preserved with this fixative; however, the morphologies of postgastrula larvae are poorly preserved.

### 3. Osmium Tetroxide–Glutaraldehyde Sequential Cocktail Fixative (Gowdy, Unpublished, Cited by Eisenman, 1980)

While it was developed for TEM of sea urchin embryos, this fixative preserves various morphologies of embryos and larvae (Fig. 2) quite well. The fixation procedure is as follows: embryos are prefixed in a solution of 4% osmium tetroxide (2.5 ml), seawater (5.0 ml), and distilled water (2.5 ml) containing 0.132 g NaCl for 5 minutes at room temperature. Then one-half of the osmium tetroxide fixative is removed and replaced with an equal volume of 5% glutaraldehyde in distilled water; fixation is continued for 1 hour at room temperature after which the embryos are postfixed (without rinsing) in a freshly prepared prefix solution of osmium tetroxide for 45 minutes at room temperature.

### 4. Weak Cocktail Prefixation/Sequential Glutaraldehyde–Osmium Tetroxide Fixation (Eisenman, 1980)

First embryos are prefixed for 5–10 minutes at 15–25°C in a fixative composed of 10% glutaraldehyde (4 ml), 0.4 *M* cacodylate buffer, pH 7.2 (5 ml), Ca–Mg supplement (1 ml), and sucrose (0.684 g). After the sucrose is dissolved, 1.0 ml osmium tetroxide is added. The Ca–Mg supplement is 0.11 *M* $CaCl_2$ and 0.56 *M* $MgCl_2$ in distilled water.

The prefixative is removed and replaced with the above fixative but without the added osmium tetroxide. After 5 minutes this fixative is removed and a fresh aliquot of this fixative is added for 90–120 minutes at 15–25°C.

The embryos are rinsed in two changes (5–10 minutes) of a wash buffer containing 0.4 *M* cacodylate buffer (5 ml), 0.175 g NaCl, and distilled water to make a total volume of 10 ml (pH 7.2).

Next the embryos are postfixed in the wash buffer containing 1 ml of 4% osmium tetroxide per 9 ml of wash buffer for 45 minutes at 15–25°C.

For further details on this procedure and a detailed evaluation of this fixative consult Eisenman (1980), who developed the fixative for the ultrastructural preservation of *Urechis* eggs and marine invertebrate tissues. Eisenman and Gowdy fixatives best preserve the morphology of pluteus larvae (Fig. 18) even though there is shrinkage and distortion of the anal arms of the specimens.

## C. Ethanol Dehydration

Following the last postfixation rinse in distilled water, specimens in a BEEM basket (see Section II) are dehydrated at room temperature in a modified 10% step series of ethanol solutions as follows. In the first step half the volume of the distilled water is removed from the snap-cap bottle with its basket (Fig. 13H); an equal volume of 10% ethanol is added (Fig. 13I) and the bottle is gently agitated by hand or on an angled rotary shaker to mix the two solutions (Fig. 13J). After 2–5 minutes the solution around the basket is gently agitated with a rubber-bulbed disposable pipet; half the volume of solution in the bottle is again removed and an equal volume of the next concentration of ethanol is added. This process is repeated until the specimens are in 100% ethanol. At this point the bottle's rim is wiped dry and a clean, dry snap-cap is placed on the bottle. After two or three 15-minute changes of 100% ethanol, the basket is quickly transferred to a shallow dish of 100% ethanol (Fig. 13K). A square of Nitex cloth is snapped onto the top of the basket (Fig. 13L) and the basket is quickly transferred to an appropriate critical point drier partially filled with ethanol (Fig. 13M).

## D. Critical Point Drying

The theory and practices of critical point drying are given by Hayat (1978), Postek *et al.* (1980), Gabriel (1982), and, especially, Cohen (1984).

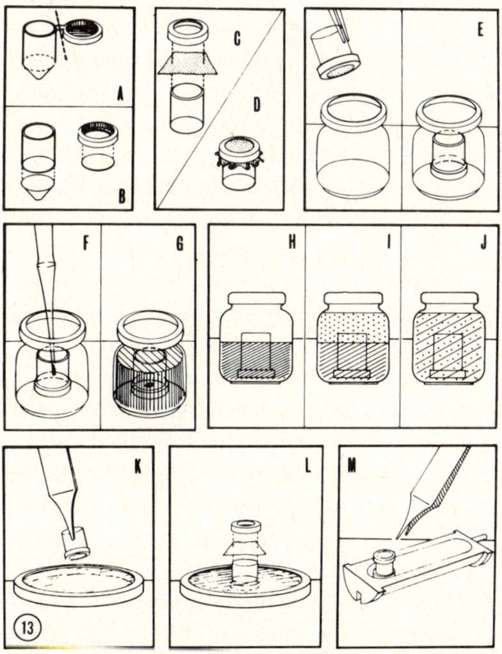


FIG. 13. Schematic drawing of preparation of embryos for critical point drying. Basket constructed from BEEM capsule (A,B) and 45-μm mesh Nitex cloth (C,D) then placed in 4-ml bottle (E). Embryos added inside basket (F) and fixative added outside to fill bottle (G). Fixative drained out (H) and sequential ethanol solutions added (I,J). Basket transferred to dish of 100% ethanol (K) and capped (L). Basket transferred to Polaron CPD boat filled with 100% ethanol (M).

We routinely critical point-dry specimens submerged in ethanol using $CO_2$ as the transitional fluid in a Polaron E3000 critical point dryer (see Hayat, 1978, for operation of this apparatus) precooled to at least 10°C below room temperature and pretested for possible pressure leaks due to faulty "O" rings. It is advisable to have two extra sets of "O" rings for the valves, door, and window of the pressure bomb and to determine if there is sufficient $CO_2$ in the gas tank for a complete run. This is conveniently accomplished by standing the gas tank on a bathroom scale and recording its weight at the end of each run. Research or hospital-grade $CO_2$ tanks with siphons contain less contaminants than industrial $CO_2$ tanks. For critical work an in-line filter between the $CO_2$ tank and the pressure bomb is desirable. Most $CO_2$ tanks have 40–50 lb of liquid $CO_2$.

Once the specimens are sealed in the pressure bomb, the bomb's chamber is quickly filled with the liquid $CO_2$. In the Polaron apparatus one can see the denser ethanol accumulate at the bottom of the chamber. With the inlet valve open, the drain valve is opened to drain the ethanol slowly. After flushing the chamber for 4 minutes, all valves are closed for 15–20 minutes to allow the exchange of $CO_2$ with ethanol in the specimens. Then the chamber is flushed with $CO_2$ again for 4 minutes. This process is repeated 3 or 4 times until all of the ethanol has been removed from the chamber. This point is determined by holding a piece of filter paper near the end of the open drain valve and collecting a small mound of frozen $CO_2$. If, upon evaporation of the $CO_2$, a wet spot remains on the paper then repeat the flushing and exchange cycling.

Prior to raising the temperature in the chamber, the level of the liquid $CO_2$ may be slowly lowered to the level of the top of the specimen boat by bleeding $CO_2$ from the drain valve. During this process the pressure will drop. If fine streams of bubbles are seen in the liquid $CO_2$, it may mean there is still ethanol present. The next step is to slowly raise the temperature of the chamber with a pressure increase of no more than 100 psi $min^{-1}$. For our water-jacketed Polaron instrument we use the hot and cold water mixing faucet in a laboratory sink. When the temperature of the chamber reaches 25–28°C, we maintain this temperature for 30 minutes before slowly raising the temperature to 40–45°C. At this point the drain valve is opened slightly to allow the gas to escape until the pressure is reduced to atmosphere. The pressure should be reduced at the rate of 100 psi $min^{-1}$ over a period of 15 minutes. At no time during this process should the temperature drop below 35°C.

After the pressure is reduced, the vent valve is opened; open the chamber and quickly transfer the specimen baskets to their clean, heat-dried snap-cap bottles and store the bottles in a desiccator over indicator silica gel or Drierite.

## E. Mounting

Specimen-mounting methods and adhesives are reviewed by Hayat (1978). For small specimens such as embryos, we use pieces of double-stick tape

(Scotch, 3M) in spite of its inherent disadvantages; it is hygroscopic and sensitive to radiant heat during sputter coating which can result in "cracking" of the tape (Fig. 23) and in settling of the embryo into the tape (Fig. 24). The surface of a trapezoidal piece of tape can be scored with a razor and embryos from up to six different stages or treatments can be mounted on one 12- to 15-mm stub.

After the tape is applied to a stub, place the stub in a SEM specimen mount support on the stage of a dissecting microscope set up for reflected light illumination (Fig. 14). Remove the protective paper from the upper surface of the tape (Fig. 15c), remove the top from a BEEM basket, and "pour" embryos from the basket onto the filter paper placed in the recessed center of the mount support (Fig. 15a). Then transfer individual embryos to the tape with an insect pin whose tip is made sticky by touching it to the edge of the tape. Where several dried embryos adhere to each other, transfer the aggregate to the tape and gently tease them apart on the tape. During this process embryos may be fractured fortuitously. Embryos may be placed on the tape in ordered rows (Fig. 15d); gastrulae and larvae may be preferentially oriented on the tape (i.e., Fig. 18). If the embryos are not dry fractured before sputter coating, then apply silver paint with a small brush to the edges of the tape (Fig. 15f) so that the paint overlaps the margins of the tape (Fig. 23) and place the stub in a stub box in a desiccator until the paint is dry (2 hours to overnight).

## F. Dry Fracturing

Once embryos are mounted on sticky tape and before the tape is rimmed with silver paint, we routinely dry fracture at least a few of the embryos in order to view the character and quality of preservation of internal structures. Three simple methods—prick, pry, and pluck—have proved successful depending on individual skills and patience. Pricking is performed free hand by touching or poking an embryo with the tip of a fine insect pin or tungsten wire needle. The resulting fracture is usually fortuitous and irregular (Fig. 6). Prying is performed with a #11 scalpel blade held at an angle to the surface of the tape in a femur clamp attached to a micromanipulator (Fig. 14). The blade is lowered until the tip penetrates the embryo; then the blade is moved one way or the other on the $X$ axis to "pry" open the embryos. Because the fracture plane is frequently 45–90° to the face of the stub (Figs. 32 and 33), the interior surfaces of the fractured halves are visible only at high angles of tilt (30–70°). Even embryos that are crushed or fragmented may yield important information with respect to bulk morphologies of the archenteron and mesenchyme cells as well as topographical features of internal surfaces of the embryos.

The cleanest fractures (Fig. 35) where the fracture plane is parallel to the face of the stub are obtained by plucking off the upper part of the mounted embryo. This is accomplished by coating the surface of a pin stub or preferably a 12-mm coverslip with an adhesive (Microstik or an adhesive tab) and lowering the pin

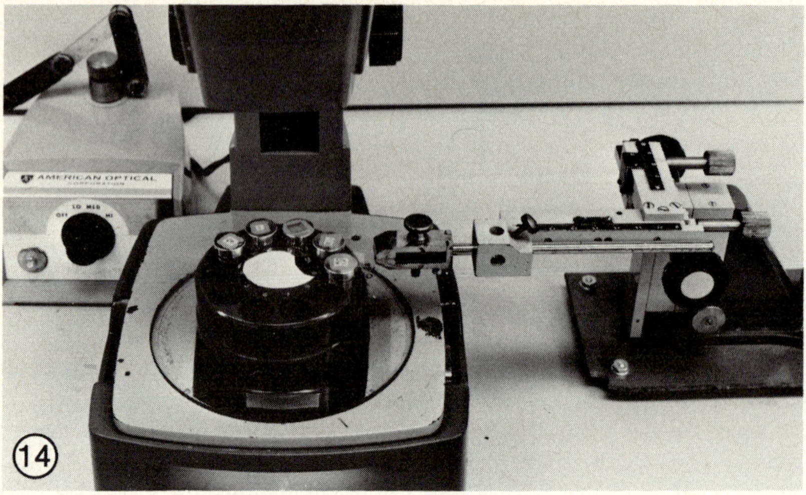

FIG. 14. Dissecting microscope with epiilluminator and stub holder for mounting specimens on stubs and a Prior X-Y-Z micromanipulator with a tendon clamp to hold a cover slip or No. 11 scalpel blade for dry-fracturing embryos on sticky tape of stubs.

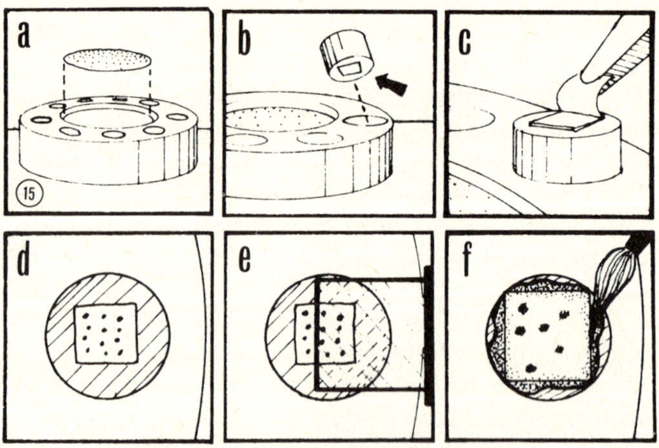

FIG. 15. Schematic drawing of mounting embryos on sticky tape of stubs and dry fracturing by plucking with cover slip mounted in micromanipulator. (a) Circular white disk of Whatmann No. 50 filter paper placed in center of stub holder to receive dried embryos in BEEM basket. (b) Small piece of double-sided sticky tape (arrow) attached to bottom of stub before stub placed in holder. (c) Square of double-stick tape placed on top of stub and protective paper peeled off. (d) Embryos transferred on surface of sticky tape with sticky tip of insect pin. (e) Dry fracturing of embryos by plucking with a cover slip with adhesive. (f) Silver paint applied to marginal region of sticky tape.

stub or cover slip with a micromanipulator (Fig. 14) until it touches and slightly compresses the embryos on the tape and then raising it slowly. The advantage of the cover slip is that you have more control over the process. With this method you can produce complementary halves of fractured embryos, selectively remove the ectodermal layer to expose the outer surface of the basal lamina (Fig. 11), and have fractures that pass between cells as well as through cells with a minimum of surface contamination (Fig. 38). After dry fracturing, the tape on the original stub is rimmed with silver paint as are the cover slip on a second stub and the adhesive area on plucking stub.

With the pluck method it is absolutely critical that the edges of the tape on the stub be trimmed with a razor blade to prevent the surface of the plucking coverslip or stub from touching the surface of the tape. This method also requires the stub to be secured in the mount support with a small piece of sticky tape (Fig. 15b).

In spite of potential surface contamination and artifacts resulting from dry fracturing, the method is simple and quick. Furthermore, whole embryos that have been metal coated and examined on the SEM scope may be dry fractured and recoated for further examination. Thus additional information of specimens on a single stub is obtainable through fracturing and coating cycles.

When the goal of fracturing whole embryos is to examine the distribution of labeled markers and the morphology of delicate internal and extracellular entities at high magnifications, the cryofracturing procedures of Humphreys *et al.* (1978) produce excellent results. Briefly, specimens in 100% ethanol in small Parafilm tubes are frozen in liquid nitrogen and fractured with a cold razor blade. The fractured, frozen pieces are transferred to 100% ethanol in a BEEM basket at room temperature, critical point dried, mounted, and coated.

## G. Coating Specimens

Because dried biological specimens and adhesives do not readily conduct electricity or heat, the specimens are routinely coated with a thin film (5–25 nm) of metal (i.e., gold–palladium) or carbon. For all but the most critical, high-resolution type of surface analysis, specimens are routinely coated with a metal film in a sputter coating instrument or with carbon in a carbon thread carbon coater. Coating in a vacuum evaporator or Penning sputtering is still the preferred method for analytical thin film, high-resolution SEM. A Triode sputter coater with a thermoelectrically cooled stage and a film thickness monitor control is preferable to the older diode-type sputter coaters in which adhesives and specimens may be stressed by excessive heating (Figs. 23 and 24). For surface topography, metal coating is used routinely; coating specimens with a thin film of carbon before metal coating can improve the stability of delicate structures and the distribution of the metal film; carbon coating specimens is most useful when exposed surfaces are labeled with electron-dense markers like colloidal gold and is necessary when backscatter electron images of metal markers and cytochemical stains or reaction products are desired.

General guidelines for sputter coating include as thin a metal film as possible to minimize damage from heating and to prevent charging of the specimen on the SEM scope. At low magnifications, structures that exhibit charging at beam voltages of 10 to 15 kV will not charge when the voltage is lowered to 2 kV. If charging of the specimens or areas of the adhesive is excessive, then recoat the sample. We have found in fractured, critical point-dried embryos that loosely attached materials, particularly in the exposed blastocoel, may be lost during sputter coating unless the pressure in the vacuum chamber is changed very slowly.

## IV. Artifacts of Critical Point-Dried Specimens

The actual and potential artifacts in wet chemical-fixed, critical point-dried specimens have been reviewed and critiqued many times (see references in the Introduction). As Cohen (1984) says, "If either microscopic or macroscopic specimens after fixation look clean, undistorted and 'natural' your chance of success is good. Critical point drying begins with the initial treatment of the specimen, and all steps are interrelated; artifacts caused by earlier processing may not show up until later." For sea urchin embryos we have observed a number of artifacts associated with each step in the critical point drying methods used. Some of the artifacts may be prevented, some minimized, and others may require special methodologies yet to be developed. Space permits only a review of the overt artifacts of the several steps in critical point-drying preparations.

### A. Fixation Artifacts

Typically, deciliation in hypertonic seawater followed by fixing in glutaraldehyde alone results in overall shrinkage of the embryo (Fig. 17) and frequently the

---

FIGS. 16–22. Shrinkage artifacts of pretreatment, fixation, dehydration, critical point drying, and air drying of embryos and larvae. Prelarval stage embryos fixed in Galileo's fixative (Fig. 16) show minimal shrinkage as compared to deciliated embryos fixed in glutaraldehyde-collidine buffer fixative (Fig. 17). Larva fixed in Eisenman's fixative exhibits distortion of the arms and overall shrinkage following critical point drying (Fig. 18). Air drying of larva on a Millipore filter in 100% ethanol results in a marked collapse of the soft tissues around the larval skeleton (Fig. 19). Figures 20–22 show the shape of the larva when it's alive (Fig. 20), after fixation in Eisenman's fixative when the embryo is in 10% ethanol (Fig. 21), and after critical point drying (Fig. 22).

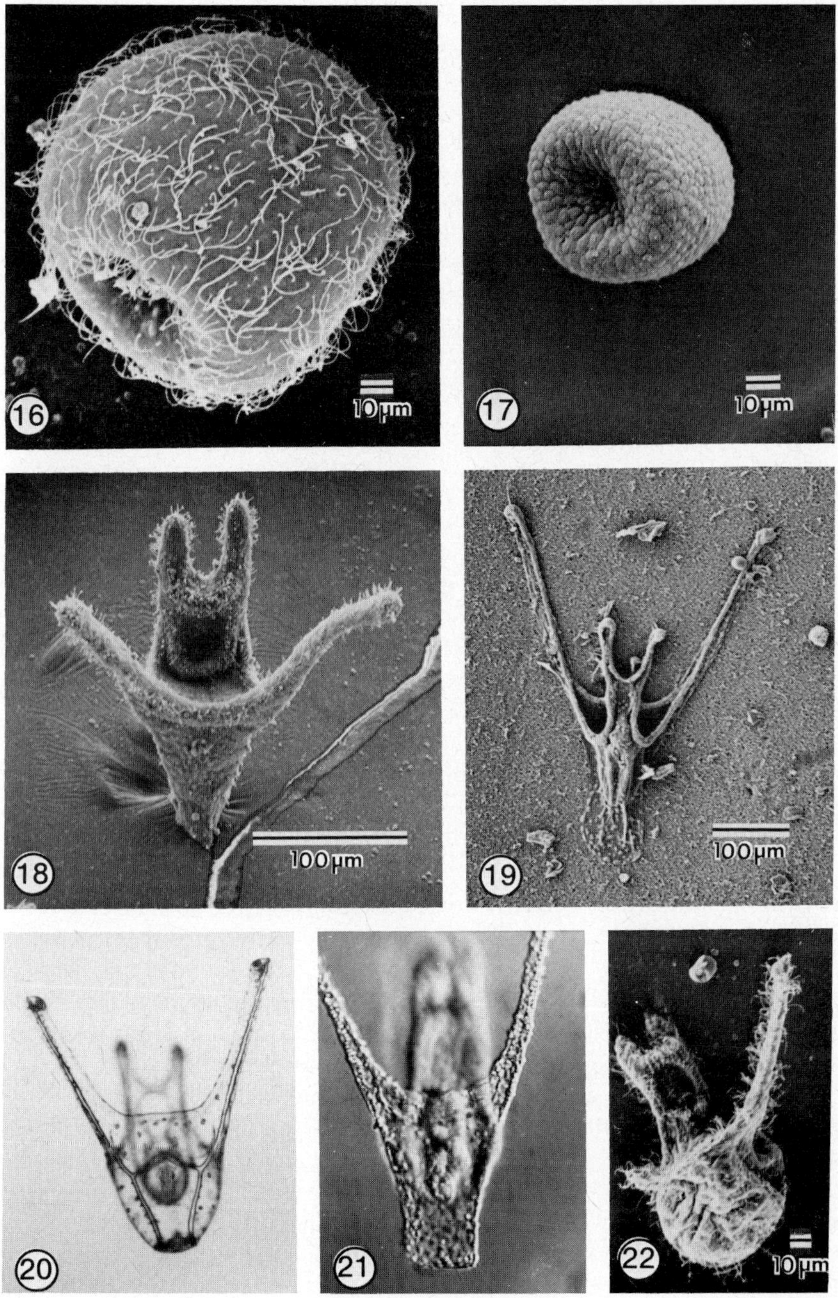
16
10μm
17
10μm
18
100μm
19
100μm
20
21
22
10μm

loss of the hyaline layer. With glutaraldehyde alone the blastocoelic compartment remains osmotically active; extracellular carbohydrate-enriched macromolecules may be subsequently lost or incompletely preserved during dehydration. These alterations may be reduced by adding primary amines to the glutaraldehyde fixative (Boyles, 1984). Alternatively, the space-filling locations and structures of these substances may be preserved by anhydrous fixation in acrolein vapors followed by dehydration in anhydrous cellusolve (Goldberg and Escaig, 1983).

While the other fixatives minimize bulk shrinkage, none preserves the natural shapes of the cilia. At high magnifications, microvilli also appear distorted. Not all of the smooth vesicle-like structures on the surfaces of cells (Fig. 9) may be secretory vesicles; some may be membrane blisters that are a frequent artifact of "double" fixation in glutaraldehyde and osmium tetroxide (Shelton and Mowczko, 1978).

## B. Dehydration Artifacts

Even embryos, especially larvae, fixed or postfixed in osmium tetroxide in which semipermeable membrane barriers are destroyed, may exhibit overall changes in size and shape during ethanol dehydration (compare Figs. 20 and 21). Potential volume changes during dehydration may be reduced by treating the embryos with aqueous 0.5% uranyl acetate after post-osmium rinsing and before dehydration (Wollweber *et al.*, 1981). Continuous gradient rather than step dehydration from distilled water to 100% ethanol (Rostgaard and Tranum-Jensen, 1980) may also reduce cellular and bulk shrinkage.

## C. Critical Point-Drying Artifacts

When monitored with a light microscope through the 100% ethanol dehydration step, embryos and larvae fixed in Galileo, Gowdy, or Eisenman fixative may retain their normal size and shape. Following critical point drying embryos, particularly larvae, exhibit measurable or severe shrinkage artifacts (Fig. 22) and distortion of their arms (Fig. 18). Recently Boyde and Tamarin (1984) have traced the origins of this artifact to when the dry samples are being "let down to air" and to the presence of residual ethanol in cells and tissues. Figure 19 is a gross example of the effects of residual ethanol in a specimen after critical point drying. In this instance there was even a layer of ethanol over the specimen that was subsequently air dried. As the authors points out ethanol can "ethanolate" proteins (probably other macromolecules) like water "hydration," that ethanol is less miscible than other intermediate solvents, and that storage of ethanol over molecular sieve does not prevent post-critical point-drying shrinkage. What does reduce this type of shrinkage is replacing ethanol by $C_2Cl_3F_3$ by refluxing in a special Soxhlet Extractor (Agar Aids, Ltd., Stansted, Essex, England). Specimens are then transferred in $C_2Cl_3F_3$ to a Polaron critical point-drying bomb and critical point dried using $CO_2$ as the transitional fluid. With the critical point drying method of Boyde and Tamarin many of the shrinkage and distortion artifacts of cellular and extracellular structures in SEM images of embryos should be reduced.

## V. Operating Parameters and Secondary Electron Imaging with the Scanning Electron Microscope

Numerous sources review the theory and practice of the interaction of the electron beam with the specimen, image formation, imaging parameters, and image artifacts. Hayat (1978) provides an excellent introduction and review for optimizing the formation of images; Gabriel (1982) discusses the photographing of SEM images, and Hayes (1980) reviews the problems of the perception and interpretation of SEM images.

With respect to the formation and quality of images one should know how beam voltage (kV), diameter of the final aperture, diameter of beam (spot size), working distance, stage tilt relative to the position of the beam and position of the secondary electron detector(s), and stigmatism may affect the resolution (sharpness), depth of field, and sources of the secondary electron signals that produce variations in contrast and brightness in the resultant image. What is frequently not appreciated or understood is that more than 70% of the secondary electron image is generated by the electron beam interacting with atoms below the spec-

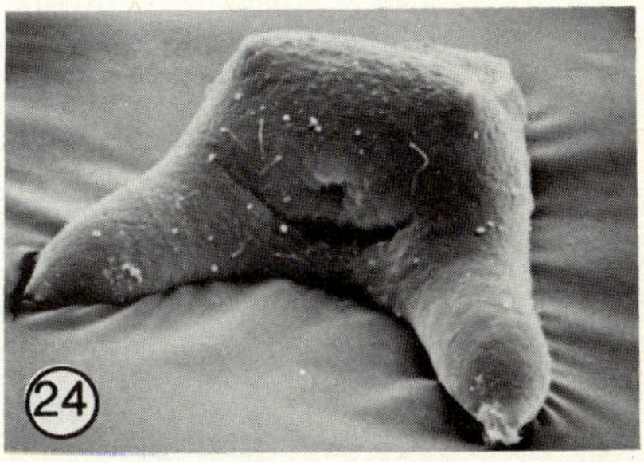


FIG. 23. Low magnification (×20) of embryos on sticky tape showing shrinkage and cracking of tape during sputter coating. The bright area of the tape is not grounded and is charging because the metal film has also cracked.

FIG. 24. Deciliated *Arbacia* early pluteus larva that has settled into the sticky tape. The curved slit beneath the stomodeum is a preparation artifact.

imen's surface to produce backscatter electrons that produce type II secondary electrons as they leave the specimen and type III secondary electrons when they collide with surfaces near the specimen such as the bottom of the objective lens and the wall of the specimen chamber. Type II and III secondary electrons impart contrast, brightness, and shadowing that enhance the three-dimensional quality of SEM images. However, these secondary electrons may also obscure the imaging of fine surface features by type I secondary electrons and cause a supersaturation of secondary electron signals from edges of structures. This is known as the brightness edge effect and is frequently interpreted as charging. With embryos the edge effect is particularly noticeable at low magnifications (< 2k×) and high beam voltages (10–25 kV) as illustrated in Figs. 6, 25–28, and 35–36.

The following guidelines are offered for improving image quality over that typically obtained with "standard operating settings" (i.e., 12- to 15 mm working distance 15–25 kV, and 100 μm final aperture) on the SEM scope:

1. *Working distance*. For whole embryos a 12- to 15-mm working distance suffices; at magnifications above 2 k× an 8- to 4-mm working distance is better because the shorter the working distance the "sharper" the surface features are even though the depth of field is less.

2. *Beam voltage*. A general rule is to use a beam voltage of 2–3 kV at magnifications less than 0.5 k×, 5 kV (0.5–2 k×), 10 kV (2–10 k×), 15–20 kV (10–20 k×) and 25–30 kV (20–100 k×). Figures 25–28 illustrate the improvement of the quality of a 500× SEM image when the beam voltage was reduced from 20 to 2 kV. In addition to causing halo-like edge effects in brightness and obscuring surface details (Fig. 25), high beam voltages may also cause surface features of the specimen to vibrate while a photo is taken. This artifact is particularly noticeable at high magnifications and may be corrected by using a lower beam voltage as illustrated in Figs. 29–31 of the surface of the hyaline layer. This image artifact is not due to charging although surface charging does create vibrations.

3. *Final aperture*. A general rule is to use a final aperture of 200 μm below 1 k×, 100 μm (1–5 k×), 50 μm (5–20 k×), and 30 μm (20–100 k×).

4. *Spot size*. To maximize the resolution of the image's features at any magnification the spot size should be as small as possible to obtain a visible or photographable image. Reducing the spot also reduces the strength of the secondary electron signal as does reducing the size of the final aperture and beam voltage. This is particularly a problem at magnifications above 10 k×.

5. *Stage tilt*. Tilting the stage and specimen toward the secondary electron detector increases the brightness of the image and the secondary electron signals. Thus by tilting the stage to 45–70° along with rotating the stage to orient a representative specimen, it is possible to overcome certain imaging restrictions

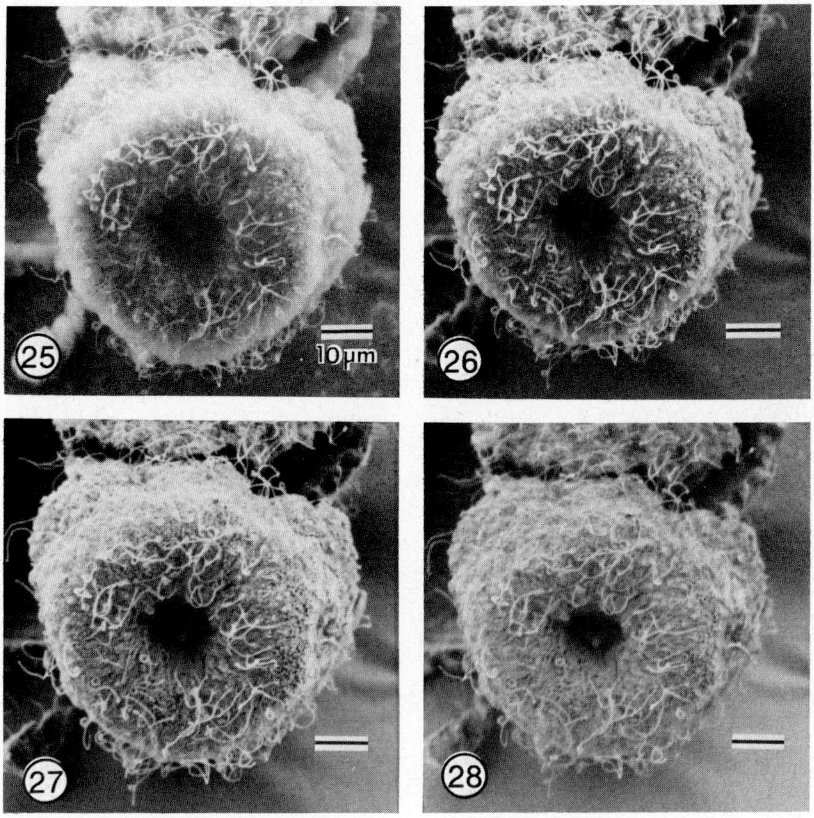


FIGS. 25–28. Effects of primary beam voltage on image quality. All photos were taken at ×700 at similar contrast-brightness settings of the vegetal pole of a mid gastrula. At 20 kV (Fig. 25) the outer edge of the embryo glows because of excess secondary electron signal and details of the surfaces of the ectodermal cells around the gastropore are obscured. Overall image quality and tonality improves when the beam voltage is decreased to 10 kV (Fig. 26), 5 kV (Fig. 27), and 2 kV (Fig. 28). The optimal beam voltage for this specimen and magnification is 5 kV.

associated with low kilovoltage, small spot size, and a small final aperture. Both tilting and rotating the stage are extremely useful in viewing an embryo from all angles and for features of embryos to be oriented for photographing (Figs. 5, 18, and 24). Two minor distractions associated with photos of specimens imaged at high tilt angles are that the background is usually brighter than the specimen and that surface features in the background may detract from the quality of the final photograph. Since embryos and their parts have complex geometries, the angle of tilt of the stage is only relative to the actual angle of the imaged portion of the specimen with respect to the detector and the incident electron beam. Finally, by tilting the stage away from detector even the $-10°$ available on some SEM instruments, the ratio of type I to other secondary electrons increases. Thus a negative tilt of the stage can improve the quality of the images of fine surface details.

## VI. Examining and Photographing Embryos

### A. Examining Embryos

When examining SEM preparations and recording images, try to examine every embryo on the stub from various angles and record the position and features of embryos to be photographed. A low-magnification (10–30×) Polaroid photo of the entire piece of sticky tape (Fig. 23) is extremely useful for orientating the observer. The stub can be marked so that it is always placed in the stub holder of the stage in the same orientation relative to the "stand-by" position of the *X* and *Y* axis controls of the stage. The locations of an embryo to be photographed as well as rephotographed at some later time may be recorded on the low-magnification photo as well by noting the position of the embryo relative to the scales for the *X*, *Y*, and rotational axis controls on the door of the stage.

In order not to waste film, try to think of every negative as being destined for publication, a poster display, or analysis from other angles or at higher magnifications. All high-magnification recordings should be accompanied by at least one low-magnification overview of the embryo.

### B. Photographing Embryos

Before photographically recording an image, it is important that the image of the embryo or the important parts of the embryo are centered on the viewing screen and at least $\frac{1}{4}$ in. from the margins of the area of the screen that appears on the negative. By leaving such a margin, then the image on the negative may be reduced, if necessary, when prints are made with an enlarger. The magnification of the image may be adjusted to one or another standard magnification if zoom control is available on the SEM instrument. We routinely record images at a few selected or preferred magnifications in order to simplify comparisons between photos.

In addition to being centered, it is especially important that whole embryos and parts of embryos be oriented with their vegetal pole lower most relative to the bottom of the camera screen. With some SEM instruments the camera screen is oriented 90° to the viewing screen. When taking full-frame stereo pair photos

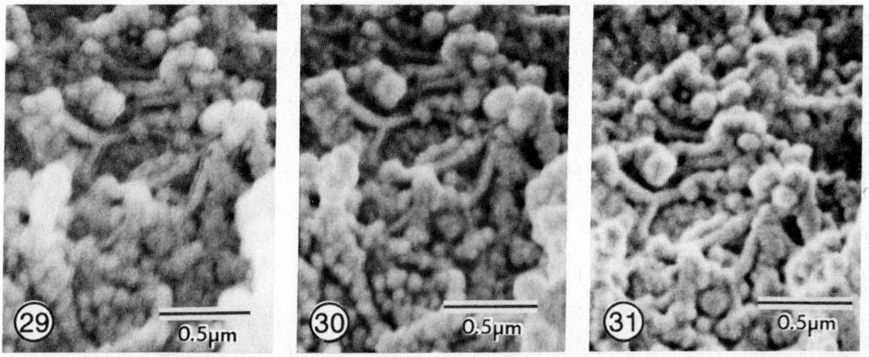


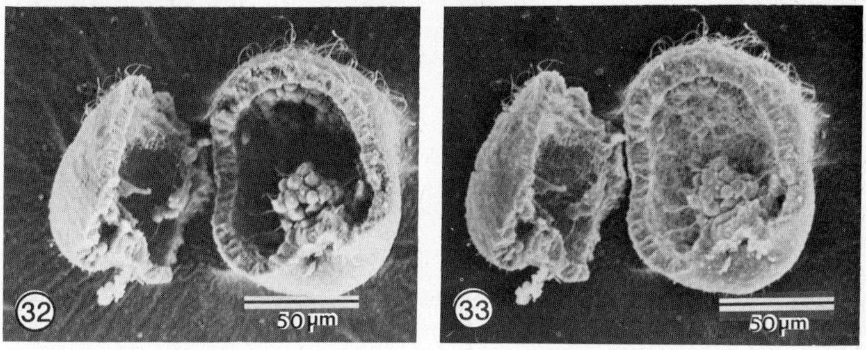


FIGS. 29–31. Effects of primary beam voltage on image stability of delicate surface features at high magnifications. Photos of the surface of the hyaline layer of a 16-cell embryo at ×30,000 taken at beam voltages of 20 kV (Fig. 29), 5 kV (Fig. 30), and 2 kV (Fig. 31).

FIGS. 32–33. Comparison of image quality and photographic information in concave regions of an early lithium-induced exogastrula with a conventional stage chamber secondary electron detector (Fig. 32) and an in-column detector in the H.R. mode (Fig. 33) at ×300.

of embryos, the plane of stage tilt relative to the orientation in the stero viewer should be considered. Figure 35 is an example where this was overlooked.

Before photographically recording the image, determine that the image is stable and not charging, is in focus and is astigmatic. It is also advisable to check that the final aperture, final lens, and gun tilt and shift are aligned. Then adjust the contrast and brightness of the image so that brightest and darkest areas in the image field are about two-thirds of the full scale settings on the instrument. The use of gamma control, if available, to attenuate the brightest relative to the darkest regions is quite useful when photographing whole embryos (Figs. 5 and 15) to reduce glare from edge effects and when photographing fractured embryos where the outer, convex surfaces are much brighter than the inner, concave regions (Fig. 32). Even with gamma control, surfaces and regions in the "shadow'' of the secondary electron detector in the stage chamber appear dark (Fig. 32). However, a detector at the bottom of the objective lens or in the column proper can reveal morphological information (Fig. 33) otherwise lost in the shadowed areas.

The dried negatives are stored in negative envelopes with the emulsion side of the negative away from the seam of the envelope; the dried, coated Polaroid prints are best attached to the outside of the envelopes.

## VII. Recording Stereo Pair Images

To avoid misinterpretations about structural features and spatial relationships that exist in biological specimens, stereo visualization (stereopsis, stereomicroscopy) of specimens should be practiced routinely and is essential for extracting both qualitative and quantitative information from SEM micrographs (Figs. 34–36). Wergin and Pawley (1980) provide an excellent, instructional review on recording and projecting SEM stereo pairs; the theory and various methods are reviewed by Chatfield (1978).

The procedure I use for taking a stereopair is as follows:

1. The image of an embryo or part of an embryo is centered on the viewing screen and oriented relative to the *X* and *Y* axis and plane of tilt of the photographed pair of images.

2. Stereo images destined for publication are taken at a magnification in which the width of the important region on the screen is less than 2.5 in. (60 mm). Stereo pairs of whole and fractured embryos are taken at 300 to 500×.

3. After the field of interest is photographed in the usual way, several prominent features of the image are outlined with a wax pencil on the plastic shield of the viewing screen. When taking stereo pairs of images at magnifications above 5000×, reduce the magnification and outline an additional reference feature that can be readily located after the specimen stage is tilted.

4. Tilt the stage, relocate, and recenter the features to coincide with the penciled outline; refocus the image; adjust the magnification to that of the first photo; adjust the contrast and brightness if necessary; rephotograph the field of interest. Ideally, refocusing is done with the *Z*-axis control for the stage to have the two images aligned; operationally the second image is refocused with the focus knobs which can result in the second image being ''skewed'' relative to the first depending on the degree of tilt, magnification, and degree of curvature of structures in the images. For qualitative stereo images the stage may be tilted 10–12° (10–100×), 6–8° (100–1000×), 6° (1000–10,000×), and 4° above 10,000×. After the second photo, check the stereo pair Polaroid prints with a

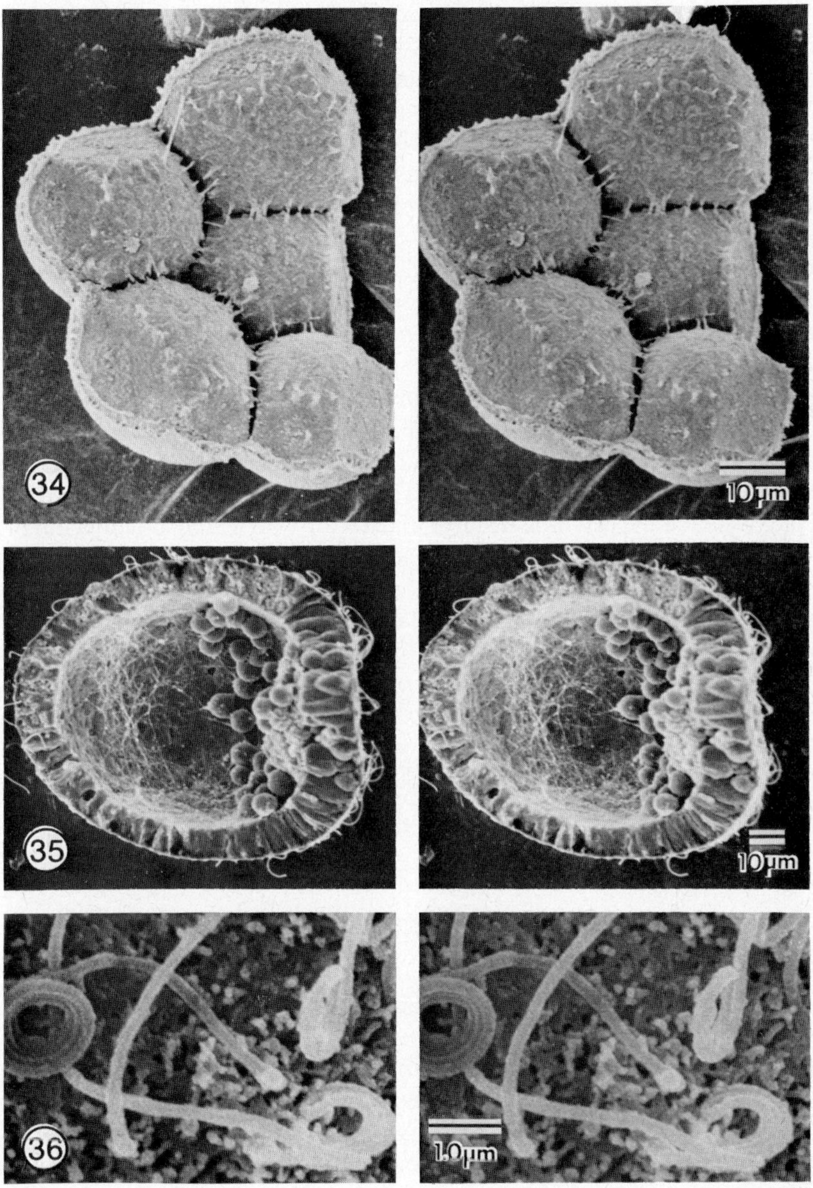


FIGS. 34–36. Stereopairs of the luminal surfaces and cell processes of blastomeres of a 16-cell embryo (Fig. 34); a dry-fractured, early gastrula (Fig. 35), and ectodermal cilia of an early gastrula (Fig. 36).

portable stereo viewer for composition and focus, then rub the wax off the plastic shield with a soft damp cloth or paper towelling.

5. When taking stereo pairs, it is convenient to take the first photo at the lower tilt angle and record on the margin of the Polaroid prints which is left and right. While half-frame stereo pairs may be taken on a single sheet film if a split screen imaging accessory is available, I prefer to take full-frame stereo pairs because printed enlargements of the original prints can be mounted on a sheet of cardboard and used for study, discussion, and display with one or more portable stereo viewers. In addition to its other attributes, a stereo image frequently provides a sharper image than either monoimage. Furthermore, each of the stereo pair may be taken with different brightness and contrast settings, kilovolts, size of the final aperture, and beam diameter, thereby providing the opportunity for additional, visual information even though the stereo pair may not be published.

6. For preparing and viewing stereo pairs for publication one needs a pocket-type stereo viewer with an interpupillary adjustment. When printing stereo pair negatives either enlarge or reduce the projected image of the negatives in an enlarger so that the essential features of each print lie within a field width of 2.5 in. and that the center-to-center distance of the trimmed and aligned prints is 2.4–2.5 in. (60–62 mm). Apply a piece of photo mount adhesive to the back of each print. Trim the vertical image of one print, overlap this print on the other print, align the images on the two prints using a pocket viewer, tape the overlapping margins of the two prints, tape the overlapping on the back of the prints, trim the outer margins of the two prints, then carefully cut the prints apart along the middle of the overlap to make the two prints equal in width. Before permanently mounting the prints side by side, check their alignment and orientation relative to one another with the stereo viewer.

## VIII. Photomontages

Photomontages of SEM images of whole embryos, particularly fractured embryos, and parts of embryos are useful where the 3× enlargement limits of Polaroid negatives have to be exceeded for poster sessions and public displays and where the enlarged in-focus morphological relationships are to be analyzed. The extra expense of time and film is more than offset by image quality of the final print, even if it is a print of a photo-reduced image (compare Figs. 37 and 38). Furthermore, negatives of parts of the montage may be enlarged individually for display purposes.

I find that photomontages of an embryo or a major morphological feature of an embryo taken at a magnification of 2.5 k× requires 12 to 15 4 × 5 in. negatives. All of the individual frames of a montage should be taken with a 5 to 10%

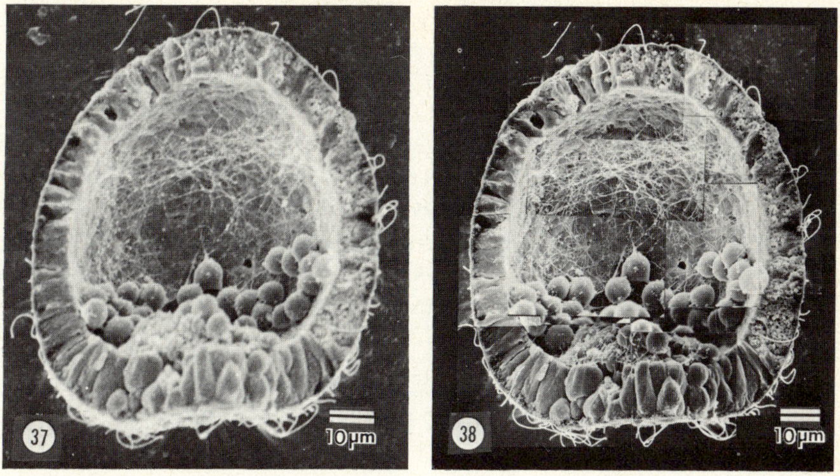


FIGS. 37–38. Comparison of the quality of the photo image of a fractured early gastrula taken at ×500 (Fig. 37) and a photomontage of the same embryo composed of 15 photos taken at ×2500.

overlap at one session (0.5–1 hour) on the scope focused at the same magnification. The Polaroid prints of all the frames should be arranged and checked for areas of the montage that may need to be rerecorded before removing the specimen from the instrument.

Once all the individual negatives have been printed, and the prints backed with pieces of photo-mounting adhesive, the prints are trimmed and fitted. An initial, if not final, photomontage may be made directly from the Polaroid prints. Once the final montage print is made, it may be photographed with a 4 × 5 in. studio copy camera; the resulting negative may be used to make enlarged prints or reduced prints for publication (Fig. 38).

## IX. SEM of Sectioned Embryos

SEM of "thick" sections of embedded specimens with or without removal of the embedding medium offer a variety of opportunities for both secondary and backscatter electron imaging of surfaces and internal structures of cells. Laane (1976) provides an introduction to the SEM of sectioned specimens. Secondary electron imaging of sections requires the attachment of sections to cover slips, the removal or partial removal (etching) of the embedding medium from the sections (Winborn, 1976), transferring the sections to an intermediate fluid, critical point drying the sections, mounting, and coating with a metal or carbon film. In addition to paraffin and epoxy resins, other embedding media for sections such as polyethylene glycol (Kondo, 1984; Wolosewick, 1984) and low-temperature acrylic resins (Armbruster *et al.*, 1984) that improve the preservation of the ultrastructure of cells and reactivity (i.e., antigenicity) of macromolecules are useful in SEM analysis of sections.

Figures 39–42 of sections of an early gastrula illustrate both the potential utility and possible artifacts of secondary electron imaging of 1.0-μm epoxy resin sections in which the resin is completely removed in an epoxy-dissolving apparatus (Fullam). Had only part of the resin been removed (etched) delicate surface or internal structures of cells might have been better preserved.

## X. Photoetching and Line Drawings

Frequently line drawings of specific features in an SEM micrograph are desired. Such drawings can be made quickly and simply with the "photoetching" method described in detail by Yob (1973). Features of a photograph printed on glossy, non-resin-coated photographic paper are outlined on the print with a Rapidograph pen. After the lines are thoroughly dry, the photographic image is bleached with an iodine solution and removed in a fixer solution. Then the print is washed and processed as a glossy print. When the print is dry, you have a line drawing suitable for publication.

## XI. Conclusions

This article's goals will be realized if the reader views and records secondary electron images at 10 kV or less at magnifications less than 10 k×, routinely records and uses stereo pair images, and appreciates and acknowledges the artifacts in once hydrated embryos, as well as cellular and extracellular entities in critical point-dried specimens.

---

FIGS. 39–42. SEM analysis of resin-embedded sections of an early gastrula. Figure 39 is a light micrograph of a 1-μm section stained with Richardson's stain. Figure 40 is a secondary electron image of another section from the same embryo in which the resin was removed. Figure 41 illustrates the utility of the split screen, dual magnification accessory of most SEM scopes. The boxed area of the ×300 image in the left panel is magnified five times in the right panel. Figure 42 shows the blastocoelic margins of the ectodermal cells in the deresinated section. ×10,000. The large intercellular spaces and the absence of a distinct basal lamina may be artifacts of embedding, sectioning, mounting, or removal of the resin.

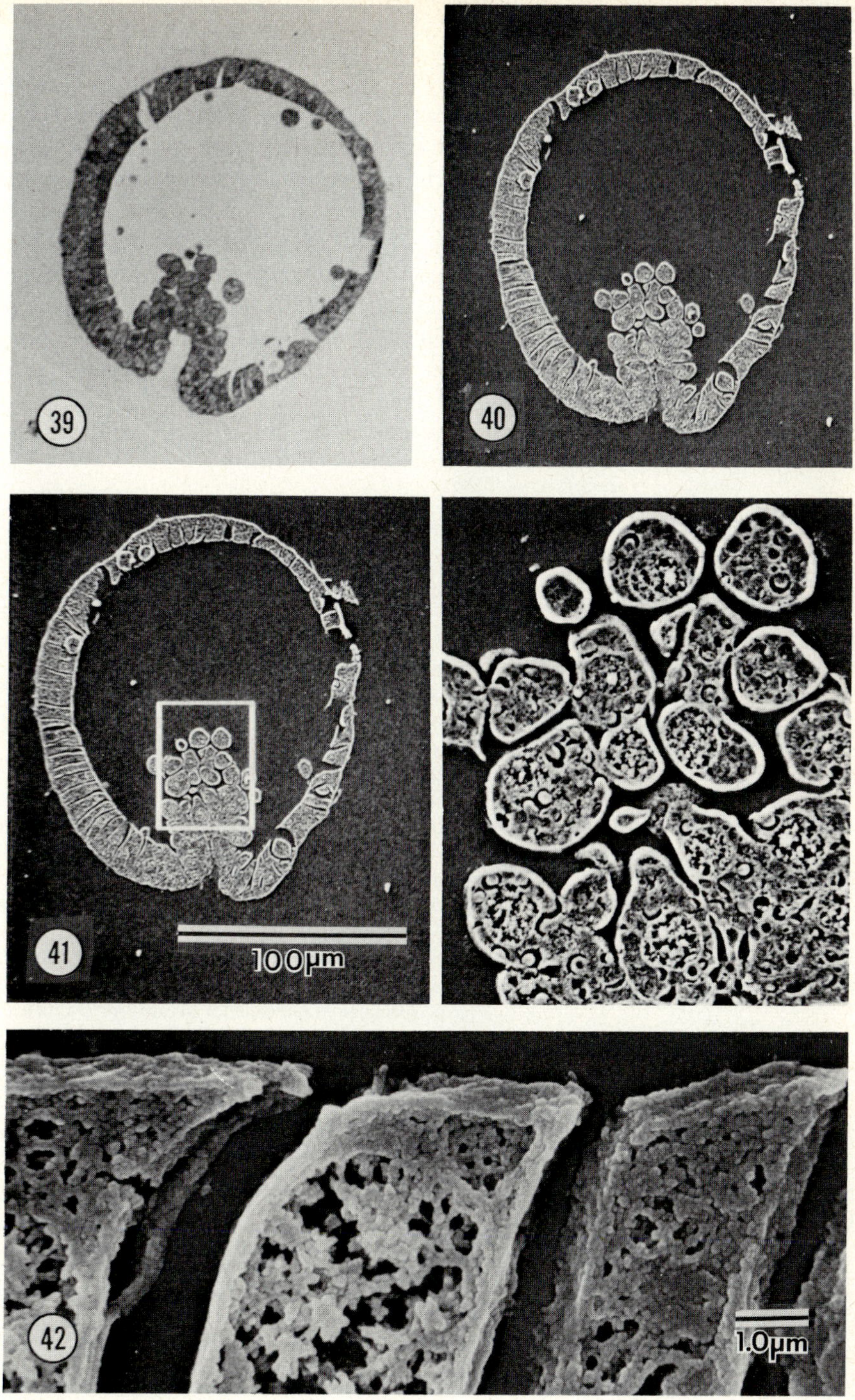
39
40
41
100μm
42
1.0μm

After the basic procedures of preparing specimens and the morphological information derived from SEM secondary electron images are obtained, then is the time to use SEM instrumentation and techniques for further bridging the gap between LM and TEM analyses. In particular, secondary electron images of the labeling of cell surfaces (Becker and Johari, 1979) via markers such as latex spheres (Molday, 1976) and colloidal gold (DeMey, 1984; Bendayan, 1984) coupled with improved techniques for preserving the integrity of hydrated molecules can vastly improve the detection and localization of a number of developmentally interesting macromolecules. Finally, backscatter electron imaging (DeNee and Abraham, 1976) of structures differentially "stained" with various heavy metals in addition to colloidal gold markers offer unexplored opportunities. The main limits of these opportunities are inertia, time, and imagination.

## References

Albrecht, R. M., and Wetzel, B. (1979). *Scanning Electron Microsc. 1980* **III,** 203–222.

Armbruster, B. L., Kellenberger, E., Villiger, E. C. W., Garavito, R. M., Hobot, J. A., Chiovetti, R., and Acetarin, J.-D. (1984). *In* "The Science of Biological Specimen Preparation for Microscopy and Microanalysis" (J.-P. Revel, T. Barnard, and G. H. Haggis, eds.), pp. 77–81. Scanning Electron Microscopy, Inc., AMF O'Hare, Illinois.

Arro, E., Collins, V. P., and Brunk, U. T. (1984). *In* "Scanning Electron Microscopy of Cells in Culture" (P. B. Bell, ed.), pp. 9–18. Scanning Electron Microscopy, Inc., AMF O'Hare, Illinois.

Becker, R. P., and Johari, O., eds. (1979). "Cell Surface Labeling." Scanning Electron Microscopy, Inc., AMF O'Hare, Illinois.

Bell, P. B., ed. (1984). "Scanning Electron Microscopy of Cells in Culture." Scanning Electron Microscopy, Inc., AMF O'Hare, Illinois.

Bell, P. B., and Revel, J.-P., (1980). *In* "Biomedical Research Applications of Scanning Electron Microscopy" (G. M. Hodges and R. C. Hollowes, eds.), Vol. 2, pp. 1–63. Academic Press, New York.

Bendayan, M. (1984). *J. Electron Microsc. Tech.* **1,** 243–270.

Boyde, A., and Tamarin, A. (1984). *Scanning* **6,** 30–35.

Boyles, J. (1984). *In* "The Science of Biological Specimen Preparation for Microscopy and Microanalysis" (J.-P. Revel, T. Barnard, and G. H. Haggis, eds.), pp. 7–22. Scanning Electron Microscopy, Inc. AMF O'Hare, Illinois.

Chatfield, E. J. (1978). *In* "Principles and Techniques of Scanning Electron Microscopy" (M. A. Hayat, ed.), Vol. 6, pp. 47–88. Van Nostrand-Reinhold, Princeton, New Jersey.

Cohen, A. L. (1984). *In* "Scanning Electron Microscopy of Cells in Culture" (P. B. Bell, ed.), pp. 49–70. Scanning Electron Microscopy, Inc., AMF O'Hare, Illinois.

DeMey, J. (1984). *EMMSA Bull.* **14,** 54–66.

DeNee, P. B., and Abraham, J. L. (1976). *In* "Principles and Techniques of Scanning Electron Microscopy" (M. A. Hayat, ed.), Vol. 5, pp. 144–180. Van Nostrand-Reinhold, Princeton, New Jersey.

Eisenman, E. A. (1980). Ph.D. thesis, University of California, Berkeley.

Gabriel, B. L. (1982). "Biological Scanning Electron Microscopy." Van Nostrand-Reinhold, Princeton, New Jersey.

Goldberg, M., and Escaig, F. (1983). *J. Microsc.* **134,** 161–167.

Hayat, M. A. (1974–1978). "Principles and Techniques of Scanning Electron Microscopy," Vols. 1–6. Van Nostrand-Reinhold, Princeton, New Jersey.

Hayat, M. A. (1978). "Introduction to Biological Scanning Electron Microscopy." Univ. Park Press, Baltimore.

Hayat, M. A. (1981). "Fixation for Electron Microscopy." Academic, Press, New York.

Hayes, T. L. (1980). *Scanning Electron Microsc. 1980* **I,** 1–10.

Humphreys, W. J., Spurlock, B. O., and Johnson, J. S. (1978). *In* "Principles and Techniques of Scanning Electron Microscopy" (M. A. Hayat, ed.), Vol. 6, pp. 136–158. Van Nostrand-Reinhold, Princeton, New Jersey.

Kondo, H. (1984). *J. Electron Microsc. Tech.* **1,** 227–241.

Laane, M. M. (1976). *In* "Principles and Techniques of Scanning Electron Microscopy" (M. A. Hayat, ed.), Vol. 5, pp. 36–52. Van Nostrand-Reinhold, Princeton, New Jersey.

Molday, R. S. (1976). *In* "Principles and Techniques of Scanning Electron Microscopy" (M. A. Hayat, ed.), Vol. 5, pp. 53–77. Van Nostrand-Reinhold, Princeton, New Jersey.

Murphy, J. A., and Roomans, G. M., eds. (1984). "Preparation of Biological Specimens for Scanning Electron Microscopy." Scanning Electron Microscopy, Inc., AMF O'Hare, Illinois.

Postek, M. T., Howard, K. S., Johnson, A. H., and McMicheal, K. L. (1980). "Scanning Electron Microscopy." Ladd Research Industries, Inc., Burlington, Vermont.

Revel, J.-P., Barnard, T., and Haggis, G. H., eds. (1984). "The Science of Biological Specimen Preparation for Microscopy and Microanalysis." Scanning Electron Microscopy, Inc., AMF O'Hare, Illinois.

Rostgaard, J., and Tranum-Jensen, J. (1980). *J. Microsc.* **119,** 213–232.

Shelton, E., and Mowczko, W. E. (1978). *Scanning* **1,** 166–173.

Wergin, W. P., and Pawley, J. P. (1980). *Scanning Electron Microsc. 1980* **I,** 239–249.

Winborn, W. B. (1976). *In* "Principles and Techniques of Scanning Electron Microscopy" (M. A. Hayat, ed.), Vol. 5, pp. 21–35. Van Nostrand-Reinhold, Princeton, New Jersey.

Wollweber, L., Stracke, R., and Gothe, U. (1981). *J. Microsc.* **121,** 185–189.

Woolosewick, J. J. (1984). *In* "The Science of Biological Specimen Preparation for Microscopy and Microanalysis" (J.-P. Revel, T. Barnard, and G. H. Haggis, eds.), pp. 83–95. Scanning Electron Microscopy, Inc., AMF O'Hare, Illinois.

Yob, C. (1973). *Peterson's Photo. Mag.* **2,** 32–36.

# Chapter 16

# *Cytological Techniques for the Study of Larval Echinoids with Notes on Methods for Inducing Metamorphosis*

ROBERT D. BURKE AND ALLAN W. GIBSON

*Department of Biology*
*University of Victoria*
*Victoria, British Columbia, Canada*



## I. Introduction

Cytological studies of echinoderm larvae have been primarily concerned with determining the structure and function of larval tissues and organs. An ancillary objective of these studies has often been to describe the development of a tissue by outlining the origin and fate of the cells that comprise it. A wide variety of techniques have been used, including various forms of light and electron microscopy, histochemistry, and immunocytochemistry. In general, function has been interpreted from structure, whereas the lineage of tissues has been determined by observing the relationships of tissues, by vital staining of embryonic cells, or by tracing specific cytological characteristics. Here we present two techniques that have proved useful in identifying structure and organization and have provided reliable cytological markers that have assisted in determining aspects of the origin and fate of certain tissues of larval echinoids. As studies of the fate of larval tissues often require a detailed analysis of the morphogenetic movements that the larva undergoes during metamorphosis, we also include in this article several methods for the induction of metamorphosis in echinoids.

## II. Monoclonal Antibodies against Tissue-Specific Antigens in Sea Urchin Embryos

### A. Introduction

Through advances in hybridoma technology it is now possible to produce unlimited quantities of antibodies with a precise specificity against any immunogenic substance. Monoclonal antibodies have acquired a wide range of uses, including immunohistochemical labeling of tissue-specific antigens. As cell-specific markers they have been used extensively in neurobiology to identify neurons (Zipser and McKay, 1981), in studies of amphibian regeneration to identify blastema cells (Kinter and Brockes, 1984), and in studies of chick development to identify neural crest cells (Ciment and Weston, 1982). In sea urchins monoclonal antibodies have been produced that label a variety of embryonic tissues (McClay *et al.*, 1983; Wessel *et al.*, 1984) and have been used to follow the lineage of mesenchymal cells (Gibson and Burke, 1985).

Techniques for raising hybridoma cells secreting monoclonal antibodies against predefined antigens were initially described by Kohler and Milstein (1975). B lymphocytes from an immunized animal secrete antibodies, but quickly die in culture. Lymphocytes fused with myeloma cells, which are adapted for permanent growth in culture, secrete antibodies and can be grown indefinitely *in vitro*. Mutant myelomas that lack the enzyme hypoxanthine guanine ribosyltransferase are used in the fusion so that hybridomas can be selected in a medium containing the nucleotide antimetabolite, aminopterin. Myelomas not fused with lymphocytes cannot utilize the nucleotide salvage pathway and thus die in a medium combining hypoxanthine, aminopterin, and thymidine (HAT medium).

Once myelomas and lymphocytes have been eliminated from the fusion culture, the resulting hybridomas can be weaned from the HAT medium and expanded for growth in culture. Individual hybridomas secreting desirable antibodies can be isolated and subsequently cloned to produce a cell line secreting antibodies of a single specificity, i.e., monoclonal antibodies.

The basic procedures for raising hybridomas secreting monoclonal antibodies have been described in detail and will not be repeated here. The reader is urged to consult Galfre and Milstein (1981) for a discussion of strategies and basic procedures, as well as Kennett *et al.* (1980) and Goding (1983) for more specific applications. In this section we will describe techniques used in our laboratory that apply specifically to aspects of raising monoclonal antibodies against tissue-specific antigens in sea urchin plutei. Modifications of the basic procedures are primarily in the preparation of the immunogen and in screening techniques for immunofluorescent localization of cell surface, tissue-specific antigens on whole-mount larvae.

## B. Preparation of Inoculum and Inoculation Procedures

Because the specificity of antibodies can be determined during the cloning and screening procedures, the antigens with which the mice are challenged need not be purified. We have been successful in producing monoclonal antibodies against antigens on various tissues and cell types by inoculating with crude preparations of fixed plutei or homogenates of adult urchin tissues.

*Materials required*

Dounce homogenizer
Two 5-cc syringes with female–female coupler
20-gauge hypodermic needles
Freund's complete and incomplete adjuvant
Phosphate-buffered saline (PBS: formula in Adams, 1980)
Three to five female BALB/c mice, 5 to 7 weeks old

*Procedure*

1. Concentrate whole larvae by gentle centrifugation, or excise adult tissues either from specific germ layers or taking various soft parts (tube feet, gut, coeloms, etc.). Cut tissues into 1- to 2-mm pieces.
2. Suspend the pelleted larvae or tissue in distilled $H_2O$ for 30 minutes on ice.
3. Dounce homogenize the suspension thoroughly and spin at 4000 *g* for 10 minutes.
4. Discard supernatant and fix pellet in 4% paraformaldehyde in PBS for 1 hour.
5. Centrifuge at 4000 *g* for 10 minutes, discard the supernatant, rinse with PBS while breaking up the pellet—repeat 3 times.
6. Emulsify 250 mg of the final pellet (enough for five mice) in 2 ml of a 1:1 mixture of Freund's complete adjuvant and PBS by repeatedly squirting the suspension between two coupled 5-ml syringes. Continue this process

until the fluid attains the consistency of thick, whipped cream (about 30 minutes).

7. Anesthetize a mouse with ether and inject 0.1 ml inoculum into each of two subcutaneous locations on the back at the base of the tail and 0.3 ml intraperitoneally.
8. Repeat the injection procedure with the other mice.
9. Give two booster inoculations at biweekly intervals, prepared as above, but with Freund's incomplete adjuvant.
10. Ten days after the second booster, bleed the mice by nicking the tail with a razor blade and holding it in a small vial containing 50 μl PBS, spin the vials at 15,000 *g* to remove cells.
11. Assay the supernatant for anti-sea urchin antibodies (procedure described below).
12. Four days prior to cell fusion the mouse showing the highest titer is given an intravenous (tail vein) injection of 25 mg unfixed homogenized plutei or adult antigen in 0.2 ml PBS.

The preparation of the immunogen can be varied to suit the type of monoclonal antibody required. We were primarily interested in raising monoclonal antibodies against cell surface antigens and thus the soluble fraction of the homogenate was not included in the inoculum. Although we had success inoculating with unfixed antigen, generally the serum titer from mice injected with fixed antigen was higher. This may be due to exposure of masked epitopes and aldehyde cross-linking of haptenic molecules to increase their antigenicity. As well, screening was carried out using fixed larvae. Fixed crude antigen can be stored at −20°C for several months.

The immunization schedule is variable but a minimum of 2 weeks between boosters should be allowed. Mice will remain primed for many months. It is critical that the cell fusion be carried out 3 or 4 days after the final intravenous boost as this is the time of maximum proliferation of antibody-secreting lymphocytes. Instructions on inoculation procedures, blood collection, and general handling of mice can be found in Weir (1973).

## C. Assay for Monoclonal Antibodies

We found a whole-mount immunofluorescent technique valuable for identifying clones secreting tissue-specific antibodies. It is simple to prepare and allows positive, three-dimensional localization of the cells without serial sectioning (Fig. 1).

*Materials required*

Five day or older plutei

4% paraformaldehyde in filtered or artificial seawater

Glass microscope slides
95% ethanol
HCl
Mactac, Mylar, or some other waterproof tape
Paper punch
Poly-L-lysine hydrobromide (300,000 $M_r$, 1.5 mg/ml in PBS)
Rabbit anti-mouse IgG conjugated with fluorescein isothiocyanate (FITC)
50% glycerol/PBS
20 × 60 mm cover slips

*Procedure*

1. Concentrate larvae by gentle centrifugation in disposable glass tubes and fix in 4% paraformaldehyde in 0.4 μm filtered seawater for 2 to 4 hours at room temperature.
2. Soak microscope slides in 3% HCl in 95% ethanol for 30 minutes, drain, rinse in distilled water, and dry.
3. Cut 20 × 50 mm strips of waterproof tape and paperpunch eight holes in each strip.
4. Apply the tape to the slides to make wells in which to carry out the immunoassay.
5. Place a drop of polylysine solution in each well and place slides in a wet chamber (a 145-mm petri dish containing a wet paper towel) for 1 hour at room temperature.
6. Remove the remaining polylysine with a Pasteur pipet (the polylysine can be reused) and air dry the slides.
7. Add a drop of fixed plutei (~400 larvae/drop) to each well and return the slides to the wet chamber.
8. Allow the larvae to settle (about 30 minutes) and aspirate off excess fluid. Adhesion seems to be enhanced by allowing the slides to dry slightly in air.
9. Wash the slides three times for 5 minutes each in PBS (Coplin jars filled with PBS are good for these washing steps).
10. Place 15 to 20 μl of spent medium from hybridoma cultures, or serum, into each well. As negative controls for hybridoma supernatant, add known, nonrelevant hybridoma supernatant, myeloma supernatant, or PBS. As negative controls for serum assays, add nonimmunized or preimmunized mouse serum. These can be included on each slide. For a positive control, with each assay run at least one antibody known to react with larval tissues.
11. Incubate the slides in a wet chamber for 1 hour at room temperature.
12. Rinse the slides once rapidly in PBS, then three times for 5 minutes each

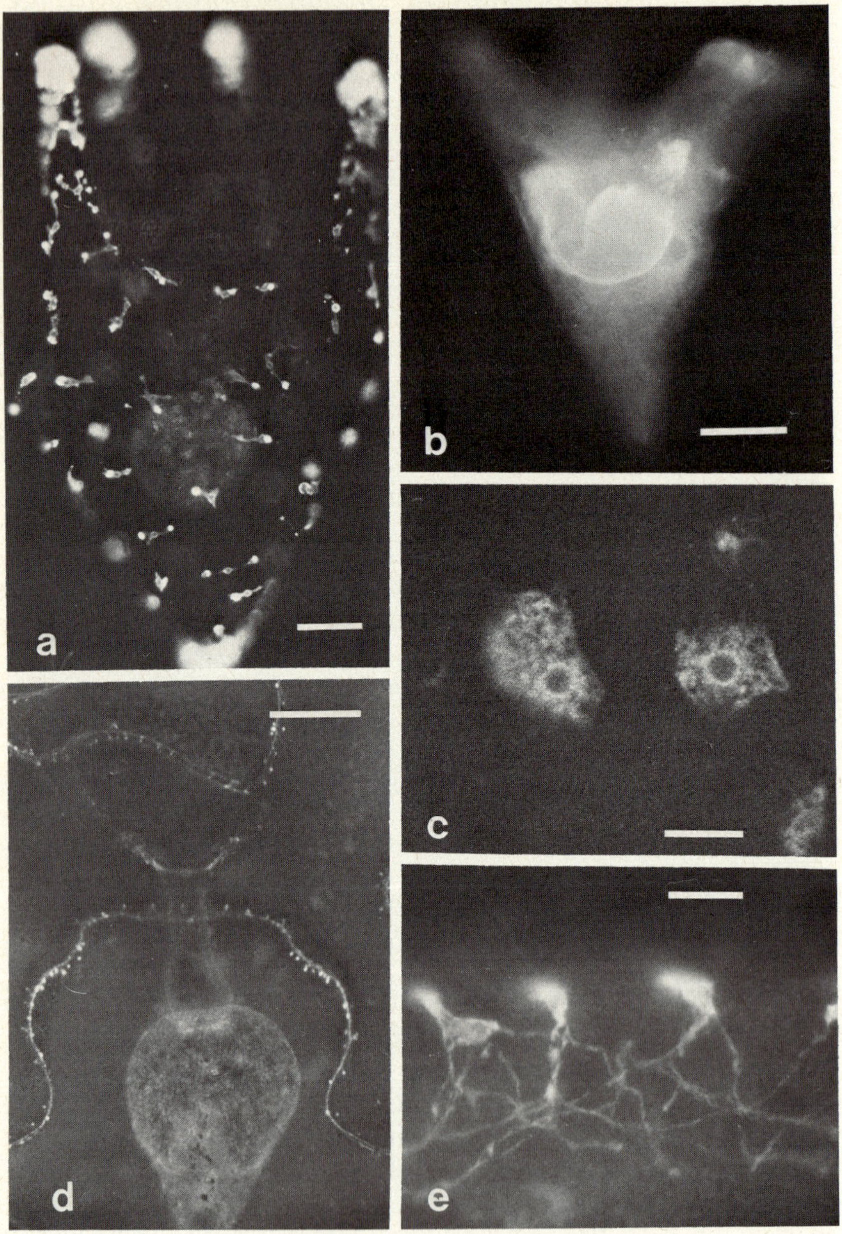
a
b
c
d
e

in PBS baths. Aspirate the excess water from around the wells. Partial drying does not appear to affect the second antibody binding.

13. Add 20 μl rabbit anti-mouse, FITC-conjugated IgG (Sigma F7506, 1:25 dilution in PBS) to each well and incubate in the wet chamber for 30 minutes.
14. Rinse the slides three times for 5 minutes each in PBS.
15. Add a drop of 50% glycerol/ PBS to each well and gently lower a 20 × 60 mm cover slip over the tape.
16. Aspirate off excess fluid and view in a fluorescence microscope.

We found that larvae fixed in 4% paraformaldehyde in filtered seawater for 2 to 4 hours provided a minimum of autofluorescence and a minimum of tissue deterioration in subsequent assay steps. Fixed larvae can be washed and stored in filtered seawater containing 0.1% $NaN_3$ for at least 6 months with little noticeable deterioration, although some antigens may be leached out. Slides with attached larvae may be stored in 0.1% $NaN_3$ in PBS for at least 2 weeks. Slides may be stored in 0.1% $NaN_3$ in filtered seawater for longer periods but should be rinsed in PBS for 15 minutes prior to primary antibody application. The polylysine solution is stored at −20°C and may be thawed and refrozen.

Some of the problems with using whole-mount preparations include the impermeability of the fertilization envelope, which must be removed for satisfactory staining of prehatched embryos, and the tendency of the hyaline layer in early embryos to bind immunoglobulins nonspecifically. The latter can be reduced by incubation of the larvae for 30 minutes in 1% bovine serum albumin before assaying and is not a problem in late gastrulae or plutei.

The simple procedure described above allows us to assay over 100 hybridoma cultures per day. However, once monoclonal antibodies have been selected, immunoassays of experimental material and embryos younger than plulei are best carried out in 10 × 75 mm disposable glass tubes. Embryos are fixed, washed, and assayed in the same tube, with appropriate controls carried out in

---

FIG. 1. (a) Indirect immunofluorescent staining of a four-armed pluteus of *Strongylocentrotus purpuratus* with a monoclonal antibody for a cell surface antigen of pigment cells (see Gibson and Burke, 1985, for details). Note that three-dimensional organization is well preserved. Bar = 20 μm. (b) Indirect immunoflorescent staining of an early plutus of *S. purpuratus* with a monoclonal antibody (Sp5/55.65). With whole-mount staining the specificity of this antibody for endodermally derived tissues as well as cells scattered over the oral field and in the preoral ciliary band can be easily determined. Bar = 50 μm. (c) With the procedures outlined, preservation of cell structure is good enough to show a reticular pattern of staining with Sp5/55.65 in the ectoderm of gastrulae. Bar = 4 μm. (d) Glyoxilic acid-induced fluorescence of nerve cells in the bipinnaria of *Orthasterias koehleri*. With some preparations it is possible to view the entire nervous system of the larva. Here the nerves form tracts underlying the ciliary bands; the periodic bright spots are nerve cells. The gut is autofluorescent. Bar = 10 μm. (e) Detail of the ciliary band showing nerve cells and axon tracts. Bar = 5 μm.

separate tubes. Incubation times are the same as with slides and rinsing the larvae between assaying steps is done by settling the larvae and discarding the supernatant.

## III. Visualization of Larval Nerves with Glyoxilic Acid-Induced Fluorescence

This simple procedure has proved effective in identifying catecholamine-containing nerve cells in larvae of all five classes of echinoderms (Burke, 1983a,b; Chia *et al.*, 1985; Burke, unpublished) and has been used with echinoid embryos to trace the lineage of nerve cells (Burke, 1983c). In various forms this technique has also been used to visualize nerves in several other kinds of animals (Goh and Davey, 1976; Sharpe and Atkinson, 1980; Grace and Bunney, 1980). The technique is based on the reaction of glyoxilic acid with catecholamines at neutral pH to produce a fluorescent product (Fuxe and Jonsson, 1973). Although there is some confusion whether in living cells it is specific for dopamine or for a wide range of catecholamines, it appears to give superior results to those obtained with formaldehyde vapor-induced fluorescence of catecholamines (Bjorlund *et al.*, 1968; Keenan and Koopowitz, 1981; Grace and Bunney, 1981).

Nerve cells in echinoderm larvae are difficult to identify and authors have differed on the issues of their existence or precise location (Mackie *et al.*, 1969; Ryberg, 1977; Burke, 1978). Histochemical techniques, such as formaldehyde-induced fluorescence (Ryberg, 1974), indogenic methods for esterases (Ryberg, 1973), and rongalit-reduced Methylene Blue (Mackie *et al.*, 1969), have produced equivocal or capricious results. The procedure outlined here is derived from the technique for staining neurons in nematodes described by Sharpe and Atkinson (1980) and has reliably and effectively produced readily interpretable images that have been corroborated by ultrastructural identification of neurons.

*Materials required*

Working solution of 1% glyoxilic acid in 0.1 *M* $NaH_2PO_4$, adjusted to pH 7.4 with concentrated NaOH and stored at 4°C

Watch glass, 5-cm diameter

Bibulous paper

Hair drier

Paraffin oil or immersion oil

Glass slides and cover slips

Clear nail polish

*Procedure*

1. Concentrate embryos by gentle centrifugation or select older larvae individually with a Pasteur pipet.
2. Transfer specimens, with as little seawater as possible, to a watch glass containing 5 ml of the glyoxilic acid solution on ice. Leave specimens in the glyoxilic acid solution for 2 to 5 min.
3. Transfer specimens with a Pasteur pipet to a clean glass slide. Draw off excess glyoxilic acid with a strip of bibulous paper.
4. Dry specimens to the slide for 5 minutes with a stream of hot air from a hair drier.
5. Place the slide on a hot plate or in an oven at 100°C for 5 minutes.
6. After cooling, place the specimens in paraffin oil or immersion oil, cover with a glass cover slip, and seal with nail polish.
7. View specimens with an epifluorescence microscope using excitation below 490 nm and emission above 500 nm, i.e., filters suitable for fluorescein fluorescence.

We have had good results with reagent-grade glyoxilic acid obtained as a 50% solution in water. In this form it is apparently stable at room temperature for several years. The 1% working solution, if stored at 4°C, can be used for at least 4 weeks. If a precipitate forms, the solution should be discarded. The intensity of staining and the levels of background fluorescence can be varied by adjusting the time in glyoxilic acid, or the concentration of the working solution. Although drying delicate larvae onto a glass slide does cause them to collapse, preservation is satisfactory for identification of tissues with Nomarski DIC or phase-contrast optics. Heating the glass slide to more than 100°C causes bubbling of the tissues and formation of crystals that ruins the specimen. Without heating, background fluorescence is high. Peak emission of the fluorescent product is 510 nm with an excitatory peak of 400 nm (Keenan and Koopowitz, 1981). Preparations are remarkably stable and can be examined repeatedly or stored for at least 3 weeks at room temperature with no significant deterioration of the image.

## IV. Induction of Metamorphosis in Echinoid Larvae

### A. Introduction

Metamorphosis in echinoids is a sequence of developmental events that rapidly transforms the planktonic pluteus larva into the benthic adult. Many aspects of metamorphosis are of interest to biologists, but studies may be hampered by difficulties in reliably initiating the process. For some species, larvae will metamorphose spontaneously when competent and left in unstirred cultures (Shott *et al.*, 1984), but, for several other species, metamorphosis in nature appears to be triggered by cues associated with adult habitats. In these forms there is a low rate of spontaneous metamorphosis when larvae are retained in culture, but if given appropriate stimuli, a large percentage of the larvae will metamorphose (Hinegardner, 1969; Burke, 1984). Researchers usually expose larvae to substrates associated with adults, such as offering sand to sand dollar larvae (Mortensen, 1921, 1938; Caldwell, 1972; Highsmith, 1982) or algal encrusted rocks to regular urchins (Marcus, 1980). Primary films of algae, diatoms, and bacteria grown on glass sides or the surface of dishes have also been used (Cameron and Hinegardner, 1974). Although these methods are satisfactory for some purposes and some species, results can be capricious and manipulations awkward. Here we describe several methods that have been used to initiate metamorphosis and have proved useful in studies in which the precise onset of metamorphosis must be controlled.

## B. Methods

### 1. Extractions

These extraction methods have been employed principally with the Pacific sand dollar, *Dendraster excentricus,* where it has been shown that larvae metamorphose preferentially in natural beds of adult sand dollars (Highsmith, 1982). A simple extraction of sand with the capacity to induce metamorphosis can be made by mixing 100 ml of sand and 100 ml of filtered seawater in a small beaker. Sand freshly collected from a sand dollar bed or sand from an aquarium containing adult sand dollars has been successfully used. The sand and seawater are mixed into a slurry with a magnetic stirrer for 5 to 15 minutes. The solution is then vacuum filtered through a medium-pore paper filter in a Buchner funnel. Competent larvae can be induced to metamorphose by transferring them directly into the filtrate, or instilling the filtrate into a small volume of seawater containing the larvae. Each batch should be tested with a number of competent larvae and only batches in which 80–90% of the larvae metamorphose within 5 minutes should be accepted. Extracts prepared in this way will retain the ability to induce metamorphosis for several days if stored at 4°C.

R. Emlet (personal communication) has been successful in inducing larvae of tropical sand dollars and heart urchins to metamorphose by filtering a slurry of sand and seawater and placing small pieces of the filter paper in dishes containing competent larvae. Highsmith (1982) put sand from a *Dendraster excentricus* bed into dialysis tubing and was able to induce metamorphosis by putting the tubing in seawater with the larvae. Although these methods take several hours to initiate metamorphosis, juveniles are more easily recovered than when they are exposed to sand directly.

It may be convenient to prepare an extract in distilled water and lyophilize it to produce a dry powder, as follows:

1. From a sand dollar bed collect sand with as little seawater as possible in 400-ml aliquots. Store at −20°C.
2. Thaw and add 300 ml distilled water.
3. Stir with an electric stirring motor for 30 minutes at 70°C in a water bath.
4. Add 1.8 ml acetic acid (0.1 *M* final concentration) and stir for an additional 10 minutes.
5. Vacuum filter with a medium-pore, paper filter in a Buchner funnel.
6. Refilter filtrate with a fine-pore paper filter.
7. Shell freeze by rotating flask in an ethanol-dry ice bath so that liquid crystallizes over the entire inner surface, and lyophilize overnight to produce an off-white powder.

The optimum concentration of powder in seawater that will induce immediate metamorphosis may be as low as 5 to 10 mg/ml and can be determined by serial dilution. The powder can be stored at −20°C for at least 1.5 years with no appreciable loss in activity. From extracts such as this, Burke (1984) has purified a peptide that appears to be the component triggering metamorphosis.

It has also been demonstrated that the peptide is produced by the adult sand dollars and that it can be extracted and purified from adult tissues. Whole sand dollars and gonads proved to have the highest titer per gram wet weight. Specimens or tissues were reduced to a slurry with a Waring blender, extracted in distilled water, and lyophilized using the procedure outlined above. Preliminary experiments indicate that similar extracts of the regular urchin, *Strongylocentrotus droebachiensis,* will induce competent larvae of that species to metamorphose.

### 2. Electrical Method

Cameron and Hinegardner (1974) and Burke (1983a) have reported that competent larvae can be induced to metamorphose by direct electrical stimulation. This technique is useful for studies of metamorphosis of individual larvae.

Suction electrodes of either polyethylene or glass can be used (Fig. 2). These can be made by cutting off the end of an 18-gauge hypodermic needle 1 cm from the plastic. Fine-gauge platinum wire is then wrapped several times around the hypodermic needle to provide an electrical contact. A 1-cm length of polyethylene tubing (P.E. 190, 1.19 mm i.d.) is forced over the platinum wire wrapping with one end of the platinum wire extending beyond the end of the hypodermic

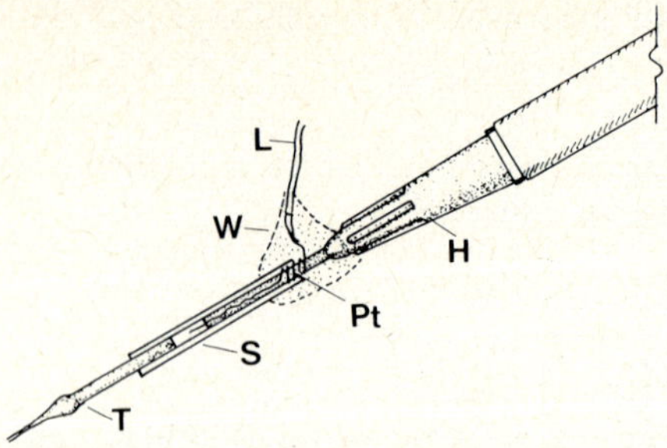


FIG. 2. Diagrammatic view of suction electrode used to induce metamorphosis in echinoid larvae. H, 18-gauge hypodermic needle; L, 22-gauge, stranded wire lead; Pt, platinum wire; S, polyethylene tubing sleeve; T, fine tip drawn from polyethylene tubing; W, tacky wax sealant.

needle into the lumen of the tubing. After soldering a connection between the end of the platinum wire and an insulated wire lead (22-gauge, stranded), the exposed wire and metal of the hypodermic needle can be sealed by dripping a low melting point wax (tacky wax) over them. This holder can then be attached to a micromanipulator or a support and connected via polyethylene tubing to a 3- or 5-ml syringe. Polyethylene tips can be drawn out to a diameter of 25–50 μm and inserted into this holder. These tips are made by taking a 2.5- to 3-cm length of polyethylene tubing (P.E. 60, 1.22 mm o.d.) and heating it over the hot tip of a soldering iron. Grasp both ends of the tubing between thumb and forefinger and slowly rotate it 5–7 mm above the hot tip. As the polyethylene becomes hot it clears and the diameter increases slightly. Carefully and slowly draw the ends apart and decrease the temperature slightly by lifting away from the tip or blowing softly on the tubing. This procedure is quite difficult to master, but with practice fine diameter tips can be produced consistently. Glass electrode tips can be drawn from glass capillaries (1.20 mm o.d.) either by hand or with an electrode puller. Although these may be drawn to a finer point, the tip often has a sharp edge that tends to pierce the tissue. Polyethylene tips are more durable than glass tips and do not need replacement as frequently.

The electrode should be positioned under a dissecting microscope so that the tip angles into a 10-cm plastic Petri dish and can be brought into focus. Seawater from the dish is drawn up into the electrode and must fill at least to the level of the metal barrel to provide electrical continuity. The positive electrode, a small length of platinum wire, is inserted into the dish of seawater opposite the suction electrode (negative). Larvae can be transferred to the dish and manipulated with small probes to the electrode tip. By drawing on the syringe, larvae are sucked onto the electrode tip and can be held in place. Burke (1983a) found there were two locations on the larvae, associated with neural centers, that are most favorable for stimulation. These are the oral ganglion, on the lower lip of the larval mouth, and the apical neuropile, between the preoral and anterolateral arms.

The source of stimulation can be a neurophysiological square-wave stimulator or a low voltage dc source such as a variable transformer. With a stimulator, precise voltages and durations of stimuli can be set. Stimulation of 12 to 20 V delivered as 5-msec pulses, 5 pulses per second for 10 seconds, has proven successful. With a dc transformer, a small switch is opened and closed to determine duration, and voltage can be set at the transformer. With this arrangement a single 12- to 20-V pulse of 2 to 3 seconds has been used.

Larvae attached to the electrode appear unaffected by being restrained. Stimulation usually initiates a ciliary reversal and activation of the tube feet in the adult rudiment. When metamorphosis is stimulated, the oral hood flexes ventrally and the rudiment begins to evert (Burke, 1983a, 1985). Larvae can be released from the electrode by relieving suction pressure, and metamorphosis procedes normally. This procedure has proved successful for inducing metamorphosis in *Dendraster excentricus, Arbacia punctulata, Lytechinus variegatus,* and *Lytechinus pictus.*

## Acknowledgments

This research was supported by a Natural Sciences and Engineering Research Council of Canada (NSERC) University Research Fellowship and Operating Grant to R.D.B. and an NSERC Post-Graduate Scholarship to A.W.G. B. W. Bisgrove and J. J. King kindly reviewed and commented on the manuscript.

## References

Adams, R. L. P. (1980). "Cell Culture for Biochemists." Elsevier, Amsterdam.
Bjorklund, B., Ehinger, B., and Falck, J. (1968). *J. Histochem. Cytochem.* **16,** 263–271.
Burke, R. D. (1978). *Cell Tissue Res.* **191,** 233–247.
Burke, R. D. (1983a). *Biol. Bull.* **164,** 176–188.
Burke, R. D. (1983b). *J. Morphol.* **178,** 23–35.
Burke, R. D. (1983c). *Cell Tissue Res.* **229,** 145–154.
Burke, R. D. (1984). *Science* **225,** 442–443.
Burke, R. D. (1985). *Cell Tissue Res.,* **239,** 589–597.
Caldwell, J. W. (1972). M.Sc. thesis, University of Florida, Gainesville.
Cameron, R. A., and Hinegardner, R. T. (1974). *Biol. Bull.* **146,** 335–342.
Ciment, G., and Weston, J. A. (1982). *Dev. Biol.* **93,** 355–367.
Fuxe, K., and Jonsson, G. (1973). *J. Histochem. Cytochem.* **21,** 293–311.
Galfre, G. and Milstein, C. (1981). *In* "Methods in Enzymology" (J. J. Langone and H. van Vunakis, eds.), Vol 73, pp. 3–47. Academic Press, New York.
Gibson, A. W., and Burke, R. D. (1985). *Dev. Biol.* **107,** 414–419.
Goding, J. W. (1983). "Monoclonal Antibodies: Principles and Practice." Academic Press, London.
Goh, S. L., and Davey, K. G. (1976). *Int. J. Parasitol.* **6,** 403–441.
Grace, A. A., and Bunney, B. S. (1980). *Science* **210,** 654–656.
Grace, A. A., and Bunney, B. S. (1981). *Science* **214,** 1152.
Highsmith, R. C. (1982). *Ecology* **63,** 329–337.
Hinegardner, R. T. (1969). *Biol. Bull.* **137,** 465–475.

Keenan, L. C., and Koopowitz, H. (1981). *Science* **214,** 1151–1152.

Kennett, R. H., McKearn, T. J., and Bechtol, K. B. (1980). "Monoclonal Antibodies—Hybridomas: A New Dimension in Biological Analysis." Plenum, New York.

Kinter, C. R., and Brockes, J. P. (1984). *Nature (London)* **308,** 67–69.

Köhler, G., and Milstein, C. (1975). *Nature (London)* **257,** 495–497.

McClay, D. R., Cannon, G. W., Wessel, G. M., Fink, R. D., and Marchase, R. B. (1983). *In* "Time, Space, and Pattern Development in Embryonic Development" (W. R. Jeffery and R. A. Raff, eds.), pp. 157–169. Liss, New York.

Mackie, G. O., Spencer, A. N., and Strathmann, R. R. (1969). *Nature (London)* **223,** 1384–1385.

Mortensen, T. (1921). "Studies of the Development and Larval Forms of Echinoderms." G. E. C. Gad., Copenhagen.

Mortensen, T. (1938). *Kg. Dan. Vidensk Selsk. Biol. Skr.* **7,** 1–59.

Ryberg, E. (1973). *Zool. Scripta* **2,** 163–170.

Ryberg, E. (1974). *Acta Zool.* **55,** 179–189.

Ryberg, E. (1977). *Cell Tissue Res.* **179,** 157–167.

Sharpe, M. J., and Atkinson, H. J. (1980). *J. Zool. (London)* **190,** 273–284.

Shott, R. J., Lee, J. J., Britten, R. J., and Davidson, E. H. (1984). *Dev. Biol.* **101,** 295–306.

Weir, D. M. (1973). "Handbook of Experimental Immunology," 2nd ed. Blackwell, Oxford.

Wessel, G. M., Marchase, R. B., and McClay, D. R. (1984). *Dev. Biol.* **103,** 235–245.

Zipser, B., and McKay, R. (1981). *Nature (London)* **289,** 549–554.

# Chapter 17

# *Embryo Dissociation, Cell Isolation, and Cell Reassociation*

DAVID R. McCLAY

*Department of Zoology*
*Duke University*
*Durham, North Carolina*



## I. Introduction

The sea urchin embryo was the first embryo to be used for studies on cell adhesion (Herbst, 1900). In the last 20 years a number of laboratories have taken advantage of several properties exhibited by this embryo. Perhaps the best asset in working with the embryo is that it can be dissociated to single-cell suspensions without using enzymes. The cells have the ability to reassociate and to sort out into ''correct'' embryonic relationships. The younger the embryo, the greater the ability of cells from that embryo to reconstitute a completely normal pluteus larva. A second advantage of the system is its simplicity. The cells will reassociate in seawater, they sort out quickly, and the transparency of embryos of many species makes morphological analysis simple. Furthermore, the ability to raise large numbers of embryos synchronously makes the sea urchin embryo a very attractive system for the study of early morphogenetic events. This article will review the methods we use for embryo dissociation, for isolation of subsets of cells, for analysis of cell–cell reassociation, and for analysis of antibody interference. The article is written to provide helpful hints where they are known to exist. We have worked with six different species and have found that optimal conditions of dissociations and reaggregation vary somewhat between species, and even between embryos from the same species during the gravid season.

## II. Embryo Dissociation

Table I gives the reagents that are used for dissociation and culture of the cells. "Hyalin extraction medium" has been found to be the best solution for removing hyalin without damaging cells. It is modified after the glycine method of Kane (1973), which utilized 1 *M* glycine, and was useful for dissociating embryos as well as for isolating hyalin, but the long-term survival of cells was compromised. Hyalin extraction medium is not as efficient in removing the hyaline membrane, leaving some hyalin behind on the surface of cells (Fink and McClay, 1985), but the cells retain a high degree of viability and ability to reaggregate and sort out.

TABLE I

FORMULA FOR DISSOCIATION AND CULTURE REAGENTS

| | Hyaline extraction medium (g/liter) | Artificial seawater | Calcium and magnesium-free seawater | Calcium-free seawater |
|---|---|---|---|---|
| NaCl | 18.5 | 28.32 | 31 | 26.5 |
| KCl | 0.7 | 0.77 | 0.8 | 0.7 |
| $MgCl_2 \cdot 6H_2O$ | | 5.41 | | |
| $MgSO_4 \cdot 7H_2O$ | 11.9 | 7.13 | | 11.9 |
| $CaCl_2$ | | 1.18 | | |
| $NaHCO_3$ | | 0.2 | 0.2 | 0.5 |
| $Na_2SO_4$ | | | 1.6 | |
| Glycine | 22.5 | | | |
| Tris | 1.21 | | | |
| EGTA | 0.76 | | | |

To dissociate embryos at any stage we use the following protocol:

1. Embryos are collected from culture and concentrated. This is done by filtering the dilute suspensions of embryos with Nitex mesh (35 μm, Tetco, New York, NY). The filters are made by cutting out the bottoms of plastic beakers and attaching the Nitex with rubber bands. One must be careful not to swirl the embryos over the Nitex because the abrasion will damage them. With the Nitex, embryos in a liter of seawater can be concentrated to 50 ml in about 30 seconds.

2. The 50-ml suspension of concentrated embryos is centrifuged in a hand certrifuge (hand centrifuges with 50-ml buckets are available from Boekel Co., Philadelphia, PA). The embryos are washed twice by resuspension in Millipore-filtered seawater or in artificial sea water (see Table I for formulation).

3. The embryos are then resuspended in hyalin extraction medium (40 ml/2 ml packed embryos). This suspension is incubated at room temperature for 5 minutes. This step is the one that enjoys the greatest variability between different laboratories. Guidice originally used 0.44% sucrose (Giudice, 1962), others have since used calcium and magnesium-free seawater, calcium-free seawater, reagents containing glycine and/or EDTA or EGTA (Spiegel and Spiegel, 1975; Okazaki, 1975; Harkey and Whiteley, 1981; McClay and Hausman, 1975). In examining the optimal conditions for separating cells from a new species we vary the time in hyalin extraction medium, and/or we dilute hyaline extraction medium with calcium and magnesium-free or calcium-free seawater to achieve more gentle conditions. The "optimal" dissociation conditions are judged by (1) ease of embryo dissociation, (2) volume of cell suspension attained per volume of starting embryo suspension, and most important, (3) the ability of the cells in suspension to aggregate and differentiate into embryoid bodies.

(4) The embryo suspension is resuspended in calcium-free seawater and dissociated with a 10-ml syringe (without a needle). The cells are fragile so the plunger of the syringe is pulled up and down slowly and the turbidity of the embryos in solution can be used as a rough indicator of extent of dissociation. The cell suspensions are then filtered twice through 28 μm Nitex in order to retain cell clumps and pass the single cells in suspension. The cell suspensions are then washed in calcium-free or artifical seawater, depending upon the experimental design that follows. It is important to keep the suspension cold at this step since the cells will begin to aggregate if they are warmed to room temperature.

## III. Germ Layer Separation

The embryo can be subdivided into a number of subsets of cells, and methods also exist for isolation of matrix material. This section will describe methods for isolation of cells of each of the three germ layers, for isolation of the hyaline layer, and for isolation of the basal lamina lining the blastocoel plus the matrix that cofractionates with the basal lamina. Again there are many species-specific differences in several steps and these will be pointed out below where they are known.

## A. Isolation of Hyalin

The hyaline layer is released from cortical granules at fertilization (McClay and Fink, 1982). Hyalin, the major $Ca^{2+}$-dependent macromolecular component in the hyaline layer, can be isolated by removal of calcium from the culture medium (Kane, 1973) or by first extracting the cells from intact hyaline bags (Citkowitz, 1971). The protocol given below removes hyalin while retaining viability in the cells. Hyalin can be isolated from any stage up to the pluteus larval stage. It is easiest to isolate from embryos 1 hour after fertilization (as given below). The same routine with a reduced yield can be used for any stage.

If hyalin is to be removed from 1-hour embryos, the fertilization membrane must be removed. This is done at fertilization by incubation of eggs in 10 m*M* para amino benzoic acid during the fertilization reaction. This inhibits the hardening of the fertilization membrane (Showman and Foerder, 1979) so that, several minutes after sperm addition, the fertilization membrane can be removed by passing zygotes through either 53- or 73-μm Nitex mesh (the optimum size varies with the species).

At 1 hour after fertilization the cells are suspended in hyalin extraction medium for 10 minutes. The zygotes are then gently swirled and the supernatant is collected and brought to 20 m*M* with $Ca^{2+}$. The solution containing $Ca^{2+}$ is stirred for 30 minutes and then is pelleted at 15,000 rpm in a Sorvall centrifuge. Hyalin pellets as a clear to translucent precipitate. In order to get more highly purified hyalin, it can be dissolved in calcium and magnesium-free seawater to which EDTA (2 m*M*) has been added to put it back into solution, centrifuged, then reprecipitated from the supernatant by dialysis against calcium. This will clean up the hyalin but will also result in more than a 50% loss of precipitable hyalin with each cycle.

## B. Isolation of Ectoderm

Ectoderm can be isolated from midgastrula stage onward. The timing must be worked out for each species and for each stage (McClay and Marchase, 1979), but it is very easy to obtain routinely cell suspensions that are greater than 90% ectoderm.

Embryos at the appropriate stage are harvested using Nitex. They are washed with seawater and resuspended in hyalin extraction medium, perhaps diluted with calcium-free seawater (the precise mixture should be determined by trial since the ease of ectoderm separation varies at this step with different species).

At 1-minute intervals collect a sample of embryos, transfer it to calcium-free seawater, and triturate it three times with a plastic 10-ml syringe. Examine under a microscope. When the proper conditions are achieved, ectoderm cells will be released from embryos leaving behind bags of basal lamina containing primary mesenchyme cells and an intact gut.

Once appropriate conditions have been established the bulk of the embryos can be treated under the same conditions as the sample. After trituration, embryos lacking ectoderm will sediment in a hand centrifuge, leaving the ectoderm cells in the supernatant. The ectoderm cells can then be pelleted at 1000 rpm in a clinical or refrigerated centrifuge.

As embryos become older the basal lamina becomes tougher, making it progressively easier to isolate ectoderm from the remaining embryo (McClay and Marchase, 1979).

## C. Isolation of Mesoderm

Harkey and Whiteley (1981) describe in detail how primary mesoderm cells can be isolated either as micromeres at the 16-cell stage, or as primary mesenchyme cells. Their methods work precisely as given for *Strongylocentrotus purpuratus*. For other species this method is more difficult to use. Recently a simple and highly efficient method has been developed for the isolation of primary mesenchyme cells (Ettensohn and McClay, 1986). This method uses wheat germ agglutinin to pan for cells.

## D. Isolation of Endoderm

The endoderm isolation protocol relies on the observation that endoderm cells are the most tightly adherent in the embryo. Whether that is due to junctional complexes (Gibbins *et al.*, 1969) or to some other adhesive property of these cells is not presently known. Nevertheless, this property makes it possible to separate endoderm from the other germ layers (McClay and Chambers, 1978; McClay and Marchase, 1979). After isolation of ectoderm, the mesoderm–endoderm pellet is resuspended in hyalin extraction medium for 5 minutes, triturated, then pelleted, and resuspended again. This sequence is repeated and the embryo is examined after each wash for complete removal of remaining ectoderm and of mesoderm. When clean, the endoderm should remain behind as a recognizable tube. A longer incubation of the endodermal tube in hyalin extraction medium (about 5 minutes, although it varies with species), followed by pipetting, yields a single-cell suspension of endoderm cells. From the prism stage onward, an alkaline phosphatase histochemical stain can be used as a diagnostic for purity of endodermal cells (McClay and Marchase, 1979).

### E. Isolation of Basal Lamina

The method of Harkey and Whiteley (1981) can be used to isolate basal lamina bags. The bags can be broken up by pipetting and then separated from primary mesenchyme cells by Nitex separation (the cells will pass through 28-μm mesh, whereas the basal lamina bag will tend to stick to the mesh).

An alternative method has been used to isolate large quantities of extracellular matrix (Wessel *et al.*, 1984). Embryos are extracted with nonionic detergent [0.1% Triton X-100, 10 m*M* $NaHCO_3$, 0.01% phenylmethylsulfonylfluoride (protease inhibitor), and 15 μg/ml DNase]. The combination of detergent and low ionic strength solubilizes cells but leaves a residue that can be pelleted (15,000 rpm, 30 minutes, 4°C, in a Sorvall centrifuge). This pellet is highly enriched for the extracellular matrix that lines the blastocoel.

## IV. Reassociation Methods

### A. Aggregation and Culture

A number of protocols exist for following the aggregation and sorting behavior of sea urchin cells. Extensive studies have been published by Guidice (1962; Giudice and Mutolo, 1970), and by the Spiegels (1975). The culture requirements for cells are minimal. Generally, Millipore-filtered seawater supplemented with streptomycin sulfate (1 μg/ml) and/or penicillin (1 μg/ml) is all that is needed to support aggregation and cell sorting. Culture of micromeres, as devised by Okazaki (26), requires a small amount of horse serum as a supplement in order for the micromeres to differentiate and make spicules. Sterile technique should be practiced in the aggregation experiments, although the cultures will not be bacteria free as there is no way to eliminate all bacteria from the embryo suspensions. This means that any insult, beyond the dissociation itself, will reduce viability of the cells. Cell crowding, introduction of dead cells, and a small surface-to-volume ratio are factors that can contribute to the death of cultures.

For long-term cultures of aggregates there are two approaches that have similar end points, as shown by Spiegel and Spiegel (1975). Cells can be incubated in stationary culture where they will aggregate, become ciliated, and then lift off the dish and swim. Alternatively, cells can be incubated in shaker culture where they will aggregate in suspension, become ciliated, and swim. In both cases cells within the aggregates, if kept in sparse culture, will sort out according to germ layer, and the embryoids that result will resemble normal embryos.

## B. Methods of Analysis of Cell Associations

A number of treatments are available for quantitating the interaction of sea urchin cells. These methods generally do not differ from the approaches used on other cells except that sea urchin cells, having been dissociated without enzymes, do not require any recovery period before regaining adhesiveness. A brief description of several approaches that have been used is given below. Following those descriptions is a detailed description of the methods used in our laboratory for quantitative analysis of adhesion.

### 1. Disappearance of Single Cells

This method was originally described by Orr and Roseman (1969), and it has been applied to sea urchins (Oppenheimer and Meyer, 1982). In practice, one counts the disappearance of single cells manually, counts with an automatic particle counter, or measures turbidity since multicellular clusters reduce the turbidity as a direct function of cells moving into aggregates. This method is quantitative though not highly sensitive and can measure adhesive properties as aggregation proceeds. The major disadvantages to this approach are that specifications between cell populations cannot be measured. Second, the assay methodology introduces an unknown fluid shear, either during sampling, or during the flow through the particle counter. This shear reduces the likelihood that the earliest steps in an interaction are detected, and since the shear forces are unknown one cannot assess biophysical properties of the interaction. As advantages, the approach is quick, and the curves that are generated do reflect an impression of cell aggregation; this can be experimentally manipulated to determine some of the properties of adhesion.

### 2. Measurement of Aggregate Size

This method was introduced by Moscona (1957) and has been widely used for a variety of cell types. For sea urchin aggregation, variations of this method have been used for a number of studies. Generally, cells are rotated in flasks and the progress of aggregation is judged by the size of aggregates that accumulate. The approach can be semiquantitated by photographing of the aggregates and then measuring their volumes. The advantages of this approach are that it is very easy to do (though not to quantitate) and that the aggregates can be further assessed for sorting out of cell types using histological approaches. A disadvantage of the approach is that the time intervals are rather long so that early events are missed. It gives one an overall impression of the ability of cells to reconstitute tissues but cannot be used if one wants to obtain information about single steps in the adhesive process.

## 3. Cluster Formation

A simple technique for following adhesion has been to add cell suspensions to microtiter wells and then simply observe (Noll *et al.*, 1979). This approach has enabled several groups to study "factors" that increase the size of the resulting clusters that result, presumably by promoting adhesiveness (Noll *et al.*, 1979, 1981). The advantages of this approach are that it is very simple, does not require large numbers of cells, and permits large numbers of replicates to be followed with ease. The disadvantage to the approach is that one must wait for extended periods of time before differences are apparent between the controls and the factor-treated cultures. This time factor means that the event being measured can be obscured by other adhesive or nonadhesive events.

## 4. Cell-Binding Assay

Several years ago we developed this assay in order to follow the earliest events in the recognition process (McClay *et al.*, 1981). Quite simply, cells are brought together and then separated using gravity and centrifugal force as a way to quantify initial binding affinities between cells. The assay is of special value when testing the effect of antibodies on putative cell-adhesive components. The assay can also be used to distinguish between the first events and later events in adhesion that increase the strength of adhesion (McClay *et al.*, 1981). The disadvantage to this assay is that later sorting out events cannot be followed. Also although the concept of the assay is simple, in practical terms it is not as easy to use as the first three techniques since there are several technical steps that must be followed. Nevertheless, the assay has provided a means for looking at the earliest events in the adhesion process. In this respect we have found the assay to be quite useful, especially for screening monoclonal antibodies for interference with adhesive components.

## 5. Protocol for the Cell-Binding Assay

*a. Overview.* Figure 1 shows the assay diagrammatically. First a monolayer of cells is placed on the bottom of a microtiter well. Radiolabeled probe cells are then added to the well, the well is sealed, and the probe cells are brought into contact with the monolayer. The plate is inverted and spun in the inverted

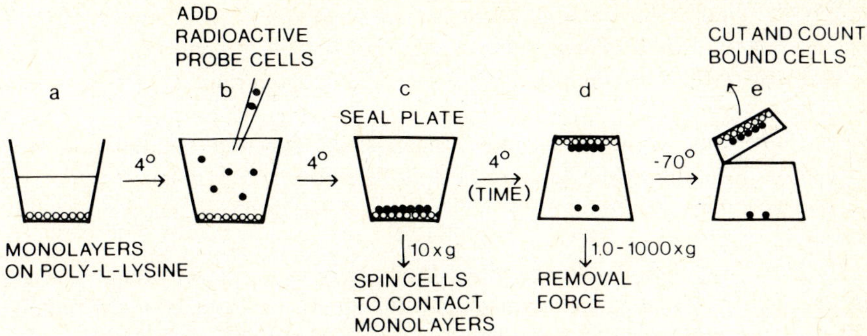


FIG. 1. Cell binding assay. (a) Monolayer cells are spread onto polylysine-treated microtiter wells. (b) Radioactive probe cells are added and each well is filled to the brim. (c) The wells are sealed with plate-sealing tape and the probe cells are centrifuged to contact the monolayers. (d) The plates are inverted (immediately or after an incubation period) and centrifuged (1–2000 *g*) to provide a force tending to remove probe cells from the monolayers. (e) After centrifuging, the plates are quick frozen in the inverted position in ethanol–dry ice and the bottom 3 mm of each well is clipped away to be counted by scintillation methods.

position in an attempt to remove the probe cells from the monolayer cells. To determine how many of the probe cells have remained in contact with the monolayer the plate is frozen, the well bottoms are cut from the plate, and the bottoms are counted in a scintillation counter. To measure initial binding only, the plate is kept at 4°C throughout the assay. To measure secondary events the plate is warmed to ambient temperature for about 10 minutes with the probe cells in contact with the monolayers. When the cells are then separated by centrifugation, there is a much greater resistance due to a presumed strengthening of the interaction between the probe and the monolayer cells. In the description that follows there is a step-by-step sequence given, including the controls that must be established.

*b. Plate Preparation.* Polyvinylchloride microtiter plates can be obtained from Dynatech Laboratories, Inc. (Alexandria, VA), Falcon, or from Costar (Corning). The soft plates are used since the well bottoms are later cut from the plates to analyze the percentage of cells binding to the bottoms. Before using the plates they must be derivatized so that monolayer cells can bind very tightly and resist centrifugal forces that tend to strip the probe cells from the monolayers. The wells are coated with three coats of poly-L-lysine plus glutaraldehyde, a treatment that has been determined empirically to provide a very sticky surface for a variety of cell types. Polylysine (70,000 $M_r$ form, Sigma, 2 mg/ml in distilled $H_2O$) is mixed 2 parts to 1 with glutaraldehyde (1% in saline, diluted from 25% stock) immediately before adding 50 μl per well. The polylysine–glutaraldehyde mixture must be added to wells within 2 minutes of mixing. After a 20-minute incubation the wells are washed and receive a second treatment of polylysine alone. Twenty minutes later they receive a third treatment of polylysine–glutaraldehyde. Plates thus prepared can be covered and stored for about a week. Just before use the plates are washed 12 times in running water and then seawater.

*c. Substrate.* The assay can be used for cell–substrate binding as well as for cell–cell binding (McClay and Fink, 1982; Fink and McClay, 1985). This section describes the addition of substrates and the controls that must be followed.

Substrate components can be added to the polylysine-treated wells or to untreated wells. Untreated wells will bind a maximum of about 20 ng of protein, whereas the polylysine-treated plates will bind about 20 μg of protein. The amount of protein bound can be quantitated immunologically if one has an antibody against the substrate material, or it can be quantitated with a micro-Bradford assay. Bradford reagent (Bio-Rad Laboratories, Richmond, CA) can be diluted 1:5 with water and added to the well bottoms (30 μl). From the color of the reagent compared with standards one can estimate how much protein was deposited on the bottom of the wells.

Before going further with the assay, the wells must be "blanked" to block nonspecific binding. The goal is to block nonspecific binding to the well bottom without disturbing the substrate. Usually, treatment with 20% fetal calf serum, bovine serum albumin (5 mg/ml), or gelatin (200 μg/ml) will block all nonspecific binding when probe cells are subsequently added to the blanked wells but it has been found that each cell type behaves differently so that wells must be blanked for each cell type tested.

In practice, after learning how to blank wells the assay includes as a control a series of wells that are blanked as a background binding control. The test wells precoated with the substrate are also blanked so that any binding above background must be due to the interaction of cells to the substrate material.

*d. Monolayers.* To make monolayers, cells are added directly to the polylysine-treated plates. To determine how many cells make a monolayer, cells at several dilutions are added to the plates and spun onto the well bottoms at 200 *g*. The plates are then examined with an inverted microscope to determine the amount of coverage of the bottom. The plates containing monolayers are incubated for about 10 minutes at room temperature and are then tested to determine whether cells have bound tightly to the well bottoms. This test is performed by inverting the plate and "flicking" the seawater out of the wells and then replacing it with fresh seawater (100 μl). After washing the monolayers twice in this way, inspection will determine whether they are tightly bound to the bottoms. In practice, we have found that when a cell monolayer survives the flicking of the plate, it will also survive gravitational forces to the limit of the centrifuge plate carrier (1500 *g*). Further, we have found that all cell types tested thus far will stick through this protocol after the well bottoms are derivatized with polylysine as above.

Several controls should be performed to determine that the cells of the monolayer remain intact throughout the washing procedures and the centrifugations. First, when radioactive cells are used for the monolayers, there is very little loss of radioactivity from the substrate after the first wash. One would expect that if cells were continuously sheared, radioactivity would appear in the supernatant. It does not. Second, if cells are double labeled with [$^3$H]amino acids (internal label), and [$^{125}$I]lactoperoxidase (surface label), the ratio of $^3$H:$^{125}$I should not change as long as the cells are intact.

After the protocol for the monolayer is established there is one additional control that must be determined. As with the substrate controls one must be assured that cells bind only to the surface of other cells and not to any substrate that happens to remain between the cells of the monolayer. Again, the blank plates can be used. Using radiolabeled cells one can determine what reagent will prevent the binding of cells to the polylysine-coated substrates. For most sea urchin cells 10% fetal calf serum in seawater will blank the well. If the blanking solution reduces binding by whole wells to less than 10%, and if the spaces between cells of the monolayers constitute less than 10% of the available space for probe cells, then one can assume that artifactual binding to substrate is less than 1%.

*e. Probe Cell Preparation.* Probe cells can be labeled with any of a number of radioactive tracers. We use [$^3$H]leucine because it is incorporated well and the pool size for this amino acid is not as large as for other amino acids. Embryos can be labeled by incubation in [$^3$H]leucine (1 μCi/ml in seawater) either overnight or from fertilization until the cells are needed. The embryos are washed twice and then dissociated, which adds about 7–10 additional washes. At this point the cells are labeled with about 0.10–0.1 cpm/cell and very little radioactivity is released from cells during the assay that follows. As a control for loss of radioactivity, cells can be precipitated at any step and the supernatant sampled. When radioactivity is detected in the supernatant it usually means the cells are damaged. As reported by Okazaki (1975), cells at very early stages in development are especially delicate and we have found them to be more easily damaged than cells at later stages.

*f. The Assay Sequence, with Controls.* Probe cells are added to monolayers in 100 μl of seawater. The wells are then brought to 315 μl with seawater in order to form a positive meniscus on the top of the well. The positive meniscus is necessary because when tape is added to seal the plate it is very important that bubbles are not introduced into the wells. About 10 μl of medium will be lost from the top of the wells when tape is rolled over the plate so, in order to prevent mixing of rows, we use only every other row of the microtiter plate. The plates are then sealed with plate sealer tape (Dynatech). The tape is added by rolling it from left to right across the rows of wells and then rolling a darkroom print roller over the plate in the opposite direction. The plates are supported by a rubber plate holder during this operation in order to keep them rigid. Sealing the plates is the most exacting aspect of the entire assay and it may take some practice in order to prevent bubbles from being introduced into the wells. A small error is introduced by the loss of the excess medium from the well, but it has been found that the error is very similar for each well and is less than the error introduced by pipetting.

After sealing, the plates are transferred to a refrigerated centrifuge equipped with microtiter plate carriers. The plates are centrifuged first to bring probe cells into contact with the monolayers (300 rpm $\simeq$ 20 *g*, 8 minutes). This can be monitored in controls by measuring the disappearance of radioactivity from samples of the medium above the monolayers. As soon as the centrifuge stops the plates are flipped, reinserted into the plate carriers, and centrifuged again. Now the centrifugal force will tend to pull the probe cells away from the monolayers. To measure the force necessary to remove 50% of the cells, one can spin replicates at different gravitational forces. The profile of cells bound vs centrifugal force can be used to calculate the force necessary to remove 50% of the cells, and this in turn can be used to calculate the force in dynes necessary to pull half the cells away from the monolayer cells (McClay *et al.*, 1981).

The assay is completed by transferring the plates from the centrifuge, still in the inverted position, to a mixture of dry ice–ethanol. The wells are frozen in seconds and then the well bottoms are cut and placed into scintillation vials. To cut the wells, we have adapted a pet toenail clipper (for small pet animals) by soldering washers onto the top of the hole in the clipper to serve as a measured stop so the bottom 3 mm of each well is clipped. In that way the same amount of medium is transferred to each scintillation vial, giving each vial approximately the same level of quench.

In each assay, controls include scintillation vials to which cells have been added directly and others containing cells spun down onto wells only as a measure of 100% binding. Because the assay is very rapid we collect eight replicates for every point. Statistically this number of replicates gives us a standard error of the mean of about 2–5% with the greatest error coming from a sampling error in the addition of cells.

*g. Initial Binding vs Secondary Steps in Binding.* The assay as described above is performed at 4°C. At the low temperature it will measure components of adhesion that do not require a temperature-sensitive step. The assay can be modified to measure temperature-sensitive steps by spinning probe cells onto the monolayers before sealing the plates, then incubating the plates at higher temperatures. After a high-temperature incubation the plates are returned to ice, then each well is brought to 315 μl and sealed. This sequence must be followed since sealed plates, if incubated at higher temperatures, accumulate bubbles due to changing vapor pressures in the media.

## V. Antibodies and Measurement of Cell Interactions

Antibodies have been used as probes for cellular interactions since Gerisch and co-workers first used Fab fragments to block cell adhesion of *Dictyostelium* cells (Beug *et al.*, 1970). The following paragraphs describe the types of measurements and the conclusions that can be made with antibodies. Often the first thing done with an antibody is to study its effects in an adhesion assay. This approach is all right but one should be prepared to have a series of controls in order to interpret any finding correctly. Sometimes antibodies have been added to embryos in order to observe effects on development. Again, interpretations of this kind of experiment must be made with care. Adhesion and morphogenesis involve complex series of events.

What is described below may seem like common sense, but must be kept in mind while doing an experiment with antibodies. The methodology described below begins with the assumption that one has antibodies against the cell surface, yet possesses no further knowledge of antibody specificity. These antibodies could be either monoclonals or polyclonals. Furthermore, it must be understood that there are two kinds of molecules on the surface of cells, those that participate in cell recognition and adhesion and others that do not (nonrelevant components). The cell-binding assay is used to establish a series of predictions that can be tested. Other assays could also be used for this exercise but it should be realized that assays measuring the end point of multiple events (aggregation assays) could be misinterpreted if antibodies interfere with an unknown step. In most cases, the tacit assumption is made that antibodies interfere with the initial binding step. This assumption is unnecessary with the cell-binding assay since only the initial steps of an adhesion are measured.

The predictions that follow have each been tested both with polyclonal and with monoclonal antibodies on the sea urchin embryo and with the chick embryo (McClay, 1984). Concentrations of antibody needed are not given since they will vary with the source of antibody. Most of our polyclonal antisera have been tested at 1:20 dilution, and the supernatants containing monoclonal antibodies have been tested without dilution but after dialysis into seawater.

*Whole polyclonal antibodies will enhance the binding of cells to monolayers.* This will be true if antibodies in the mixture recognize nonrelevant components as well as adhesion components. Antibodies against nonrelevant antigens will cross-link the two cells adding to existing adhesive cross-linking components. Antibodies against adhesion components should simply replace adhesive cross-links with antibody cross-links. Obviously there will be differences in binding constants that could affect the relative affinity but at the level of current resolution, these properties, if they exist, have not been noticed.

*Fab fragments to the polyclonal antibody will inhibit binding.* This will be true if there are Fab fragments directed against the adhesion component(s).

*Whole antibodies whose second sites are blocked will behave like Fab fragments.* Springer and Barondes (1980) introduced this approach and it works especially well with the cell-binding assay. After addition of polyclonal antibodies to the monolayer, excess antibody is washed away. Then an anti $F(ab')_2$ is added to block the second site of the original antibody. If there is an antibody in the mixture that recognizes adhesion components, adhesion should be inhibited.

*Monoclonals to cell surface components will enhance initial binding.* Again this has been true for a large number of monoclonal antibodies, even those that have been shown in later steps to interfere with adhesion. Many of the monoclonals tested had no noticeable affect on aggregation of cells by other assays.

*Monoclonals to "known" adhesion antigens will block adhesion if they are added in combination with anti* $F(ab')_2$. This was observed with one monoclonal antibody (against a component in the neural retina system, McClay, 1984). It is difficult, however, to use this as a good criterion at present since there are so very few molecules for which enough data verify that they are adhesion antigens.

*Monoclonals against nonrelevant antigens will not block binding when added in combination with anti* $F(ab')_2$. With one known surface molecule, this has been true (McClay, 1984). Again this criterion is difficult to satisfy at present since so few molecules are characterized.

*Some monoclonal antibodies against cell surface components will inhibit initial binding if the second site of the monoclonal is blocked by anti* $F(ab')_2$. This has been true for a small subset of monoclonals that recognize the surface of sea urchin cells. Again, it should be noted that these monoclonals had no apparent affect on aggregation by other methods.

*Fab fragments of monoclonals in the previous prediction will, by themselves, inhibit binding.* This has been true for several of the surface antigens. The reason for including $F(ab')_2$ fragment predictions is that they will permit screening of supernatants without having to produce Fab fragments from every monoclonal antibody produced.

*Monoclonals against two independent sites, both with inhibitory activity in the presence of* $F(ab')_2$, *will have additive effects when added together.* This too has been confirmed both in the sea urchin and in the neural retina system.

# VI. Summary

In resolving the role of cell recognition events in the process of morphogenesis it is necessary to focus on single events against a background of many complex interactions. This article presents a series of approaches that are designed to do just that. It should be noted that with simplification there is a danger of oversight. Nevertheless, in learning about mechanisms of cellular movement at gastrulation we have found that cell separation techniques, simplified adhesion assays, and predictable antibody activities are helpful for approaching the complex mechanisms of morphogenesis.

## References

Beug, H., Gerisch, G., Kempff, S., Riedel, V., and Cremer, G. (1970). *Exp. Cell Res.* **63,** 147–158.

Citkowitz, E. (1971). *Dev. Biol.* **24,** 348–362.

Ettensohn, C. A. and McClay, D. R. (1986). *Dev. Biol.* (in press).

Fink, R. D., and McClay, D. R. (1985). *Dev. Biol.* **107,** 66–74.

Gibbins, J. R., Tilney, L. G., and Porter, K. R. (1969). *J. Cell Biol.* **41,** 201–226.

Giudice, G. (1962). *Dev. Biol.* **5,** 402–411.

Giudice, G., and Mutolo, V. (1970). *Adv. Morphog.* 8, 115–158.

Harkey, M. A., and Whiteley, A. H. (1981). *Wilhelm Roux's Arch. Dev. Biol.* **189,** 111–122.

Herbst, C. (1900). *Arch. Entwicklungsmech. Org.* **9,** 424–463.

Kane, R. E. (1973). *Exp. Cell Res.* **81,** 301–311.

McClay, D. R. (1984). *J. Cell Biol.* **99,** 116a.

McClay, D. R., and Chambers, A. C. (1978). *Dev. Biol.* **63,** 179–186.

McClay, D. R., and Fink, R. D. (1982). *Dev. Biol.* **92,** 285–293.

McClay, D. R., and Hausman, R. E. (1975). *Dev. Biol.* **47,** 454–460.

McClay, D. R., and Marchase, R. M. (1979). *Dev. Biol.* **71,** 289–296.

McClay, D. R., Wessel, G. M., and Marchase, R. M. (1981). *Proc. Natl. Acad. Sci. U.S.A.* **78,** 4975–4979.

Moscona, A. (1957). *Proc. Natl. Acad. Sci. U.S.A.* **43,** 184–194.

Noll, H., Matranga, V., Cascino, D., and Vittorelli, L. (1979). *Proc. Natl. Acad. Sci. U.S.A.* **75,** 288–292.

Noll, H., Matranga, V., Palma, P., Cutrono, F., and Vittorelli, L. (1981). *Dev. Biol.* **87,** 229–241.

Okazaki, K. (1975). *Am. Zool.* **15,** 567–581.

Oppenheimer, S. B., and Meyer, J. T. (1982). *Exp. Cell Res.* **137,** 472–476.

Orr, C. W., and Roseman, S. (1969). *J. Membr. Biol.* **1,** 109–115.

Showman, R. M., and Foerder, C. A. (1979). *Exp. Cell Res.* **120,** 253–256.

Spiegel, M., and Spiegel, E. S. (1975). *Am. Zool.* **15,** 583–606.

Springer, W., and Barondes, S. (1980). *J. Cell Biol.* **87,** 703–714.

Wessel, G. M., Marchase, R. M., and McClay, D. R. (1984). *Dev. Biol.* **103,** 235–245.

# Chapter 18

# *Manipulative Methods for Analyzing Embryogenesis*

YOSHIHIKO K. MARUYAMA, KENYA YAMAMOTO[1], AND IZUMI MITA-MIYAZAWA

*Department of Zoology, Faculty of Science*
*Kyoto University*
*Kyoto, Japan*

TETSUYA KOMINAMI

*Department of Biology, Faculty of Science*
*Ehime University*
*Matsuyama, Japan*

SHIN-ICHI NEMOTO

*Tateyama Marine Laboratory*
*Ochanomizu University*
*Tateyama, Japan*

---



[1]Present address: Department of Biology, Faculty of General Education, Gifu University, Gifu, Japan.

## I. Introduction

Manipulative approaches have become valuable for studying the temporal and spatial organization of embryos of diverse species (Chambers and Chambers, 1961; Hörstadius, 1973). In this article we describe a number of manipulative methods adopted to the egg cells and embryos of the sea urchin *Hemicentrotus pulcherrimus* and the starfish *Asterina pectinifera*. With minor modifications, they will be applicable to other species.

## II. Stripping Eggs

Stripping eggs of the fertilization membrane or the vitelline coat is frequently required for manipulative experiments, such as egg bisection, embryo fusion, blastomere isolation, and chromosome staining.

Mature oocytes of starfishes are obtained by immersing oocytes in 1-methyladenine in seawater (Kanatani, 1969). Spermatozoa are obtained from testis fragments placed in a dry Petri dish. Insemination is made about 40 minutes after the start of the 1-methyladenine treatment or at any time from germinal vesicle breakdown up to polar body extrusion. Insemination should be made with a considerably diluted sperm suspension to avoid polyspermic fertilization.

One-molar urea (dissolved in distilled water) is used to remove the elevating fertilization membrane of the starfish egg (Dan-Sohkawa, 1976; Ikeda *et al.*, 1976). Eggs 3–5 minutes after insemination are treated with 1 *M* urea solution: In these eggs, fertilization membranes have been elevated from entire egg surfaces. Using a hand centrifuge, 1 ml (or less) of eggs 2–3 minutes after insemination are collected in a 15-ml centrifuge tube, and washed once with $Ca^{2+}$-free or $Ca^{2+}$- and $Mg^{2+}$-free seawater, and then are suspended in about 10 ml of 1 *M* urea solution. Eggs in the tube that is sealed with one's thumb are shaken by slowly tipping it back and forth several times. In this solution, the fertilization membrane will be highly elevated. The eggs are immediately collected by using a hand centrifuge and resuspended in 10 ml of fresh 1 *M* urea solution. The eggs in the tube are similarly shaken several times until they are denuded. Denuded eggs will settle rapidly in the urea solution. If eggs are not yet denuded, the urea solution is changed once more using a hand centrifuge, and the above procedure is repeated. Then, the tube is left standing until denuded eggs settle to the bottom. After discarding the supernatant, denuded eggs are washed two or three times with seawater. They are then pipetted into seawater in a 1% agar-based dish. Denuded eggs are cultured in agar-based dishes to prevent adhesion to the substratum. These dishes are prepared as follows. Agarose is dissolved at 1% in boiling seawater and this solution is layered on bottoms of plastic dishes. After gelation, the dishes are used as agar-based dishes. If necessary, $Ca^{2+}$-free or $Ca^{2+}$- and $Mg^{2+}$-free seawater is also used as the solvents for agarose.

Sodium thioglycolate removes fertilization membranes of starfish eggs or embryos at *any* stage. It is dissolved at a concentration of 1% in seawater and is adjusted to pH 10 with 1 *N* NaOH. Eggs or embryos are suspended in the solution and treated for 3 minutes. During the first 1–2 minutes of the treatment, fertilization membranes become weakened. Then, eggs in the solution are gently passed through a Pasteur pipet several times to remove the envelopes. The denuded eggs are washed two or three times with seawater, and cultured in agar-based dishes filled with seawater.

Vitelline coats of starfish oocytes are removed by using the thioglycolate solution. On the other hand, trypsin dissolved in seawater at 0.1% is useful for removing vitelline coats from immature starfish oocytes (Nemoto *et al.*, 1980). Immature oocytes are first treated with trypsin solution for 1–2 minutes to elevate the vitelline coat from the oocyte surface, and then oocytes with elevated envelopes are treated with 1 *M* urea solution to remove the envelopes, as described previously. Pronase dissolved at 0.01% in seawater is also used for dissolving vitelline coats from starfish oocytes (Nemoto *et al.*, 1980). Denuded oocytes can be induced to mature by 1-methyladenine, and, upon fertilization, they will develop. They are cultured in agar-based dishes filled with seawater. These methods may be modified for other starfishes. A combined use of $Ca^{2+}$ ionophores and proteases has been developed for *Pisaster ochraceus* (Schroeder and Stricker, 1983).

Sea urchin eggs can be denuded by similar procedures as in starfishes. Here, a technique for denuding a single sea urchin egg (or egg fragment) of its fertilization membrane and hyaline layer is briefly described. The following procedures are performed under a dissecting microscope. An egg is inseminated with a small amount of diluted sperm suspension. One minute after insemination, the egg with an elevating fertilization membrane is transferred to $Ca^{2+}$- and $Mg^{2+}$-free seawater in an agar-based dish. Two or 3 minutes later, the egg with a highly elevating fertilization membrane is passed through a micropipet whose tip diameter is similar to the egg diameter (cf. Section III). The fertilization membrane is removed. The egg is left in the solution for an additional 10–20 minutes, and then transferred to $Ca^{2+}$-free seawater in an agar-based dish. Formation of the hyaline layer is prevented in these media. This egg can be used for manual operations. If necessary, the egg is transferred to seawater. Similarly, in the starfish *Asterina pectinifera*, the fertilization membrane can be removed by passing fertilized eggs through a micropipet whose tip diameter is slightly smaller than that of the fertilization membrane. This method of denuding can easily be performed on eggs before the first cleavage.

# III. Bisection of Sea Urchin and Starfish Eggs

The normal interaction of the nucleus and cytoplasm can be altered by removing the nucleus or some regions of the cytoplasm (including the cortex) from the egg. Sea urchin or starfish eggs can be separated into egg fragments. By using egg fragments, we can examine possible roles for the nucleus in cytoplasmic activities or, in turn, possible roles for cytoplasmic materials in embryogenesis. There are two ways to obtain egg fragments: manual bisection and centrifugal separation. Manual bisection enables us to divide eggs in any desired direction so that we can compare a single egg fragment with its complementary partner. On the other hand, centrifugal separation makes large enough numbers of nonnucleate and nucleate egg fragments for biochemical analyses.

## A. Manual Bisection

Bisection of sea urchin eggs in a definite direction in relation to the animal–vegetal axis can be achieved by using the jelly canal as a marker of the animal pole. In this section, a technique for visualizing the jelly canal is described first. Procedures of egg bisection of sea urchins and starfishes follow.

### 1. Staining the Jelly Canal

The jelly canal, which was also called the micropyle (Boveri, 1901), is a narrow channel in the transparent jelly coat which surrounds the egg. This structure occurs commonly among sea urchin eggs and marks the animal pole of the egg and embryo (Boveri, 1901; Morgan and Spooner, 1909; Hörstadius, 1937; Schroeder, 1980a,b). For visualizing the jelly canal, eggs just emerging from the genital pores or obtained directly from ovaries are used, because ''staining'' of the jelly canal becomes difficult as the jelly coat swells in seawater (Tennent *et al.*, 1929; Schroeder, 1980a).

Sumi-ink solutions are prepared from commercial liquid sumi-ink called bokuju (for example, from Fueki Nori Co., Japan). This concentrated sumi-ink is used after twice or more dilution with seawater. Alternatively, sumi-ink solutions are prepared by rubbing a dry cake of commercial sumi-ink in a small amount of seawater on an inkstone or frosted glass surface (Schroeder, 1980a,b).

Female sea urchins are induced to spawn by injecting 0.54 *M* KCl into the body cavity and placed with the aboral surface downward (not immersed in seawater). Eggs emerging from the genital pores are transferred into a small amount of the sumi-ink solution, using a Pasteur pipet. Alternatively, a piece of an ovary is placed into a sumi-ink solution and dissected with forceps to release eggs. One to 3 minutes later, the sumi-ink solution containing eggs is gently pipetted into seawater in a plastic dish (15 × 60 mm). They are left there for 5 minutes or more until jelly coats swell and jelly canals are stained. Just before use, they are washed two or three times with seawater. A single jelly canal filled with sumi-ink particles will be seen in each egg (Fig. 1A). These eggs with stained jelly canals should be used immediately for the next step, bisection or fertilization.

Jelly canals stained with sumi-ink in eggs with thick jelly coats such as *Mespilia globulus* and *Strongylocentrotus droebachiensis* are particularly long and are retained for a much longer period in seawater than the jelly canals of eggs with thin jelly coats such as *Anthocidaris crassispina* and *Hemicentrotus pulcherrimus*. Culturing the eggs in diluted sumi-ink solutions may prevent diffusing out of sumi-ink particles from the jelly canal. It should be noted that the jelly coat is fragile, so the egg with the stained jelly canal is handled carefully, if necessary, with a pipet of a relatively wide orifice.

## 2. Procedure of Manual Bisection

Manual bisection is carried out by the free-hand use of a fine glass needle. A dissecting microscope or an inverted microscope is suitable for the free-hand operation.

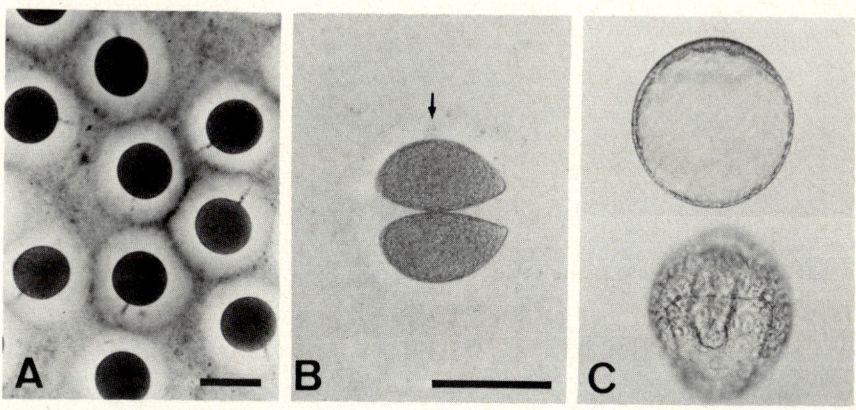


Fig. 1. Manual bisection of sea urchin (*H. pulcherrimus*) eggs with stained jelly canals. (A) Unfertilized eggs treated with a sumi-ink solution. The jelly canals are stained. (B) One pair of egg halves just after bisection in relation to the animal–vegetal axis, as defined by the jelly canal. The upper is the animal half with the jelly canal, as indicated by an arrow, and the lower is the vegetal half. (C) One pair of egg halves 43 hours after fertilization. An egg has been bisected into the animal and the vegetal halves, inseminated, and separately cultured at 16–18°C. The animal half (upper) develops to a permanent blastula, and the vegetal half (lower) develops to a prism larva with the gut and skeletons. Bars, 100 μm. The bar in (B) is for (B) and (C).

Fine glass needles for manual bisection can be hand pulled in two steps from glass rods (or glass tubes) of about 4 mm in diameter. First, the glass rod is pulled by using the flame of an ordinary Bunsen burner to make a needle whose tip diameter is about 1 mm (Fig. 2A). Second, one portion of the needle tip is melted on a flame of a micro-gas burner (Fig. 2B), and is quickly pulled apart out of the flame by using a fine forceps so as to make a fine needle forming a slight angle with the glass rod axis (Fig. 2C). A micro-gas burner can be made of a commercial Bunsen burner (see Fig. 2B). A suitable tip diameter of the fine needle is 10–15 μm for eggs or oocytes of 100–170 μm in diameter. A micro-needle whose fine portion is short (1–2 cm long) and tough is convenient. Among needles, an appropriate one is selected by trial. The tip of the needle should be kept clean by frequent sonication, since an egg (fragment) is apt to stick to a dusty (or sticky) needle.

A fine pipet is frequently used to transfer an egg or egg fragment. Such a micropipet can be easily made from a commercial Pasteur pipet (7 × 146 mm; Corning, NY) or from a hand-pulled pipet with a tip diameter of about 1 mm (cf. Fig. 3). The tip of the pipet is heated for several seconds on the flame of an alcohol lamp or of a micro-gas burner and is quickly hand pulled out of the flame by using a fine forceps so as to obtain a suitable tip diameter. If a glass tube with

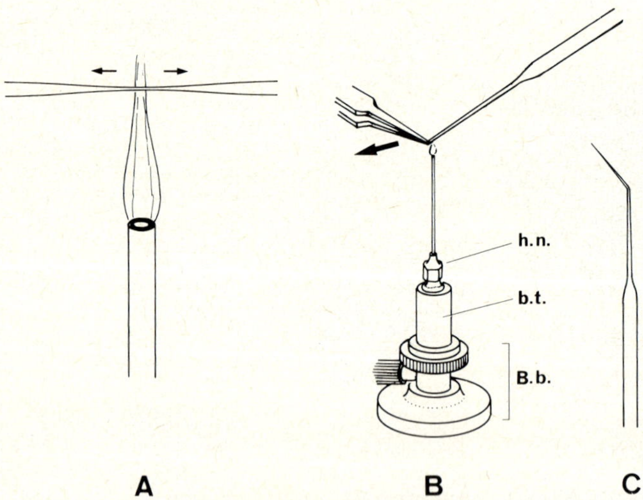


FIG. 2. Schematic drawing of the procedure for making micro-glass needles (A–C) and of a micro-gas burner (B). The micro-gas burner is assembled from a hypodermic needle with its end filed flat (h.n.), a connecting piece of brass tubing (b.t.), and a commercial Bunsen burner whose air inlet has been removed (B.b.). A piece of brass rod (35 × 17 mm) is drilled and tapped so as to screw into the burner while a small hole is bored at the other end that has been shaped to match the base of the needle. Each connecting portion is sealed with silicon grease against gas leakage. The size of the microflame is controlled by adjusting the gas pressure regulator of the burner.

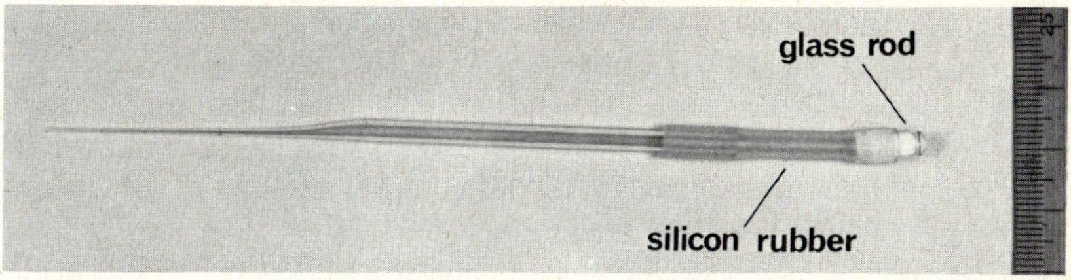


FIG. 3. A micropipet for transferring egg fragments. The micropipet is connected to a piece of rubber tubing whose end is stoppered with a piece of glass rod. A scale (1 mm unit) is inset.

a thin diameter (4 mm) is available, such a micropipet can be made directly of it. While supporting the pulled-out portion on a finger, the glass is scratched with the edge of a piece of sand paper and broken to obtain a micropipet with a proper diameter. The micropipet thus prepared can be connected to silicon rubber tubing and controlled by mouth (Hörstadius, 1973). Alternatively, the micropipet can be connected to a short piece of silicon rubber tubing stoppered with a piece of glass rod and manipulated by hand (Fig. 3). The micropipet is used after drawing in a bit of seawater just before use. The micropipet should be kept clean by frequent washing with distilled water; otherwise, new micropipets should be made before each experiment.

Egg bisection is carried out on an agar-based dish to prevent slippage of eggs during bisection. This agar-based dish is prepared as follows. Boiled seawater containing 1% agarose is layered on the bottom of a plastic dish (15 × 60 mm), and left to cool and slightly dry in air for several hours. Then, the dish is covered until use.

The procedure for bisecting sea urchin eggs in a definite direction in relation to the animal–vegetal axis is now described. Using a pipet, one drop of seawater containing a few eggs with stained jelly canals is placed on an agar-based dish. An egg whose animal–vegetal axis lies horizontally is found under a dissecting microscope. The fine tip of the micro-glass needle is dipped into the seawater drop and then placed horizontally on the egg along the line of the desired bisection. The needle is slowly shifted downward while still horizontal. The egg beneath the needle is first squeezed and eventually will be separated into two halves (Fig. 1B). If any damage is detected on the surfaces, the egg fragments should be discarded. Using a micropipet whose tip diameter is larger than the egg fragment, one of the egg fragments is picked up and transferred to one drop of seawater in a plastic dish. The counterpart is then transferred to another seawater drop in the dish. By bisecting another egg in a fresh drop of egg suspension on a different part of the agar-based dish, an additional pair of egg fragments is obtained. These egg fragments should be used immediately for fertilization or artificial activation.

Using a micropipet, a small amount of diluted sperm suspension is added to each egg fragment. Both egg fragments derived from an egg form fertilization membranes and begin development. About 5 minutes later, each egg fragment should be washed with seawater to remove excess sperm and transferred to a small culture vessel or a microwell of a multiwell plate (96 wells; Falcon) filled with seawater. They are cultured up to desired developmental stages (Fig. 1C).

In the case of fertilized sea urchin eggs, removal of both the fertilization membrane and the hyaline layer is necessary before bisection. Eggs denuded according to the method described in Section II can be easily bisected by the same procedure as just described. Thus, one pair of nonnucleate and nucleate egg fragments can be obtained from an egg. These egg fragments should be handled and cultured in a 1% agar-based dish to prevent adhesion to the substratum.

Techniques for bisecting starfish oocytes or eggs will now be described. A small number of oocytes are placed into a 1% agar-based plastic dish filled with seawater. The oocyte is bisected using a micro-glass needle. In contrast with the sea urchin egg, bisected halves of a starfish oocyte remain oval and are often joined together by the vitelline coat (Fig. 4A). In some batches of occytes, two glass needles manipulated by both hands are required to separate the fragments completely. Such difficulties in bisecting intact starfish oocytes are probably due to their stubborn vitelline coats. If the vitelline coat is removed from the oocyte beforehand, the bisection can be done more easily or quickly and the bisected halves are separated immediately after bisection (Fig. 4B). Halves derived from immature oocytes become round more slowly than those derived from maturing oocytes. By 1-methyladenine treatment, halves derived from immature oocytes can be induced to mature (Yamamoto and Yoneda, 1983). Maturing starfish oocytes can be bisected in a definite direction relative to the animal–vegetal axis by using the polar body as the marker of the animal pole (Mita, 1983). Oocytes

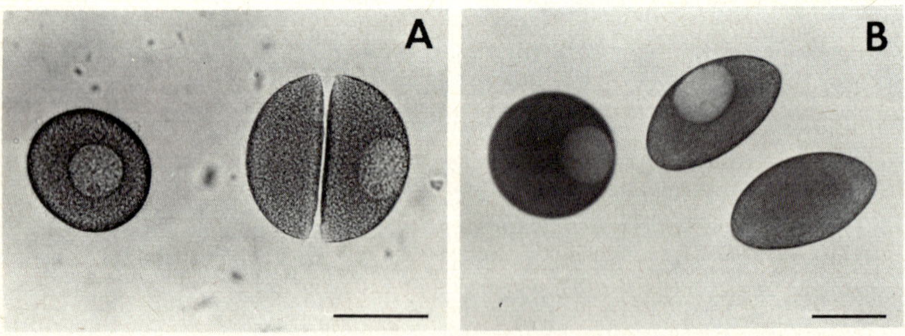


FIG. 4. Manual bisection of starfish (*A. pectinifera*) oocytes. (A) An intact oocyte is bisected 16 minutes after commencement of 1-methyladenine treatment. A normal oocyte is seen at the left. (B) A denuded oocyte at the immature stage is bisected. Nonnucleate fragment, right; nucleate fragment, middle; unbisected control oocyte, left. Bars, 100 μm.

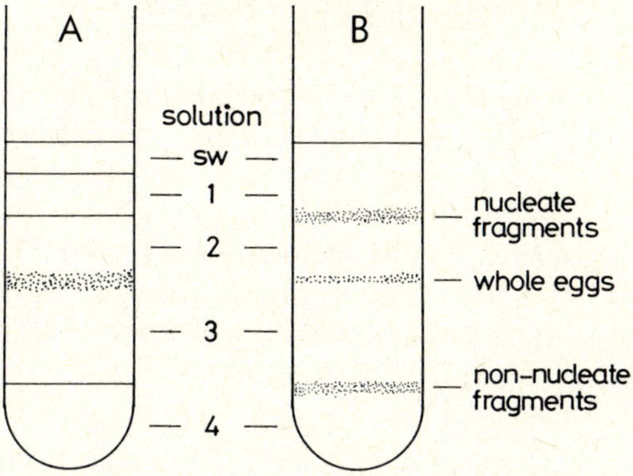


FIG. 5. Schematic representation of layered solutions in the tube before (A) and after (B) centrifugal separation.

just after the first polar body extrusion are used for this purpose. The bisected halves are immediately inseminated with a diluted sperm suspension and cultured. In the case of fertilized eggs, the fertilization membrane is first removed according to the methods in Section II. The denuded egg can be bisected easily. Denuded halves should be handled and cultured in agar-based dishes.

## B. Centrifugal Separation

A large number of nucleate and nonnucleate fragments can be obtained by a single run of centrifugation, avoiding the laborious repetition of manual bisection. By centrifuging eggs, the particulate materials in the endoplasm are stratified according to differences in their densities, so that in a medium isopycnotic with the whole egg the heavier part of the egg is pulled downward while the lighter part tends to float upward until the two parts are pulled apart.

Practically, rather than solutions with a continuous density gradient, four layers of solutions with different densities in a centrifuge tube are suitable for successfully separating eggs and efficiently recovering the resulting fragments (Fig. 5). Mixtures of seawater and a sucrose solution have often been used for this purpose. Ficoll dissolved in seawater may also be appropriate. The composition of each solution is selected by trial, so that intact eggs come to the boundary between the solutions 2 and 3, the heavier parts between 3 and 4 and the lighter parts between 1 and 2 (see Fig. 5). Use of centrifuge tubes made of transparent materials such as polycarbonate is preferable, since these tubes allow us to sequentially layer the solutions, to monitor the separation of eggs, and to collect resulting fragments.

### 1. Sea Urchin Eggs

The composition of the four layers of solutions for centrifugally separating unfertilized eggs of *Hemicentrotus pulcherrimus* (Yoneda *et al.*, 1978) is as follows:

| | |
|---|---|
| 1. (top) | 1 *M* sucrose:seawater = 45:55 |
| 2. | 1 *M* sucrose:seawater = 50:50 |
| 3. | 1 *M* sucrose:seawater = 55:45 |
| 4. (bottom) | 0.65 *M* sucrose |

Solutions 4 (7 ml), 3 (13 ml), 2 (8 ml), and 1 (5 ml) are sequentially layered in a 50-ml centrifugate tube (10.3 × 2.7 cm). Eggs in seawater are loaded on the top layer. They come to rest at the boundary between the solutions 2 and 3 by centrifuging the tube at 5600 *g* for 25–30 minutes. Eggs are elongated as the constituents of their cytoplasm are stratified. By subsequent gradual increase in centrifugal acceleration up to 11,000 *g* and additional centrifugation for 2–5 minutes at this force each egg is separated into two fragments, a light nucleate fragment (at the 1–2 boundary) and a heavy nonnucleate fragment (at the 3–4 boundary) (Fig. 5). This separation will be achieved successfully when the amount of loaded eggs forms a layer one or two eggs thick. During centrifugation the temperature is kept at 10°C. Fragments are collected by withdrawing the layered solutions from the top of the tube. Collected fragments are rinsed with seawater to remove sucrose. Before use the fragments are kept standing in seawater for at least 1 hour, so that fragments will recover from possible aftereffects of centrifugation.

Depending on batches of eggs, the centrifugation may fail to separate the eggs into fragments in spite of the stratification of the endoplasm. This is mostly due to the stiffness of the jelly coat. In such cases, the eggs should be washed several times with seawater to partly remove the jelly coat. Complete removal of the jelly coat with either acidified seawater or isotonic (0.54 *M*) NaCl should be avoided, since it causes aggregation of eggs during centrifugation. By the present procedure, a nucleate fragment is slightly larger than the nonnucleate partner. With higher centrifugal forces, the latter may become larger than the former.

### 2. Starfish Oocytes

*a. Maturing Oocytes.* Jelly coats of starfish oocytes are very tough, so that oocytes are washed once or twice with acidified sea water (pH 4.5 by HCl) or isotonic NaCl in order to partly remove the jelly coat. Care should be taken to avoid complete removal of the jelly coat. As in sea urchin eggs, maturing starfish oocytes thus treated can be separated into nucleate and nonnucleate fragments on four layers of solutions (Nemoto *et al.*, 1980). However, proper compositions of the four layers of solutions must be determined beforehand for each batch of oocytes since densities of starfish oocytes are quite variable among batches. We have found that the following six layers of solutions with diverse densities are satisfactory for separating oocytes of all batches.

| | |
|---|---|
| 1. (top) | 1 *M* sucrose:seawater = 40:60 |
| 2. | 1 *M* sucrose:seawater = 50:50 |
| 3. | 1 *M* sucrose:seawater = 55:45 |
| 4. | 1 *M* sucrose:seawater = 60:40 |
| 5. | 1 *M* sucrose:seawater = 65:35 |
| 6. (bottom) | 0.75 *M* sucrose |

Whole oocytes will either go to boundaries between the solutions 2 and 3 or 3 and 4. Resulting nucleate fragments invariably come to the 1–2 boundary, while the nonnucleate ones are found at either the 4–5 or 5–6 boundary. A typical schedule of the centrifugation is (1) 6000 *g* for 30 minutes, (2) gradual increase up to 12,000 *g* over a 2- to 5-minute period, and (3) 12,000 *g* for 5 minutes (Fig. 6A, B, and C). During these procedures, the temperature is kept at 5°C in order to slow the progress of maturation (such as formation of the polar body), which may otherwise proceed during centrifugation.

Both nucleate and nonnucleate fragments develop at least to the gastrula stage, if fertilized.

*b. Immature Oocytes (Germinal Vesicle Stage).* Intact immature oocytes of *A. pectinifera* are neither elongated nor separated into fragments even by centrifugation at 14,000 *g*, although the endoplasm is well stratified as in maturing oocytes (Nemoto *et al.*, 1980). Martindale and Brandhorst (1984) used 10,000–12,000 rpm in a Beckman JS-13 swinging bucket rotor for separating *Asterias forbesi* oocytes. According to Dorée (1982), a small population of *Marthasterias glacialis* oocytes were broken into halves by centrifugation at 10,000 rpm in a Spinco swinging bucket rotor SM 27. This high resistance to centrifugation must be due to the extreme stiffness of immature oocytes (Nakamura and Hiramoto, 1978; Nemoto *et al.*, 1980; Nemoto, 1982).

A method to obtain fragments in a high yield, although requiring some care, is to soften the cortex by treating the oocytes with 1-methyladenine for a period shorter than the length of the "hormone-dependent period," which is defined as a minimal time of exposure to a threshold concentration for 50% germinal vesicle breakdown to occur. Such an abbreviated treatment does not trigger maturation, but definitely decreases the stiffness (Nemoto, 1982). The low stiffness is held for 1 hour or more, so that the oocytes can be easily separated into fragments.

The length of the hormone-dependent period is first determined in small num-

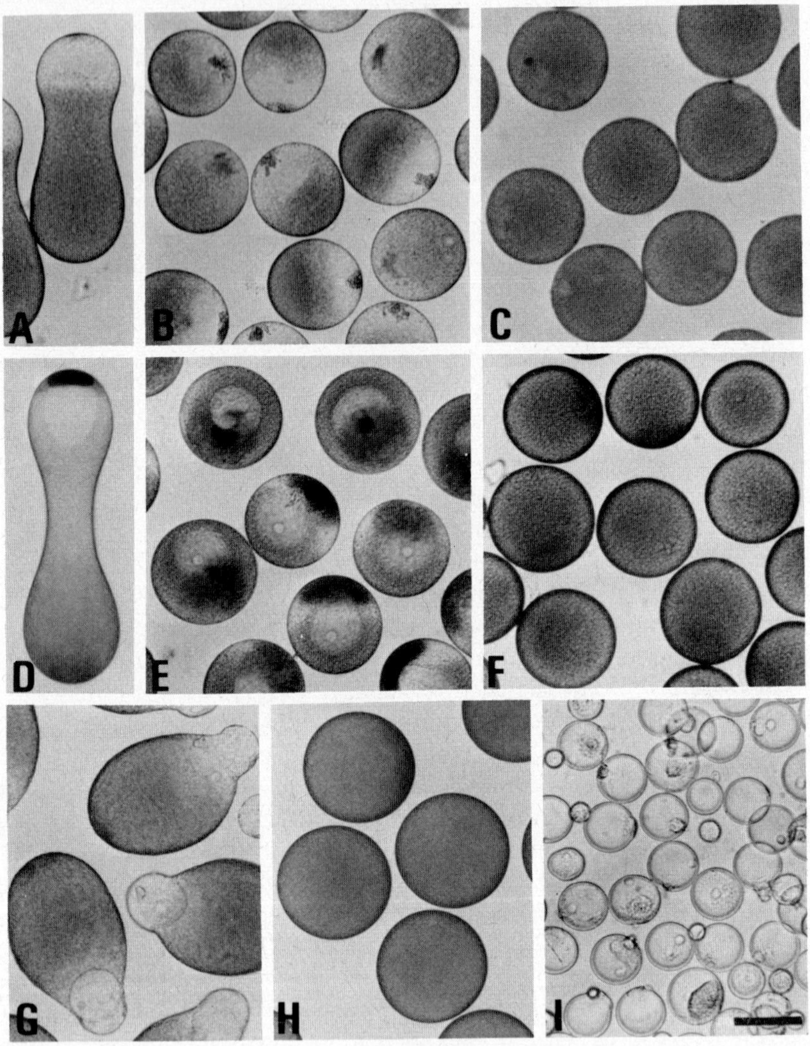


FIG. 6. Centrifugal separation of starfish (*A. pectinifera*) oocytes. Maturing oocytes centrifuged at 6000 *g* for 30 minutes are stratified and elongated (A), and further centrifugation at 12,000 *g* for 5 minutes separates the oocytes into nucleate fragments (B) and nonnucleate fragments (C). Immature oocytes pretreated with 1-methyladenine are centrifuged at 9000 *g* for 30 minutes (D), and further centrifugation at 12,000 *g* for 20 minutes separates them into nucleate fragments (E) and nonnucleate fragments (F). Immature oocytes treated with cytochalasin B are centrifuged at 14,000 *g* for 10 minutes (G), and additional centrifugation at this force for 10 minutes produces cytoplasts (H) and karyoplasts (I). Bar, 100 μm.

bers of oocytes taken out from the batch (cf. Guerrier and Dorée, 1975; Nemoto, 1982; for details of the determination). Oocytes are treated with 1-methyladenine for a slightly shorter period. After washing with seawater to eliminate 1-methyladenine, the oocytes are washed with acidified seawater or isotonic NaCl to partly remove the jelly coat. They are then centrifuged at 10°C in the same set-up as used for maturing oocytes. A typical schedule of the centrifugation is (1) 7000–9000 *g* (according to the batch) for 30 minutes, (2) gradual increase up to 12,000 *g*, and (3) 12,000 *g* for 20 minutes (Fig. 6D, E, and F).

Pairs of nucleate and nonnucleate fragments of comparable sizes are obtained by the methods described above. On further treatment with 1-methyladenine, nucleate fragments undergo germinal vesicle breakdown and form first and second polar bodies. On fertilization they develop normally.

*c. Enucleation of Immature Oocytes.* When one wishes to separate the germinal vesicle from the whole cytoplasm, the cytochalasin B method is also recommended (Prescott and Kirkpatrick, 1973; Nemoto *et al.*, 1980).

Immature oocytes are pretreated with 10 μg/ml cytochalasin B in seawater for 10 minutes. The four-layered solutions are as follows.

| | |
|---|---|
| 1. (top) | 1 *M* sucrose:seawater = 35:65 |
| 2. | 1 *M* sucrose:seawater = 50:50 |
| 3. | 1 *M* sucrose:seawater = 60:40 |
| 4. (bottom) | 0.75 *M* sucrose |

Solutions 2 and 3 contain 3 μg/ml cytochalasin B. In this procedure it is not necessary to remove the jelly coat. Centrifugation at 14,000 *g* for 20 minutes at 10°C is enough to separate the oocytes into nucleate fragments (karyoplasts) and enucleated fragments (cytoplasts) (Fig. 6G, H, and I). Collected fragments are rinsed with seawater to remove cytochalasin B and sucrose. Within 1 to 2 hours after washing, the enucleated fragments regain the high stiffness characteristic of the immature oocyte, and resume the spherical shape.

## IV. Formation of Fused Starfish Embryos

A fused starfish embryo is produced by simply placing two denuded eggs side by side so as to bring them into contact (Dan-Sohkawa, 1977; Mita and Satoh, 1982). These two eggs continue to cleave and some of them eventually develop to a double embryo with a confluent blastocoel (Fig. 7). The fusion can be made between eggs at different developmental stages as well as the same stages as shown in Fig. 7A. Therefore, the fused embryo can be used as a system for

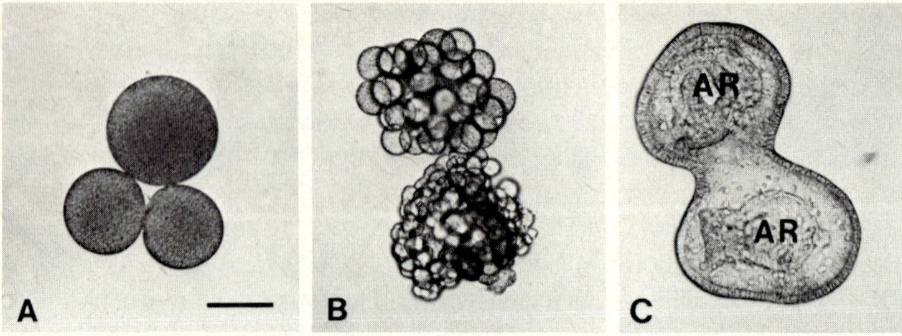


FIG. 7. The process of embryo fusion in starfish (*A. pectinifera*). (A) One denuded egg (upper, nonstained) is laid near to a two-cell stage embryo (lower, stained with Nile Blue Sulfate). They differ by 2 hours in the timing of 1-methyladenine application and insemination. (B) The lower embryo reaches the 256-cell stage while the upper one is still at the 32-cell stage. (C) A fused embryo fixed with 2% acetic acid at the late gastrula stage. The lower member began to gastrulate about 2 hours earlier than the upper. AR, archenteron. Bar, 100 μm.

analyzing the timing mechanism in development (Mita and Satoh, 1982) as well as the mechanism of regional specification.

Eggs denuded of the fertilization membrane (see Section II) are cultured in a 1% agar-based plastic dish containing seawater. Some eggs are vitally stained with Nile Blue Sulfate dissolved in seawater at a concentration of 0.005% for 3 minutes to mark their descendant cells. Only those eggs that divide on schedule at the first cleavage are used. As soon as the cleavage furrow appears, two eggs (one stained and one nonstained) are placed in close contact side by side in a corner between the slant wall and the bottom of the microwell (Terasaki plate, 60 wells, Falcon) filled with seawater. If two eggs with different developmental stages are to be fused, a batch of immature oocytes are split into two groups. They are induced to mature by 1-methyladenine with a time lag, and inseminated at definite times after 1-methyladenine treatments. After removing fertilization membranes, eggs from one group are vitally stained. Pairs of eggs from the two groups are successively transferred with an interval equal to the time lag between the groups. They are left undisturbed during the first several cleavages; during that period, cell–cell adhesion is weak. Septate desmosomes begin to develop at the $2^9$-cell stage (Dan-Sohkawa and Fujisawa, 1980). Adjacent blastomeres derived from either stained or nonstained eggs (Mita and Satoh, 1982) become attached, so that the whole population of blastomeres gives rise to a single fused embryo.

The cleavage and the fusion process of embryos in the microwell can be observed with an inverted microscope. After the fused embryo begins swimming, it is transferred to a microaquarium (3 × 15 mm) filled with seawater. The microaquarium is covered with a cover slip against evaporation and then mounted on a microscope for observation. A vitally stained half of the fused embryo can be distinguished from the unstained member during these observations. By this procedure, about 5 pairs out of 60 pairs of denuded eggs develop into fused embryos with confluent blastocoels.

## V. Marking Starfish Blastomeres with Injected Horseradish Peroxidase

Intracellular injection of horseradish peroxidase (HRP) has been used for marking blastomeres of various animal embryos; leeches (Weisblat *et al.*, 1978), amphibians (Hirose and Jacobson, 1979), mice (Bałakier and Pedersen, 1982), starfishes (Kominami, 1983), and ascidians (Nishida and Satoh, 1983). In the case of starfish embryos, normal organization of blastomeres is preserved only in the presence of the fertilization membrane. This section describes how to inject HRP into blastomeres of starfish embryos through the intact fertilization membrane.

### A. Set-Up of Manipulator and Microscope

An ordinary microscope and a micromanipulator are mounted on an iron base plate (53 × 38 cm) for micromanipulation and fastened with screws (Fig. 8A). A micromanipulator with a long working distance (for example, Model MK-2, Narishige Sci. Inst. Lab., Tokyo) is recommended. A pipet holder (Instrument Collar, Leitz, West Germany) is mounted on the manipulator and is connected to a syringe (Injecting Device, Leitz) with a polymer tube made of a relatively stiff and transparent material such as nylon or Toalon-G (Toa-Kagaku, Gunma, Japan). One portion of the tube is attached to the iron base plate (Fig. 8A). The entire system, from the holder through the tubing and to the cylinder of the syringe, is filled with distilled water, leaving an air bubble a few centimeters long in the attached portion of the tube (Fig. 8B). This air bubble serves as a buffer and indicator of the hydraulic pressure generated by the movement of the syringe. A millimeter scale set along the fixed portion is useful for monitoring the pressure. Expensive commercial injecting devices can be substituted by an ordinary medical syringe for this sort of microinjection as described below.

### B. HRP Solution and Injection Pipet

HRP (type VI, Sigma) is dissolved at a concentration of 5% in half-strength $Ca^{2+}$-free artificial seawater. Depletion of calcium ions from the solvent is

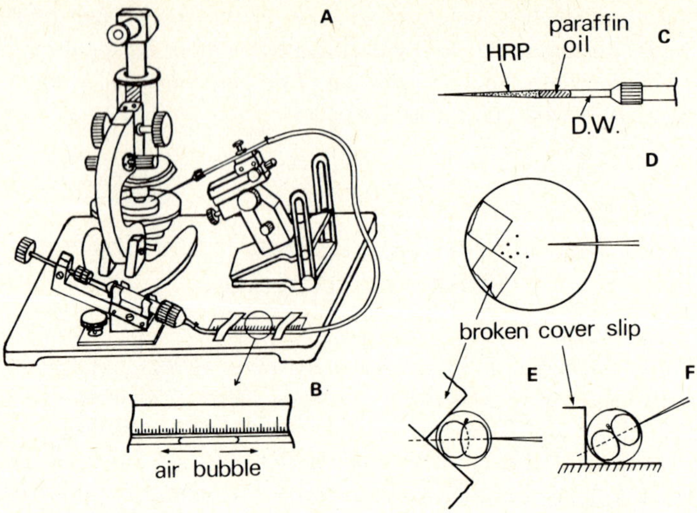


FIG. 8. Schematic drawing of the HRP injection procedure. (A) Experimental set-up, (B) pressure indicator, (C) injection pipet mounted in a pipet holder, (D) arrangement for confining an egg to be injected, (E) upper view and (F) side view of an egg to be penetrated by an injection pipet.

necessary to prevent coagulation of the injected solution with cytoplasm. The solution of HRP is drawn into 5-μl disposable glass micropipets and stored in a freezer (−20°C): It is not necessary to seal the ends. HRP is stable for several months or more.

Glass tubes (10 cm long and 1 mm in diameter) with glass rod cores, used conventionally by electrophysiologists, are necessary for manufacturing injection pipets. An injection pipette is hand pulled from the glass tube. First, its central portion is attenuated to less than one-half of its initial diameter on a small gas flame of a micro-gas burner. Second, on a diminished gas flame, the attenuated portion is pulled apart. Micropipets with very long tapered tips (1 cm long), which tremble in ordinary air currents, are suitable. The tapered portion is broken with watchmaker's forceps at a point where it is about 3–5 μm thick (by checking in a microscope) to make a pipet with the tip of this size. The tapered portion works as a brake for the injection medium and, moreover, it ensures that the pipet can penetrate the elevated fertilization membrane.

The pipet thus prepared is loaded with the HRP solution by using a capillary, as follows. Five μl of the HRP solution is gently dropped on a piece of cleaned glass slide and drawn into a fine glass capillary with a diameter smaller than the inner diameter of the intact end of the injection pipet. The HRP solution is loaded into the back of the pipet. If small air bubbles develop in the solution, they should be removed completely by tapping the pipet with a finger. A small amount of paraffin oil is then layered on the HRP solution and the remaining part of the pipet is filled with distilled water (Fig. 8C). The pipet is mounted into the pipet holder.

Estimation of the volume of injected HRP solution is undertaken as follows. The tip of the pipet is dipped into a drop of paraffin oil and then a definite amount of hydraulic pressure (as monitored by the air bubble) is imposed for a certain period. By measuring the resulting diameter of the droplet of HRP solution in the paraffin oil, the amount of the pressure and the duration necessary for injecting a definite volume of solution are roughly determined beforehand to each pipet.

## C. Injection Procedure

To hold an embryo to be injected, two pieces of broken cover slip are placed at the periphery of the bottom of a plastic Petri dish (15 × 60 mm) so as to form a right-angled corner that opens toward the central portion of the dish (Fig. 8D). The dish is then filled with seawater and placed on the microscope stage, and the corner formed by the cover slip fragments is brought into the field of the microscope. Using a pipet, a small number of embryos are placed near the corner. The tip of the injection pipet is then brought into the field of the microscope. Using the tip end as a bar, an embryo to be injected is moved step by step into the corner by moving the dish with a hand. The embryo is rotated so that the blastomere to be injected is properly oriented (Fig. 8E and F).

For successful penetration by the injection pipet, it is necessary to adjust the slant of the pipet axis so that it is normal to the surface of the embryo (Fig. 8E and F). HRP solutions near the tip end, possibly diluted with seawater, are discarded by being ejected into water. Using the micromanipulator, the pipet is then pushed against the embryo until the fertilization membrane and the underlying cell surface are indented, forming a conical depression (one-third to one-half of cell diameter). Penetration into the blastomere is accomplished by a light tap on the arm of the manipulator or the iron base plate on which the manipulator is mounted. An adequate volume of HRP solution to be injected into a single blastomere is 5–10 pl, 2–4 pl, and 1–2 pl at the 2-cell, 4-cell, and 8-cell stages, respectively, when the total volume of the uncleaved egg is 2500 pl (165–175 μm in diameter, in the case of *Asterina pectinifera*). These amounts of HRP solution correspond to droplets 21–27 μm, 16–20 μm, and 12–16 μm, respectively, in diameter. HRP solution is injected by a predetermined pressure for a definite duration. To withdraw the injection pipet from the embryo, the pipet is gently raised. The embryo coming up with the pipet is released from the pipette at the surface of water. The floating embryo is found by spot lighting the water surface with a flashlight. The embryo is drawn into a micropipet, transferred to a Petri dish (10 × 35 mm) filled with seawater, and cultured up to the desired developmental stages.

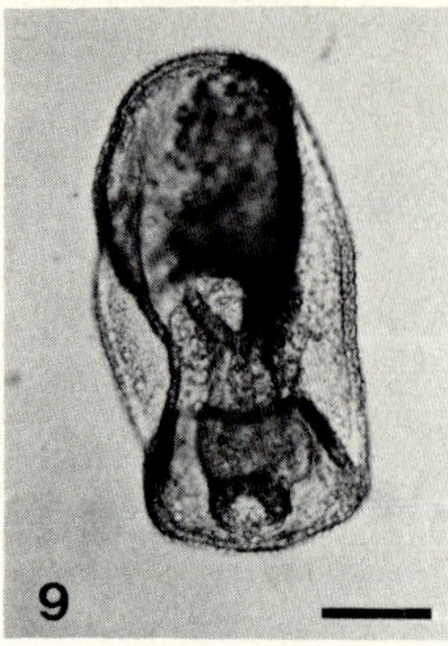


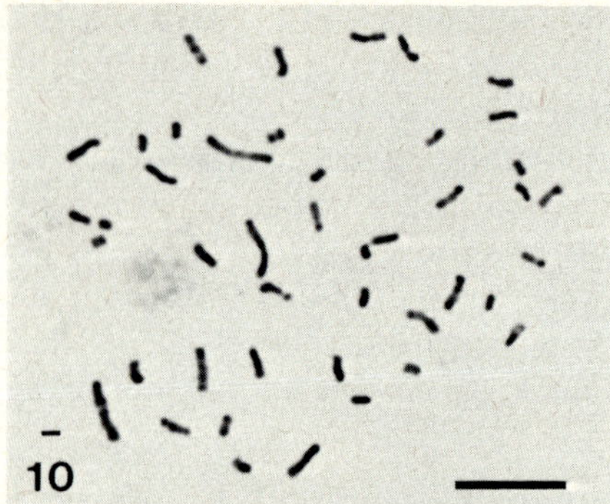


FIG. 9. A starfish (*A. pectinifera*) embryo injected with HRP. HRP has been injected into one of the blastomeres at the animal hemisphere of an eight-cell stage embryo and detected at the early bipinnaria stage. Reaction products are distributed at the anterior ventral side of the ectoderm. Bar, 100 μm.

FIG. 10. Chromosomes of a starfish (*A. pectinifera*) embryo stained with aceto-orcein. $2n = 44$. Bar, 10 μm. (From Obata and Nemoto, 1984).

### D. Detection of HRP and Observation

HRP-injected embryos are fixed with 2.5% glutaraldehyde dissolved in seawater for 1 hour at room temperature (about 20°C). They are then washed three times (10 minutes each) in 0.1 *M* phosphate-buffered saline (pH 6.4), and preincubated in about 5 ml of 0.03% 3,3′-diaminobenzidine dissolved in 0.1 *M* phosphate buffer for 20 minutes at room temperature. By adding a few drops of 0.1% $H_2O_2$ to the above solution, the enzyme reaction starts. By monitoring under a microscope, the enzyme reaction is allowed to progress until an adequate contrast is obtained, and is then stopped by promptly washing the specimen three times with 0.1 *M* phosphate buffer (Fig. 9). Since the reaction product is rather stable, the specimens can be observed not only as whole mounts but also in sections after paraffin embedding. Clarification in xylene is also possible.

## VI. Chromosome Staining

Chromosome preparations of starfish embryos are made according to the following method (Mita and Obata, 1984; Obata and Nemoto, 1984).

A fresh mixture of acetic acid and absolute ethanol (1:3) is used as the fixative. Chromosomes are stained with aceto-orcein: 1 g of orcein (Merck) is dissolved in 45 ml of hot glacial acetic acid and, after cooling, 55 ml of distilled water is added. This solution is paper filtered and stored. Further filtration through a filter paper is made just before use.

Fertilized eggs are denuded (see Section II) and cultured in an agar-based dish. When denuded eggs have developed to the 16- or 32-cell stage, they are suspended in $3\text{–}5 \times 10^{-4}$ *M* colchicine in seawater for 10–15 minutes. They are then transferred to about 5 ml of diluted seawater (seawater:distilled water = 3:7) to swell the blastomeres. As soon as a small number of blastomeres begin to burst, a few drops of the fixative are added. Blastomeres sedimented to the bottom are pipetted and suspended in the full-strength fixative in small test tubes with round bottom (35 × 6 mm) and fixed for several hours at room temperature (about 20°C). The fixative is replaced by the aceto-orcein solution and the materials are stained for 2–3 days: The top of the test tube is sealed with a piece of Parafilm M (American Can Co.) against evaporation. Stained blastomeres in one drop of the aceto-orcein solution are placed on a glass slide, and one drop of concentrated lactic acid is added. A cover slip is put on the materials, and excess solutions around the cover slip are blotted with filter paper. Another piece of filter paper is then placed on the cover slip and a rubber eraser is put on it. By pressing the eraser with one's thumbs, stained blastomeres are squashed between the glass slide and the cover slip. Preparations are sealed and cemented with a finger nail lacquer and then observed using a 100× oil-immersion objective (Fig. 10). The number of metaphase chromosomes is counted in photographs and/or sketches. A high-contrast film (for example, Minicopy HR II, Fuji Photo Film Co., Tokyo) is used for the photography. The glass surfaces must be dust free, since any dust may be mistaken for chromosomes.

## Acknowledgment

We would like to thank Dr. Mitsuki Yoneda for suggestions and critical reading of the draft.

## References

Bałakier, H., and Pedersen, R. A. (1982). *Dev. Biol.* **90,** 352–362.

Boveri, T. (1901). *Verh. Phys.-Med. Ges. Würzburg* **34,** 145–176.

Chambers, R., and Chambers, E. L. (1961). "Explorations into the Nature of the Living Cell." Harvard Univ. Press, Cambridge, Massachusetts.

Dan-Sohkawa, M. (1976). *Dev. Growth Differ.* **18,** 439–445.

Dan-Sohkawa, M. (1977). *Dev. Growth Differ.* **19,** 233–239.

Dan-Sohkawa, M., and Fujisawa, H. (1980). *Dev. Biol.* **77,** 328–339.

Dorée, M. (1982). *Exp. Cell Res.* **139,** 127–133.

Guerrier, P., and Dorée, M. (1975). *Dev. Biol.* **47,** 341–348.

Hirose, G., and Jacobson, M. (1979). *Dev. Biol.* **71,** 191–202.

Hörstadius, S. (1937). *Biol. Bull.* **73,** 295–316.

Hörstadius, S. (1973). "Experimental Embryology of Echinoderms." Clarendon, Oxford.

Ikeda, M., Nemoto, S., and Yoneda, M. (1976). *Dev. Growth Differ.* **18,** 221–225.

Kanatani, H. (1969). *Exp. Cell Res.* **57,** 333–337.
Kominami, T. (1983). *J. Embryol. Exp. Morphol.* **75,** 87–100.
Martindale, M. Q., and Brandhorst, B. P. (1984). *Dev. Biol.* **101,** 512–515.
Mita, I. (1983). *J. Exp. Zool.* **225,** 293–299.
Mita, I., and Obata, C. (1984). *J. Exp. Zool.* **229,** 215–222.
Mita, I., and Satoh, N. (1982). *J. Exp. Zool.* **223,** 67–74.
Morgan, T. H., and Spooner, G. B. (1909). *Arch. Entwicklungsmech. Org.* **28,** 104–117.
Nakamura, S., and Hiramoto, Y. (1978). *Dev. Growth Differ.* **20,** 317–327.
Nemoto, S. (1982). *Dev. Growth Differ.* **24,** 429–442.
Nemoto, S., Yoneda, M., and Uemura, I. (1980). *Dev. Growth Differ.* **22,** 315–325.
Nishida, H., and Satoh, N. (1983). *Dev. Biol.* **99,** 382–394.
Obata, C., and Nemoto, S. (1984). *Biol. Bull.* **166,** 525–536.
Prescott, D. M., and Kirkpatrick, J. B. (1973). *Methods Cell Biol.* **7,** 189–202.
Schroeder, T. E. (1980a). *Exp. Cell Res.* **128,** 490–494.
Schroeder, T. E. (1980b). *Dev. Biol.* **79,** 428–443.
Schroeder, T. E., and Stricker, S. A. (1983). *Dev. Biol.* **98,** 373–384.
Tennent, D. H., Taylor, C. V., and Whitaker, D. M. (1929). *Carnegie Inst. Washington Publ.* (391), 1–104.
Weisblat, P. A., Sawyer, R. T., and Stent, G. S. (1978). *Science* **202,** 1295–1298.
Yamamoto, K., and Yoneda, M. (1983). *Dev. Biol.* **96,** 166–172.
Yoneda, M., Ikeda, M., and Washitani, S. (1978). *Dev. Growth Differ.* **20,** 329–336.

# Chapter 19

# *Simple Methods and Devices for Handling Echinoderm Eggs*

**R. RAPPAPORT**

*Department of Biological Sciences*
*Union College*
*Schenectady, New York*
*and The Mount Desert Island Biological Laboratory*
*Salsbury Cove, Maine*

---



## I. Introduction

Many investigators who use echinoderm eggs and embryos in their research travel to a marine laboratory during the season when the echinoderms are reproductively active. Although the seasonal migration may yield a more reliable source of material in better condition and superior holding facilities, the investigator who uses bulky, complex equipment may find the chores of packing and unpacking, the risk of damage in transit, and the limitations of space which are sometimes encountered at marine laboratories to be restricting. The ability (which can be cultivated) to improvise simple, effective experimental equipment can help alleviate some of those difficulties. More importantly, the capacity to work independently of large pieces of major equipment sometimes helps the investigator to pursue the solution of the biological problem rather than the method.

The following limitations in the author's background have shaped this article: (1) All experience with echinoderm eggs has been related to investigations on cleavage mechanisms. (2) Techniques used in handling eggs and in micromanipulation have been acquired by personal trial and error, and by reading the literature. (3) All investigations have involved the use of inverted microscopes which provide easy access to the operation chamber and permit simple solutions to the problems of adequate working distance and evaporation of the medium.

Most of the tools and techniques used today for the manipulation of echinoderm eggs were devised and described, at least in principle, by previous generations of investigators. The purpose of this article is to describe several simple, potentially useful inexpensive variations on those techniques. Two very useful general references are de Fonbrune's "Technique de Micromanipulation" (1949) and a chapter by Chambers and Kopac in "McClung's Handbook of Microscopical Technique" (1950). Although the former emphasizes the use of the de Fonbrune microforge, it also contains much valuable information concerning the properties and handling of glass, and the ways in which an incandescent filament can be used to shape glass tools. The Chambers and Kopac chapter contains information on all aspects of "micrurgical technique." The simplicity of their approach is especially appropriate for this article. Their extensive discussion of freehand methods is valuable, although few students should expect to appreciate that a usable, hand-pulled capillary separates with ". . . a feeling much like that experienced when a taut thread, held in the fingers, is parted in a small flame." In the same handbook, Horstadius describes in a brief chapter how to make certain instruments used in freehand manipulations.

## II. Equipment for Making Instruments

### A. Alcohol Burner

A small, cool, clean flame is very useful for the shaping of glass instruments. Although small gas burners fashioned from or based in principle upon hypodermic needles are usually recommended for this purpose, an alcohol burner works just as well. It is also more portable and does not require careful regulation of gas pressure. The heart of the alcohol burner is the wick, which is shaped and held by the nozzle and lower 4–6 cm of a broken or chipped borosilicate measuring pipet. A disk-type powered glass tubing cutter can be used to cut off the pipet and to cut back the nozzle if the opening must be enlarged. Wicking (2077-H55, Thomas Scientific, Swedesboro, NJ) must be trimmed with scissors to match the contour of the inside of the pipet end and then stuffed into the pipet as far as possible. A small portion of the wick should also be pulled out through the nozzle with forceps or a needle. A 50-ml Erlenmeyer flask is an adequate alcohol reservoir. The overall height of the burner should be such that your wrists can rest on the bench top when you are working glass in the flame (Fig. 1). A hole large enough to admit the pipet end is drilled in a cork fitted to the neck of the flask. After the pipet end and wick are inserted through the cork, the flask is vented by cutting a slot along the entire length of the cork. The burner is fueled

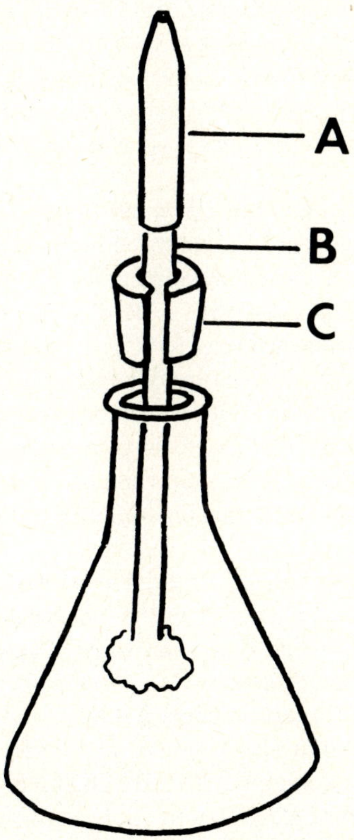


FIG. 1. Alcohol burner. (A) Cut off measuring pipet; (B) wick; (C) split cork.

with 95% alcohol. Progress of alcohol up a dry wick can be accelerated by shaking the flask. A small alcohol flame emits very little light, and it may be necessary to work against a dark background in order to observe carefully the relation between the flame and the glass. The flame of a burner of this design is hot enough to melt a 1-mm rod of borosilicate, as well as soft glass.

By working freehand with an open flame, it is possible to pull a preliminary taper in a glass needle that is to be finished in a microforge, or to make the bends in a finished needle that may be necessary to work around optics, the walls of the operation chamber, and temperature control devices. For this purpose, it is useful to make a full-sized pattern out of stiff wire to which the glass tool can be compared for the suitability of dimensions and angles of the shaft as it is made.

## B. Microforge

Although there is a certain satisfaction in working freehand, there are occasions when a mechanism to control the position of the glass stock, as well as the position and temperature of an incandescent filament, and to permit observation under significant magnification, can save time and material and require less skill. Ideally, it should be possible to inspect the functional tip of a tool under the same magnification that will be used in the subsequent operation before the tool is removed from the forge. The ideal is often not realized in commercial microforges.

A simple substitute for a commercial microforge, consisting of an incandescent filament with current control mounted on a microscope mechanical stage, can, in conjunction with a current control and micromanipulation equipment, yield excellent results (Cook, 1968). A length of platinum–iridium wire is mounted on a suitably insulated holder having the dimensions of a conventional microscope slide (Fig. 2). A power supply consisting of the transformer from a microscope illuminator, in series with a variable autotransformer (Variac), has proved perfectly satisfactory in emergencies, but a more compact substitute can be devised. A foot switch would be a convenience but has not proved essential. A similar device was at one time manufactured by the Stoelting Co., Chicago, IL.

In practice, this forge functions most satisfactorily when it is used to shape the operational tip of the tool. If the shank of the tool must be bent to accommodate the operation chamber of optics, it is usually easier and faster to work freehand with a small flame, at which time it is also possible to taper the end that will be worked with the filament. After the general dimensions of the needle have been established, it is mounted in a needle holder or pin vise and then inserted in the instrument holder of a micromanipulator. By combining movements of the manipulator and the mechanical stage it is possible to execute complex maneuvers, and these in combination with adjustment of filament temperature permit fairly

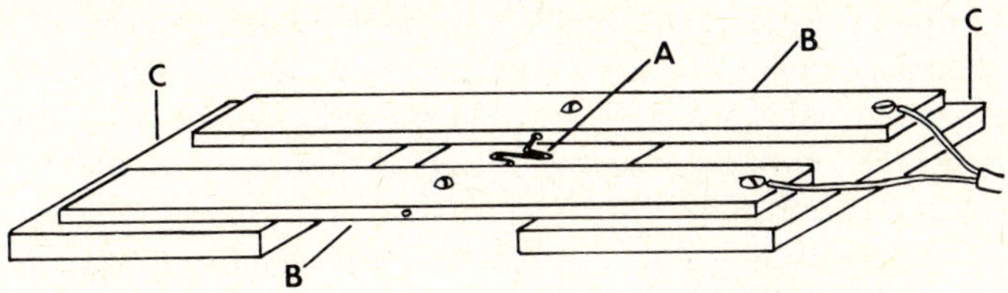


FIG. 2. Device for holding an incandescent filament on a microscope mechanical stage. The overall length and width are the same as those of a conventional microscope slide. (A) Platinum–iridium wire; (B) brass bar; (C) plastic block (Based on W. F. Cook, *J. Royal Micr. Soc.* **88,** 607–609 (1980).]

precise control of the form and dimensions of the tip. Because the arrangement is so simple and easy to install and remove, it can be mounted on the same microscope and manipulators that are used for work on biological material. Possible overheating of the objective lenses is a hazard, and it may be necessary to use a long working distance objective or work with a 10× objective and then inspect the tip after the filament cools with a shorter working distance, higher power objective.

## III. Fabrication and Care of Instruments

### A. Cutting Glass

Despite its fragility, glass is still the material of choice for most tools. Soft glass is easiest to work. Borosilicate glass is more durable. No matter how experienced the operator or durable the glass, tools will be accidentally broken before they wear out, so that learning how to work glass is an essential part of the art.

A good diamond point (such as Thomas Scientific 5680-D60, glass marker) can be very useful. Diamond points are fragile and should be protected when not in use. Even a well-made point may not cut equally well in all orientations, and it is worth learning to hold it so that the best face of the point is used in the customary cutting motion.

Thin glass rod or capillary also breaks cleanly and precisely when it is fused to a metal filament and then subjected to tension after it has cooled. De Fonbrune (1949) described and explored the possibilities of this method and showed how it can be used for various purposes. The technique requires that the filament be hot enough to fuse the glass to its surface but not hot enough to distort it. When the current through the filament is interrupted, the cooling results in the hardening of the glass–metal union and the overall contraction of the filament. The tension caused by filament contraction causes the glass to fracture at the point of fusion. The method requires that the glass not yield when it is subjected to tension. The glass being worked must therefore be straight and oriented parallel to the principal direction of contraction.

The methods for cutting rods and capillary just described presume that the smaller piece is to be discarded and the larger is to be retained. If the opposite is desired, additional steps are necessary. Short pieces of capillary (about 80 μm or longer) may be cut from a longer piece with a diamond point and saved if the capillary is placed on a short length of transparent tape that is coated with adhesive on both sides and which, in turn, is stuck to the surface of a microscope slide. With the work on the stage of a stereomicroscope, the capillary is scored with a diamond point and then flexed at the scored location by using watchmaker's forceps to grasp the capillary just distal to the score point. As each short length is broken, it may be left embedded in the adhesive material. The entire assembly of capillary, tape, and slide can be stored until required. Pieces of capillary embedded in adhesive for as long as a year separate freely from it when grasped and pulled with forceps. When short capillary segments are broken from a longer piece using de Fonbrune's method, they usually separate from the filament when it recoils, and fly rapidly beyond the control of the investigator. The trajectory of the segments is, however, predictable, and they can be caught by pieces of transparent tape placed in the line of flight. These "catcher's mitts" together with their embedded capillary segments and the stock from which they were broken can be stored indefinitely. This method has been used to make segments of 80 μm i.d. capillary that are 40 μm long.

Cover slips can be cut into small rectangles on the stage of a stereomicroscope if a thin straight edge is used to guide the diamond point. After the glass is scored, it must be turned over and flexed and broken by holding it next to the score line with wide-tipped forceps (like cover slip forceps) and then pushing downward against the microscope stage insert.

## B. Cements

Cements that contact eggs and media should adhere to glass, be nontoxic, be waterproof or water resistant, and be convenient to handle in very small amounts. It is also desirable that they harden rapidly, resist cleaning measures, and be transparent.

Hard waxes, including sealing wax, histological embedding paraffin and beeswax, are satisfactory in many respects, and the rapidity with which they form a rigid joint can be a great convenience. It is, however, sometimes difficult to use them in small amounts. Soft waxes, which are plastic at room temperature, form joints that are not as rigid as hard waxes but are sufficient to withstand the forces usually involved in cell manipulations. Mixtures of beeswax and colophony are available under various brand names. One of the more convenient and inexpensive cements is the soft paraffin used to seal domestic gouda cheese and other processed cheeses marketed in the United States. Soft waxes are useful as sealants as well as cements. They share the disadvantage with other waxes of being difficult to handle in very small amounts in the molten state.

Of the modern cements, epoxy glues form strong joints, but preliminary mixing and slow hardening can be inconveniences. Silicone rubber adhesives (e.g., Dow Corning's "silicone adhesive/sealant" or General Electric's "clear household glue and seal") have all the desirable qualities except rapid hardening. The joints that they form are elastic rather than rigid, but, depending on the specific application, elastic joints can be convenient and reduce breakage. Silicone rubber can also be cast in convenient shapes (see Section IV,C).

### C. Ultrasonic Cleaning

After repeated use, both tools and operation chambers become coated or plugged with hyaline layer material and cellular debris. Tools and chambers that withstand normal wear and tear usually withstand ultrasonic cleaning very well. The glass parts are rarely damaged and silicone adhesives as well as hard and soft waxes are unaffected. After treatment, the detergent can be rinsed off with running tapwater or by repeated immersion in a beaker of water.

The smallest commercially available cleaners are adequate for this purpose. Small units have the added advantage of making it possible to clean tools and large-bore pipets without detaching them from the micromanipulator or breaking a connection to the syringe. In practice the instrument holder can be swivelled away from the microscope stage and pointed downward. The operating cleaner is then held under the tool and raised until the tip is submerged in the detergent solution. The tip can also be rinsed by repeated submersion in a beaker of water.

## IV. Instruments and Devices

### A. Braking Pipet

The movement of fluid into a braking pipet is controlled by a small aperture that restricts the movement of air into and out of the part of the pipet that contains the fluid. When the size of the aperture is right, the inrush of fluid which would otherwise occur when the nozzle is submerged is prevented, yet small volumes of fluid can be aspirated and expelled from the pipet by changing the air pressure in the pipet by mouth.

Both capillary filaments and segments of thermometer tubing have been used for apertures. The pipets described here are based on capillary filaments. A chronic problem encountered in the use of the braking pipet is the clogging of the air capillary by droplets of moisture. When the pipet is used with seawater, the problem is compounded by the formation of salt crystals. It is ofter easier to make a new pipet than to properly clean an old one. This fact, plus the frequent need for a special purpose pipet with a nozzle or tip of a special form, makes the semidisposable braking pipet attractive.

The braking pipet described here consists of a modified piece of thin-walled capillary tubing mounted in a disposable Pasteur pipet with soft wax. One end of the capillary tubing is pulled by hand or with a needle puller, and the suitability of the aperture at the tip of the pulled out part is determined by dipping the unmodified end of the capillary in water. If the aperture is too large, the water rises rapidly. If the aperture is too small, the water does not rise at all. With a little experience, it is possible to judge the rate of rise that indicates a suitable aperture. If the aperture is too large, the capillary can be repulled by hand. If it is too small, it can be cut back with scissors. When the aperture is suitable, the other end of the capillary can be reshaped either freehand or with a microforge, according to the use to which the pipet will be put. The capillary is then inserted in the tapered end of a Pasteur pipet which has, if necessary, been cut back to provide a snug fit. This step should be done under a stereomicroscope in order to insure that the tapered end of the capillary is neither broken off nor clogged during the process. After the capillary is in the pipet, it can be sealed in place with either hard or soft wax, or silicone rubber adhesive, which is less vulnerable to heat (Fig. 3).

Braking pipets can be used to transfer or measure volumes of fluid or to manipulate individual eggs. They can also be used to reshape eggs by extrusion if the nozzle diameter is appropriately reduced. The gum rubber (latex) tubing used to convey pressure changes from mouth to pipet often tends to assume a particular curvature or "set." When performing delicate freehand operations with the

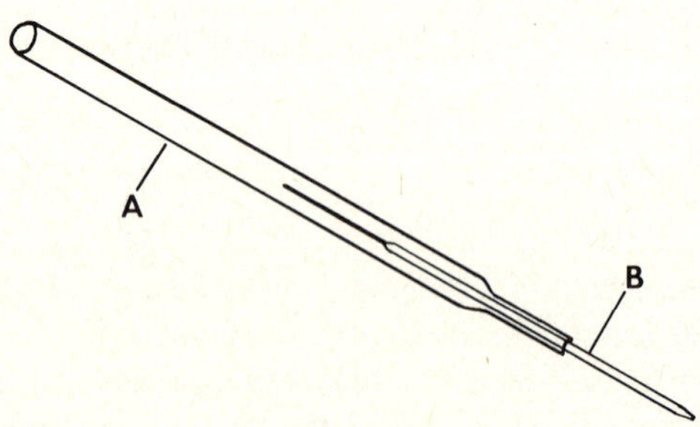


FIG. 3. Braking pipet. (A) Disposable Pasteur pipet; (B) modified length of capillary tubing.

pipet, it is worthwhile to ascertain the way in which the tubing is predisposed to curve and then adjust the pipet and other circumstances of the operation so that it is not necessary to overcome the resistance of the tubing while the manipulation is in progress.

## B. Operating Chamber

A complicated operating chamber is unnecessary when an inverted microscope is used. In its simplest form a chamber can consist of a glass ring (Thomas Scientific, Micro slide ring 6705-r24, 5 × 18 mm) to which is attached a cover slip, that serves as the bottom through which observations are made (Rappaport, 1968). If the chamber bottom is to be changed after each use, it can be secured with petrolatum or stopcock grease (silicone grease retains its viscosity and shape in an autoclave). Silicone rubber (which is autoclavable) or wax can be used for more permanent chambers. If wax is used, an incandescent light bulb can serve as a heat source to neatly secure the bottom. Two or three small slugs of soft wax are rolled between the fingers and placed on the edge of the ring to which the bottom will be secured. A cover slip is gently pressed into the wax and, with the cover slip upmost, the ring and cover slip are placed on the light bulb and the current is turned on. As soon as the wax melts and runs around the circumference of the wall, the current is turned off. With a little practice one can learn the amount of wax necessary to form a good seal without any surplus.

If it is necessary to insert the finished chamber in a small space that does not admit the fingers (like the well of a microscope stage temperature control device); it can be grasped by the edge with insect pinning forceps (Carolina Biological Supply Co., Burlington, NC, 65-5120) and lowered into place.

This simple operation chamber has several desirable features. It fits in the well of a thermoelectric temperature control device (Midland Ross, Electronic Connector Division, Cambridge, MA, formerly Cambion). It can be sealed with oil (silicone, halocarbon, and paraffin are possibilities), or oil can be used to flatten eggs while operations are performed with tools inserted through the oil (Rappaport, 1968). It can readily be cleaned ultrasonically and the observation surface can be replaced quickly and easily. Its principal disadvantage is that the height of the wall sometimes requires that one or two bends be made in the tool so that it clears the optics and the chamber wall.

## C. Silicone Rubber Capillary

Sometimes it is desirable to reshape echinoderm eggs into cylinders for experimental purposes. Although the fertilization membrane and hyaline layer appear to retain the form imposed on the egg immediately after fertilization (Lindahl, 1932; Rappaport, 1975; Presnov and Isaeva, 1983), the shapes obtained in this way are difficult to control, and the substances that are retained around the egg may make it difficult to penetrate. Eggs may also be reshaped into cylinders by inserting them in short lengths of glass capillary (Rappaport, 1981). Although the method is effective, glass capillaries are tedious to make and to maintain in a usable state. These problems can be avoided by using capillaries cast in a block of transparent silicone rubber and cemented to the bottom of an operation chamber (Rappaport, 1985).

The pattern for casting a capillary is a filament of polypropylene that has been checked for roundness and approximate diameter with a stereomicroscope. The source of the polypropylene can be a disposable micropipet tip that is melted in a shallow metal container (like an inverted metal jar cover) on a hot plate. A small amount of the molten mass is grasped with cool forceps and pulled. Select filament lengths with the desired diameters, mark them with a felt-tipped pen, and cut them from the remainder with a razor blade.

Before casting, place the polypropylene filament on a clean cover slip and position it so that the part with the desired diameter is centered on the cover slip. Pick up a lump-free droplet of transparent silicone rubber adhesive (such as Dow Corning) on the tip of a toothpick and spread it over the selected part of the filament, ensuring that the space between the filament and the cover slip is filled. Press a small sheet of Parafilm (American Can Co., Greenwich, CT) onto the surface of the droplet, smooth side down. If Parafilm is unavailable, a smooth piece of any other substance that will not bond to the adhesive should suffice. The Parafilm should be oriented as nearly parallel to the plane of the cover slip as possible.

After the rubber has cured (several hours or overnight), peel off the Parafilm. Gently grasp the filament and pull it out of the casting. Since the filament is unaffected by the process, it can be labeled, stored, and reused many times. The rubber block can then be trimmed into regular form with a razor blade. In order to faciliate insertion of the egg, the face of the block containing one end of the capillary should be perpendicular to the surface of the cover slip and the face of the block containing the other end of the capillary should be slanted at about 45° (Fig. 4). The method for inserting an egg in a capillary will be described in Section V,B.

## D. Micromanipulator

R. J. Goldacre (1954) described a simple, inexpensive, small micromanipulator that allows fine movement of the tool, eliminates hand tremor, and fixes the position of the tool. In this ingenious device the muscles of the hand work against uniform resistance; the resistance is provided by a viscous layer of grease. It may be fashioned from pieces of glass (microscope slides) cemented together with sealing wax, or a more permanent version can be made of sheet brass with

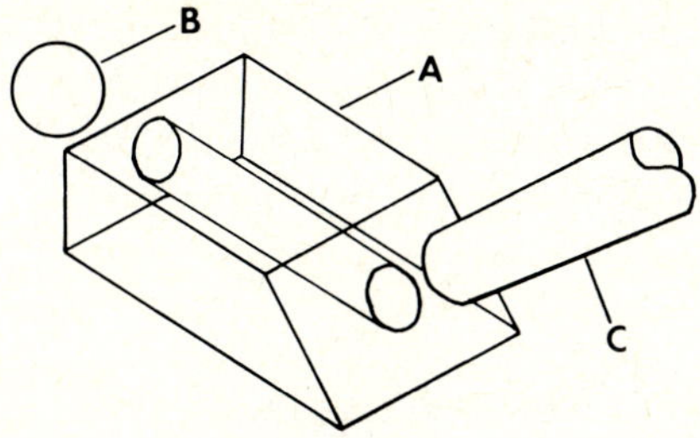


FIG. 4. Capillary cast in transparent silicone rubber. (A) Silicone block; (B) egg; (C) tip of braking pipet. (Based on R. Rappaport, 1985.)

soldered joints. The horizontal motion is achieved by sliding an inverted T-shaped assembly of glass pieces in a small quantity of grease on one end of a microscope slide. The vertical motion is achieved by sliding a piece of glass in grease on the surface of the vertical portion of the inverted T. The tool is cemented to the glass piece which moves in the vertical plane. Goldacre cemented the pieces of the T with sealing wax, which forms a strong bond with clean glass and is bulky enough to be mechanically self-reinforcing. In the original description, the viscosity of the grease was varied by combining petrolatum with different proportions of paraffin wax. Silicone greases are now available in which the viscosity is not changed by the heat of the hand (Fig. 5).

In use, one or two devices can be put on a single slide that is then mounted on the microscope stage, or more devices can be mounted on a larger piece of glass. The sliding components can be used to achieve motion in all planes or, with a somewhat different arrangement, the sliding components can be used to center the tip of the tool in the field and to position the tip at the described height above the stage, and then the mechanical stage of the microscope can be used for horizontal motions. This technique was used in some of the very early micro-

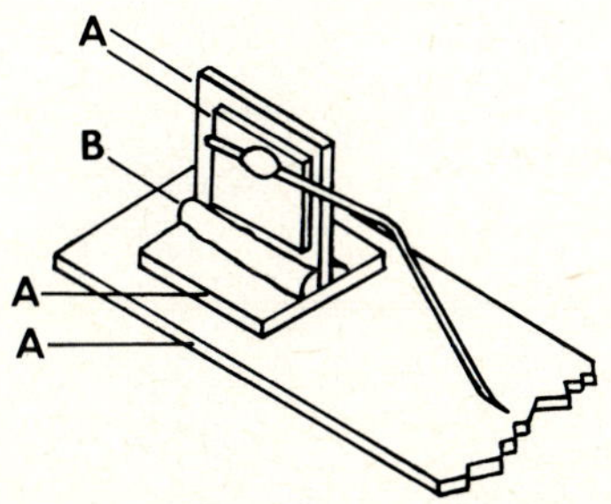


FIG. 5. Simple micromanipulator. (A) Pieces of glass microscope slide; (B) sealing wax. (Based on Goldacre, 1954.)

manipulation studies; it ameliorates the problem of image reversal and tool movement. It may also be possible to cope with the problem of image inversion by placing a simple image erector prism (available from Edmund Scientific Co., Barrington, NJ) in the optical axis of the microscope.

# V. Manipulations

## A. Mechanical Denuding

Physical removal of jelly and fertilization membrane can be used to avoid problems associated with chemical treatment and to conserve experimental material. The following method, which is slightly modified from that described by Showman and Foerder (1979), has proved very satisfactory for denuding small batches of eggs for micromanipulation studies. The materials required are a 2-ml syringe with a Luer lock tip, a disposable syringe needle with a plastic hub compatible with the Luer lock tip, and a small piece of Nitex bolting cloth with a mesh size slightly smaller than the size of the egg. To assemble the device, cut the tip off the syringe needle and retain the hub. Lay a small square of Nitex across the syringe nozzle and then push the hub onto the nozzle over the Nitex and lock it in place (Fig. 6). To prepare the device for use, wet the plunger and the inside of the barrel and then position the barrel so that the tip is under seawater in a small dish. The side of a $2\frac{1}{2}$ in. culture dish (Carolina Biological Supply Co., No. 74-1000) is high enough to hold the barrel at an angle of about 30° with the tip resting on the bottom. Shortly after fertilization (about a minute), pipette the eggs into the upper opening of the barrel and then insert the plunger

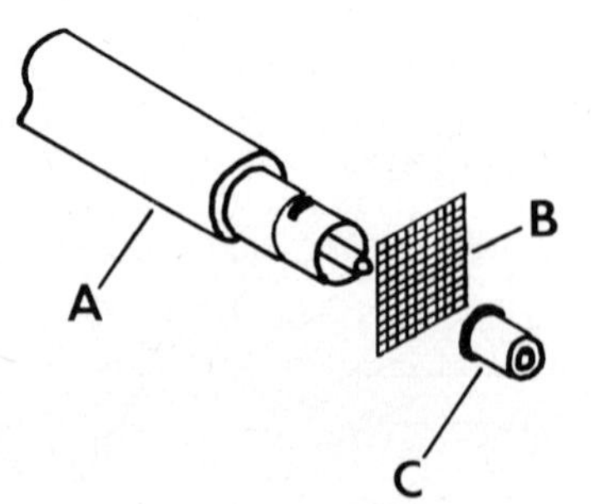


FIG. 6. Denuding syringe. (A) Barrel of 2-ml syringe with Luer lock tip; (B) Nitex; (C) plastic hub of disposable syringe needle.

and push, keeping the Nitex submerged. The denuded eggs can be centered by gently swirling the culture dish in small circles.

The effect of this treatment on the eggs depends upon the rate at which they are expelled from the syringe, as well as the relation between the mesh size and the toughness and diameter of the fertilization membrane and the dimensions and properties of the jelly. The method yields a satisfactory proportion of denuded, undistorted *Echinarachnius parma* eggs (diameter about 135 μm) when they are gently forced through 132-mesh Nitex. The syringe and Nitex should be back-flushed with tap water after each use.

## B. Inserting Eggs in Short Capillaries and Glass Loops

Experimental design sometimes requires that the egg be reshaped and then held in the imposed form. The method can be used to change the geometrical relations between some of the parts, as well as to hold the eggs in place during an operation. Imposition of cylindrical form has the virtue of altering internal geometrical relations and fixing certain dimensions, while at the same time leaving the egg accessible for operations if the confining mold is open at both ends.

The technique used to insert eggs in glass and silicone rubber capillaries requires a braking pipet. For glass capillaries it is necessary to bend the tip of the pipet so that it is parallel to the bottom of the operation chamber during the procedure (Rappaport, 1982). After the capillary has been submerged in the chamber and the air bubble that is usually trapped in the lumen has been sucked out with the pipet, place an egg next to the opening at one end of the capillary. Then place the pipet nozzle against the opening at the opposite end and apply suction. If the capillary diameter is suitable, the flow of medium will carry the egg into the capillary. If silicone rubber capillaries are shaped as described above, it is not necessary to bend the tip of the pipet because the end of the capillary to which the pipet should be applied is slanted (Fig. 4).

For the purposes of this description, glass loops may be considered to be very short glass capillaries. They can be used to hold eggs and change their dimensions while at the same time leaving most of the surface accessible for manipulation. Glass loops can be made by the technique of de Fonbrune in which a glass rod is pushed against a filament that is hot enough to soften it but cool enough to discourage adhesion. The finished loop is oriented so that the loop is perpendicular to the chamber bottom and then secured to it with a droplet of silicone rubber adhesive (Rappaport and Rappaport, 1984). The egg which is to be inserted in the loop is placed next to it on one side and a braking pipet with a bent tip and an orifice of about the same diameter as the loop is applied to the other side (Fig. 7). It is usually necessary to suck the egg part way through the loop and into the pipet nozzle and then gently expel it as the pipet is moved away.

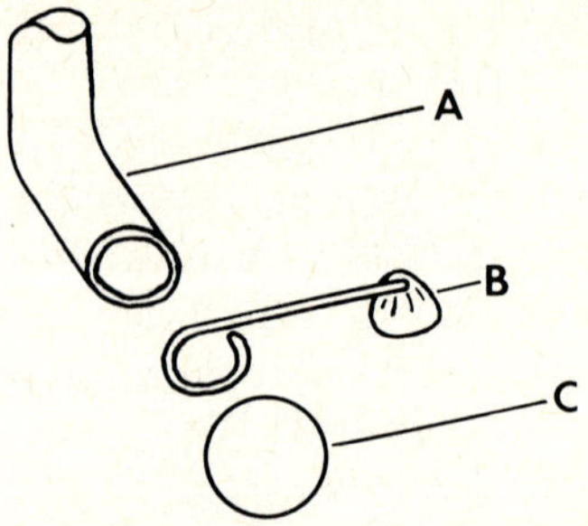


FIG. 7. Arrangment for inserting an echinoderm egg in a glass loop. (A) Tip of braking pipet bent parallel to chamber bottom; (B) glass loop secured to chamber bottom at the end of the shank with a droplet of silicone rubber; (C) egg.

## VI. Summary

This article describes several simple methods and easily made devices that have proved useful in experimentation on echinoderm eggs. It may be that some of the suggestions will prove useful for teaching purposes as well as for research. Beyond providing detailed instructions for a few methods, it is just as important to stress that simple, direct methods can still yield valuable data in important areas of experimentation.

### ACKNOWLEDGMENTS

This work was supported by Grant #PCM-8404174 from the National Science Foundation. The skilled assistance of Barbara N. Rappaport is gratefully acknowledged.

### REFERENCES

Chambers, R., and Kopac, M. J. (1950). *In* "McClung's Handbook of Microscopical Technique" (R. McC. Jones, ed.), pp. 492–543. Hoeber, New York.

Cook, W. F. (1968). *J. R. Microsc. Soc.*, **88,** 607–609.

deFonbrune, P. (1949). "Technique de Micromanipulation." Masson, Paris.

Goldacre, R. J. (1954). *Nature (London)* **173,** 45.

Lindahl, P. E. (1932). *Wilhelm Rouxs Arch. Entwicklungsmech. Org.* **127,** 300–322.

Presnov, E. V., and Isaeva, V. V. (1983). *Ontogenesis* **3,** 319–323.

Rappaport, R. (1961). *J. Exp. Zool.* **148,** 81–89.

Rappaport, R. (1968). *Embryologia* **10,** 115–130.

Rappaport, R. (1975). *In* "Molecules and Cell Movement" (S. Inoué and R. E. Stephens, eds.), pp. 287–303. Raven, New York.

Rappaport, R. (1981). *J. Exp. Zool.* **217,** 365–375.

Rappaport, R. (1982). *Cell Differ.* **11,** 359–361.

Rappaport, R. (1985). *J. Exp. Zool.*, **234,** 167–171.

Rappaport, R., and Rappaport, B. N. (1984). *J. Exp. Zool.* **231,** 81–92.

Showman, R. M., and Foerder, C. A. (1979). *Exp. Cell Res.* **120,** 253–255.

# Chapter 20

# *Induced Fusion of Echinoderm Oocytes and Eggs*[1]

S. G. VASSETZKY AND M. N. SKOBLINA

*Institute of Developmental Biology*
*USSR Academy of Sciences*
*Moscow, USSR*

G. G. SEKIRINA

*Experimental Station "Gorki Leninskiye" of the*
*Institute of Applied Molecular Biology and Genetics*
*Lenin Academy of Agricultural Sciences*
*Moscow, USSR*

---



[1]With an Appendix, "Methods for Maintaining Starfish and Obtaining Gametes and Embryos Year Round," by P. V. Davydov, O. I. Shubravyi, and S. G. Vassetzky.

# I. Introduction

The eggs of echinoderms, especially sea urchins, are widely used in cell and developmental biology due to the homogeneity of their population, well-developed methods of obtaining gametes and artificial insemination, synchrony of embryonic development, etc. (for reviews see Harvey, 1956; Giudice, 1973). At the same time, the small size of eggs complicates the application of microsurgery, which allowed very important advances to be obtained in experimental embryological studies on the eggs of lower vertebrates (for a review see Diberardino, 1980). In studies on echinoderm eggs, microsurgery can in some cases be replaced by the method of induced cell fusion and/or by a combination of cell fragmentation by density gradient centrifugation with subsequent induced fusion. Similar experiments have already been conducted on mammalian eggs and blastomeres (for a review see Vassetzky and Sekirina, 1985).

The possibility of egg fusion in echinoderms was corroborated by the spontaneous and rare appearance of fused sea urchin eggs after treatments designed to remove the fertilization membrane (see Bennett and Mazia, 1983). The first successful experiments on induced fusion of echinoderm eggs were carried out in 1981: Hybrids between mature (fertilized and nonfertilized) eggs of sea urchins (intra- and interspecific) were obtained (Bennett and Mazia, 1981a,b, 1983; Richter *et al.*, 1981).

We have first used starfish oocytes for induced fusion in order to obtain mass materials for studying the role of the nucleus and nucleocytoplasmic relations during oocyte maturation and early development. The methods of obtaining gametes in starfishes and of inducing oocyte maturation were already developed in detail (for reviews see Kanatani, 1973, 1979; Meijer and Guerrier, 1984).

All experiments were carried out at the Biological Station "Vostok" of the Institute of Marine Biology, Far East Science Center of the USSR Academy of Sciences (Vostok Bay, Sea of Japan). On the other hand, the potential for inland research of this type now exists; it has recently been shown that starfishes can be maintained in a closed cycle aquarium with artificial seawater and that gametes and embryos can be obtained from them year round (Davydov *et al.*, 1986; see also Appendix).

# II. Fusion between Starfish Oocytes and between Oocytes and Cytoplasts

## A. Characteristics of the Object

Our experiments were conducted on a starfish *Aphelasterias japonica* Bell, which is easily available in the Vostok Bay. In this region *A. japonica* spawns in

July–August at water temperatures of 14–18°C (our observations). The gonads are bright orange in females and cream-colored in males. Fecundity attains 1.5 million eggs in *A. japonica* (Kasyanov *et al.*, 1980). During the season of spawning, the oocytes are uniform in size (111–125 μm), and mature under the influence of 1-methyladenine in virtually 100% of cases. The percentage of "spontaneously" matured oocytes is low (no more than 10% in most cases).

## B. Preparation of Oocytes for Fusion

The technique of induced cell fusion calls for the solution of three problems: (1) the removal of all envelopes peripheral to the plasma membrane, i.e., the removal of the follicular envelope and vitelline membrane; (2) the agglutination of the oocytes; and (3) the fusion of the apposed membranes (Bennett and Mazia, 1981a).

All procedures described below were aimed at the solution of these problems (Skoblina *et al.*, 1982; Sekirina *et al.*, 1983). They were conducted at temperatures of 13–20°C, in an air-conditioned room when necessary, with the use of artificial seawater. The use of artificial seawater allows various surprises to be avoided, such as pollution (both from the shore and from the open sea) or dilution (as a result of abundant rains). There are many recipes for artificial seawater, and we used one of them often utilized in experiments on the unicellular alga *Acetabularia* (Betina *et al.*, 1973):

26.05 g/liter NaCl, 5.2 g/liter $MgCl_2 \cdot 6H_2O$, 4.48 g/liter $MgSO_4 \cdot 7H_2O$, 1.43 g/liter $CaCl_2$, 0.69 g/liter KCl, 0.198 g/liter $NaHCO_3$, pH 7.8.

The artificial seawater was prepared in distilled water using a mixture of weighed portions of the ingredients (except $NaHCO_3$). All solutions for further procedures were prepared in this artificial seawater.

About 10 starfishes collected in the sea were brought to the laboratory. Each of them was cut with scissors along one of the arms (small cut). Forceps with curved tips were used to tear off a small portion of the gonad. If the sperm was active, one gonad was taken out of this male, placed in a dry Petri dish, and preserved in a refrigerator. In these conditions the sperm retained their fertilizing capacity for 24 hours. All males were returned to the sea. All pieces of ovaries were examined under the microscope. Ovaries containing many small-sized or damaged oocytes were sorted out. From the chosen females (usually two or three), the gonads were taken out from one arm. All females were returned to the sea.

Oocyte suspensions were prepared from the gonads of the chosen females. The ovaries were minced with scissors in a small volume of seawater and the suspension was filtered through gauze with a mesh size of 0.2 mm. The suspension containing the fewest damaged oocytes was chosen for the experiments.

The oocytes remained viable in the suspension for 10–12 hours if kept in loosely closed vessels in a refrigerator at 4–6°C. The vessels, e.g., 200-ml beakers, usually contained 1 cm of settled oocytes in 100 ml of seawater. After such long storage, many oocytes matured under the influence of 1-methyladenine, could be fertilized, and endured various treatments preceding and accompanying the induced fusion.

We washed the oocytes with seawater and other solutions, usually in a hand centrifuge with four 10-ml test tubes since this, on one hand, markedly accelerated all procedures (e.g., the three-fold washing of 40–50 ml of oocytes usually required less than 3 minutes) and, on the other hand, ensured the delicate handling of the oocytes and eggs.

The follicular envelopes and vitelline membranes of the oocytes were removed as follows. The initial oocyte suspension (~5% by volume) was centrifuged in a hand centrifuge and about 5 ml of Ca-free seawater was added to a ~ 1 ml of settled oocytes.

The oocytes were resuspended in Ca-free seawater and centrifuged, the supernatant was discarded, and a new portion of Ca-free seawater was added to the oocytes. The state of the oocytes in the suspension was monitored under the microscope. The follicular cells lost their contact with the oocytes usually within a few minutes after the beginning of treatment with Ca-free seawater. Later, within 10–15 minutes, the vitelline membrane began to break away. At this stage the suspension was centrifuged, the supernatant was discarded, and 1 ml of 0.25–0.5% trypsin (Spofa, Czechoslovakia) was added. The duration of trypsin treatment varied from 10 to 25 minutes. The treatment was terminated usually when the oocytes began to form small aggregates and a few cells lost their normal form and became spindle shaped. During both these treatments the test tube containing the oocyte suspension was regularly shaken. After the trypsin treatment the oocytes were washed three or four times in 10 ml of seawater.

The first stage of treatment, removal of follicular envelope and vitelline membrane, was often conducted on all previously chosen oocyte suspensions and the best of them was chosen for further treatments. Such a procedure complicates experimentation but provides better results in the end.

The denuded oocytes were then exposed to 1 *M* urea (Sigma) with 1 m*M* $CaCl_2$ for 30 seconds. The oocytes were washed twice with seawater and, without any delay (!), treated with polyethylene glycol (see Section II,E).

The treatment with urea was proposed by Bennett and Mazia (1981a) to enhance the adhesion of the eggs to each other at the next step. In our experiments this procedure restored the cell turgor, led to the disappearance of aggregates, and, apparently, reduced the damaging effect of the subsequent treatment with polyethylene glycol dissolved in a Ca- and Mg-free seawater.

## C. Obtaining Mature Eggs

The maturation of oocytes was induced by 1-methyladenine (Sigma) at concentrations of $1 \times 10^{-4}$ to $1 \times 10^{-5}M$. The stock solution of 1-methyladenine ($1 \times 10^{-2}M$) was made up in distilled water and kept at $-10°C$. Before use, the stock solution was thawed and diluted with seawater. The oocyte suspension can be treated with equal success with 1-methyladenine before and after removal of external membranes. In the former case, all above described procedures were conducted after the germinal vesicle breakdown in the oocytes.

## D. Obtaining Cytoplasts

Cytoplasts were obtained by a modified method of Dyban *et al.* (1983). After removing the follicular envelope and vitelline membrane the oocytes were preincubated for 30–45 minutes in 5% Ficoll solution with 5 μg/ml cytochalasin B (Sigma). The preincubation in cytochalasin B for this duration is quite sufficient, so that one should already have the Ficoll density gradient and a centrifuge prepared for experiments. The centrifugation procedure should be modified for every new species to be used in such experiments (Vassetzky *et al.*, 1984). In our case, to determine the zone of sedimentation of the starfish oocytes and to choose the optimal form of the Ficoll gradient, the suspension of denuded oocytes (in 5% Ficoll, $M_r$ 400, Pharmacia, Sweden) was layered on a linear Ficoll gradient (15–30%, by weight) and centrifuged (K-23, East Geramny) at 2500 rpm for 20 minutes. The oocytes formed a compact layer at the 22% Ficoll zone.

A stepwise Ficoll density gradient was prepared in 35-ml nitrocellulose (transparent) centrifuge tubes. The total volume of the gradient was roughly 13 ml. The gradient was prepared using a programmable gradient pump "Mikrotechna" (Czechoslovakia) or a standard peristaltic pump: 4.5 ml of 22% + 6.6 ml of 17% Ficoll solutions with 5 μg/ml cytochalasin B. Such a gradient was chosen for the oocytes to pass rapidly across the upper layer of 17% Ficoll and to be sharply slowed down in the zone of sedimentation at the 17–22% interface, thus promoting oocyte fragmentation. This gradient can also be prepared easily with a standard 10-ml syringe fixed on a support.

The preincubated oocyte suspension (1–2 ml) in 5% Ficoll was layered onto this gradient and diluted with seawater to 35 ml. The centrifugation was performed in a VAC-602 (East Germany) in a bucket rotor with three tubes at 13,000–14,000 rpm (21,000 *g*) for 25 minutes. Upon ultracentrifugation of a suspension of oocytes in this gradient, two fractions, one with germinal vesicles (near the 5–17% interface) and one with cytoplasts (near the 17–22% interface) were obtained (Fig. 1). A small number of germinal vesicles was found above the layer of cytoplasts and mixed with them upon separating the cytoplasts. The mixture was, however, readily washed out. The majority of the cytoplasts did not

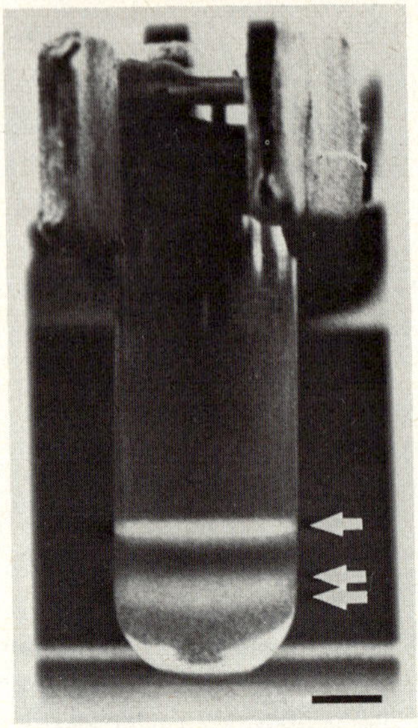

FIG. 1. The test tube after the ultracentrifugation (VAC-602, 21,000 g, 25 minutes) of the suspension of denuded oocytes of *Aphelasterias japonica*, pretreated in cytochalasin B (5 μg/ml) in the stepwise Ficoll gradient (17/22%). Single arrow: layer of germinal vesicles; double arrow: layer of cytoplasts. Scale bar, 1 cm.

differ markedly in diameter (97–105 μm) from the whole oocytes (111–125 μm), but smaller fragments (28–70 μm) were also observed. The cytoplasts differed markedly from the whole oocytes in color intensity and were, hence, easily identified both in the mixed suspensions and products of fusion.

Cytoplasts can also be obtained from oocytes (with intact germinal vesicles) by centrifugation in a sucrose density gradient. We will briefly outline here two sucrose gradients developed for two starfish species. Nemoto *et al.* (1980) preincubated the oocytes of *Asterina pectinifera* with jelly coats in seawater with 10 μg/ml cytochalasin B for 10 minutes, layered this suspension on a sucrose density gradient which consisted of 0.7 *M* sucrose + 1 *M* sucrose/seawater mixture (2:1 + 3:2 + 1:1) with 3 μg/ml cytochalasin B and centrifuged it at 14,000 *g* for 20 minutes. The obtained enucleated oocytes (cytoplasts) were somewhat stretched but rounded again within hours.

Malchenko *et al.* (1984) preincubated intact, full-grown oocytes of *Aphelasterias japonica* in seawater with 10 μg/ml cytochalasin B, and layered 0.2–0.3 ml of this suspension on a sucrose density gradient: 1.5 ml 0.95 *M* sucrose + 15 ml of 0.95 *M* sucrose/seawater mixtures (2:1 + 3:2 + 1:1) with 3 μg/ml cytochalasin B and centrifuged it at 15,000 *g* for 20 minutes. The cytoplasts accumulated in the lower layer of the gradient, while the "karyoplasts" were located in the upper layer.

After centrifugation was complete, the upper layers of the Ficoll gradient with the layer of germinal vesicles were carefully removed with a peristaltic pump or a Pasteur pipet so that the layer of cytoplasts was left intact. The layer of cytoplasts was transferred by Pasteur pipet in a cytochalasin B-free 5% Ficoll solution and the cytoplasts were washed three times in this solution.

While the oocytes were fragmented into cytoplasts and germinal vesicles and the cytoplasts were isolated from the Ficoll density gradient after centrifugation, the denuded oocytes were kept in 5% Ficoll solution at 10°C. After a pure suspension of cytoplasts was obtained, the oocytes and cytoplasts were mixed in about equal ratio.

## E. Treatment with Polyethylene Glycol

We have used polyethylene glycol (PEG) of different molecular mass produced by different companies, first of all a "6000 PEG," since such a PEG was effective on sea urchin eggs, and a 1450 PEG since this effectively induced the fusion of mammal oocytes and eggs (Dyban *et al.*, 1981; Sekirina, 1985). In the preliminary experiments, the oocytes fused under the influence of the 6000 PEG (Loba, Austria, or Merck, West Germany) whereas the 1450 PEG (Serva, West Germany) destroyed the oocytes. In most experiments we used the 6000 PEG (Loba).

We used 50% (w/w) PEG solution prepared on a Ca- and Mg-free seawater. To preserve the ionic strength of the solution, $MgCl_2$ and $CaCl_2$ were replaced by 2.25 g/liter NaCl and $MgSO_4$ was replaced by 2.58 g/liter $NaSO_4$.

About 0.25 ml of the prepared PEG solution was poured into a 5-ml plastic test tube [Flow Lab., Falcon #4-2058-O-(K)] and an approximately equal volume of 4–6% (by volume) suspension of oocytes or of a mixture of oocytes and cytoplasts was layered over it. The tube was then carefully inclined to a nearly horizontal position, so that the PEG solution occupied three-quarters of the tube length with the suspension above it. In this orientation the tube was carefully rotated by hand for 1 minute during which time the oocytes settled down into the PEG solution. The tube was then returned to the vertical position and rotated by hand for 30 seconds and the suspension was gradually diluted for 30 seconds to a final volume of 5 ml. The suspension was washed three times with seawater, transferred into a 5% Ficoll solution, and poured out in a plastic Petri dish (Flow Lab., Falcon #3001). On examining the suspension of oocytes just after PEG treatment, a few fusion products were usually found. At this stage the process of fusion was easily observed under the dissection microscope (magnification up to ×50) (Fig. 2).

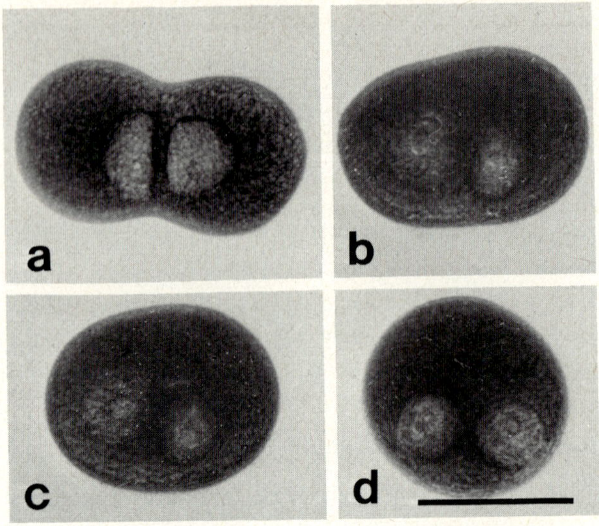


FIG. 2. Consecutive stages of fusion of two oocytes of *Aphelasterias japonica*. (a) Advanced contact of oocytes; (b and c) formation of a fusion product; (d) definitive fusion product (note its spherical form). Scale bar, 100 μm.

## F. Products of Fusion[2]

The time taken from the initial contact between oocytes until a single spherical fusion product is formed can vary from several minutes to an hour or more. Soon after PEG treatment and during the subsequent 4–5 hours, the fusion products showed neither marked morphological changes or discernible mixing of cytoplasms of the fused oocytes, and they remained viable.

The frequency of fusion in the above-described conditions attained 12%[3] for the oocytes (Skoblina *et al.*, 1982; Sekirina *et al.*, 1983) and 16% for the oocytes and cytoplasts (Vassetzky *et al.*, 1984).

The frequency of fusion in our first experiments with starfish oocytes was much less than that indicated above (Skoblina *et al.*, 1982; Sekirina *et al.*, 1983). It is possible our later experiments (Vassetzky *et al.*, 1984) benefited from the

[2]In our previous publications (Skoblina *et al.*, 1982; Sekirina *et al.*, 1983; Vassetzky *et al.*, 1983, 1984) we referred to the products of fusion between oocytes, oocytes and eggs, and oocytes and cytoplasts as cell hybrids. But since the cells of the same type were involved in these procedures we prefer to use the term "fusion products."

[3]The incidence of fusion indicated in our first papers (Skoblina *et al.*, 1982; Sekirina *et al.*, 1983) did not exceed 6% but these values were calculated per total number of cells rather than per total number of pairs of cells and cell fragments as was done by us later (Vassetzky *et al.*, 1984) and in the most other publications on induced cell fusion.

addition of Ficoll to seawater to a final concentration of 5%, leading to an improved morphological condition of the cells both during preincubation and after fusion procedures and subsequent observations. The stabilizing effect of Ficoll, as well as of some other high-molecular-weight compounds, on mammalian cells is well known. In particular, Ficoll was successfully introduced in place of albumin in the medium for preimplantation mouse embryos (Brinster, 1968). In our experiments we likewise preferred to use gradients containing Ficoll for fragmenting of the oocytes, rather than the sucrose, because of the stabilizing effect of Ficoll.

Selection of fusion products is one of the most complex problems when using the method of induced fusion, especially if morphologically similar cells are induced to fuse. In somatic hybridization studies modified media are used in which only cell hybrids can survive. In our experiments we selected the fusion products manually.

The products of fusion of two or more oocytes are easily distinguished from the unfused oocytes by the number of germinal vesicles and overall size. The number of germinal vesicles in fusion products varies from 0 to 4 (Fig. 3), but more than 80% of the fusion products contained two, i.e., the majority of fusions take place between pairs of oocytes (Table I).

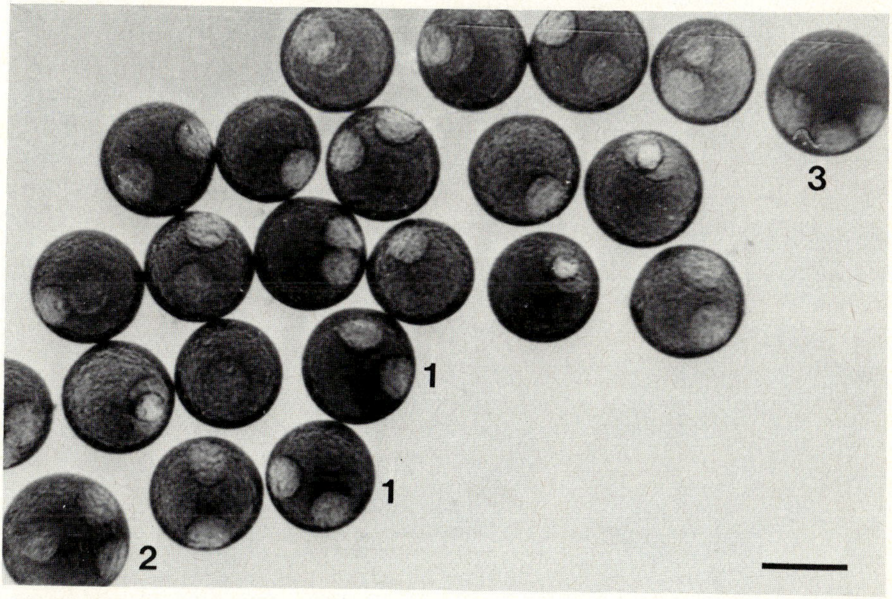


FIG. 3. Fusion products taken from a suspension of *Aphelasterias japonica* oocytes after treatment with PEG. (1) Fusion products with two germinal vesicles; (2) fusion product with three germinal vesicles; (3) fusion product with four germinal vesicles. Scale bar, 100 μm.

TABLE I

PRODUCTS OF FUSION BETWEEN THE *Aphelasterias japonica* OOCYTES CONTAINING VARIOUS NUMBERS OF GERMINAL VESICLES (GV) (SKOBLINA *et al.*, 1982)

| Number of experiment | Number of fusion products: Without GV | With 1 GV | With 2 GV | With 3 GV | With 4 GV | Total |
|---|---|---|---|---|---|---|
| 4 | 1 | 1 | 7 | — | — | 9 |
| 5 | — | 2 | 18 | — | — | 20 |
| 7 | — | — | 14 | — | — | 14 |
| 8 | — | — | 82 | 8 | 1 | 91 |
| 9a | 1 | — | 71 | 3 | 1 | 76 |
| 9b | — | — | 14 | — | — | 14 |
| 10 | 3 | 34 | 185 | 13 | 1 | 236 |
| Total: | 5 | 37 | 391 | 24 | 3 | 460 |
| Percentage: | 1.1 (0.3–2.3) | 8 (4.6–9.0) | 85 (83.1–88.9) | 5.2 (3–6.7) | 0.7 (0.1–1.7) | |

[a] Note: Confidence intervals for $p = 0.05$ are indicated in parentheses (Yanko, 1961).

As already noted, no marked mixing of the cytoplasms of the fused oocytes was observed, and this was clearly seen both just after fusion (Fig. 4a) and during 4–5 hours of observation (Fig. 4b and c). One can see both the boundary between the respective cytoplasms and differences in the color or texture of the cytoplasm. This is especially important for selecting the products of fusion between the oocytes and the cytoplasts or between the cytoplasts. The fact is that the cytoplasts are, as a rule, lighter in color than the oocytes, and upon fusion of

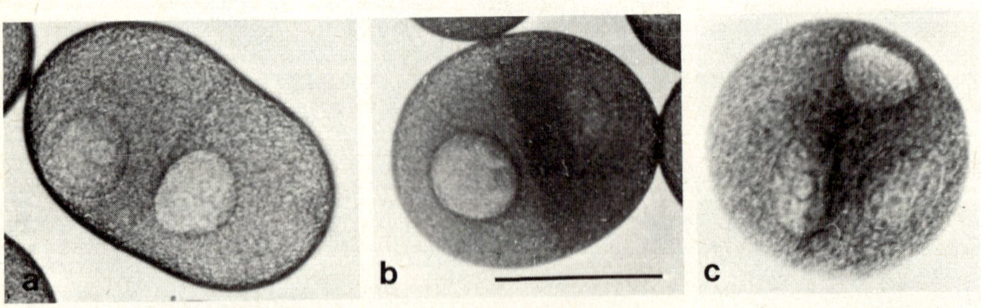


FIG. 4. Fusion products just after PEG treatment of the *Aphelasterias japonica* oocytes (a) and 4 to 5 hours after fusion (b and c). (a and b) Fusion products with two germinal vesicles; (c) fusion product with three germinal vesicles. No discernible mixing of cytoplasms is seen. Scale bar, 100 μm.

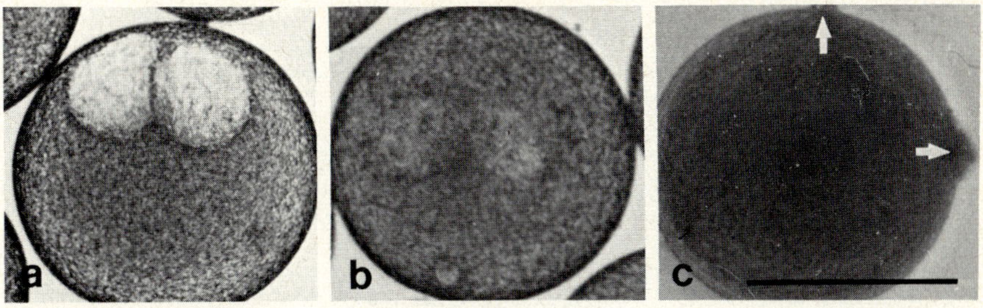


FIG. 5. The maturation of a product of fusion of two oocytes of *Aphelasterias japonica* under the effect of 1-methyladenine ($1 \times 10^{-5}$ *M*). (a) fusion product prior to the treatment with 1-methyladenine; germinal vesicles and nucleoli are clearly seen; (b) several minutes after the beginning of 1-methyladenine action; the germinal vesicle membrane breaks down; (c) a mature fusion product; one can see polar bodies (arrows) on the plasma membrane. Scale bar, 100 μm.

the oocytes with the cytoplasts the boundary between the respective cytoplasms is clearly discernible.

Obviously, from the above description the fusion products are easily distinguishable from the oocytes or cytoplasts. With a certain experience, their selection becomes quite easy. To select the fusion products we used a dissection microscope with magnification of up to ×50 and a micropipet prepared from a Pasteur pipet. In such a way a few hundred fusion products can be selected within 15–20 minutes.

The oocytes and fusion products matured under the influence of 1-methyladenine (Vassetzky *et al.*, 1983): germinal vesicle breakdown took place (Fig. 5a and b). The process was completed in less than 30 minutes by the extrusion of polar bodies (Fig. 5c). It can be seen (Table II) that the percentage of matured fusion products is close to 100% in most experiments, except in the first three where treatment with PEG resulted in a marked decrease in the reaction to 1-methyladenine both in the oocytes and fusion products due, apparently, to the time of the experiments (very beginning of the spawning season). The percentage of oocytes without germinal vesicles (so called "spontaneously" matured) in the initial suspension of oocytes of a given female did not reflect their ability to react to 1-methyladenine (Table II). The induced fusion of oocytes did not cause them to undergo maturation, nor did it affect the capacity of oocytes and fusion products to respond to 1-methyladenine.

The matured fusion products cleaved after insemination (Fig. 6a and b). In four experiments 434 cleaving fusion products were observed, i.e., an average of 53.6% (40–90%). Development of fusion products began within 60–90 minutes after insemination and, during 4–5 hours of observation, proceeded to the early morula stage (Fig. 6c) (Vassetzky *et al.*, 1983).

1-Methyladenine induced the maturation of the products of fusion between the

TABLE II

CAPACITY FOR MATURATION OF OOCYTES AND FUSION PRODUCTS OF *Aphelasterias japonica* (VASSETZKY *et al.*, 1983)

| | | Percentage of oocytes and fusion products without GV | | | |
|---|---|---|---|---|---|
| | | | Oocytes and fusion products after 1-methyladenine ($1 \times 10^{-5}$ *M*) treatment | | |
| Experiment number | Date (day/month) | Intact | Oocytes prior to PEG treatment | PEG-treated oocytes | Fusion products[a] |
| 3a | 21/7 | 3 | 71 | 0 | 0 |
| 3b | 21/7 | 5 | 52 | 10 | 16 (4) |
| 5 | 27/7 | 27 | 91 | 35 | 33 (10) |
| 6 | 29/7 | 13 | 92 | 97 | 98 (64) |
| 7 | 30/7 | 7.4 | — | 100 | 100 (121) |
| 8 | 5/8 | 10 | — | — | 99 (90) |
| 9 | 6/8 | 9 | — | — | 98 (88) |
| 10 | 6/8 | 10 | — | — | 99 (198) |
| 11 | 7/8 | — | — | — | 99 (220) |
| 12 | 8/8 | — | — | — | 94 (34) |

[a] Figures in parentheses refer to the number of fusion products which underwent maturation under the effect of 1-methyladenine; —, no counts were made.

oocytes and cytoplasts, and germinal vesicle breakdown was observed in them approximately at the same time as oocytes, i.e., the altered (reduced) nucleocytoplasmic ratio in the fusion products appeared not to affect the time course of maturation.

The oocytes and fusion products can easily be observed in a low-power dissection microscope or a light microscope with magnification of up to ×300. In order to ensure better conditions for photography, the oocytes and fusion products can be placed in a small drop of seawater under a layer of paraffin oil. In these conditions the visibility is very clear and the objects remain in the same position. Besides, all processes occurring in the living cells take their normal course, so that one can record the fusion between oocytes or of oocytes with cytoplasts, the maturation of oocytes and fusion products, and their subsequent cleavage.

## III. Fusion of Mature Unfertilized Sea Urchin Eggs

### A. Characteristics of the Object

Our experiments (Vassetzky *et al.*, 1986) were conducted on a sea urchin *Strongylocentrotus nudus* (Agassiz), which is a common representative of the

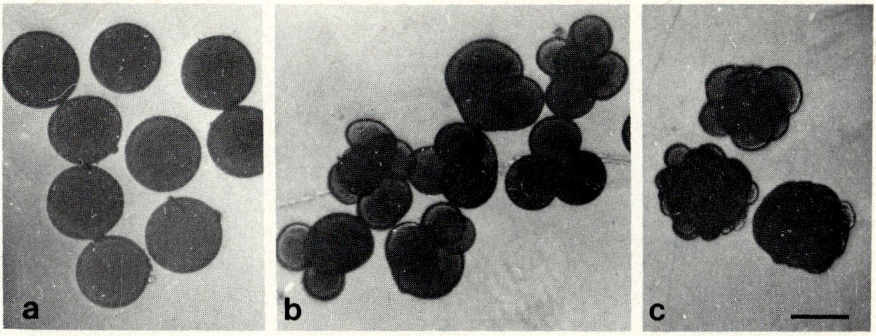

FIG. 6. The development of products of fusion of the oocytes of *Aphelasterias japonica* following their maturation under the effect of 1-methyladenine ($1 \times 10^{-5}$ *M*) and subsequent insemination. (a) Mature fusion products; (b) beginning of cleavage; (c) early morula stage. Scale bar, 100 μm.

coastal fauna of the Sea of Japan. In the Vostok Bay it spawns from July to September at a water temperature of 14 to 22°C. The first peak of spawning is usually observed in the second half of July and the second one in September. The color of gonads varies from cream to dark brown in females and from orange to dark red in males. Its fecundity amounts to 10 million eggs, on the average. During the period of spawning the oocytes are uniform in size (~100 μm) and readily fertilizable (Kasyanov *et al.*, 1980).

## B. Preparation of Eggs for Fusion

The mature eggs were obtained by injecting the females with 2–3 ml 0.5 *M* KCl. The egg suspension was washed with artificial seawater two to three times. In order to remove the jelly coat 1 ml of seawater adjusted to pH 4.5–5.0 by HCl was added to the ~1 ml of the initial suspension sediment. The eggs in the test tubes were periodically shaken for 2 minutes and washed three times with seawater. In order to remove the vitelline membrane, 1 ml 0.25% trypsin (Spofa, Czechoslovakia) was added to the acid water-pretreated sediment of the egg suspension. The treatment usually lasted 15 minutes and the test tube with the eggs was periodically shaken. The treatment was terminated when egg aggregates appeared in the suspension, the suspension was thereafter washed three times with seawater, and treated with 1 *M* urea with 1 m*M* $CaCl_2$ and washed twice with seawater.

## C. Treatment with Polyethylene Glycol

Immediately after washing the urea-treated eggs they were layered on a PEG solution. All other procedures were the same as for starfish oocytes (see Section II,E).

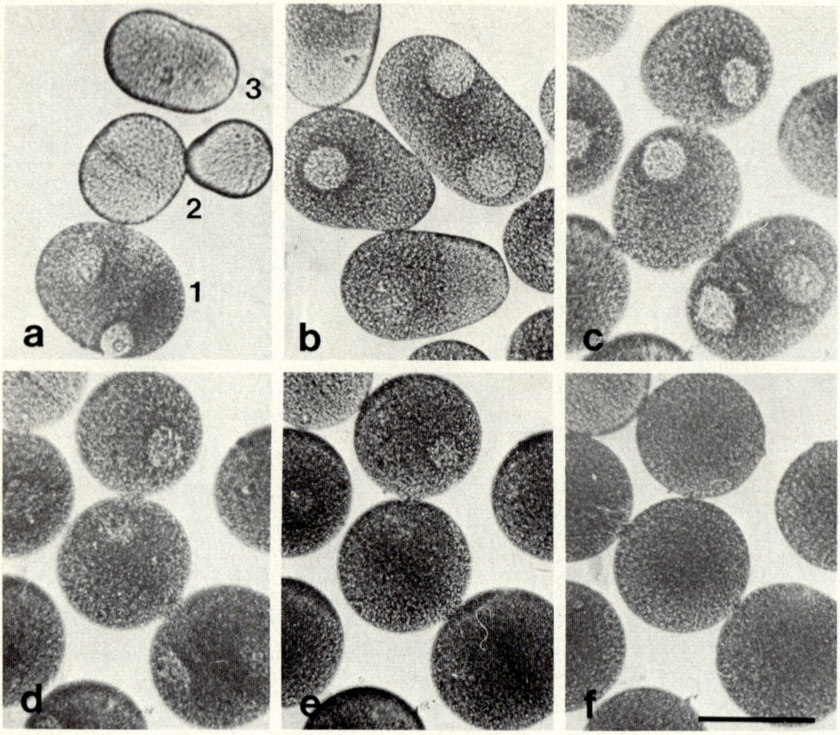


FIG. 7. Products of fusion between the oocytes and cytoplasts of *Aphelasterias japonica*. (a and b) Fusion products soon after PEG treatment: (1) oocyte/oocyte; (2) oocyte/cytoplast; (3) cytoplast/cytoplast; (c–f) consecutive stages of the maturation of fusion products induced by 1-methyladenine ($1 \times 10^{-5}$ *M*). Scale bar, 100 μm.

The frequency of fusion was up to 44%. In these experiments we used a 1450 PEG (Serva, West Germany). Fusion was also observed when using 4000 and 6000 PEG (Loba, Austria, and Merck, West Germany). After the PEG treatment the eggs were washed 3–4 times with seawater. The fusion products were not activated as a result of all these treatments and remained viable for a few hours. Just after the fusion one can see one to four pronuclei in the fusion products (Fig. 7). Within a few hours one larger nucleus was sometimes observed.

## D. Other Methods of Induced Egg Fusion

As already mentioned, Bennett and Mazia (1981a,b) were the first to induce fusion between the unfertilized and fertilized eggs of two species of sea urchins, *Strongylocentrotus purpuratus* and *Lytechinus pictus*. To remove the external membranes of the unfertilized eggs, they were treated first with seawater at pH 5 and then with a Ca-free seawater containing 0.2% cysteine and 1.2 mg/ml Pronase (Pronase C, Calbiochem) at pH 8 for 5 minutes. The fertilized eggs were incubated for 2 minutes in a Ca- and Mg-free seawater containing 0.2% cysteine and 1.2 mg/ml Pronase at pH 8. These treatments were followed by 30 seconds exposure to 1 *M* urea with 1 m*M* $CaCl_2$, and thereafter the eggs were returned to seawater.

Fusion was induced in a Ca-free seawater diluted with distilled water to 75% of its original volume and supplemented with $CaCl_2$ to a final concentration of 25 m*M* (total osmolarity 800 mOsm). Equal volumes of 5% (by volume) suspensions were mixed and diluted 10-fold with a 0.0075 mg/ml polyarginine (Sigma) in the medium described above and incubated for 40 minutes. The eggs were then washed twice with a seawater containing 1.5 mg/ml arginase and, within 5 minutes, washed twice with seawater. Successful fusions were also obtained using a 6000 PEG (Koch-Licht, UK): 50 μl of the egg suspension were layered over an equal volume of 50% (w/w) PEG solution, and the eggs were allowed to settle by gravity within 2 minutes.

Unlike Bennett and Mazia (1981a,b), we used another method for removing external membranes. When using cysteine and Pronase on *S. nudus* eggs, we

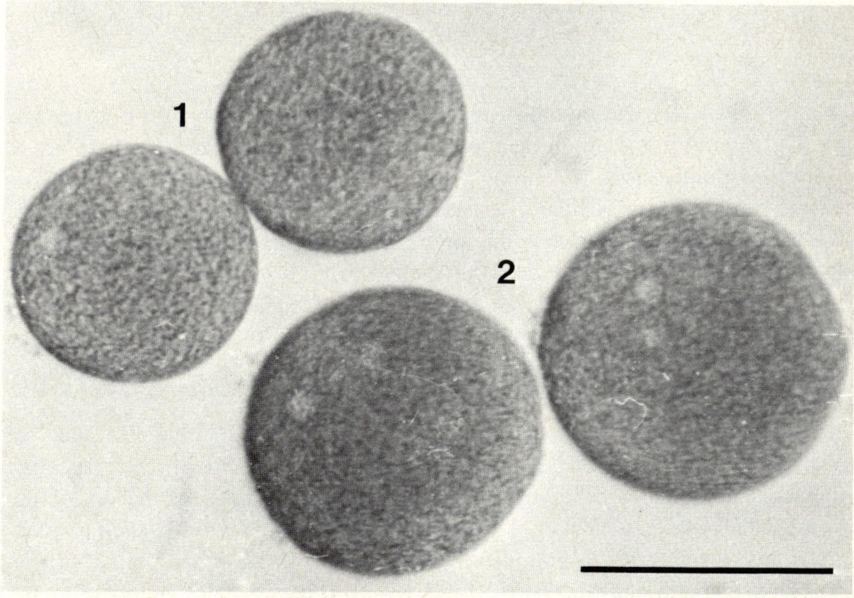


FIG. 8. Eggs (1) and products of fusion (2) between eggs of a sea urchin (*Strongylocentrotus nudus*). Two pronuclei are clearly seen in the fusion products. Scale bar, 100 μm.

observed the membranes' removal in a small percentage of eggs. Besides, we did not use the special medium for fusion since we found no advantages of such a medium compared with seawater.

Richter *et al.* (1981) reported that fusion of eggs of the sea urchin *Paracentrotus lividus* could be induced by an electric field. The eggs were dejellied by pouring 1 ml of eggs into 10 ml of seawater at pH 4.5. The dejellied eggs were brought back within 2 minutes to pH 8 with seawater. The eggs were then treated with 2 mg/ml Pronase (Pronase E, Merck, or Pronase P, Serva) to remove the vitelline membrane.

Egg fusion was induced on a gold-sputtered microslide with a 1-mm wide gold-free slot. The edges of the gold film served as electrodes. A function generator type 7404P (Toellner Electronic, West Germany) was used as a voltage source and a Hewlett–Packard pulse generator 214 B as a source of square pulses. An alternating voltage of 10–22 V and 2 MHz applied for 1–2 minutes resulted in the formation of "egg chains." Fusion was initiated with, first, a short electric field (voltage 40 V, duration 50–400 μsec, and an electrode distance of 1 mm) followed 2 minutes later by a second field pulse. The alternating voltage was then reduced to 10 V and fusion was completed within 5 minutes, although it occasionally took place just after the first pulse.

## IV. Concluding Remarks

We reviewed here the methods of induced fusion of echinoderm oocytes and eggs. Up to now this method has been used in only a few studies; there are at present nine publications, including four by the present authors. In all of these studies different fusogens were used: polyarginine (Bennett and Mazia, 1981a,b, 1983) and polyethylene glycol on the sea urchin eggs (Bennett and Mazia, 1981a,b, 1983; Vassetzky *et al.*, 1986), electric field pulse on sea urchin eggs (Richter *et al.*, 1981), and polyethylene glycol on starfish oocytes (Skoblina *et al.*, 1982; Sekirina *et al.*, 1983; Vassetzky *et al.*, 1983, 1984, 1986; Bulet *et al.*, 1985). In the latter case, we undertook the first successful attempts to reconstitute new cells from whole cells and their fragments.

Using this method, it has been shown that the fused cells retain their respective properties until the beginning of further development: Not only do the respective cytoplasms not mix together but the unfertilized and fertilized halves of the fusion product of sea urchin eggs preserve their specific features, such as presence or absence of cortical granules in the cytoplasm and of long microvilli on the surface (Bennett and Mazia, 1981a,b, 1983). It has also been shown that the procedures leading to the fusion of oocytes or eggs do not affect the normal condition of the cells: They retained the capacity for maturation, fertilization, and subsequent development (Richter *et al.*, 1981; Bennett and Mazia, 1983; Vassetzky *et al.*, 1983).

The use of the induced fusion method in studies on the echinoderm oocytes and eggs makes it possible to obtain mass quantities of oocytes and eggs combined at different developmental stages or to alter (qualitatively or quantitatively) the nucleocytoplasmic ratio. At present it is difficult to foresee all cases when this method might become useful, nevertheless it is important to note that the methodology now exists.

## V. Appendix: Methods for Maintaining Starfish and Obtaining Gametes and Embryos Year Round

P. V. DAVYDOV,[4] O. I. SHUBRAVYI,[5] AND
S. G. VASSETZKY[6]

Many echinoderm species have already been introduced in culture. In some cases the echinoderm larvae have been reared in artificial conditions until settlement and metamorphosis (Costello and Henley, 1971; Kinne, 1977), but in most cases natural seawater has been used. We have attempted to obtain the whole cycle of development of a starfish *Patiria pectinifera* in a closed cycle aquarium with artificial seawater.

*Patiria (Asterina) pectinifera* (Muller et Troschel) (Asterinidae) is one of the common species of the Sea of Japan which has a planktotrophic larva with a short pelagic life. Its embryonic and larval development in nature has already been described (Kasyanov *et al.*, 1983; Kasyanov, 1984).

Starfishes were collected near the Popov Island (Peter the Great Bay, Sea of Japan) at depths of 1–3 m in 1981 and kept in Moscow in 250-liter closed cycle aquaria made of plastic with bottom biofilter, skin separation, and heat exchanger. Artificial seawater was prepared according to an "Instant Ocean" recipe with a somewhat altered ratio of microelements (Shubravyi, 1983) (Table III). The density of animals was 30 per 250 liters. Food was provided at about 0.5 g/animal/week. Food composition: squid filet, 500 g; sweet pepper powder (as a source of carotinoids), 2 g; calcium glycerophosphate, 1.5 g. This mixture was ground and kept in a freezer at −10°C. Given abundant food and suitable conditions, the starfishes grew rapidly: Animals of 1.5–2 cm in size increased almost twice within 2–3 months. To avoid spontaneous spawning, we kept the starfishes at low temperatures (up to 9°C) with moderate illumination.

[4]Biological Faculty, State University of Moscow, Moscow, USSR.
[5]Moscow Zoo, Moscow, USSR.
[6]Institute of Developmental Biology, USSR Academy of Sciences, Moscow, USSR.

TABLE III

COMPOSITION OF ARTIFICIAL SEAWATER OF 35.00‰ (SHUBRAVYI, 1983)[a]

| |
|---|
| Basic constituents |
| $NaCl$—2640, $MgCl_2 \cdot 6H_2O$—491, $MgSO_4 \cdot 7H_2O$—674, $KCl$—65.5, $H_3BO_3$—2.48, $NaHCO_3$—13.6, $KBr$—9.7, $Na_2CO_3$—3.34, $SrCl_2 \cdot 6H_2O$—2.38, $NaF$—0.27, $CaCl_2$—111 |
| Complexing agent |
| EDTA ($Na_4$–salt)—1 |
| Trace elements |
| $KIO_3$—0.009, $MnSO_4 \cdot 5H_2O$—0.397, $NaH_2PO_4 \cdot 2H_2O$—0.397, $LiCl$—0.099, $Na_2SiO_3 \cdot 9H_2O$—0.099, $Na_2MoO_4 \cdot 2H_2O$—0.099, $Al_2(SO_4)_3 \cdot 18H_2O$—0.172, $CoSO_4 \cdot 7H_2O$—0.0089, $CuSO_4 \cdot 7H_2O$—0.0009, $ZnSO_4 \cdot 7H_2O$—0.0096, $Rb_2SO_4$—0.0149, $BaCl_2$—0.009 |

[a] Note: All values in g/100 liters.

To obtain gametes, starfish were taken one by one and injected with 0.3–0.5 ml of 1-methyladenine solution (see Section II,C) per arm. The starfishes were then placed into separate 1-liter glass vessels. Spawning started within 15–90 minutes after injection. Gametes were shed through the paired gonopores. Non-spawning and spent starfishes were returned to aquaria after being segregated by sex.

Eggs were inseminated at different times after the beginning of spawning, from 10–15 minutes (active spawning) to 60 minutes (weak spawning). Spermatozoa were obtained using either the method of gonad isolation (Costello and Henley, 1971) or 1-methyladenine stimulation. In the latter case the spawning starfishes (female and male) were placed together in a broad vessel with low walls (2–3 liters), where embryonic development proceeded. Water was changed every 30 minutes during the first 2–3 hours; the intervals were later increased. Developing embryos were kept in subdued light. Blastulae were characterized by a positive phototaxis, and this facilitated their transfer into vessels for rearing larvae.

Larvae were reared in 1-liter tall battery jars in a room with the constant air temperature (23–24°C). The jars were placed on a continuously working shaker. The density of larvae was about 100/liter. Water was changed once every 2–3 days. The larvae were fed on a mixture of microalgae *Platymonas viridis, Pavlova lutheri* and *Bodo marina;* late brachiolaria were given diatoms a few days prior to settlement and transferred into the fresh artificial seawater every 5 days. Three days prior to settlement the brachiolaria were transferred into clean jars containing oyster shells that were taken from the aquaria and were covered by the primary microbial films. Settlement of the larvae took place on the thirtieth day of development. Brachiolaria settled on the lower sides of the oyster shells in grooves and splits of the shell surface (Fig. 9).

Using the stimulation with 1-methyladenine we obtained gametes and embryos

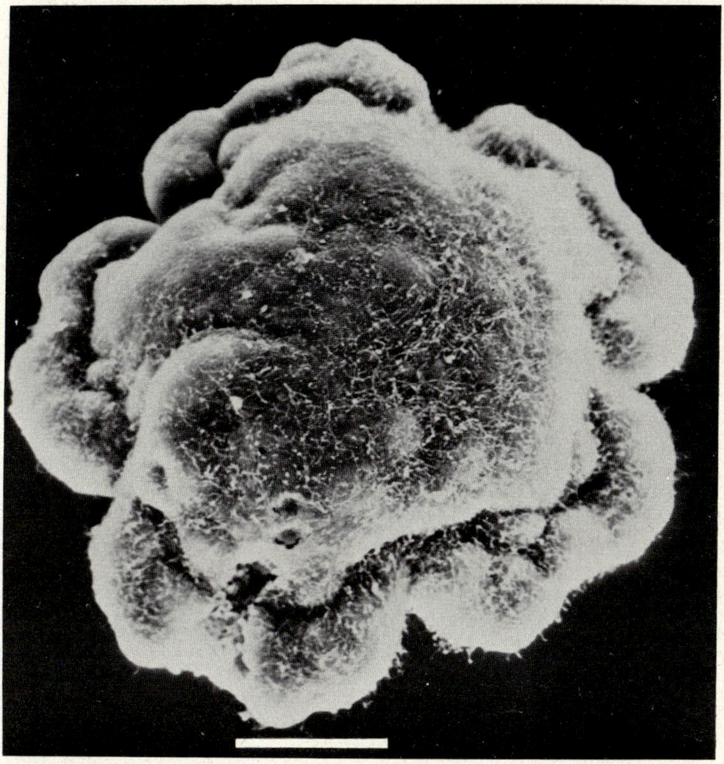

FIG. 9. Scanning electronmicrograph of *P.(A.) pectinifera* on the second day after settlement. View from aboral side. ×260. Scale bar, 100 μm.

of starfishes in winter and spring, i.e., beyond the natural season of spawning. From the results of these experiments we have described embryonic and larval development in *P.(A.) pectinifera* under artificial conditions (Davydov *et al.*, 1985). Using the same methods, we repeatedly obtained gametes, embryos, and larvae of another starfish species, *Aphelasterias japonica,* brought from the Sea of Japan in 1984 and kept in the same aquaria.

## REFERENCES

Bennett, J., and Mazia, D. (1981a). *Exp. Cell. Res.* **131,** 197–207.

Bennett, J., and Mazia, D. (1981b). *Exp. Cell. Res.* **134,** 494–498.

Bennett, J., and Mazia, D. (1983). *In* "Time, Space and Pattern in Embryonic Development," (W. R. Jeffrey and R. A. Raff, eds.), pp. 1–27. Liss, New York.

Betina, M. I., Yazykov, A. A., Yanushevich, I. V., and Zubarev, T. N. (1973). *Plant Physiol.* (in Russian) **20,** 1075–1082.

Brinster, R. L. (1968). *J. Reprod. Fertil.* **17,** 521–525.

Bulet, P., Kishimoto, J., and Shirai, M. (1985). *Dev. Growth Differ.* **27,** 243–250.

Costello, D. P., and Henley, C. (1971). "Methods for Obtaining and Handling Marine Eggs and Embryos," 2nd ed. Marine Biological Laboratory, Woods Hole, Massachusetts.

Davydov, P. V., Shubravyi, O. I., Tsetlin, A. B., Dautov, S. Sh., and Vassetzky, S. G. (1986). *Mar. Biol.* (in Russian), in press.

Diberardino, M. A. (1980). *Differentiation* **17,** 17–30.

Dyban, A. P., Sekirina, G. G., and Ignatova, T. N. (1981). *Cytology* (in Russian) **23,** 1200.

Dyban, A. P., Veisman, B. L., and Golinsky, G. F. (1983). *Ontogenesis* (in Russian) **14,** 420–426.

Giudice, G. (1973). "Developmental Biology of the Sea Urchin Embryo." Academic Press, New York.

Harvey, E. B. (1956). "The American Arbacia and Other Sea Urchins." Princeton Univ. Press, Princeton, New Jersey.

Kanatani, H. (1973). *Int. Rev. Cytol.* **35,** 253–298.

Kanatani, H. (1979). *In* "Hormones and Evolution" (E. J. W. Barrington, ed.), Vol. I, pp. 273–307. Academic Press, London.

Kasyanov, V. L. (1984). *Mar. Biol.* (in Russian) **1,** 1–13.

Kasyanov, V. L., Medvedeva, L. A., Yakovlev, S. N., and Yakovlev, Yu. M. (1980). "Reproduction of Echinoderms and Bivalve Molluscs" (in Russian). Nauka, Moscow.

Kasyanov, V. L., Kryuchkova, G. A., Kulikova, V. A., and Medvedeva, L. A. (1983). "Larvae of Marine Bivalve Molluscs and Echinoderms" (in Russian). Nauka, Moscow.

Kinne, O. (1977). *In* "Marine Ecology" (O. Kinne, ed.), Vol. 3, Part 1, pp. 936–967.

Malchenko, L. A., Buznikov, G. A., Gordeev, E. N., Kudryashova, N. I., Mndzhoyan, O. L., Ovsepyan, T. R., and Khromov-Borisov, N. V. (1984). *Ontogenesis* (in Russian) **15,** 211–215.

Meijer, L., and Guerrier, P. (1984). *Int. Rev. Cytol.* **86,** 129–196.

Nemoto, S.-I., Yoneda, M., and Nemura, I. (1980). *Dev. Growth Differ.* **22,** 315–325.

Richter, H.-P., Scheurich, P., and Zimmerman, U. (1981). *Dev. Growth Differ.* **23,** 479–486.

Sekirina, G. G. (1985). *Ontogenesis* (in Russian) **16,** 583–588.

Sekirina, G. G., Skoblina, M. N., Vassetzky, S. G., and Bilinkis, A. A. (1983). *Cell Differ.* **12,** 67–71.

Shubravyi, O. I. (1983). *Zool. J.* (in Russian) **62,** 618–621.

Skoblina, M. N., Sekirina, G. G., Vassetzky, S. G., and Bilinkis, A. A. (1982). *J. Gen. Biol.* (in Russian) **43,** 847–853.

Vassetzky, S. G., and Sekirina, G. G. (1985). *Cell Differ.*, **16,** 77–82.

Vassetzky, S. G., Sekirina, G. G., Skoblina, M. N., and Bilinkis, A. A. (1983). *Cell Differ.* **12,** 73–76.

Vassetzky, S. G., Sekirina, G. G., Veisman, B. L., Skoblina, M. N., and Bilinkis, A. A. (1984). *Cell Differ.* **14,** 47–52.

Vassetzky, S. G., Sekirina, G. G., Skoblina, M. N., and Bilinkis, A. A. (1986). *Bull. Moscow Soc. Nat.* (in Russian), in press.

Yanko, Ya. (1961). "Mathematical-Statistical Tables" (in Russian). Gosstatizdat, Moscow.

# Chapter 21

# *Microinjection and Cytoplasmic Transfer in Starfish Oocytes*

TAKEO KISHIMOTO

*Department of Developmental Biology*
*National Institute for Basic Biology*
*Okazaki, Japan*



## I. Introduction

Introduction of materials into the interior of a single living cell is an invaluable technique in studies of cellular processes. A variety of methods have been devised: volume-controlled pressure microinjection (Hiramoto, 1962, 1974; Kiehart, 1982), simple pressure microinjection (Chambers and Chambers, 1961; Gurdon, 1977; Kopac, 1964; Yamamoto and Furusawa, 1978), microelectrophoresis (Curtis, 1964), lipid vesicle fusion (Poste *et al.*, 1976), erythrocyte ghost fusion (Furusawa *et al.*, 1976), and cell-to-cell fusion with an electric field

(Richter *et al.*, 1981), with Sendai virus (Balakier, 1978; Poste, 1972), with polyethylene glycol (Bulet *et al.*, 1985; Vassetzky *et al.*, 1983, this volume), or with polyarginine (Bennett and Mazia, 1981). Among these, however, volume-controlled pressure microinjection is the only method which satisfies all of the following necessary criteria: (1) precise control over the amount of material to be introduced, (2) specified timing of introduction, (3) introduction into specified cells, (4) high success rate of introduction, (5) few restrictions on the nature of the extracellular medium, and (6) removal of cell constituents. Thus, volume-controlled pressure microinjection is a versatile way to add various kinds of substances directly into the inside of individual cells.

The microinjection technique has made remarkable contributions in elucidating the mechanism of oocyte maturation (Kanatani, 1985; Kishimoto, 1982; Masui and Clarke, 1979). This article provides the technical details of a volume-controlled pressure microinjection system which has been applied for more than 10 years to the study of starfish oocyte maturation. The principle of the system was devised by Hiramoto (1956, 1962, 1974, 1984). The present description should enable other workers to carry out microinjection and cytoplasmic transfer in echinoderm oocytes and eggs.

## II. Apparatus for Microinjection

### A. Micromanipulator

The microinjection set-up consists of an observation microscope and a micromanipulator; I use a single right-handed manipulator unit (Fig. 1). A micropipet is fitted to a needle holder, which is attached to the micromanipulator, and is connected to a screw-driven syringe by plastic tubing. The recipient cell is held in a manipulation chamber (see Section II,C), and the micropipet approaches the cell from a horizontal direction (Fig. 2). The injection volume is controlled by the pressure, which is caused by advancing the screw drive on the injection syringe.

A Leitz micromanipulator is used to position the micropipet. Although any micromanipulator with similar features (for example, manufactured by Narishige, Tokyo) can be used, the Leitz is preferable, since various movements of the micropipet are controlled by a single micromanipulator. Coarse adjustment of the micropipet in three dimensions is achieved by screw-driven knobs, and fine adjustment is achieved in the two horizontal coordinates by a joy stick and by a fine-drive knob in the vertical dimension. Further, the microinstrument holder, which spring clamps the needle holder and is attached directly to the micromanipulator, allows additional adjustments in all directions before manipulation.

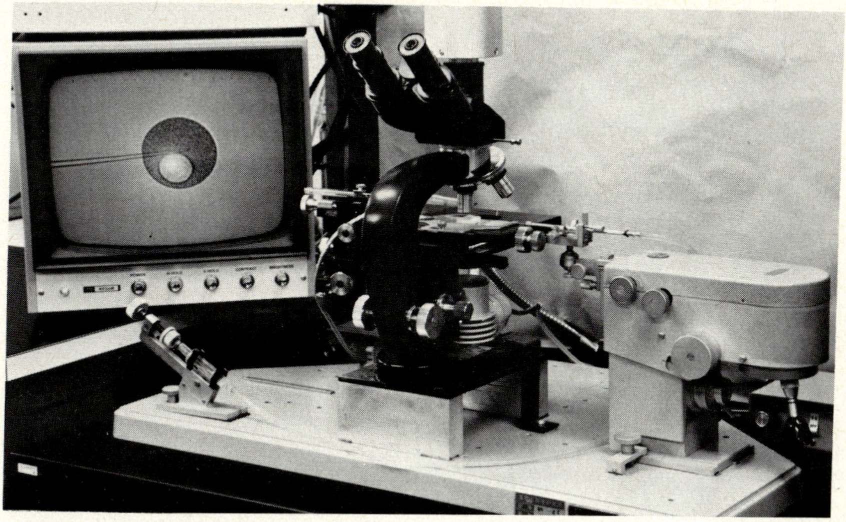

FIG. 1. The microinjection set-up. The complete system includes a microscope and a right-handed Leitz micromanipulator, both of which are clamped on a common base plate. Microinjection procedures into an immature oocyte of *Asterina pectinifera* are monitored by video recording.

The micromanipulator and microscope are mounted and clamped on a common base plate. The right-hand manipulator alone is sufficient for the microinjection of starfish oocytes.

Since this microinjection system depends upon hydraulic pressure, a sufficiently pressure-tight system is constructed as follows. A Leitz instrument collar is used as the needle holder. The micropipet is firmly gripped into the instrument collar by the captive nut. The shaft of the micropipet is friction fitted into the inside of the needle holder by means of the brass collar and silicon rubber bushing. This ensures against pressure leaks at the joint of the micropipet and needle holder. The rear end of the instrument collar is connected to thick, flexible, transparent, plastic tubing (approximately 2.5 mm i.d.) leading to the glass syringe (2 ml) mounted in a Leitz screw-advance syringe holder. The piston of the syringe is greased with high-vacuum silicon grease to prevent pressure leakage. The space from the tip of the instrument collar to the injection syringe is filled thoroughly with distilled water before fitting the micropipet (Fig. 2). Air is allowed only in the shaft of the micropipet. In order to keep everything pressure tight, the joints between the instrument collar, plastic tubing, and glass syringe are sealed with dental wax. Practically, small pieces of the wax placed on the joint are melted by heating and then allowed to set by air cooling. Changes in pressure caused by the injection syringe are delivered to the micropipet through a pressure-tight system.

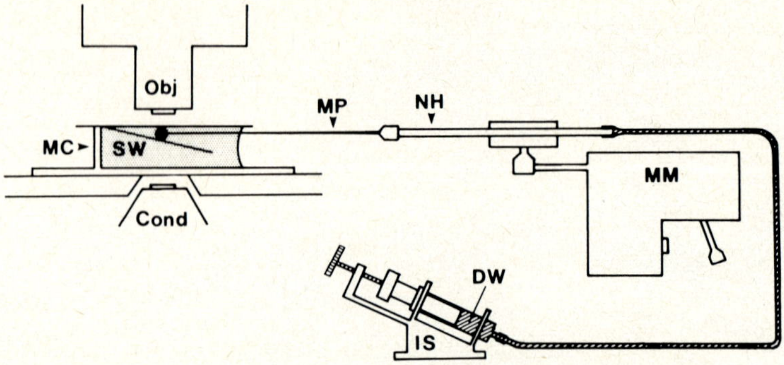


FIG. 2. Schematic of microinjection system depicting the relationship between the microscope, recipient oocyte (black ball), manipulation chamber (MC), micropipet (MP), needle holder (NH), micromanipulator (MM) and injection syringe (IS). The space from the tip of needle holder to injection syringe is filled with distilled water (DW). Obj, objective; Cond, condenser; SW, seawater.

A microburner shown in Fig. 5 is convenient for melting small pieces of wax as mentioned above. Gas is connected through a flexible rubber tubing to a glass pipet, the tip of which is fitted to a hypodermic needle. The flame is controlled by a hosecock clamp on the rubber tubing.

For work with the micromanipulator, a microscope with a fixed stage and body tube focusing is recommended, since the relative position between the recipient cell and micropipet does not change when adjusting focus in this type of microscope. A movable mechanical stage is useful for positioning the manipulation chamber in the horizontal plane. Normal microscope objectives (10 or 20×) for bright-field observation are generally useful for starfish oocytes or eggs. It is important that the microscope is clamped tightly to the base plate of the micromanipulator, otherwise when the focus is adjusted the microscope will move a little. This leads to a change of position of the recipient cells on the stage with respect to that of the micropipet, resulting in a failure of injection.

A microscope with stage focusing and a fixed tube can also be used with the micromanipulator, although the sequence of necessary movements, which have to be mutually adjusted, must be strictly adhered to. This type of microscope is suitable when the moving element of the micromanipulator, separated from its control unit, is clamped to the stage of microscope. An example of such a manipulator is Narishige type MO-102.

## B. Micropipet

Although the micropipet is connected to a screw-controlled syringe, the syringe is not suitable for regulating the injection volume directly, since the syringe is far larger than the volume of the injected material. In order to control the speed of flow in and out of the micropipet, Hiramoto (1974) devised a braking micropipet with a constriction near the tip, where it is filled with silicon oil (see Fig. 6). The appropriate combination of constriction size and viscosity of the silicon oil functions to attenuate the passage of fluid. Picoliter volumes can be injected by this method. Such a braking micropipet is made by the following procedure.

1. A glass micropipet is drawn from glass capillary tubing (Drummond Microcaps, 50-μl capacity in 10-cm lengths, Drummond Scientific Co., Broomall, PA) with a Livingston-type horizontal needle puller (Narishige, type PG-1). The outer diameter of the glass capillary should be 1.0 mm in order to seal properly in the instrument collar of the Leitz needle holder.

2. Since the lumen of the micropipet as pulled is too fine for this particular kind of microinjection, the tip must be broken to an inner diameter of approximately 5 μm. Practically, the tip of the micropipet (fitted to the needle holder of the manipulator) is gently pushed straight against to the edge of a piece of cover slip cemented onto a pedestal on a glass slide. This maneuver is carried out using the micromanipulator and is observed under the microscope. In this manner, the extreme tip is broken off. Micropipets with tips of 3–10 μm are suitable for injecting various materials, including test liquids and cytoplasm, into starfish oocytes. Instead of breaking the tip, the tip can be bevelled by sharpening with a microneedle grinder (Narishige, type EG-1). Alumina abrasive film (A. E. Thomas Co., Philadelphia, PA, mesh size 0.3 μm) is suitable for this purpose. Fine glass powder adhering to the tip after abrasion can be removed by washing with concentrated sulfuric acid followed by deionized water and ethanol. However, a broken tip is completely suitable for starfish oocytes and is more readily manufactured.

3. A constriction is made in the micropipet approximately 3–5 mm from the tip, as shown in Fig. 3. A glass micropipet with the correct size of tip is fixed into the needle holder of a micromanipulator (not attached to the syringe set-up) and positioned for viewing with a microscope. The micropipet is held horizontally, and its tip is inserted about 3–5 mm into a 1-mm diameter loop of platinum wire (0.1 mm thick) which is then heated by applying an electric current using a variable transformer (less than 2 V). A thickening in the wall of the micropipet produces a constriction so that the bore diameter is reduced to about 10 μm (Fig. 3c).

The platinum wire is attached to a glass or plastic rod that is then attached to an independent, simple three-dimensional micromanipulator that can be positioned temporarily on the left side of the microscope. The ends of the platinum wire are soldered to leads to the transformer. The loop is positioned in the field of the microscope, and the micropipet is inserted from the right (Fig. 3d).

Since heating the micropipet causes it to bend in an undesirable way, construc-

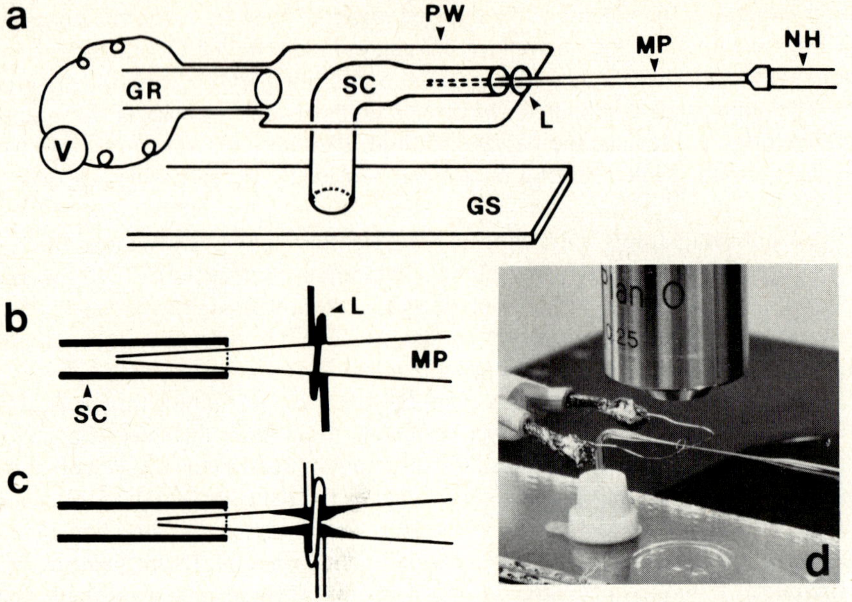


FIG. 3. Procedures for making a constriction in a braking micropipet. Schematic diagram (a) and photograph (d) of the apparatus for making a constriction. Platinum wire (PW), with a loop (L) of 1 mm diameter, is attached to a glass rod (GR). This glass rod is positioned by a simple three-dimensional micromanipulator. The micropipet (MP) is held horizontally, and its tip is inserted into the loop and enters a supporting glass capillary (SC) with horizontal direction fixed on the glass slide (GS). (b and c) A constriction is formed by heating a part (3 to 5 mm from the tip) of the shaft of the micropipet with applying an electric current to the loop of platinum wire. (b) and (c) were adapted from Hiramoto (1974).

tion of a support for the tapered end (Fig. 3a and d) is recommended. This can be done by mounting a short segment of capillary tubing having a right-angle bend onto a glass slide so that its position can be regulated by the stage mechanism of the microscope. The open tip of this supporting capillary (preferably about 200 μm in diameter) is positioned directly to the left of the platinum wire loop. Thus, when the micropipet is inserted into the loop, its extreme tip enters the capillary, where it is supported and prevented from bending by gravity as the loop is heated.

4. Silicon oil is put into the braking micropipet from the rear end using a long slender glass pipet or a long hypodermic needle. Silicon oil with a viscosity of 500 centistokes is generally used (Toshiba Silicone, Tokyo), although paraffin oil is also appropriate. The tapering part of the micropipet, including the constriction, is thoroughly filled with silicon oil, although it is not necessary to fill the whole shaft of the micropipet. Any small bubbles of air should be avoided in

the tapering part of micropipet, since they could interfere with the precise controlling action of the braking pipet. The micropipet thus formed is fitted to the instrument collar and syringe set-up. Now, the microinjection can be carried out by applying pressure to the micropipet with the screw-controlled syringe.

## C. Manipulation Chamber

In microinjection experiments, it is important to immobilize the recipient cells without threatening their viability. In our system, starfish oocytes are immobilized in a wedge-shaped space as shown in Fig. 2. Such a manipulation chamber is composed of a wedge-shaped egg holder and its supporting glass slide (Fig. 4). In this chamber, starfish oocytes undergo maturation normally. Each oocyte can be removed easily from the chamber for subsequent processing by sweeping it out using the side of the micropipet.

The support for the egg holder is made by attaching a U-shaped piece of acrylic plate (less than 4 mm in height) to a glass microscope slide with dental wax (Fig. 4). The wedge-shaped egg holder is made by the following procedure.

1. Cover slips (24 × 50 mm, No. 1 thickness) are scored with a diamond glass scriber and cut to make rectangular fragments of two sizes: approximately 24 × 12 mm (large fragment) and approximately 8 × 10 mm (small fragment).

2. The long edge of a large fragment is inserted 2 to 3 mm into an ordinary paper clip (approximate size, 28 × 7 mm; made of stainless wire) as shown in Fig. 5. One long edge of a small fragment is positioned on the paper clip so that it rests at an angle on the large fragment. Thus, small and large fragments form a wedge shape supported by the paper clip.

3. A small piece (less than 1 × 1 × 1 mm) of dental wax is placed on the large fragment adjacent to the line touching the small fragment. When this wax is gently heated with a microburner as shown in Fig. 5a and b, it melts and spreads along the long edge of small fragment by capillarity. The paper clip is removed

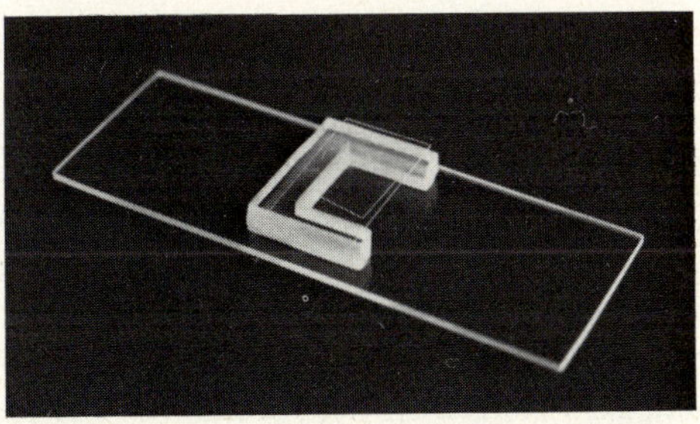

FIG. 4. Manipulation chamber. A wedge-shaped egg holder rests on a U-shaped acrylic plate attached to a supporting glass slide.

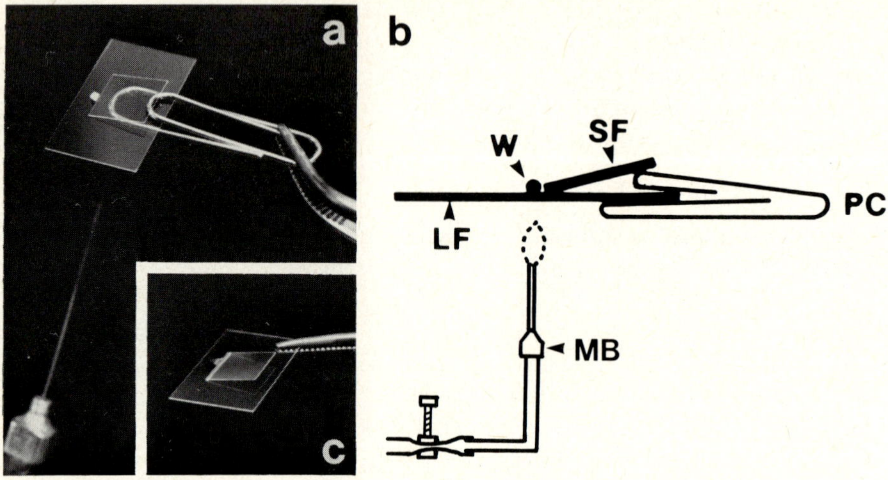


FIG. 5. Preparation of a wedge-shaped egg holder. Photograph (a) and schematic side view (b) of the procedure for making a wedge-shaped egg holder. The edge of a small cover slip fragment (SF) is placed on a paper clip (PC) which clasps a large cover slip fragment (LF). A small piece of wax (W) is placed on the large fragment adjacent to the small fragment. The tip of a microburner (MB) is seen underneath the cover slip fragments. (c) A completed wedge-shaped egg holder. When the wax is melted by heating, it spreads along the edge of the small fragment. After the wax has cooled and hardened, it cements one edge of the small fragment at an angle to the face of the large fragment, even after the paper clip is removed.

after the wax has cooled and hardened. One edge of the small fragment is thereby cemented to the large fragment, and all other sides remain open (Fig. 5c). Thus, a wedge-shaped egg holder is completed. Since it is very easy to construct such holders, they can be used disposably.

4. An oocyte suspension (30 to 40 μl containing 20 oocytes) is introduced with a capillary micropipet into the open end of the wedge-shaped egg holder, which is held so the open end is directed upward. Oocytes are allowed to settle by gravity to the vertex of the wedge-shaped space, where they are sandwiched. Thus, oocytes that settle in a line near the wax are flattened slightly and immobilized in the egg holder.

5. Such an egg holder is placed downward (small fragment side down) over the U-shaped plate on the supporting glass slide, which has been filled beforehand with about 0.4 ml of seawater (see Figs. 2 and 4). The open end of the wedge faces the open end of the U-shaped plate. Surface tension maintains the seawater in this chamber, and the open side of the chamber provides access to the oocytes for microinjection.

# III. Procedures of Microinjection

## A. Microinjection

The fundamental principle in microinjection is that the tip of the micropipet and the periphery of the recipient cell are adjusted to the same focal plane in the same field of view in the microscope. Under these circumstances, the micropipet can enter the recipient cell. Practically, the following procedures are performed.

1. The micropipet, mounted in the micromanipulator, is angled slightly higher than horizontal.
2. The focal plane of the microscope is adjusted on the recipient cell mounted in the manipulation chamber.
3. Without changing the focus, the manipulation chamber is removed from the field of view (to the left) with the drive of the mechanical stage.
4. The tip of the micropipet is brought into focus in the center of the microscope field, then moved downward a little, slightly out of focus, in order to avoid damaging the tip in the next step.
5. The recipient cell in the manipulation chamber is brought back into the microscope field with the drive of the mechanical stage. After the chamber comes over the micropipet tip, the micropipet is brought back into focus. Thus, the micropipet and the recipient cell are both in focus in the same microscope field.

After the micropipet penetrates the cell following the drive of the micromanipulator and the mechanical stage, the material in the micropipet is injected by applying positive pressure with the syringe. In general, the injected aqueous solution is observed to disperse readily into the cytoplasm. When pressure is reduced within the syringe, materials in the cell, including the cytoplasm or the nucleus, are drawn into the micropipet. The micropipet is kept in an almost horizontal position. This makes the manipulation of the micropipet easier than if it is inclined at an angle, since the micropipet remains in focus and therefore the movement of the meniscus of injected material can be monitored.

Although it is easy to penetrate the surface of starfish oocytes with a micropipet, the surface membrane of some eggs is so extensible that the cell surface merely depresses without allowing the micropipet to actually penetrate; in such cases, a long, funnel-shaped sleeve of the cell surface surrounds the micropipet. When the aqueous solution is injected, it merely passes out along the shaft of the micropipet toward the periphery of the egg without entering. In such cases, it is necessary to suck a small amount of egg cytoplasm from the outside of the surface membrane into the micropipet before injection in order to break the surface membrane of the egg. After this, the micropipet may be assumed to be inside the egg. A micropipet with a bevelled tip (see Section II,B) is also recommended in these cases.

When the material to be injected is an aqueous solution, it can be introduced into the micropipet through the tip from a short segment of glass capillary containing the injection solution by the following procedure (Fig. 6). A small amount of the injection solution (approximately 1 μl) is drawn by capillary action into the end of a glass capillary (for example, a 10-μl Drummond Micro-cap), which is cut and then serves as a reservoir. This reservoir is immersed in silicon oil filling the U-shaped space of a support slide (see Section II,C) covered by a cover slip. This prevents evaporation of the solution in the reservoir during handling. Under the microscope, the tip of the micropipet is advanced into the reservoir, and the injection solution is sucked up into the micropipet from the tip by applying negative pressure in the syringe. Then, the tip is sealed with a small amount of silicon oil surrounding the reservoir (Fig. 6c). Such a plug of silicon oil prevents loss of the injection solution before microinjection. It also prevents contamination of the injection solution with the medium surrounding the recipient cells in the manipulation chamber. The silicon oil plug, which is micro injected into the recipient cell together with the injection solution, is observed in the cell as a minute drop and serves as a landmark of the injected cell.

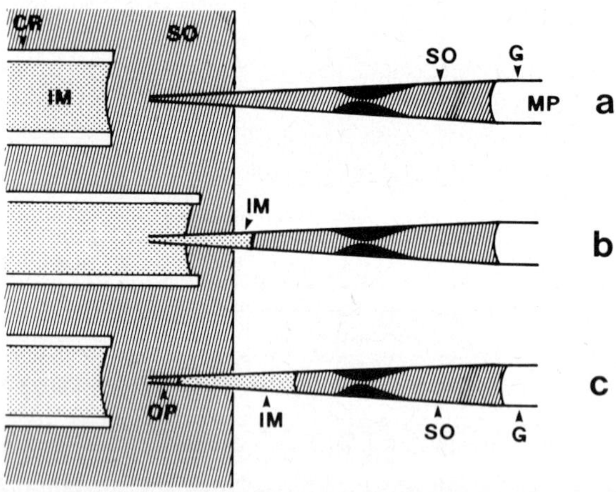


FIG. 6. Loading a micropipet with an aqueous solution to be injected. (a) A micropipet (MP), which has been loaded with silicon oil, is positioned under a microscope near the capillary reservoir (CR) which contains an aqueous solution to be injected and has been completely immersed in silicon oil (SO). (b) Injection material (IM) is sucked up from the capillary reservoir into the micropipet through its tip. (c) A silicon oil plug (OP) is next introduced from the surrounding silicon oil into the tip to seal the injection solution in the micropipet. G, gas phase (air). (a), (b), and (c) were adapted from Hiramoto (1974).

## B. Determination of Volume Injected

In order to measure the volume of material injected into a cell, the volume of the material sucked up into the micropipet is calculated by assuming that the part of the micropipet containing the material is a frustum of a cone. In Fig. 7a, the volume of material in the micropipet ($V$) is calculated as follows: $V = \pi L(D_0^2 + D_0 D + D^2)/12$, where $D_0$ and $D$ are the internal diameters of the micropipet at the two meniscuses of the material to be injected and $L$ is the height of cone. When one dose of material is loaded into the micropipet, $D_0$ is approximately 5 μm, since this is the conventional tip diameter for injecting into starfish oocytes. Since the error introduced by deviations in this assumption in $D_0$ is so small as to be negligible, the volume is determined in practice by referring to a table indicating relationships among $D$, $L$, and $V$ in the ranges typically encountered.

In precisely estimating the internal diameter, the error owing to the refraction of light at the surface of the micropipet is taken into consideration (Hiramoto, 1962, 1974) (Fig. 7b). The ratio of the true internal diameter to the apparent internal diameter measured under the microscope is assumed to be identical to the ratio of the refractive index of the surrounding medium to that of the glass of

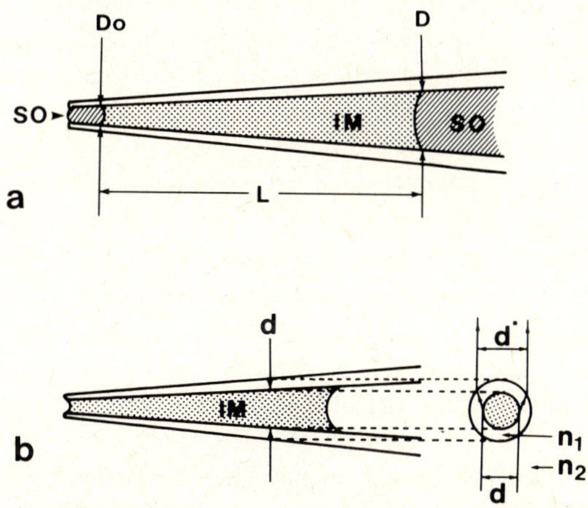


FIG. 7. Calculation of the injection volume in the micropipet. (a) The volume ($V$) of the material sucked up into the micropipet is calculated by assuming its shape as a frustum of a cone: $V = \pi L(D_0^2 + D_0 D + D^2)/12$, where $D_0$ and $D$ are the internal diameters of the micropipet corresponding to the top and end of introduced material, respectively, and $L$ is the height of the cone. IM, injection material; SO, silicon oil. (b) Correction of the error in the estimation of the internal diameter owing to the refraction of light rays at the surface of glass micropipet: $d = d' n_2/n_1$, where $d$ and $d'$ are the true and apparent internal diameters of micropipet, respectively, and $n_1$ and $n_2$ are the refractive indices of glass and the surrounding medium, respectively (see Hiramoto, 1962). (b) was adapted from Hiramoto (1984).

the micropipet: $d = d'n_2/n_1$, where $n_1$ and $n_2$ are the refractive indices of glass and the surrounding medium, respectively, and $d$ and $d'$ are the true and apparent internal diameters of the micropipet, respectively. When the surrounding medium is an aqueous solution, the correction due to refraction is more than 10%.

# IV. Application of Microinjection Techniques in Starfish to the Study of Oocyte Maturation

## A. Cytoplasmic Transfer

In starfish, fully grown ovarian oocytes are arrested at the prophase of first meiosis. Such immature oocytes are characterized by the presence of a large nucleus called the germinal vesicle. The resumption of meiosis, which is called oocyte maturation, is induced by a hormone, 1-methyladenine (Kanatani, 1973; Meijer and Guerrier, 1984). This substance is produced in the follicle cells surrounding each oocyte under the influence of gonad-stimulating hormonal peptide of neural origin (see Shirai, this volume). When added to the immature oocytes isolated in seawater, 1-methyladenine induces breakdown of the germinal vesicle and subsequent processes of meiotic maturation, including formation of two polar bodies and the egg pronucleus. In studying the mechanism of action of 1-methyladenine in inducing oocyte maturation, the microinjection technique has fully exhibited its power.

Microinjection of 1-methyladenine into the cytoplasm of immature oocytes of *Asterina pectinifera* failed to induce germinal vesicle breakdown in recipient oocytes even at a concentration higher than that which is effective when applied externally (Kanatani and Hiramoto, 1970). This result shows that 1-methyladenine acts on the external oocyte surface. In other words, this fact indicates that the hormonal message of 1-methyladenine is conveyed to the cytoplasm, after the binding of 1-methyladenine with its receptor in the oocyte surface, to promote the production of a cytoplasmic factor which directly induces germinal vesicle breakdown. Evidence for the production of such a "maturation-promoting factor" in the oocyte cytoplasm has been generated by the microinjection technique as follows (Fig. 8). Cytoplasm taken from an oocyte of *A. pectinifera*, which is maturing after the treatment with 1-methyladenine, induces germinal vesicle breakdown when injected into an intact immature oocyte, whereas injection of cytoplasm from nontreated immature oocytes has no effect (Kishimoto and Kanatani, 1976). Cytological examination of oocytes injected with maturation-promoting factor revealed that meiosis is completed normally and that the time course of meiosis in injected recipients is similar to that of oocytes treated with 1-methyladenine.

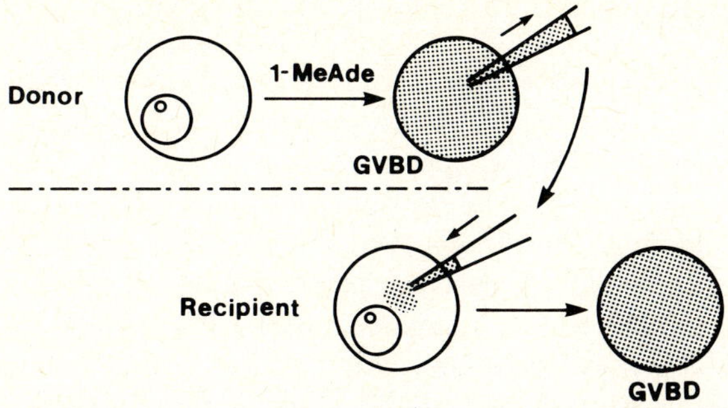


FIG. 8. Cytoplasmic transfer in starfish oocytes. Maturation-promoting factor-containing cytoplasm taken from a 1-methyladenine-treated oocyte is injected into an intact immature oocyte. This causes maturation in the recipient oocyte. GVBD, germinal vesicle breakdown.

The frequency with which germinal vesicle breakdown is induced in recipient oocytes by injecting cytoplasm from maturing oocytes is proportional to the volume of injected cytoplasm. Almost 100% germinal vesicle breakdown is obtained when the volume of cytoplasm containing maturation-promoting factor is 220 pl, which corresponds to about $\frac{1}{15}$ to $\frac{1}{20}$ of the volume of the *Asterina* oocyte. Donor oocytes of maturation-promoting factor are treated with $10^{-5}$ *M* 1-methyladenine for 10 minutes and washed with two changes of seawater in order to dilute 1-methyladenine in the medium. At room temperature, the content of maturation-promoting factor in the maturing cytoplasm reaches its maximum level about 20 minutes after 1-methyladenine addition, at which time germinal vesicle breakdown generally occurs. After this, it persists for a period of about 20 minutes, which corresponds to the ''M-phase'' of the first meiotic division, and then declines at the time of the first polar body formation. The maturation-promoting activity in donor oocytes appears again at the second metaphase of meiosis, and then decreases, showing that maturation-promoting activity in maturing oocytes fluctuates along with the meiotic cell cycle (Kishimoto and Kanatani, 1976; Doree *et al.*, 1983). Further, maturation-promoting activity is already found in the cytoplasm of some oocytes 13 minutes after 1-methyladenine treatment, when the germinal vesicles are still intact, although it is not detected at 7 minutes (Kishimoto *et al.*, 1984a). The activity thus does not seem to be the result of germinal vesicle breakdown.

In general, recipient oocytes of *A. pectinifera* are placed in 80% seawater before and during injection with cytoplasm containing maturation-promoting factor. This treatment allows the recipient oocytes practically to be more tolerant to a large volume of injected material than normal seawater. In 80% seawater, the volume of *Asterina* oocyte is about 4000 pl.

## B. Enucleation

The role of germinal vesicle material in the production of maturation-promoting factor was investigated with *Asterina* oocytes (Kishimoto *et al.*, 1981). The results show that the amount of maturation-promoting factor produced in enucleated oocytes is small compared with that in nucleated oocytes. Production of maturation-promoting factor in enucleated oocytes is restored to a considerable extent by injecting germinal vesicle material. On the other hand, the amount of maturation-promoting factor is amplified during maturation when maturation-promoting factor is injected into intact oocytes. In contrast, such amplification is not observed in enucleated oocytes. Thus, the germinal vesicle is considered to have some role in production of maturation-promoting factor by amplification. For such study, enucleated oocytes and oocyte fragments with or without germinal vesicles were obtained as follows.

Immature oocytes with germinal vesicles are put into the wedge-shaped egg holder. After the micropipet is inserted into the germinal vesicle, germinal vesicle material, including nuclear membrane, is sucked up into the micropipet. Thus, enucleation is performed. For this purpose, a micropipet with an internal diameter of more than 10 μm is recommended, since the germinal vesicle material is sticky and will plug a smaller tip. The entire contents of the germinal vesicle (about 400 pl), or part of it, can be removed as desired. This procedure is applied to the transfer of germinal vesicle material. In another way to obtain an enucleated oocyte, the germinal vesicle is squeezed out from the oocyte through a perforation made with a fine glass needle in the oocyte surface near the germinal vesicle under a dissecting microscope (Hirai *et al.*, 1971).

Oocyte fragments with and without a germinal vesicle can be obtained from a single oocyte using two fine glass needles attached to separate micromanipulators. Micropipets for injection serve as glass needles. An oocyte in a small drop of seawater (about 0.1 ml) is placed on a glass slide under a microscope. The two glass needles are laid across the oocyte from opposite sides and then are separated (Fig. 9), cutting the oocyte into two halves, one enucleated and the other nucleated.

# V. Concluding Remarks

This article describes in concrete detail the equipment and procedures used for microinjection that I use with starfish oocytes. In this system, I cannot only inject material into the cell but can also remove and transfer the constituents of the cell. This system has been applied successfully to oocytes and eggs of amphibia and mice in addition to echinoderms. Most of the system was devised by Hiramoto (1974), especially the elegant and simple method for controlling the injection

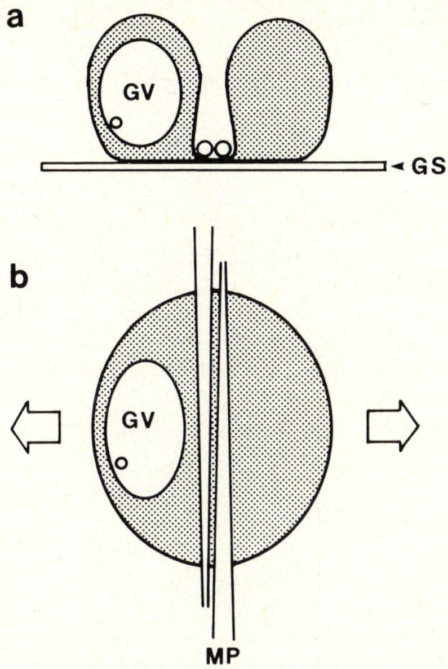


FIG. 9. Schematic procedure for making oocyte fragments with and without a germinal vesicle (GV). (a) Side view. (b) Bottom view. Two glass micropipets (MP) are laid across the oocyte on a glass slide (GS) and then are separated laterally as shown with arrows.

volume by constricting the micropipet. My further modifications, including the procedure for constructing the wedge-shaped egg holder, permit everyone to master microinjection techniques easily after a little practice.

Using these procedures, it has been found that maturation-promoting activity occurs in a wide variety of eukaryotic cells such as mitotic cells as well as maturing oocytes (see Kishimoto *et al.*, 1982, 1984b). Maturation is induced in frog or starfish oocytes when injected with cytoplasm taken from maturing oocytes or cleaving blastomeres and also with the M-phase extracts of mammalian cultured cells, including HeLa, CHO, and V-79 cells. Thus, maturation-promoting factor seems to play a general role in initiating cell division both in meiosis and in mitosis and to act nonspecies specifically throughout the animal kingdom. Considering these facts, the microinjection techniques described in this article will make a contribution not only to the study of starfish oocyte maturation but also to the study of cell division itself, which attracts interest more generally in cell biology.

## ACKNOWLEDGMENTS

This article is dedicated to the late Professor Haruo Kanatani, who provided me the opportunity to learn microinjection and passed away on February 13, 1984. I thank Professor Yukio Hiramoto for introducing me to the microinjection technique in 1974 and for his continual encouragement and valuable discussion. Thanks are also due to Mrs. K. Noda for line drawings. Supported in part by grants-in-aid from the Ministry of Education, Science and Culture of Japan.

## REFERENCES

Balakier, H. (1978). *Exp. Cell Res.* **112,** 137–141.

Bennett, J., and Mazia, D. (1981). *Exp. Cell Res.* **131,** 197–207.

Bulet, P., Kishimoto, T., and Shirai, H. (1985). *Dev. Growth Differ.* **27,** 243–250.

Chambers, R. W., and Chambers, E. L. (1961). "Explorations into the Nature of the Living Cell." Harvard Univ. Press, Cambridge, Massachusetts.

Curtis, D. R. (1964). *In* "Physical Techniques in Biological Research" ( W. L. Nastuk, ed), Vol. 5, pp. 144–190. Academic Press, New York.

Doree, M., Peaucellier, G., and Picard, A. (1983). *Dev. Biol.* **99,** 489–501.

Furusawa, M., Yamaizumi, M., Nishimura, T., Uchida, T., and Okada, Y. (1976). *Methods Cell Biol.* **14,** 73–80.

Gurdon, J. B. (1977). *Methods Cell Biol.* **16,** 125–139.

Hirai, S., Kubota, J., and Kanatani, H. (1971). *Exp. Cell Res.* **68,** 137–143.

Hiramoto, Y. (1956). *Exp. Cell Res.* **11,** 630–636.

Hiramoto, Y. (1962). *Exp. Cell Res.* **27,** 416–426.

Hiramoto, Y. (1974). *Exp. Cell Res.* **87,** 403–406.

Hiramoto, Y. (1984). *In* "Jikken Seibutsugaku Koza" (T. Niitsu and T. Okigaki, eds.), Vol. 8, pp. 277–297. Maruzen, Tokyo (in Japanese).

Kanatani, H. (1973). *Int. Rev. Cytol.* **35,** 253–298.

Kanatani, H. (1985). *In* "Biology of Fertilization" (C. B. Metz and A. Monroy, eds.), Vol. 1, pp. 119–140. Academic Press, New York.

Kanatani, H., and Hiramoto, Y. (1970). *Exp. Cell Res.* **61,** 280–284.

Kiehart, D. P. (1982). *Method Cell Biol.* **25,** 13–31.

Kishimoto, T. (1982). *Cell Differ.* **11,** 255–257.

Kishimoto, T., and Kanatani, H. (1976). *Nature (London)* **260,** 321–322.

Kishimoto, T., Hirai, S., and Kanatani, H. (1981). *Dev. Biol.* **81,** 177–181.

Kishimoto, T., Kuriyama, R., Kondo, H., and Kanatani, H. (1982). *Exp. Cell Res.* **137,** 121–126.

Kishimoto, T., Usui, N., and Kanatani, H. (1984a). *Dev. Biol.* **101,** 28–34.

Kishimoto, T., Yamazaki, K., Kato, Y., Koide, S. S., and Kanatani, H. (1984b). *J. Exp. Zool.* **231,** 293–295.

Kopac, M. J. (1964). *In* "Physical Techniques in Biological Research" (W. L. Nastuk, ed.), Vol. 5, pp. 191–233. Academic Press, New York.

Masui, Y., and Clarke, H. J. (1979). *Int. Rev. Cytol.* **57,** 185–282.

Meijer, L., and Guerrier, P. (1984). *Int. Rev. Cytol* **86,** 129–196.

Poste, G. (1972). *Int. Rev. Cytol.* **33,** 157–252.

Poste, G., Papahadjopoulos, D., and Vail, W. J. (1976). *Methods Cell Biol.* **14,** 33–71.

Richter, H. P., Scheurich, P., and Zimmermann, U. (1981). *Dev. Growth Differ.* **23,** 479–486.

Vassetzky, S. G., Sekirina, G. G., Skoblina, M. N., and Bilinkis, A. A. (1983). *Cell Differ.* **12,** 73–76.

Yamamoto, F., and Furusawa, M. (1978). *Exp. Cell Res.* **117,** 441–445.

# Chapter 22

## *Rapid Repetitive Microinjection*

ANITA M. COLIN

*Department of Zoology*
*University of Washington*
*Seattle, Washington*



## I. Introduction

Microinjection provides a rapid and certain means of changing the internal environment of the cell, both throughout the cytosol and in local regions. Various cell types have been injected with a great diversity of materials, including macromolecules, subcellular fractions, and chemical reagents (for reviews, see Baserga *et al.*, 1980; Celis *et al.*, 1980). Microinjection of genes and nuclei has been used to study transcription in amphibians (Gurdon and Brown, 1977; Birnstiel *et al.*, 1979; DeRobertis *et al.*, 1979; Gurdon *et al.*, 1976). Microinjection has also been used successfully to study the regulation of translation in amphibian oocytes and eggs (Gurdon *et al.*, 1971; Laskey *et al.*, 1977; Richter *et al.*, 1982, 1983) and as a means of blocking translation of specific mRNAs in amphibian and mammalian cells (Izant and Weintraub, 1984; Melton, 1985).

Echinoderm eggs have been injected to answer a variety of biological questions. Hiramoto (1962) injected sperm into eggs to study the mechanism of fertilization. The concentration of free calcium at fertilization was measured by the injection of aequorin (Steinhardt *et al.*, 1977). Maturation-promoting factor was identified by the injection of the cytoplasm of starfish eggs into starfish oocytes (Kishimoto and Kanatani, 1976; Kishimoto *et al.*, 1981; Picard and Doree, 1984; Kishimoto, this volume). The injection of anti-myosin antibodies revealed a role for myosin in cytokinesis, but not in chromosome movement (Mabuchi and Okuno, 1977; Kiehart *et al.*, 1977), and the injection of calcium into local regions of zygotes illuminated its effect on the mitotic apparatus (Kiehart, 1981). McMahon *et al.* (1984) have injected DNA into sea urchin eggs to analyze the regulation of gene transcription.

In 1966, Spirin observed messenger ribonucleoprotein particles, which he called "informosomes", in fish eggs. He proposed that the proteins associated with these mRNAs prevent their translation in eggs. In order to test whether mRNAs are masked in amphibian oocytes, i.e., are the rate-limiting component of protein synthesis, Laskey *et al.* (1977) injected globin mRNA into *Xenopus* oocytes. They observed that the globin mRNA was translated by the *Xenopus* oocytes, but that the synthesis of endogenous proteins decreased in proportion to the increase in globin synthesis. They concluded that the globin mRNA competed with the endogenous mRNAs for a rate-limiting component of the protein synthesis machinery. Several laboratories have proposed that mRNAs are masked in sea urchin eggs (Humphreys, 1971; Davidson, 1976; Jenkins *et al.*, 1978; Ilan and Ilan, 1978; Raff, 1980; Winkler and Steinhardt, 1981). The controlled release of the masking factor could account for the gradual recruitment of mRNAs after fertilization (Dolecki *et al.*, 1977; Raff *et al.*, 1981).

I have refined the methods for injecting mRNA to test directly whether mRNAs limit the rate of protein synthesis in sea urchin eggs, or whether, as in *Xenopus,* some other component of the translational machinery is rate limiting. Since sea urchin eggs are much smaller than *Xenopus* oocytes (80 to 180 μm in diameter as compared to 1.2 mm), a large number of eggs are required to synthesize sufficient peptides for an accurate statistical analysis. Thus, it was necessary to devise methods for injecting large numbers of eggs in a relatively short time. Although the following methods are designed for the rapid injection of large numbers of echinoid eggs or zygotes, they should also be useful for injecting a variety of cell types or for fewer cells. McMahon *et al.* (1985) have adapted the method described in this article for injecting DNA into sea urchin eggs.

## II. General Method

Dejellied eggs or zygotes are placed in a ring in small culture dishes treated with protamine sulfate (Steinhardt *et al.*, 1971). The cells attach to the dish and can be quickly injected under a stereomicroscope by turning the dish and penetrating them with a micropipet attached to a standard micromanipulator. I have used the eggs of the green sea urchin, *Strongylocentrotus droebachiensis,* and the sand dollar, *Dendraster excentricus,* as they are among the largest echinoid eggs available (180 and 120 μm in diameter, respectively) and thus, are excellent for rapid injection with simple equipment. Smaller eggs can be injected successfully with similar methods if an inverted phase microscope is used to view the cells (McMahon *et al.*, 1985).

Microinjection is performed with a glass micropipet that is attached to a syringe via polyethylene tubing. The entire system is filled with fluid, the tip of the pipet containing the injection solution. Constant pressure is applied to the syringe via a micrometer screw to produce a continuous flow of solution from the micropipet tip. A continuous flow prevents clogging and facilitates the rapid injection of large numbers of cells. I have used two methods for applying pressure to the syringe. A motor-driven syringe is useful for injecting relatively large cells. This method allows for good control of the flow rate, but requires a relatively large micropipet tip and extra time to fill the pipet. A high-pressure syringe is more useful when injecting smaller cells and is easier to set up. In this method, a micropipet with a small-diameter tip is necessary. The equipment for high-pressure injection is shown in Fig. 1.

The proteins synthesized by the injected cells are labeled with a mixture of tritiated amino acids that is added to the injection dish. After incubation, the cells are collected with a braking pipet and analyzed for the incorporation of [$^3$H]amino acids into total proteins and into specific peptides.

## III. Preparations

### A. Making the Micropipets

Motor-driven and high-pressure injection require different types of micropipets. For motor-driven injection, the micropipets can be made from 50-μl disposable pipets (Curtin Matheson Scientific). The glass is siliconized with Prosil-28 (PCR, Inc., Gainesville, FL), then autoclaved (or baked for 4 hours at 160°C) and stored in a foil package. Siliconizing the glass prevents macromolecules from sticking to it, and baking destroys ribonuclease. The micro-

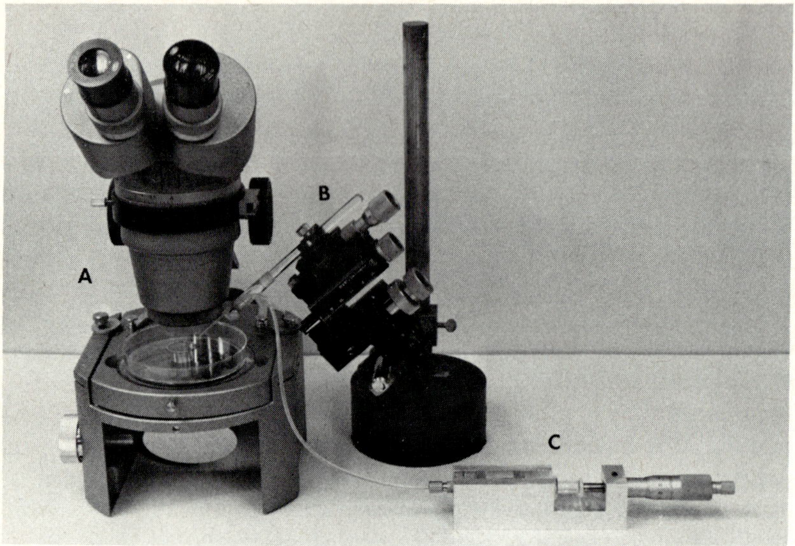


FIG. 1 Microinjection apparatus. The equipment shown includes a zoom stereomicroscope (A), a micromanipulator (B), and a fluid-filled pressure injection system (C). The pressure system includes a micrometer screw mounted in an aluminum block with a glass syringe attached. The syringe is connected to the micropipet with polyethylene tubing.

pipets are produced as needed on a vertical pipet puller (David Kopf Instruments, Model 700C) with a one-coil heating element, with the solenoid on 60 and current at 21 mA. Three pipets can be made from a single 10-cm length of glass. Longer pipets are more difficult to fill. For high-pressure injection, a capillary glass with a filament must be used (Frederick Haer and Co., borosilicate, with Omega Dot, 1.0 mm o.d., 0.75 mm i.d.). The glass is cleaned by placing it in hot 35% nitric acid for 30 minutes, followed by rinsing with distilled water. The glass is siliconized, then rinsed in distilled water that has been filtered through a 0.2-μm sterilization filter (type S, Sybron/Nalge). The pipets are pulled as above except that a two-coil heating element is used. They are best prepared on the day of the injection and stored on a clay strip in a small box.

## B. Filling the Micropipets

For motor-driven injection, polyethylene tubing (PE 50, Clay Adams) is attached to a 10-μl Hamilton syringe which has been mounted in a wooden or aluminum block so it can be driven by a micrometer screw. A fine, but inexpensive, micrometer screw can be obtained from any tool company. I use one manufactured by Mitutoyo with a range of 0–25 mm in 1 μm divisions. The end of the micrometer screw is attached to the plunger of the Hamilton syringe with a piece of rubber tubing. The micrometer screw is coupled to and driven by a reversible, variable-speed ($\frac{1}{8}$–4 rpm) motor. The micrometer is attached to the motor shaft with another piece of rubber tubing. The filling of the micropipet and assembly of the apparatus is as follows. First, the polyethylene tubing is attached via a 25-gauge needle to a syringe containing Fluorinert (FC-77, 3M Co.) and filled with the fluid. A 1-cm piece of silicone tubing (0.03 in. i.d., 0.09 in. o.d.) is attached to the other end of the polyethylene tubing and also filled. The micropipet is filled from the back using another syringe filled with Fluorinert. The pipet is connected to the exposed end of the fluid-filled polyethylene tubing by pushing the back of the pipet snugly into the silicone tubing, avoiding air bubbles. The pipet is then clamped securely onto the micromanipulator. A plastic Petri dish is filled with distilled water and the pipet is gently touched to the bottom of the dish while viewing it under the microscope, so as to break the extreme tip. Water will enter the broken tip. The tip should be 5–10 μm in diameter. Fluorinert is pushed through with the syringe, which is still attached to the polyethylene tubing, to rid the tip of water and air. Next, the polyethylene tubing is disconnected from the syringe and connected to an Fluorinert-filled 10-μl Hamilton syringe. There should be no air in the Hamilton syringe or in the connection. (I keep the Hamilton syringe attached to the micrometer screw and motor for easy use.) To fill the micropipet with injection solution, a drop of the solution (2–5 μl) is placed on a small piece of Parafilm under the microscope, the tip of the micropipet is positioned in the drop, and the motor is run in reverse at high speed (4 rpm). The solution can be covered with Fluorinert to prevent evaporation. Clogging can be a problem at this stage—if so, the injection solution may require centrifuging. After filling the tip, the motor is run forward while watching the Fluorinert–water interface until the solution just begins to flow out. This step eliminates any vacuum in the system. The pipet is filled with injection solution only after the cells are ready, as the filled tip will dry out and clog in air. During the injections, the motor is run forward at a low speed (about $\frac{1}{8}$ rpm).

For high-pressure injection, the connections between the pipet and the pressure syringe must be leak-proof and the solution path free of air bubbles. Polyethylene tubing (PE 240, Clay Adams) is attached through a Clay Adams adapter to a 2-ml glass syringe with a metal Luer-Lok tip. The syringe and micrometer screw are mounted on an aluminum block, and the micrometer screw is held firmly in place with a set screw (Fig. 1). The polyethylene tubing and syringe should be prefilled with Fluorinert. The micropipets are filled from the back using a separate filler pipet. To make a filler pipet, a siliconized and baked 50-μl disposable pipet is hand pulled over a low Bunsen burner flame. The injection solutions must be dust free. Buffers can be filtered through a 0.2-μm Nalgene sterilization filter unit, and RNA-containing solutions can be centrifuged at 12,000 *g* for 3 minutes in an Eppendorf microfuge. About a microliter of injection solution is drawn into the filler pipet with a metal-barreled, screw-type pipet filler (Clay Adams) and introduced through the back of the micropipet into its tip. If the injection solution is not too viscous, it will flow from the filler pipet into the very tip of the micropipet. All bubbles in the tip must be gently tapped out. The rest of the pipet is filled with Fluorinert using a syringe with a 2-in., 30-gauge needle. To mount the micropipet, a leak-proof, air bubble-free connection between the pipet and the polyethylene tubing is made in the following manner. The polyethylene tubing is filled with Fluorinert using a syringe. A 1-cm piece of silicone tubing (0.04 in. i.d., 0.07 in. o.d.) is inserted into the polyethylene tubing and also filled with Fluorinert. The micropipet is gently inserted into the silicone tubing and the micrometer screw is backed off. The pipet is then pushed into the silicone tubing simultaneously as the silicone tubing is pushed into the polyethylene tubing to form a tight connection. Backing off on the micrometer avoids the high pressures brought about by this connecting process. (If solution can be seen flowing from the tip at this time, the pipet is already broken and must be replaced.) The micrometer screw is turned to create the pressure necessary for the injections. After attempting to inject the first few cells, the pipet tip will break slightly and a steady flow will soon be established.

## C. Preparing Injection Dishes and Braking Pipets

Linbro tissue culture wells (96 flat-bottom wells per tray, 1.0 × 1.5 cm, Flow Laboratories) make excellent injection dishes. I find it useful to mark the wells on the bottom with a fine-tipped water-proof pen. A cross divides the dish into quadrants of cells which can be injected with different solutions. The wells are cut into single dishes and 1% protamine sulfate (Sigma) is added. After 30–60 seconds, the dishes are rinsed with distilled water. The 1% protamine sulfate solution can be reused many times if refrigerated. Seawater that has been filtered through nitrocellulose with 0.45-μm pores and contains 10 m*M* glycylglycine (pH 8.0) and 1 m*M* streptomycin sulfate is then added to each dish.

A good braking pipet is essential for placing the cells in the injection dish and removing them to assay. A braking pipet is made by pulling one end of a 50-μl glass pipet over a Bunsen burner to make a tip slightly larger than the egg diameter. The other end is then pulled very long and thin. The quality of this tip can be tested by putting the larger end into water. Water should flow in slowly. If water does not enter, a bit of the thin end should be broken off and the tip retested. If it enters too quickly, the thin end should be pulled again. The larger end of the good tip is carefully fire polished and the opposite end is inserted into a shortened Pasteur pipet and sealed with a soft wax, e.g., from Gouda cheese. A piece of rubber tubing attached to the Pasteur pipet can be used to control the fluid flow by mouth. With a good braking pipet, one can easily pick up single eggs. To avoid clogging, care is taken not to suck water up near the braking end of the pipet.

## D. Isolating and Treating Cells

Echinoid gametes are spawned by injection of 0.55 *M* KCl into the coelomic cavity of the adults. Sperm are collected dry by inverting the male in a Petri dish. The eggs are collected at 10°C in filtered seawater and washed by settling in filtered seawater containing 1 m*M* streptomycin sulfate. To dejelly the eggs, a small number is put into a used, but cleaned (not with soap) Petri dish and washed in pH 4–5 filtered seawater. (Acid-washed cells tend to stick to a new Petri dish.) Since different batches of eggs will dejelly under different conditions, I make very acid seawater (pH 3, or one drop of concentrated HCl in 50 ml filtered seawater) and add it gradually to the eggs while observing them under the microscope. The eggs are swirled in the dish under the microscope until they touch each other. Then they are quickly rinsed in filtered seawater containing glycylglycine and streptomycin. Fertilized eggs are prepared by adding diluted sperm to the eggs in filtered seawater containing 5 m*M* *p*-amino benzoic acid (PABA). PABA prevents hardening of the fertilization envelop (Hall, 1978). After 5 minutes, the zygotes are washed well and dejellied as described for the eggs. Zygotes with intact fertilization envelopes will sometimes attach well to the protamine sulfate-coated injection dish, but often the zygotes will attach better and are more easily injected if the envelop is removed. The envelopes can be removed by flushing the cells in and out of a pasteur pipet in acid seawater. Demembranated zygotes must be rinsed quickly with normal seawater as they become sticky in acid seawater. One to two hundred eggs or zygotes are collected with a braking pipet in a minimal amount of seawater and placed in a ring in the prepared dish. Placing the cells in a ring allows one to reach all of the cells easily with the micropipet and to keep track of the injected cells. After placing them in rings, the eggs are allowed to settle for about 20 seconds before moving the dish. The dish of cells can be temporarily stored if it is suitably chilled.

# IV. Injecting the Cells

## A. Techniques

During the injection, the small dish of cells is kept cool (5–10°C) by placing it in a larger Petri dish with some ice chips. To begin injections, the dish with cells is positioned under a stereomicroscope, illuminated with transmitted light

(Olympus Zoom Stereomicroscope, model SZ-III, or any suitable stereomicroscope, 80× magnification). The micropipet, positioned at an angle of about 45°, is placed in the dish and located near the cells under 20× magnification. During injection, the cells are viewed at 80× magnification and moved sequentially under the pipet by turning the Petri dish. The cell to be injected is ''jabbed'' with the coarse axial advance control of the micromanipulator. The cells must be jabbed carefully, however, as the tip of the pipet can be broken against the bottom of the dish. Echinoid eggs are very soft, so some skill is necessary to penetrate them without hitting the dish. The flow of solution from the tip is a great aid to penetration. If the flow is too slow, the pipet will not penetrate the cell. However, if the flow is too fast, it will cause much damage. As a last resort, the cells can be made more turgid by adding some distilled water to the dish. The pipet tip must be kept clean during the injections. A tip coated with cytoplasm and membrane will do a great deal of damage to the cells. It can be easily and frequently cleaned by pulling it through the air–water interface. If necessary, the tip can be gently scraped along the bottom of the dish to dislodge adherent matter. A good micropipet can be used to inject several hundred cells at the rate of 10 cells/minute.

During the injection of a cell, the emergence of solution into the cell is seen as a clearing of cytoplasm. If the tip is just under the membrane, a clear blister will form. If the tip is actually in the cytoplasm, a less distinct clearing will form. The tip of the pipet is pulled out as soon as the clearing forms. The volume injected will depend on the flow rate but is generally 10–20% of the cell volume. The volume can be estimated by measuring the diameter of the clear area appearing in the cell just after injection. The amount of substance injected is more easily changed by varying the concentration of the solution than by trying to inject different volumes. Although an injection solution as simple as 10 m*M* Tris, pH 6.8, and 1 m*M* EGTA will work, I generally use a more physiological solution containing 300 m*M* glycine, 10 m*M* NaCl, 2 m*M* $MgCl_2$, 10 m*M* glycylglycine (adjusted to pH 6.8 with KOH), and 0.1 m*M* EGTA. For globin mRNA injection, rabbit reticulocyte poly(A)$^+$ RNA is dissolved in the glycine buffer.

## B. Incubating and Collecting the Cells

If the cells are to be labeled with radioactive amino acids, the amino acids are added to the dish after the injections. The injection dishes are placed in covered Petri dishes containing wet paper and are rotated gently with a clinical rotator (Thomas) at 60 rpm in an incubator. At the end of the incubation, normal-looking cells are collected individually by ''scooping'' and sucking with the braking pipet. Although whole cells can be removed with the braking pipet, they are often damaged at their attachment point and cannot be cultured further. Sand dollar eggs are particularly difficult to remove intact. The cells can be stored in ethanol at −20°C in small microfuge tubes (Eppendorf). For gel electrophoresis, the samples are centrifuged in an Eppendorf microfuge, the ethanol is removed, and the cells are dissolved in sample buffer. Alternatively, the cells can be placed directly in sample buffer.

## V. Results of Microinjection and Gel Electrophoresis of Proteins

The major difficulty I have encountered with injecting unfertilized eggs is that of artificially activating them. Activation occurs more frequently at the end of the reproductive season and for certain batches of eggs. Some solutions, such as those with pH below 6, will cause more activation than others. EGTA appears to decrease the frequency of egg activation. The appearance of a full or partial fertilization envelope after injection is always accompanied by an increase in amino acid uptake and incorporation into proteins. Activation is reduced by using appropriate solutions and sharp, clean micropipets. Injected zygotes can develop normally, although cleavage is delayed and rate of protein synthesis reduced. I have been able to fertilize injected eggs of *S. droebachiensis* and *D. excentricus,* but the fertilization rate has been low even with eggs which are merely attached to the protamine sulfate-coated injection dish. Eggs of the sea urchins *Lytechinus pictus* and *S. purpuratus* are more easily fertilized after attachment to dishes.

Echinoid eggs and embryos translate injected globin mRNA. Fig. 2 is an example of the results I have obtained from injecting *D. excentricus* eggs with globin mRNA. The eggs were injected with 8 mg/ml reticulocyte poly(A)$^+$ RNA in glycine buffer. Control eggs, in the same dish, were injected with glycine buffer. After injection, 40 μCi/ml [$^3$H]amino acid mix (146 mCi/mg, ICN Pharmaceuticals) was added to the seawater in the dish. The eggs were incubated on a rotator for 3 hours at 12°C. Forty uninjected eggs, 40 buffer-injected eggs, and 40 RNA-injected eggs were collected and stored in ethanol at −20°C. The eggs, dissolved in SDS sample buffer, were electrophoresed on a 12.5% polyacrylamide gel (Laemmli, 1970). The gel was treated with an autoradiography enhancer (Enlightening, New England Nuclear), dried, and exposed to preflashed XRP-1 film for 6 days. The new band in the lane containing RNA-injected eggs comigrates with rabbit globin (Fig. 2, lane C). The film can be cut and scanned in a Gilford spectrophotometer to quantify the amount of [$^3$H]-labeled peptides in the globin and endogenous bands. Several exposure times are needed to accurately measure dark bands. The method has been successfully used to determine the contribution of mRNAs to the rates of protein synthesis in *S. droebachiensis* eggs and zygotes (Colin and Hille, 1986).

FIG. 2. Fluorograph of mRNA-injected sand dollar eggs. Lane A: Uninjected eggs. Lane B: Buffer-injected eggs. Lane C: Globin-injected eggs. The dark band in lane C comigrates with rabbit globin. The other, fainter protein bands are from the translation of endogenous mRNAs.

## VI. Summary

This article describes methods to inject large numbers of echinoid eggs and zygotes with globin mRNA and label them with tritiated amino acids. The products can be quantitatively analyzed to determine the amounts of peptide synthesis. The injections can be performed quickly and require relatively simple and inexpensive equipment. Although I designed my methods to inject large numbers of echinoid eggs and zygotes, these methods should also be useful for injecting many different molecules and ions into smaller numbers of cells as well as into a variety of cell types.

## ACKNOWLEDGMENTS

I would like to thank Dr. Andrew P. McMahon for useful discussions and Drs. Merrill B. Hille and Robert E. Maxson for careful reading and criticism of this manuscript. This research was supported by National Institutes of Health Training Grant GM 07270 to A.C. and by National Institutes of Health Grants HD 11070 to Dr. Merrill B. Hille.

## REFERENCES

Baserga, R., Croce, C., and Rovera, G., eds. (1980). "Introduction of Macromolecules into Viable Mammalian Cells." Liss, New York.

Birnstiel, M., Portmann, R., Busslinger, M., Schaffner, W., Probst, E., and Kressmann, A. (1979). *Alfred Benzon Symp.* **13,** 117–129.

Celis, J. E., Graessman, A., and Loyter, A., eds. (1980). "Transfer of Cell Constituents into Eukaryotic Cells." Plenum, New York.

Colin, A. M., and Hille, M. B. (1986). *Dev. Biol.,* in press.

Davidson, E. H. (1976). "Gene Activity in Early Development," 2nd Ed. Academic Press, New York.

DeRobertis, E. M., Laskey, R. A., and Gurdon, J. B. (1977). *Trends Biochem. Sci.* **2,** N250–N253.

Dolecki, G. J., Duncan, R. F., and Humphreys, T. (1977). *Cell* **11,** 339–344.

Gurdon, J. B., and Brown, D. D. (1977). *In* "Molecular Biology of the Mammalian Genetic Apparatus" (P. O. P. T'so, ed.), Vol. 2, pp. 111–123. Elsevier, Amsterdam.

Gurdon, J. B., Lane, C. D., Woodland, H. R., and Marbaix, G. (1971). *Nature (London)* **233,** 177–182.

Gurdon, J. B., DeRobertis, E. M., and Partington, G. (1976). *Nature (London)* **260,** 116–120.

Hall, H. G. (1978). *Cell* **15,** 343–355.

Hiramoto, Y. (1962). *Exp. Cell Res.* **27,** 416–426.

Humphreys, T. (1971). *Dev. Biol.* **26,** 201–208.

Ilan, J., and Ilan, J. (1978). *Dev. Biol.* **66,** 375–385.

Izant, J. G., and Weintraub, H. (1984). *Cell* **36,** 1007–1015.

Jenkins, N. A., Kaumeyer, J. F., Young, E. M., and Raff, R. A. (1978). *Dev. Biol.* **63,** 279–298.

Kiehart, D. P. (1981). *J. Cell Biol.* **88,** 604–617.

Kiehart, D. P., Inoue, S., and Mabuchi, I. (1977). *J. Cell Biol.* **75,** 258a (Abstr.).

Kishimoto, T., and Kanatani, H. (1976). *Nature (London)* **260,** 321–322.

Kishimoto, T., Hirai, S., and Kanatani, H. (1981). *Dev. Biol.* **81,** 177–181.

Laemmli, U. K. (1970). *Nature (London)* **227,** 680–685.

Laskey, R. A., Mills, A. D., Gurdon, J. B., and Partington, G. A. (1977). *Cell* **11,** 345–351.

Mabuchi, I., and Okuno, M. (1977). *J. Cell Biol.* **74,** 251–263.

McMahon, A. P., Novak, T. J., Britten, R. J., and Davidson, E. H. (1984). *Proc. Natl. Acad. Sci. U.S.A.* **81,** 7490–7494.

McMahon, A. P., Flytzanis, C. N., Hough-Evans, B. R., Katula, K. S., Britten, R. J., and Davidson, E. H. (1985). *Dev. Biol.* **108,** 420–430.

Melton, D. A. (1985). *Proc. Natl. Acad. Sci. U.S.A.* **82,** 144–148.

Picard, A., and Doree, M. (1984). *Dev. Biol.* **104,** 357–365.

Raff, R. A. (1980). *In* "Cell Biology: A Comprehensive Treatise" (D. M. Prescott and L. Goldstein, eds.), Vol. 4, pp. 107–136. Academic Press, New York.

Raff, R. A., Brandis, J. W., Huffman, C. J., Koch, A. L., and Leister, D. E. (1981). *Dev. Biol.* **86,** 265–271.

Richter, J. D., Jones, N. C., and Smith, L. D. (1982). *Proc. Natl. Acad. Sci. U.S.A.* **79,** 3789–3793.

Richter, J. D., Evers, D. C., and Smith, L. D. (1983). *J. Biol. Chem.* **258,** 2614–2620.

Spirin, A. S. (1966). *Curr. Top. Dev. Biol.* **1,** 1–38.

Steinhardt, R. A., Lundin, L., and Mazia, D. (1971). *Proc. Natl. Acad. Sci. U.S.A.* **68,** 2426–2430.

Steinhardt, R., Zucker, R., and Schatten, G. (1977). *Dev. Biol.* **58,** 185–196.

Winkler, M. M., and Steinhardt, R. A. (1981). *Dev. Biol.* **84,** 432–439.

# Chapter 23

## *Electrophoresis of Eggs and Blastomeres*


KIYOSHI SANO

*Developmental Biology Laboratory*
*Department of Plastic Surgery*
*School of Medicine*
*Kitasato University*
*Sagamihara, Kanagawa, Japan*


---



## I. Introduction

Surface membranes of eggs and blastomeres seem to play important roles in the cell-to-cell interactions during development, as observed in fertilization and the morphogenetic processes of embryos. Cell electrophoresis is one of the useful methods to analyze the surface properties of cells in the living state. The surface potential (zeta potential) of cells is revealed by measuring their mobility relative to an applied electric field using a cell electrophoresis apparatus.

Only small numbers of cells are required for measurements of their electrophoretic mobility. In addition, even a mixture of the cells consisting of different cell types can be measured at the same time, if each type of cell can be distinguished from the others by its morphology under the light microscope.

This article describes the methods to make a rectangular, horizontal type of cell electrophoresis apparatus and to measure the electrophoretic mobility of the eggs and blastomeres of sea urchins.

# II. Theory

## A. Electrophoretic Mobility

When an electrically charged particle is suspended in a medium under a defined electric field, the particle begins to move owing to the electrical force. Electrophoretic mobility is defined as the velocity of the charged particle per unit strength of the electric field, and is expressed by $\mu m \cdot sec^{-1} \cdot V^{-1} \cdot cm$. Under certain conditions the electrophoretic mobility of a particle is proportional to its zeta potential, which is a potential at its surface with respect to the bulk of the medium. When a uni-univalent electrolyte is used as the electrophoresis medium in the concentration ranges of $10$–$10^{-3}$N, the following Helmholtz–Smoluchowski equation applies for a particle whose radius is more than several micrometers, including almost all cells (Brinton and Lauffer, 1959).

$$u = \frac{\zeta D \cdot 10^4}{4\pi\eta \cdot (300)^2} \quad (1)$$

where $u$ is the electrophoretic mobility of the particle ($\mu m \cdot sec^{-1} \cdot V^{-1} \cdot cm$)
$\zeta$ is the zeta potential of the particle (V)
$D$ is the dielectric constant of the electrophoresis medium (80 for water)
$\eta$ is the viscosity of the electrophoresis medium in cgs units (poise)

## B. Stationary Level

When an electric field is applied to the medium in the electrophoresis chamber, the medium streams or flows by electroosmosis for the following reason. The glass surface of an electrophoresis chamber has electrically negative charges. Therefore, the cations in the medium are attracted toward the negatively charged glass surface and are preferentially accumulated near the surface. On the other hand, the anions in the medium tend to accumulate in the center of the chamber. When a certain voltage difference is applied to the chamber, the cations near the glass surface flow toward the cathode. The anions in the center of the chamber flow toward the anode. Thus, the medium in the chamber as a whole flows as shown in Fig. 1.

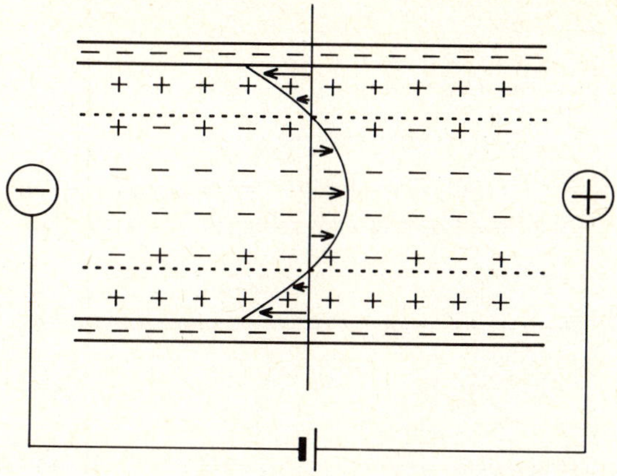

FIG. 1. Schematic cross-section of the inner space of the electrophoresis chamber. Arrows indicate the flow of the electrophoresis medium due to electroosmosis. The dotted line indicates the stationary levels, where the net flow of the medium is zero. +, Positive charge; −, negative charge.

Figure 1 reveals that there is a level within the chamber where the net flow of the medium is zero. This level is called the "stationary level." To measure the mobility of cells without effects of the medium flow, only the cells located at the stationary level must be measured.

The two stationary levels of an electrophoresis chamber are determined as follows, on the basis of Komagata's theory (Brinton and Lauffer, 1959)

$$a = \frac{1}{2} + \sqrt{\frac{1}{12} + \frac{32}{\pi^5 K}} \qquad (2)$$

where $K$ is the ratio of width to the thickness of the inner space of the chamber
$a$ is the position of the stationary levels in terms of the fractional depth from the top or bottom of the chamber

For example, when $K$ is equal to 21.5, $a$ is 0.202. When the width of the inner space of the chamber is far greater than its thickness, the stationary levels are located at 0.21 and 0.79 of the depth of the chamber.

# III. Apparatus

## A. Electrophoresis Chamber

A simple electrophoresis chamber, after a design by Mori (1968), is made by adhering glass plates with the adhesive of epoxy resin as shown in Fig. 2. This

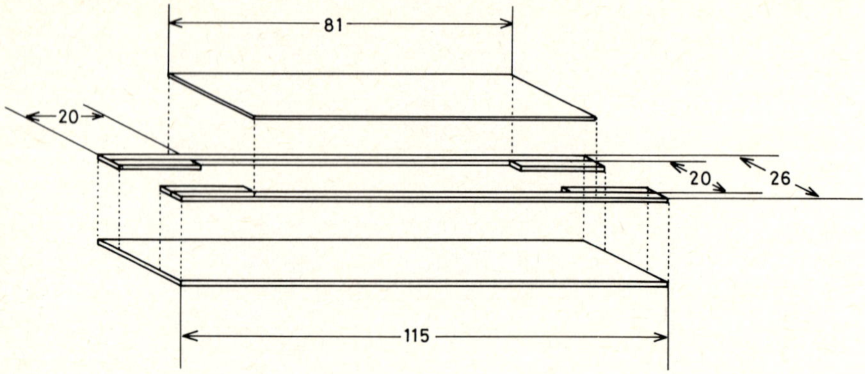


FIG. 2. The composition of the simple electrophoresis chamber. The upper glass plate is 0.2 mm thick. All other parts are of 1.0-mm thick glass. The units of the figures are in millimeters.

chamber has no temperature-control system but is useful for simple, short-range measurements of the electrophoretic mobility of cells. Using this type of chamber, Oshima (1975) has measured the electrophoretic mobility of sea urchin eggs during the division cycle.

In the following, a modified chamber with a water jacket for circulating water of constant temperature is described. The electrophoretic mobility of particles is influenced by changes of temperature of the medium; therefore, the chamber with the circulating system is useful for maintaining a constant temperature during a series of measurements that continue for several days.

This chamber is made of three types of glass plates of different thickness (0.2, 1.0, and 3.0 mm). The detailed composition of the chamber is shown in Figs. 3 and 4. The size of the inner space for electrophoresis is 20 × 1 × 74 mm. The horizontality and the thickness of the inner space of the chamber must be confirmed to be even on the stage of a light microscope by focusing the microscope on both its "floor" and "ceiling."

A holder for the electrophoresis chamber is made by sticking together glass plates (Fig. 5). The holder is used to fix the chamber to the mechanical stage of the light microscope. I have used a Nikon microscope with a 20× objective and 10× eyepiece.

## B. Circuit and Electrode

The electric current is supplied by a dry-cell battery of 90 V. The circuit includes an ammeter, a variable resistance, and a reversing switch (Fig. 6). The amount of the current is controlled by changing the resistance and is checked with the ammeter. A piece of platinum wire about 0.3 mm in diameter is coiled and soldered to the tip of the conducting wire of the circuit. The platinum

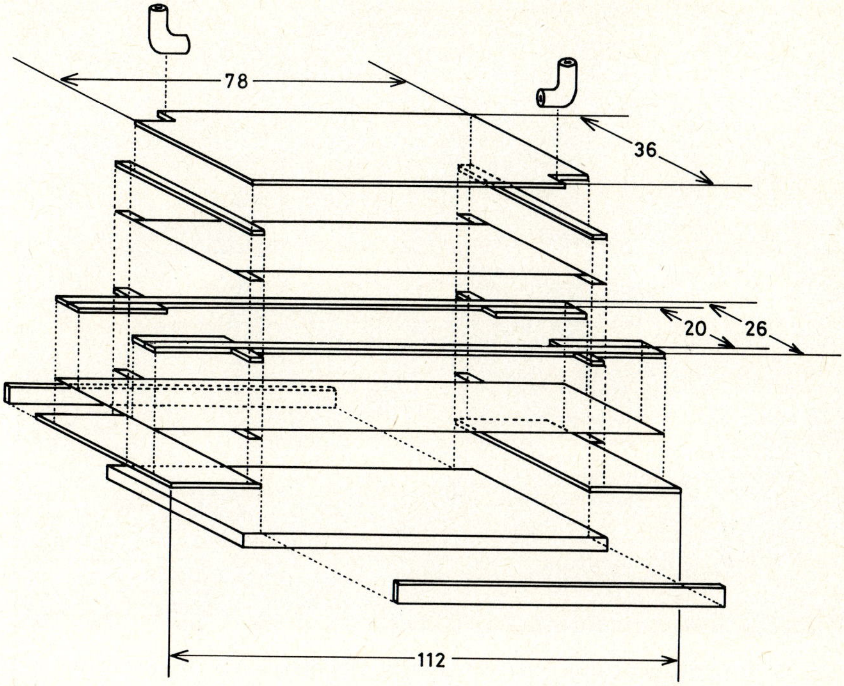


FIG. 3. The composition of the water-jacketed electrophoresis chamber with a circulation system for maintaining constant temperature. Three types of glass plates, with thicknesses of 0.2, 1.0, and 3.0 mm, are used. The units for the figures are millimeters.

electrode is immersed in a small beaker containing a saturated solution of $KNO_3$ and is connected to the agar block on the electrophoresis chamber by an agar bridge. The agar blocks and agar bridges are made of 3% agar (powdered agar; Wako Chemicals, Osaka) dissolved in a mixture of Ca-free seawater and distilled water in a 7:93 ratio (v/v). The ratio of the Ca-free seawater is the same as used for the electrophoresis medium.

## C. Circulation System for Constant Temperature

Water of a constant temperature is made to flow by gravity from a thermoregulated bath to the outer space (water jacket) of the electrophoresis chamber through vinyl tubing. The water is then allowed to flow from the chamber to a reservoir. The reservoir is set below the thermoregulator so that the water flows from the regulator to the reservoir by the difference of the height. Such a flow of water does not generate pulsatory vibrations as observed when a pump is used. The water accumulated in the reservoir is pumped back to the thermoregulator.

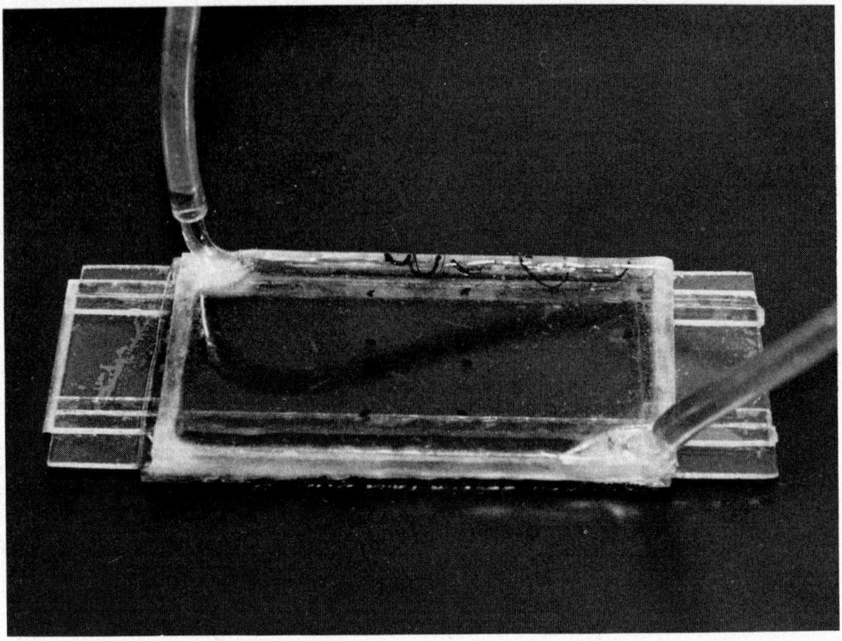

FIG. 4. A photograph of a complete water-jacketed electrophoresis chamber with the circulation system for maintaining constant temperature.

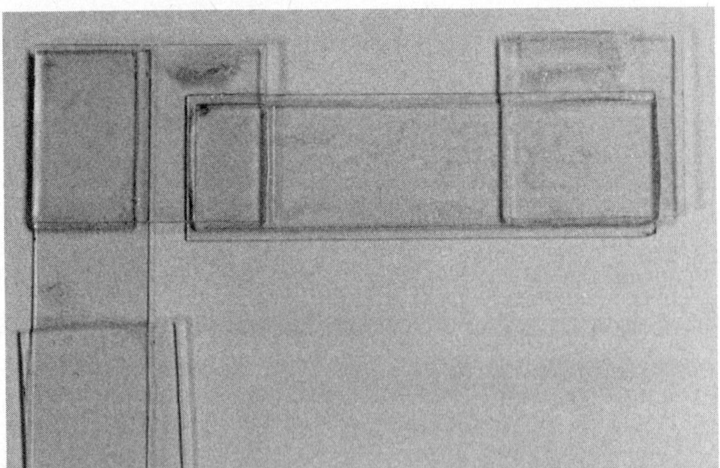

FIG. 5. A platform for holding the electrophoresis chamber. The holder is adhered to the moving guide of the mechanical stage of a light microscope, attaches the chamber to the mechanical stage, and supports two small beakers containing $KNO_3$ solutions.

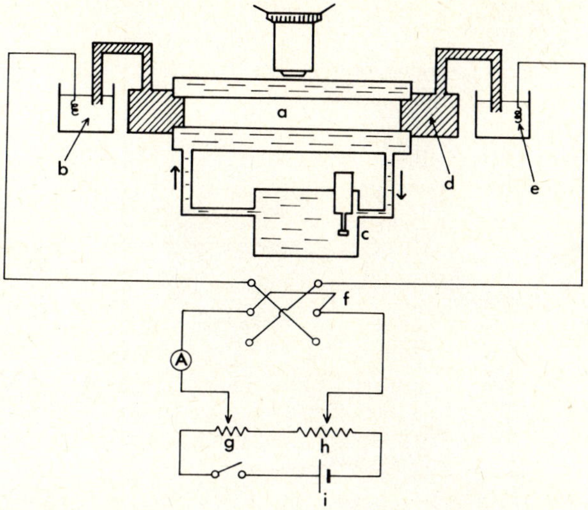


FIG. 6. Schematic outline of the electrophoresis apparatus and the electrical circuit. (a) Inner space of electrophoresis chamber; (b) saturated $KNO_3$ solution; (c) thermoregulator; (d) agar block and agar bridge; (e) platinum electrode; (f) reversing switch; (g, h) resistance, 10 and 100 kΩ; (i) 90-V dry battery. (From Sano, 1977.)

A survey view of the complete cell electrophoresis apparatus is shown in Fig. 7.

## IV. Measurement of Electrophoretic Mobility

### A. Determination of the Electrophoresis Medium

In order to avoid rapid sedimentation of eggs or blastomeres during the measurement of their electrophoretic mobility, a solution whose density is nearly equal to the eggs or blastomeres must be used as an electrophoresis medium. In addition, to reduce the thermal convection caused by the electric current, the ionic strength of the medium must be low.

To determine the composition of the electrophoresis media that fulfil the above conditions, media composed of 0.7 ml of Ca-free seawater and 9.3 ml of the mixed solutions of 0.8 *M* mannitol and 0.75 *M* sucrose in 20 m*M* Tris(hydroxymethyl)aminomethane (Tris)/HCl (pH 8.05 at 25°C) in various ratios are prepared in centrifuge tubes. The eggs or cells are layered over these mixed media. After centrifuging the tubes at 1250 *g* for 15 to 20 minutes with a swinging-bucket rotor, the distribution of the eggs in the tubes is examined. A medium is selected as the electrophoresis medium for the eggs or blastomeres if most of the cells are suspended in it, yet some sediment to a pellet. For the eggs of *Hemi-*

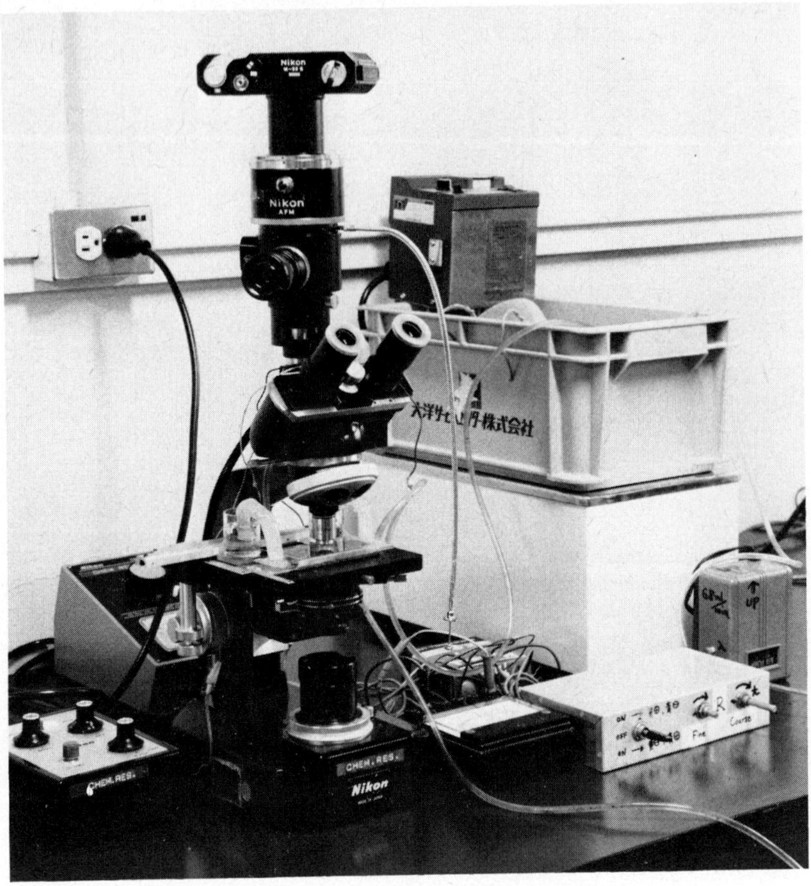


FIG. 7. Survey view of the complete cell electrophoresis apparatus.

*centrotus pulcherrimus,* a medium composed of 7 vol of Ca-free seawater, 45 vol of 0.8 *M* mannitol in 20 m*M* Tris/HCl (pH 8.05 at 25°C), and 48 vol of 0.75 *M* sucrose in 20 m*M* Tris/HCl (pH 8.05 at 25°C) was used as the electrophoresis medium. It may be possible to replace mannitol with sorbitol.

In some cases, 0.5 m*M* $CaCl_2$ was added to the above medium to minimize the surmised release of surface materials from cells caused by calcium deficiency after dissociation into single cells.

## B. Determination of the Stationary Level

Water at 25°C is circulated around the chamber, and the chamber is filled with the electrophoresis medium. The thickness of the inner space of the electrophoresis chamber is measured by changing the focus level of the light microscope using the calibrated fine adjustment knob. Then, from Komagata's theory [Eq. (2)], the location of the stationary level can be determined.

## C. Preparation of Eggs and Blastomeres

To measure the electrophoretic mobility of eggs or blastomeres before the hatching stage, the removal of the fertilization membranes is required. To obtain eggs deprived of the fertilization membranes, concentrated eggs are fertilized with freshly diluted sperm (final 2000-fold dilution of dry sperm in *H. pulcherrimus*) in a mixture of Ca-free and natural seawater in a 4:1 ratio. After 40–60 seconds, the egg suspension is diluted with 180–200 vol of Ca/Mg-free seawater to raise the fertilization membranes as high as possible. After the fertilization membranes have elevated high, the egg suspension is passed twice through a nylon mesh (with 74 μm of openings for *H. pulcherrimus* eggs whose diameter is about 90 μm). After washing with Ca-free seawater, the fertilized eggs are cultured in Ca-free or natural seawater in Petri dishes coated with 1.5% agar. At a desired developmental stage, the embryos are collected by centrifugation and dissociated into single cells in a dissociation medium. One of the useful dissociation media for sea urchin embryos is composed of 0.4 *M* sorbitol, 0.5 *M* glycerol, and 20 m*M* Tris/HCl (pH 8.2 at 4°C) (Sano, 1977). After incubating the embryos in the dissociation medium for 5 minutes at 4°C, the embryos are dissociated into single cells by pipetting. After centrifugation, the dissociated single cells are washed once with a washing solution, which is a mixture of 7 vol of Ca-free seawater and 93 vol of a solution composed of 0.4 *M* sorbitol, 0.5 *M* glycerol, and 20 m*M* Tris/HCl (pH 8.2 at 4°C). After centrifugation at 500–1250 *g* for 5 minutes with a swinging-bucket rotor, the pellet of cells is suspended in the electrophoresis medium of 25°C. The cell suspension is incubated in the thermoregulator at 25°C until the cells are transferred to the electrophoresis chamber.

## D. Loading the Chamber with Cells

Three percent agar solution is prepared by dissolving agar powder in a mixture of Ca-free seawater and distilled water in a 7:93 ratio. A "U"-shaped glass tube, whose inner diameter is about 10 mm, is filled with the hot 3% agar solution, and is used as an agar bridge between an agar block at the end of the chamber and a beaker containing $KNO_3$.

Before use, the electrophoresis chamber is filled with a detergent solution, such as "Haemosol" (Haemo-Sol Ltd.), and is rinsed well with water. This washing is conducted every time before use. Then the chamber is rinsed once with the electrophoresis medium. Drops of the hot agar solution are placed on one end of the electrophoresis chamber and pressed by a glass slide to make an agar block whose thickness is equal to that of the inner space of the chamber. Two irregular sides of the agar block are trimmed to make it rectangular. A small amount of the electrophoresis medium is introduced into the chamber, and the agar block is pushed several millimeters into the chamber. It is necessary not to produce any air bubbles at this time. Then, additional drops of the hot agar solution are placed on the agar block to seal one entrance of the chamber completely.

On the other end of the chamber drops of the dissolved hot agar are pressed with a glass slide, and a rectangular agar block is made by trimming it. The sample cells suspended in the electrophoresis medium at 25°C are introduced into the chamber using a pipet until the chamber is filled. The agar block is pushed several millimeters into the chamber without producing bubbles. Then, the entrance of the chamber is sealed completely by dripping an additional hot agar. The sealed chamber is set on the stage of the light microscope. The electrophoresis chamber is fixed to the mechanical stage of the microscope by the holder. Then, both ends of the chamber are connected to the circuit by the agar bridges. One end of an agar bridge is fixed to an agar block of the chamber using hot agar, and the other end of the bridge is immersed in the beaker containing saturated $KNO_3$ as shown in Fig. 8. The platinum electrodes are also immersed in the $KNO_3$ solution.

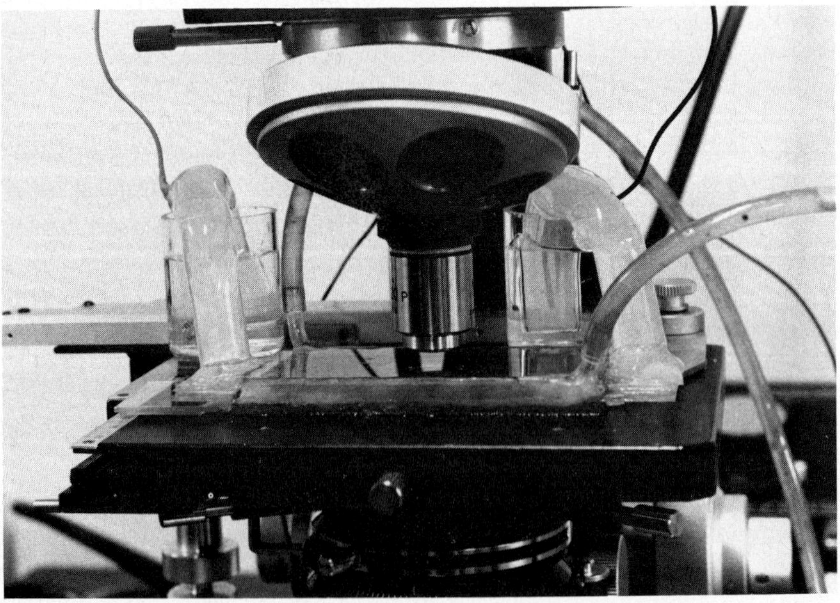

FIG. 8. Close-up view of the electrophoresis chamber mounted on a microscope stage, with agar bridges in place.

## E. Measurement

### 1. Measurement of the Velocity of Cells

First, the focus of the light microscope is adjusted to the bottom surface of the inner space of the electrophoresis chamber. Then, the body tube of the microscope is lifted with the fine adjustment knob the exact distance to the stationary level as determined in Section IV,B.

The switch of the circuit is on, and the flowing current is adjusted to a certain constant value to get the voltage gradient of about 5 V/cm. When the electrophoresis medium for *H. pulcherrimus* was used, the voltage gradient was 4.8 V/cm at a current of 5.0 mA.

Under an adequate voltage gradient, the time required for a cell observed at the stationary level to traverse the markings of an eyepiece micrometer is measured using a stop watch. In my set-up the space between the major markings is about 25 μm. After the cell traverses one marking, the polarity of the current is immediately reversed, and the time required for the same cell to traverse the marking toward the opposite direction is measured. It is also possible to measure the velocity of the cells by taking photographs.

For each sample, more than 20 cells are measured. After every five measurements, the location of the focus and the amount of the current are checked and reset.

When a set of measurements of one sample is complete, the voltage difference between the agar blocks is confirmed to be constant using a voltmeter during a series of measurements. Most of the cells were viable after electrophoresis when examined by a dye-exclusion test using 0.25% Trypan Blue in Ca-free seawater.

### 2. Voltage Gradient and Electrophoretic Mobility

The voltage gradient applied during electrophoresis is determined as follows.

First, the specific conductance of the electrophoresis medium is measured. Using both a conductivity cell with platinum electrodes and a Wheatstone bridge, the resistance ($R$) of 0.1 $M$ KCl solution is measured. Since the specific conductance (conductivity) ($\kappa$) of 0.1 $M$ KCl at 25°C is known to be 0.01288 $\Omega^{-1}\cdot\text{cm}^{-1}$, the cell constant ($k$) of the conductivity cell is determined from $k = \kappa\cdot R$.

Then, the resistance of the electrophoresis medium ($R_m$) is measured using both the above conductivity cell and the Wheatstone bridge. Then, the specific conductance of the electrophoresis medium ($\kappa_m$) is obtained by

$$\kappa_m = k/R_m \tag{3}$$

The resistance of the electrophoresis chamber filled with the electrophoresis medium ($R_c$) is obtained by

$$R_c = l/\kappa_m \cdot a \tag{4}$$

where $l$ is the length of the chamber

$a$ is the cross-sectional area of the inner space of the chamber

The voltage difference applied to the chamber ($V_c$) is calculated from

$$V_c = R_c \cdot I_c = l \cdot I_c/\kappa_m \cdot a \tag{5}$$

where $I_c$ is the current flowing the circuit (A). Therefore, the voltage gradient is expressed as follows

$$V_c/l = I_c/\kappa_m \cdot a \qquad \text{(V/cm)} \tag{6}$$

The electrophoretic mobility of a cell ($\mu m \cdot sec^{-1} \cdot V^{-1} \cdot cm$) is obtained by dividing the average velocity ($\mu m/sec$) of a cell for both directions by the voltage gradient (V/cm).

## V. Application

Using this electrophoresis chamber, the electrophoretic mobilities of the micromere- and mesomere-derived cells of sea urchin embryos were examined during their development (Sano, 1977), showing that the negative charges of the cell surface increase during the migration as primary mesenchyme cells of the micromere-derived cells *in vitro*.

Since the size of sea urchin eggs is large, it is rather difficult to measure their electrophoretic mobility accurately at the stationary level. This difficulty is reduced to some degree by coating the inner wall of the chamber with agar as shown by Oshima (1975). Since this coating reduces the streaming of the electrophoresis medium due to electroosmosis, it will reduce the errors in the measurements caused by the displacement of eggs from the stationary level. For coating the chamber with agar, a hot 1% agar solution is introduced into the chamber for 2–3 seconds and is then allowed to flow out. The newly formed surface of the coated agar becomes a new wall with fewer surface charges for cell electrophoresis. Oshima (1975) has reported using the agar-coated chamber; a peak of negative potential on egg surface occurs 15 minutes before the formation of a cleavage furrow.

Enzymatic treatment can alter the charged groups at the surface. Kojima and Maekawa (1972) combined enzymatic pretreatment with cell electrophoresis studies of an island-forming strain of rat hepatoma cells. Sialic acid is present in a cryptic form but exposed on the cell surface after pretreatment with chondroitinase-ABC, whereas it is present in an exposed form in the free-cell strain of hepatoma without treatment with chondroitinase-ABC.

Measurements of the surface charges of cell organelles, such as nuclei, are also possible using an electrophoretic method (Sato *et al.*, 1975).

As reviewed briefly, cell electrophoresis is a unique method for detecting changes that occur at the outer surface of living cells and organelles. Although this method is limited by the fact that the only property it detects is electric charge or potential, it seems to provide a useful tool for analyzing cell surface architecture when combined with other methods that modify or dissect the molecules on the cell surfaces.

## Acknowledgment

The author thanks Dr. N. Oshima of Toho University for her kind comments on the agar-coating method of the electrophoresis chamber.

## References

Brinton, C. C., Jr., and Lauffer, M. A. (1959). *In* ''Electrophoresis; Theory, Methods, and Applications'' (M. Bier, ed.), pp. 427–492. Academic Press, New York.

Kojima, K., and Maekawa, A. (1972). *Cancer Res.* **32,** 847–852.

Mori, Y. (1968). *Jpn. J. Bacteriol.* **23,** 32–33 (in Japanese).

Oshima, N. (1975). *Dev. Growth Differ.* **17,** 19–25.

Sano, K. (1977). *Dev. Biol.* **60,** 404–415.

Sato, C., Kojima, K., Matsuzawa, T., and Hinuma, Y. (1975). *Radiat. Res.* **62,** 250–257.

# Chapter 24

# *The Compression Method for Determining the Surface Force*

MITSUKI YONEDA

*Department of Zoology*
*Faculty of Science*
*Kyoto University*
*Kyoto, Japan*

---



## I. Introduction

The rounding up of cultured cells at mitosis is a familiar observation. Denuded amphibian eggs flattened on the substratum by gravity also round up shortly before cytokinesis. This phenomenon of rounding up reflects physiological changes in cytoplasm that can be monitored continuously in living materials. Since intact echinoderm (and most other marine) ova are spherical even in interphase and do not change their form during the first mitotic cycle, their stiffness has been assessed by applying defined external forces and measuring the resulting deformation. One such method is to compress the egg, as originally developed by Cole (1932).

When a spherical egg with the initial diameter $Z_0$ is compressed between parallel plates with a constant load $F$, the egg will be gradually flattened over several minutes until it attains an equilibrium thickness $Z$. The ratio $Z/Z_0$ will serve as an estimate of the stiffness of the egg. The flattening of course causes a stretching of the egg surface, as well as global deformation of the bulk endoplasm.

This compression method does not by itself distinguish between the stiffness of the cortical layer and possible contributions from structures in the inner cytoplasm. Indeed, measurements of the viscoelasticity of the cytoplasm of sea urchin eggs (Hiramoto, 1970) indicate that the inner cytoplasm may temporarily resist deformation when subjected to external forces. Nevertheless, the resistance will soon decay without leaving a permanent stress after the egg shape comes to equilibrium. In terms of mechanics, therefore, the egg cytoplasm in equilibrium will be practically fluid so that we can assume that the resistance of sea urchin eggs to the force of compression is mainly due to the stiffness of the egg cortex which opposes stretching. Thus, the value of $Z/Z_0$ will represent a measure of the mechanical property, or the "stiffness," of the egg cortex. Changes in this value of $Z/Z_0$ have been taken as indicators of certain activities of egg cytoplasm in various conditions (Hiramoto, 1963b; Yoneda *et al.*, 1978; Schroeder, 1981; Yamamoto and Yoneda, 1983; Yoneda and Schroeder, 1984). Moreover, the measured stiffness affords clues to the nature of possible structures that support the integrity of the egg cortex (Usui and Yoneda, 1982).

## II. The Bending Balance as a Monitor of the Force of Compression

The conventional procedure adopted by Hiramoto (1963a,b, 1976) and Yoneda (1964, 1973, 1980) is to compress a single egg between a pair of parallel pieces of cover glass (Fig. 1). One of them is cemented to the end of a slender glass fiber which works as a bending balance to monitor the force of compression. The other piece, which will be called the "stage," is held by the short, stiff rod.

If commercial glass wool is available, a straight glass fiber suitable for the bending balance will be easily found. If not, one can be made by pulling molten soft glass in the flame of a microburner. The thickness of the fiber suitable for the balance depends on the materials to be studied. The proper compliance (ratio of force to yielding) is $10^{-9}$ N/μm ($10^{-4}$ dyn/μm or about 100 ng/μm) for immature oocytes of starfish (which are very stiff) and $10^{-10}$ N/μm ($10^{-5}$ dyn/μm or about 10 ng/μm) for unfertilized sea urchin eggs (which are very weak). A fiber with the proper thickness may be selected by consulting Table I, which shows the compliance of glass fibers 3 cm long calculated by the formula for the bending of a cantilever. Fibers 3–4 cm in length are thus prepared. The

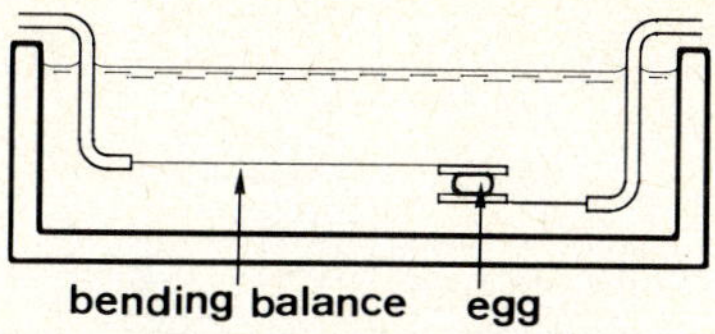


FIG. 1. Compression of a single egg between a pair of plates. One of the plates is attached to a flexible glass fiber working as a bending balance. The lower plate acts as a stage for mounting the egg. The dimensions of the plates and the egg are exaggerated in the figure.

glass rod for supporting the stage should be about 5–7 mm long and 0.2–0.3 mm thick. Two glass tubes, thinned to a diameter of 1–2 mm and bent (Fig. 2A) so as to fit into an observation chamber (cf. Fig. 4), serve as holders for the fiber and the rod. The latter are cemented to the thinned ends of the vertically held tubes with epoxy resin (Fig. 2A).

Small pieces of cover glass measuring about 1 × 2 mm are suitable as plates for compressing eggs. Long, parallel scratches 1 mm apart are made on a sheet of cover glass (0.17–0.2 mm thick) with a diamond pencil in order to make long rectangular fragments. After wrapping these strips in a small sheet of black paper, they are broken by bending the wrap between one's thumb nails. Among the random fragments, nearly rectangular pieces (1 × 2 mm) are selected.

Cementing a piece to the free end of the bending balance requires some practice: The piece is placed on a clean glass slide with its scratched side facing upward (so that the irregular scratched edges are away from the egg during measurements). The face of the piece is cleaned by applying a drop of alcohol. The slide is mounted on a dissecting microscope. The glass fiber attached to its holder is also cleaned in alcohol. The holder is mounted on a micromanipulator and, by monitoring with the microscope, the free end of the fiber is brought above the piece of cover glass on the slide (Fig. 2B). The piece is oriented to the

TABLE I

COMPLIANCE OF GLASS FIBERS 3 cm IN LENGTH

| Diameter (2$R$) (μm) | Compliance ($\times 10^{-10}$ N/μm)[a] | Suitable for |
|---|---|---|
| 20 | 0.4 | |
| 30 | 2 | Unfertilized sea urchin eggs |
| 40 | 7 | |
| 50 | 17 | Immature starfish oocytes |

[a] The compliance is estimated by the formula, force/bending = $3\pi R^4 E/4L^3$. ($L$ = 3 cm, and Young's modulus $E$ of soft glass is $5 \times 10^{10}$ N/m$^2$).

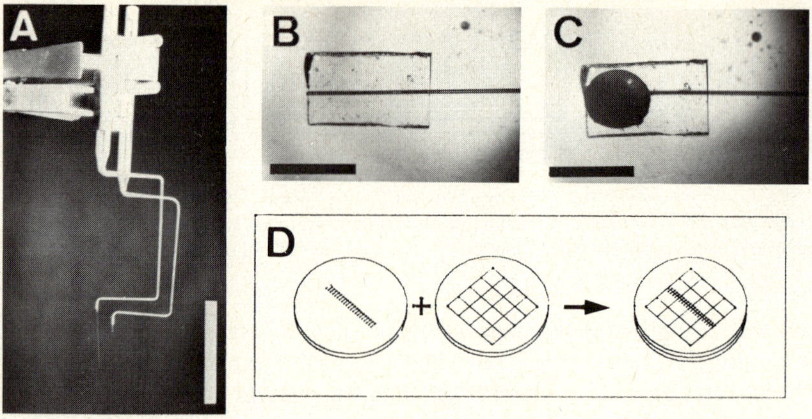


FIG. 2. (A) A pair of glass holders held vertically. A flexible fiber and a short rod are attached to the ends of the holders with epoxy resin. Bar is 5 cm. (B) An end of the glass fiber is positioned above the rectangular piece of cover glass. Bar is 1 mm. (C) The tip of the fiber with a droplet of adhesive is placed on the piece. Bar is 1 mm. (D) Combining an ocular micrometer and a square reticle.

fiber by moving the glass slide on the microscope stage so that the longer dimension of the piece is parallel to the glass fiber. After fine positioning, the fiber is raised by a vertical movement of the manipulator away from the piece. A very small drop of adhesive is then attached to the tip of the fiber, and the bending balance is lowered again very gradually until the tip comes into contact with the upper side of the piece (Fig. 2C). The adhesive will gradually spread over the piece. Wait until the adhesive hardens. During frequent use of the bending balance, the glass plate may become detached from the tip of the fiber. A proper choice of adhesive will reduce the occurrence of detachment: The two-component type of epoxy resin for repairing porcelain is excellent.

Similarly the piece of cover slip serving as the stage is attached to the rod. The pair of holders, one bearing the bending balance and the other supporting the stage, is held by two micromanipulators.

## III. Experimental Set-Up

The observation chamber, which is to be mounted on the stage of a horizontal microscope, should be large enough to allow ample clearance for movement of the balance. The simplest one is just a box of the size of a standard glass slide and several millimeters depth. In Fig. 3, the back wall is a piece of acrylic sheet 5 mm thick. A block of acrylate 7 mm thick is carved into a U-shaped piece for the

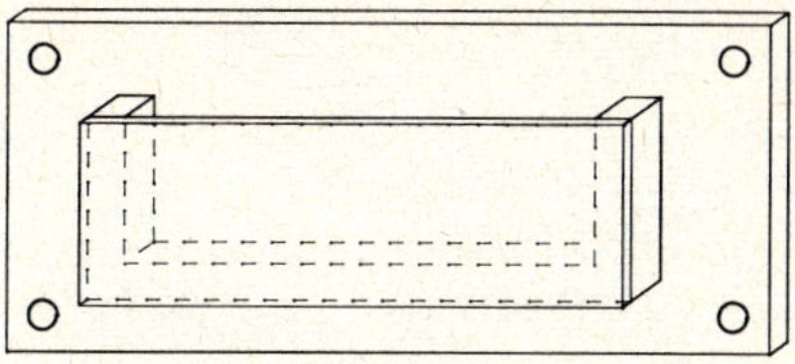

FIG. 3. Design of a simple observation chamber made of a glass slide, a U-shaped piece of acrylic, and an acrylic back wall with holes for attaching to the microscope stage.

bottom and sides of the box. The U-shaped piece and the back wall are fused together with a viscous solution of acrylate in ethylene dichloride. A glass slide as the front face of the box is finally cemented to the U-shaped piece with epoxy resin. Probably due to different thermal expansion coefficients of glass and acrylate, the bond between these two materials may often break. To minimize this trouble, the use of epoxy resin for repairing procelain is again recommended. This simple chamber suffices for water temperatures close to room temperature. If water temperatures substantially different from the surrounding temperature are needed, it will be necessary to construct a special chamber surrounded by a water jacket. The reason for this will be explained in Section IV. As seen in Fig. 4, such a chamber can be easily constructed by assembling two pieces of U-shaped acrylic sheet.

A classic microscope whose arm breaks 90° at the joint is convenient for the present purpose. Its arm is turned to the horizontal plane, and the observation chamber is fastened to the vertical stage with screws. Either a 10 or 20× objective with a long working distance (more than 2 mm) is used for observing the egg through the front face of the chamber. For convenience a prism star diagonal, available at a low price from suppliers of telescope accessories, is inserted between the ocular lens and the microscope body in order to bend the optical path vertically. Its sleeves are either thinned (by carving them) or thickened (by wrapping them with paper) to adapt them to the dimensions of the microscope tube and the ocular lens.

While an ocular micrometer is indispensable for measuring the displacement of the balance, a square reticle is very helpful to align the compressing plates. With their etched faces apposed, the two reticles are joined with Scotch tape (Fig. 2D). The assembled reticles are placed at a focal plane of the ocular lens.

The compliance of the bending balance is calibrated *in air*. The balance is placed deep inside the empty observation chamber by moving the manipulator and positioning the tip in the field of the horizontal microscope. Small pieces of weighing paper are suitable as loads of known weights for the calibration. Four triangular pieces (1–3 mm) are cut from the four corners of a weighing paper with a sharp razor. The area of each piece is determined by measuring the sides

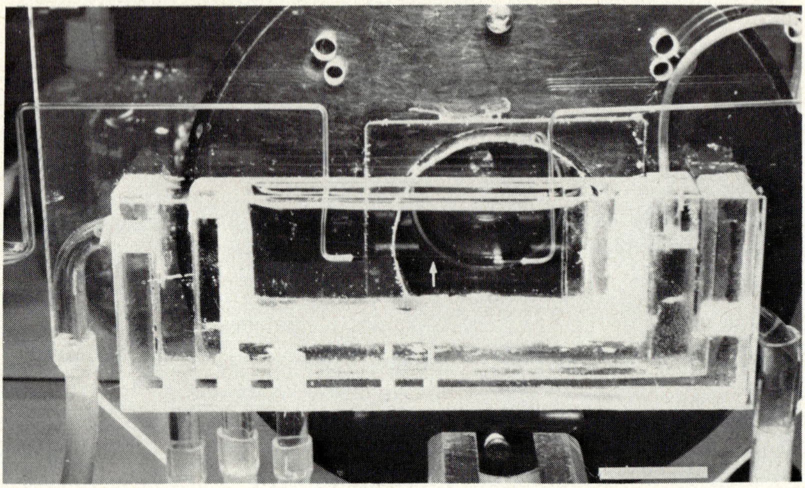

FIG. 4. A special water-jacketed chamber, 11 × 4 × 1.5 cm. Water for perfusion comes from the right side and is drained by a pair of tubes opening at the top and bottom of the water jacket. Seawater in the main chamber can be exchanged (see a fine tube for inflow to the right and two tubes for drainage to the left). Arrow points to the bending balance. Bar is 2 cm.

with a microscope equipped with an eye-piece micrometer. Weights are calculated on the basis of the weight/area ratio of the original sheet of weighing paper. One piece at a time is taken up with the tip of a sharpened sewing needle and placed onto the plate of the bending balance. Vertical displacement of the plate is measured in the horizontal microscope at a low magnification (4× objective). A fairly linear relation between the weights and displacements will be found. Table II shows an example of the calibration. Hiramoto (1963a) calibrated the balance by putting minute drops of mercury in water. Compliance was determined by the diameter of each drop and the difference in densities between mercury and water.

Calibrated balances are numbered and stored in a box. Sonication is the best way to remove any dirt from the balance, especially to clean the surface of the plate.

## IV. Practicals of Measurements

The observation chamber is filled with seawater. The bending balance and the stage are gently dipped into water by movement of the manipulator. Both plates are brought into the field of the microscope (4× objective) so as to make the surfaces of the plates strictly horizontal with the aid of the square reticle. The bending balance is moved off to one side.

TABLE II

SAMPLE CALIBRATION OF A BENDING BALANCE USING PIECES OF WEIGHING PAPER[a,b]

| | Piece | | | | |
|---|---|---|---|---|---|
| | 1 | 2 | 3 | 4 | Total |
| Area ($mm^2$) | 0.33 | 0.54 | 0.89 | 1.30 | 3.06 |
| Weight ($10^{-7}$ N) | 0.80 | 1.31 | 2.15 | 3.14 | 7.40 |
| Displacement (μm) | 170 | 314 | 493 | 792 | 1769 |

[a] Weight/area = 362 mg/146.5 $cm^2$ = 2.47 mg/$cm^2$ = $2.42 \times 10^{-5}$ N/$cm^2$.
[b] Compliance = $7.40 \times 10^{-7}$ N/1769 μm = $4.18 \times 10^{-10}$ N/μm.

Small numbers of eggs suffice for the experiments. To work on fertilized eggs, the fertilization membrane is of course removed beforehand. Similarly, unfertilized eggs should be deprived of the jelly coat.

Seawater containing a small number of eggs is drawn into a pipet. The tip of the pipet is dipped vertically into the water of the chamber and the pipet is held in position without expelling its contents. The eggs will slowly drop out of the pipet and some of them will come to rest on the stage in the observation chamber. The plate of the bending balance is brought above the stage. If more than one egg rests on the stage, all but one is swept away from the stage by using the balance as a broom. Now turn to the 10 or 20× objective. The balance is brought level with one of the major divisions of the ocular micrometer (Fig. 5A), which is recorded. This marks the position of zero force. The initial diameter ($Z_0$) of the egg (white arrow) is then measured using the micrometer. If the egg is not perfectly spherical, measure the diameters in both vertical and horizontal directions (use the square reticle) and average them. The egg is then compressed against the balance by raising the stage (Fig. 5B and C) until the balance is displaced by a definite distance (read by the micrometer) which means a definite force. The amount of force effective for a given experiment may be determined by trials. A force that reduces the thickness of the egg to 70 to 40% of the initial diameter will be appropriate. The equilibrium thickness ($Z$) of the egg, read by the micrometer, will be attained within 5 minutes, unless the egg is severely compressed to about 20% of the initial diameter (Yoneda, 1980). Usually the compression itself does not interfere with the normal course of development, so continuous records of the egg thickness ($Z$) under a constant force can be obtained. When the egg tends to push up the balance by an increase in the stiffness, the stage may be lowered so that the position of the balance stays at a certain level (i.e., a constant force), and vice versa. After taking a series of records for the thickness ($Z$) under a constant force of compression, the stage is lowered to detach the egg from the balance. At this time, the position of the balance at zero force should be checked.

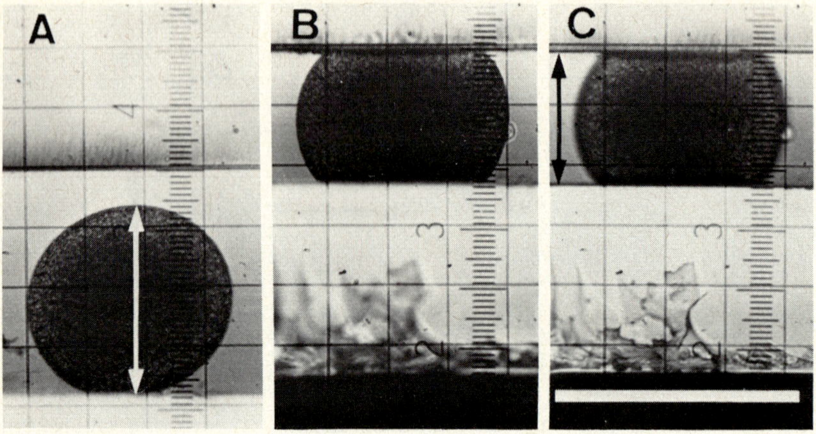


FIG. 5. A secondary oocyte of *Asterina pectinifera* under measurement. Bar is 200 μm. (A) The oocyte resting on the stage. The initial diameter (white arrow) is 156 μm. (B) The oocyte is compressed by raising the stage. The level of focus is on the contour of the oocyte. A polar body is seen at the right. (C) By focusing on the edges of the plates, the position of the plate (bending balance) shifts by 20 divisions of the micrometer scale (= 100 μm). The compliance of the balance ($1.59 \times 10^{-9}$ N/μm) × 100 μm gives the force of compression as $1.59 \times 10^{-7}$ N. The thickness (black arrow) is 109 μm ($z = Z/Z_0 = 0.699$). The surface force is calculated to be $1.23 \times 10^{-3}$ N/m.

The balance may not return to the original position attained before the compression. Such a slow drift of the sensitive balance is undoubtedly due to thermal turbulence in the water. If the positions of zero force before and after compression are close, the average position may be taken as the point of zero force, but otherwise all the measurements may have to be invalidated. This is a major trouble with the compression method. The drift is often encountered when the room temperature is unstable or the chamber with seawater is much cooler than the room. In such cases we must frequently check the position of zero force after each measurement, and readjust the zero position. To minimize such a drift, constructing a chamber with a water jacket (see Fig. 4) is recommended. Perfusing water through the insulated reservoir of the jacket is quite effective in reducing the drift.

## V. Calculation of the Surface Force

A relative value of the thickness ($Z/Z_0$) suffices when the temporal change in the cortical stiffness is a main concern. But this value depends of course on the force of compression ($F$) and size of individual eggs and so cannot be used for comparative purposes. It is advisable to convert the apparent value of $Z/Z_0$ into a physically defined term, that is, the amount of force generated over a unit length of egg surface that tends to reduce the surface area. This force has the dimension of Newtons/meter (=N/m), the same as the surface tension of liquid, and has been called either "tension at the surface" (Harvey, 1931; Yoneda, 1964), "surface force" (Cole, 1932; Hiramoto, 1963a,b), or "cortical tension" (Schroeder, 1981). Here the term "surface force" will be used. Exact calculation of the surface force involves tedious measurement of the curvatures and other parameters of the shape of compressed eggs, as done by Cole (1932) and Hiramoto (1963a,b, 1976), but if we assume that the surface force is uniform all around the surface of the compressed egg, as is practically the case in unfertilized and fertilized sea urchin eggs (Yoneda, 1973), the surface force can be calculated easily by using $Z$, $Z_0$, and $F$. The relation between the force of compression ($F$), the surface force ($T$), and the thickness of the egg ($Z$) may be expressed as $F\,dZ = -T\,dS$, or

$$T = F/\pi Z_0 (-ds/dz) \tag{1}$$

where $s$ is the surface area ($S$) of the compressed egg divided by the initial surface area ($S_0$), and $z$ is the relative thickness of the egg, which is $Z/Z_0$. By assuming a uniform surface force, we can derive the theoretical contour of the compressed egg by which the value of $(-ds/dz)$ can be expressed as a function of $z$ (cf. Yoneda 1964, 1973). Table III shows the calculated values of $-ds/dz$ and the relative surface area $s$ $(=S/S_0)$ as a function of $z$ $(=Z/Z_0)$. In practical terms, $-ds/dz$ is approximated by the empirical expression

$$-ds/dz = (1-z)(0.3212 + 0.412z - 0.78z^2 + 1.08z^3 - 0.705z^4)/z^2 \tag{2}$$

When $z$ ranges from 0.88 to 0.17, the error due to this approximation is less than 0.2%.

One example is shown here. A fertilized egg of *Dendraster excentricus* measuring 123.8 μm $(=Z_0)$ was flattened to 52.3 μm ($Z$) under compression by $3.32 \times 10^{-8}$ N $(=3.32 \times 10^{-3}$ dyn). The relative thickness, $z=Z/Z_0$, is 0.422, resulting in $-ds/dz$ of 1.347. The surface force ($T$) is therefore $3.32 \times 10^{-8}$ N/$(3.14 \times 1.347 \times 1.238 \times 10^{-4}$ m$) = 6.34 \times 10^{-5}$ N/m (0.063 dyn/cm).

The calculated surface force is independent of the degree of flattening of the egg. Several pairs of $Z$ and $F$ measured on a single egg have been found to give similar values of the surface force (Yoneda, 1973).

Some cautions are necessary when the surface force is to be calculated from the continuous measurement on dividing eggs. These will be explained in reference to Fig. 6. It should be noted that Eq. (1) applies only to spherical eggs whose surface is uniform. Therefore the surface force cannot be calculated for eggs during cleavage. Assuming that there are no constraining extracellular

TABLE III

CALCULATED PARAMETERS OF COMPRESSED EGGS AS A FUNCTION OF $z=Z/Z_0$[a]

| $z$ | $-ds/dz$ | $2R/Z_0$ | $S/S_0$ | $z$ | $-ds/dz$ | $2R/Z_0$ | $S/S_0$ |
|---|---|---|---|---|---|---|---|
| 0.99 | 0.003 | 1.001 | 1.000 | 0.59 | 0.505 | 1.173 | 1.071 |
| 0.98 | 0.006 | 1.003 | 1.000 | 0.58 | 0.535 | 1.180 | 1.076 |
| 0.97 | 0.010 | 1.005 | 1.000 | 0.57 | 0.566 | 1.188 | 1.081 |
| 0.96 | 0.014 | 1.007 | 1.000 | 0.56 | 0.600 | 1.196 | 1.087 |
| 0.95 | 0.019 | 1.009 | 1.000 | 0.55 | 0.635 | 1.204 | 1.093 |
| 0.94 | 0.024 | 1.012 | 1.001 | 0.54 | 0.672 | 1.213 | 1.100 |
| 0.93 | 0.029 | 1.014 | 1.001 | 0.53 | 0.712 | 1.222 | 1.107 |
| 0.92 | 0.035 | 1.017 | 1.001 | 0.52 | 0.754 | 1.231 | 1.114 |
| 0.91 | 0.041 | 1.020 | 1.002 | 0.51 | 0.799 | 1.240 | 1.122 |
| 0.90 | 0.047 | 1.022 | 1.002 | 0.50 | 0.846 | 1.250 | 1.130 |
| 0.89 | 0.054 | 1.025 | 1.003 | 0.49 | 0.897 | 1.260 | 1.139 |
| 0.88 | 0.061 | 1.028 | 1.003 | 0.48 | 0.951 | 1.270 | 1.148 |
| 0.87 | 0.068 | 1.032 | 1.004 | 0.47 | 1.009 | 1.281 | 1.158 |
| 0.86 | 0.076 | 1.035 | 1.005 | 0.46 | 1.070 | 1.292 | 1.168 |
| 0.85 | 0.084 | 1.038 | 1.005 | 0.45 | 1.136 | 1.304 | 1.179 |
| 0.84 | 0.092 | 1.042 | 1.006 | 0.44 | 1.207 | 1.316 | 1.191 |
| 0.83 | 0.101 | 1.045 | 1.007 | 0.43 | 1.282 | 1.329 | 1.203 |
| 0.82 | 0.110 | 1.049 | 1.008 | 0.42 | 1.364 | 1.342 | 1.217 |
| 0.81 | 0.120 | 1.053 | 1.009 | 0.41 | 1.452 | 1.355 | 1.231 |
| 0.80 | 0.130 | 1.057 | 1.011 | 0.40 | 1.546 | 1.369 | 1.246 |
| 0.79 | 0.141 | 1.061 | 1.012 | 0.39 | 1.648 | 1.384 | 1.262 |
| 0.78 | 0.152 | 1.065 | 1.013 | 0.38 | 1.759 | 1.399 | 1.279 |
| 0.77 | 0.164 | 1.070 | 1.015 | 0.37 | 1.879 | 1.415 | 1.297 |
| 0.76 | 0.176 | 1.074 | 1.017 | 0.36 | 2.009 | 1.432 | 1.316 |
| 0.75 | 0.188 | 1.079 | 1.019 | 0.35 | 2.151 | 1.450 | 1.337 |
| 0.74 | 0.202 | 1.083 | 1.020 | 0.34 | 2.306 | 1.468 | 1.359 |
| 0.73 | 0.216 | 1.088 | 1.023 | 0.33 | 2.476 | 1.487 | 1.383 |
| 0.72 | 0.231 | 1.093 | 1.025 | 0.32 | 2.662 | 1.507 | 1.409 |
| 0.71 | 0.246 | 1.098 | 1.027 | 0.31 | 2.867 | 1.529 | 1.437 |
| 0.70 | 0.262 | 1.104 | 1.030 | 0.30 | 3.093 | 1.551 | 1.466 |
| 0.69 | 0.279 | 1.109 | 1.032 | 0.29 | 3.344 | 1.575 | 1.499 |
| 0.68 | 0.297 | 1.115 | 1.035 | 0.28 | 3.622 | 1.600 | 1.533 |
| 0.67 | 0.316 | 1.120 | 1.038 | 0.27 | 3.932 | 1.626 | 1.571 |
| 0.66 | 0.335 | 1.126 | 1.042 | 0.26 | 4.280 | 1.654 | 1.612 |
| 0.65 | 0.356 | 1.132 | 1.045 | 0.25 | 4.670 | 1.684 | 1.657 |
| 0.64 | 0.378 | 1.139 | 1.049 | 0.24 | 5.112 | 1.716 | 1.706 |
| 0.63 | 0.401 | 1.145 | 1.053 | 0.23 | 5.612 | 1.750 | 1.759 |
| 0.62 | 0.425 | 1.152 | 1.057 | 0.22 | 6.184 | 1.786 | 1.818 |
| 0.61 | 0.450 | 1.159 | 1.061 | 0.21 | 6.840 | 1.825 | 1.883 |
| 0.60 | 0.477 | 1.166 | 1.066 | 0.20 | 7.598 | 1.867 | 1.955 |

[a] Note: A similar table presented in an earlier paper (Yoneda, 1973) contains errors in the last digits in some figures of $-ds/dz$. All of them are corrected in the present table.

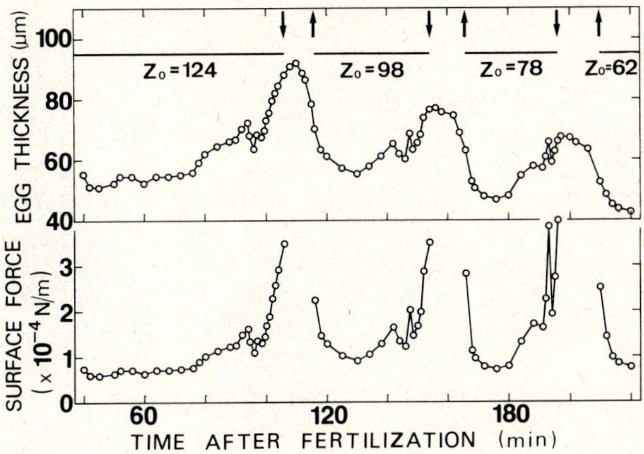


FIG. 6. Continuous measurement of the thickness of a fertilized egg of *Dendraster excentricus* and the calculated surface force. Force of compression is $3.32 \times 10^{-8}$ N. Upward and downward arrows indicate the start and end of cleavages. The surface force is not calculated during cleavages. Initial diameters $Z_0$ are shown in micrometers for each stage of development.

materials, blastomeres between cleavages should be essentially spherical. To calculate the surface force after first cleavage, two blastomeres of reduced diameter are compressed simultaneously. To calculate the surface force in each blastomere, the value of $F$ should be divided by two, and the initial diameter $Z_0$ should be replaced by $Z_0/\sqrt[3]{2}$. This conversion of $F$ and $Z_0$ is repeated after each cleavage.

## VI. Weight Loading Method

A method of compressing eggs that is simpler than using the bending balance is to load a known weight directly onto a number of eggs resting upon the bottom of a chamber (Danielli, 1952; Wolpert *et al.*, 1971; Schroeder, 1981). Though requiring some skill and practice, this method does not require a micromanipulator or a horizontal microscope. The procedure adopted by Schroeder (1981) will be described here. Technical details of the procedure were provided by T. E. Schroeder (personal communication).

A minute glass plate (about $0.6 \times 0.4 \times 0.075$ mm each) is prepared by blowing a 3-cm globe from a Pasteur pipet, fracturing it, and selecting bits having the appropriate dimensions. The volume of the plates is calculated from the area and thickness which are determined from photomicrographs. The reduced weight of the plate in water is the volume multiplied by the difference in

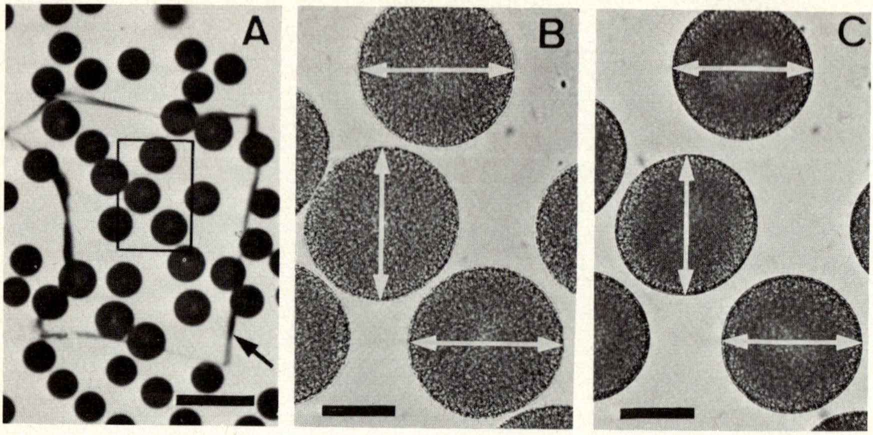


FIG. 7. The weight-loading method applied to fertilized eggs of *Strongylocentrotus purpuratus*. The observations were made by T. E. Schroeder. (A) A minute glass plate (marked by arrow) is resting on several eggs at prophase. Flattening of the eggs under the plate is indicated by the increased diameters. Bar is 200 μm. (B) Enlarged view of the rectangular field in (A). Arrows are diameters (2*R*) of eggs under the load 99 minutes after fertilization. Bar is 50 μm. (C) The same field 40 minutes later than (B), shortly before the onset of cleavage. Note reduced diameters. Bar is 50 μm.

density of the glass (2.48 g/cc) and the medium (1.023 g/ml). The density of the glass is determined by drying several clean pipets, weighing them, and measuring the displacement volume when immersed in water. The force of compression ($F$) is $0.98 \times 10^{-2}$ N/g × (weight in grams).

Eggs are transferred in a small drop of water to a prechilled glass slide to which an open square composed of thin glass spacers have been cemented.

Under a dissecting microscope, the plate is immersed in the droplet and carefully balanced on a group of about 15 eggs using very fine forceps (Fig. 7A). A cover glass is placed upon the spacers so that the droplet contacts it and is sealed against evaporation. The preparation is then placed upon a thermostatted cooling stage of a vertical microscope. At intervals (every 1–5 minutes), photomicrographs at 100× are taken of about five compressed eggs in a field of view (Fig. 7B and C). Diameters ($2R$) of the eggs, as marked by arrows in Fig. 7B and C, are measured from negatives with a 7× magnifier and the average diameter is obtained for each time. The initial diameter of the egg ($Z_0$) is determined by averaging the diameters of several uncompressed eggs. The degree of flattening of the egg is expressed by the ratio of the diameter of $2R$ under load to the initial diameter $Z_0$. From the ratio $2R/Z_0$, the corresponding values of $z$ and $-ds/dz$ are read from Table III. The average thickness of the egg $Z$ is calculated to be $z \times Z_0$. The load per egg ($F$) is the reduced weight of the plate divided by the number of eggs supporting it in the medium. Using $F$ and $-ds/dz$, the surface force ($T$) is calculated by Eqs. (1) and (2).

An example of the calculation is shown here. A new set of data and the photographic records (shown in Fig. 7) assembled by T. E. Schroeder will be used. A plate with the reduced weight of $3.78 \times 10^{-7}$ N rests on 18 eggs, so the weight per egg ($F$) is $2.10 \times 10^{-8}$ N (= $2.10 \times 10^{-3}$ dyn). The average diameter ($Z_0$) of uncompressed eggs of this batch was 84.5 μm. The eggs at prophase are very flattened (Fig. 7B) and the average diameter ($2R$) was 107 μm. Shortly before the onset of cleavage (Fig. 7C) the diameter ($2R$) was reduced to 96 μm. For the two times, $2R/Z_0 = 1.266$ and 1.136. From Table III, $-ds/dz =$ 0.929 and 0.369 ($Z = 0.484 \times 84.5$ μm and $0.644 \times 84.5$ μm). By calculation using Eq. (1), the surface force ($T$) = 0.85 and $2.14 \times 10^{-4}$ N/m (0.085 and 0.214 dyn/cm).

## VII. Critical Comments on the Methods

The compression method is applicable to any kind of spherical egg if a semi-quantitative estimation of the egg stiffness is a main concern. However, when the absolute value of the surface force is to be calculated, it should be kept in mind that the calculation assumes the mechanical uniformity of the egg surface. When local difference in mechanical property is evident, as in furrow and polar regions of dividing eggs, the apparent value of $Z/Z_0$ is no longer a reliable measure of the surface force. Obviously the suction or elastimeter method (Hiramoto, this volume) is preferable in such a case. The calculation will safely apply to oocytes during formation of polar bodies, since nearly all of the oocyte's surface remains spherical.

Another problem is the possible contribution of the endoplasm in opposing the force of compression. When immature oocytes of starfish are bisected into halves, each half remains oval for some period of time (Yamamoto and Yoneda, 1983), unlike bisected maturing oocytes or sea urchin eggs. Endoplasmic structure may not be negligible in immature oocytes. Accordingly, the marked change in $Z/Z_0$, and hence the calculated "surface force" that occurs during maturation of starfish oocytes (Nemoto, 1982; Nemoto *et al.*, 1980), may reflect endoplasmic changes and not be solely due to changes in the cortex. Despite these limitations, the compression method is nevertheless useful for detecting changes in the mechanical properties of the cytoplasm of marine ova.

### Acknowledgments

I express my deep gratitude to Dr. Thomas E. Schroeder for informing me of technical details of the weight-loading method by which Section VI has been accomplished.

## References

Cole, K. S. (1932). *J. Cell. Comp. Physiol.* **1,** 1–9.
Danielli, J. F. (1952). *Nature (London)* **170,** 496.
Harvey, E. N. (1931). *Biol. Bull.* **61,** 273–279.
Hiramoto, Y. (1963a). *Exp. Cell Res.* **32,** 59–75.
Hiramoto, Y. (1963b). *Exp. Cell Res.* **32,** 76–89.
Hiramoto, Y. (1970). *Biorheology* **6,** 201–234.
Hiramoto, Y. (1976). *Dev. Growth Differ.* **18,** 377–386.
Nemoto, S. (1982). *Dev. Growth Differ.* **24,** 429–442.
Nemoto, S., Yoneda, M., and Uemura, I. (1980). *Dev. Growth Differ.* **22,** 315–325.
Schroeder, T. E. (1981). *Exp. Cell Res.* **134,** 231–240.
Usui, N., and Yoneda, M. (1982). *Dev. Growth Differ.* **24,** 453–465.
Wolpert, L., Marsland, D., and Hirshfield, M. (1971). *J. Cell Sci.* **8,** 87–92.
Yamamoto, K., and Yoneda, M. (1983). *Dev. Biol.* **96,** 166–172.
Yoneda, M. (1964). *J. Exp. Biol.* **41,** 893–906.
Yoneda, M. (1973). *Adv. Biophys.* **4,** 153–190.
Yoneda, M. (1980). *Dev. Growth Differ.* **22,** 39–47.
Yoneda, M., and Schroeder, T. E. (1984). *J. Exp. Zool.* **231,** 367–378.
Yoneda, M., Ikeda, M., and Washitani, S. (1978). *Dev. Growth Differ.* **20,** 329–336.

# Chapter 25

# *Determination of Mechanical Properties of the Egg Surface by Elastimetry*

YUKIO HIRAMOTO

*Biological Laboratory*
*Tokyo Institute of Technology*
*Tokyo, Japan, and*
*Department of Cell Biology*
*National Institute for Basic Biology*
*Okazaki, Japan*



## I. Introduction

Elastimetry is a method for measuring mechanical properties of the cell surface based on the relation between its deformation and a negative pressure applied to a part of the cell surface through a micropipet. Such a measurement was first attempted by Vlès (1933) in sea urchin eggs. Mitchison and Swann (1954a) refined this method and discussed the physical meaning of the negative pressure–deformation relation. Since then, elastimetry has been used by many investigators for measuring mechanical properties of the cell surface, especially in echinoderm and amphibian eggs and in blood cells. For echinoderm eggs, the following references should be consulted: Mitchison and Swann (1954a,b, 1955),

Mitchison (1956), Wolpert (1963, 1966), Nakamura and Hiramoto (1978), Hiramoto (1979), and Ohtsubo and Hiramoto (1985). For the mechanical properties of echinoderm eggs in general, see reviews by Hiramoto (1970, 1981) and chapters by Yoneda and Hiramoto and Yoneda in this volume.

## II. Principles of Elastimetry

In elastimetry, a micropipet whose tip is cut at right angles to the miropipet axis is brought into contact with the cell surface (Fig. 1), and the deformation of the cell surface and the negative pressure applied through the micropipet are measured. Because the deformation is restricted to a part of the cell surface, this method is suitable for measurement of cells with mechanical properties that differ at different regions of the cell surface. The size of the micropipet can be chosen by the purpose of measurement. Properties of as little as 1% of the total surface can be measured by using a micropipet with a diameter that is one-fifth the diameter of the cell.

The height of bulge ($x$ in Fig. 1) is suitable measure for the deformation of the cell surface. It depends on the size of the micropipet and the magnitude and duration of the negative pressure. It has been found that a linear relation exists between the height of the bulge and the applied negative pressure in sea urchin eggs (Mitchison and Swann, 1954a) and in starfish oocytes (Nakamura and Hiramoto, 1978). Mitchison and Swann referred to the slope of the pressure–deformation curve as the "stiffness." The stiffness, of course, depends on the

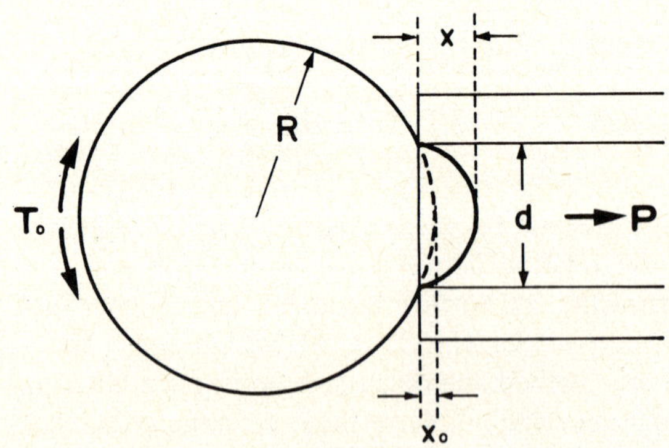


FIG. 1. Principle of elastimetry. $d$, Inner diameter of the micropipet; $P$, negative pressure; $R$, radius of the cell surface; $T_0$, surface force; $x$, height of the bulge formed by the negative pressure; $x_0$, $R - \sqrt{R^2-d^2/4}$ corresponding to $x$ at nil negative pressure.

size of the micropipet and the duration of the application of the negative pressure, as will be mentioned later.

## III. Experimental Set-up

Figure 2 shows a typical experimental set-up for elastimetry. An egg suspension is put into a chamber on the stage of a microscope. It is more convenient to use an inverted microscope, rather than an ordinary upright one, because there is easier access for the micropipet and the cell resting at the bottom of the chamber is readily visible. In this case, the bottom of the chamber should be made of a flat glass plate (e.g., a cover slip) allowing good visibility. The micropipet is held with a micromanipulator (MM1 or MM2) so as to keep the axis of the micropipet

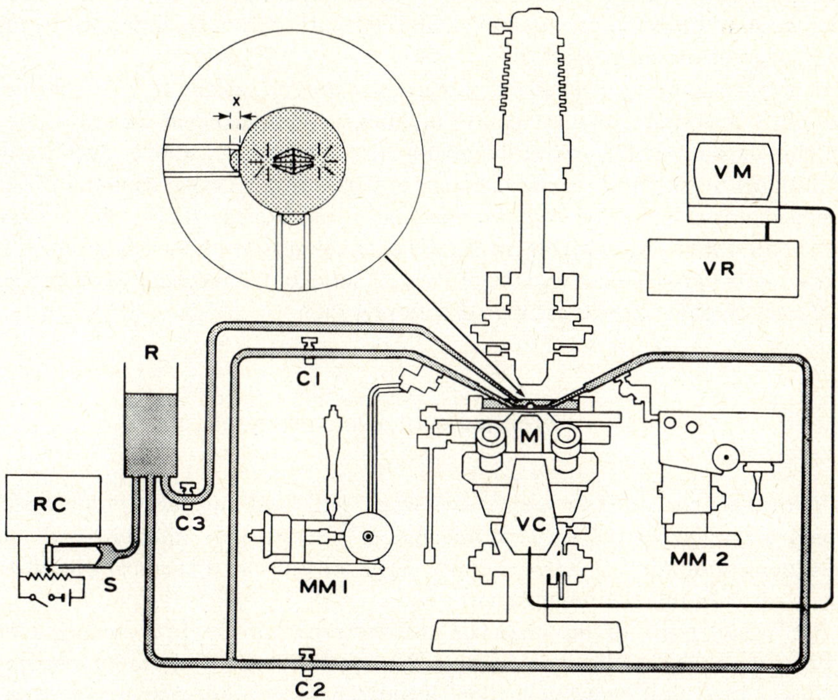


FIG. 2. Experimental set-up. $C_1$, $C_2$, and $C_3$, stop cocks; M, inverted microscope; MM1 and MM2, micromanipulators; R, reservoir; RC, voltage recorder; S, syringe; VC, video camera; VM, video monitor; VR, video-tape recorder; *x* represents the height of the bulge formed by applied negative pressure.

near the tip horizontal. The micropipet is connected through a tubing to a reservoir (R), and all are filled with seawater or Ca-free seawater. The water level in the reservoir can be controlled by moving the plunger of a syringe (S) connected to the reservoir (R). The sensitivity of this control can be regulated by choosing a syringe with a suitable capacity. The displacement of the plunger is read directly from a screw micrometer connected to the plunger, or automatically by a voltage recorder after transducing the plunger's displacement to voltage (cf. Fig. 2). The fluid in the reservoir can be equalized to the level in the chamber by opening a stop cock ($C_3$) set along a tubing that directly connects the chamber with the reservoir.

The deformation of the cell surface is measured directly through the microscope with an ocular micrometer or measured after recording the deformation by photomicrography or a video system (Fig. 2). The micropipet is made of glass capillary and the tip is cut at right angles to the pipet axis. This is important in determining the height of the bulge with exactitude and in avoiding leakage of seawater around the tip. The orifice of the micropipet can be rounded off with a microburner, if it is required to prevent the cell surface from being injured by the sharp edge of the micropipet.

In order to compare mechanical properties of the cell surface at two regions of the cell, it is desirable to use a pair of micropipets with identical sizes, since the stiffness depends on the size of the micropipet, as mentioned above. Such a pair of identical micropipets can be made by cutting a single drawn capillary of the correct diameter and using both portions (Sawai and Yoneda, 1974; Ohtsubo and Hiramoto, 1985). By applying an equal negative pressure to the micropipets in contact with two different regions of the cell surface, the mechanical properties of the two regions can be compared with precision.

## IV. Practicals of Measurement

Extraneous coats and/or membranes (e.g., jelly coat and fertilization membrane) are removed before experimentation, because they interfere with the measurement of the cell surface proper. Eggs are put into a chamber placed on the stage of an inverted microscope.

The water levels of the chamber and the reservoir are first equalized by opening the cock $C_3$. Negative pressure is applied to the cell surface by lowering the water level in the reservoir after placing the tip of a micropipet against the cell surface and closing $C_3$. A constant negative pressure can be applied by displacing the plunger of the syringe by a definite distance. The height of bulge formed in the micropipet increases rapidly at first and then gradually approaches a steady level, as shown in Fig. 3, owing to the viscoelasticity of the cell surface. Instead of maintaining a constant negative pressure, it is possible to adjust the

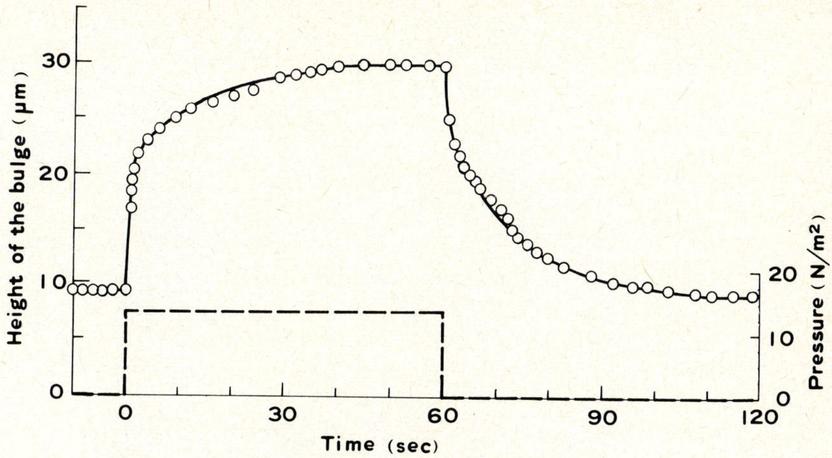


FIG. 3. Change in the height of the bulge (solid line) when a constant negative pressure is suddenly applied to a part of a starfish egg and then terminated.

applied negative pressure so as to keep the height of the bulge constant throughout the measurement. In this case, the negative pressure is large at the beginning of the application and then must be decreased toward a steady level during the first several minutes. The negative pressure can be adjusted manually when the change is not too fast. A more sophisticated adjustment system may be made by using photoelectronic feedback devices that measure and control the deformation of the cell surface automatically.

The above two procedures, the "constant-pressure run" and the "constant-deformation run," are suitable for analyzing the viscoelasticity of the cell surface. They are also useful in analyzing temporal changes in mechanical properties of the cell surface accompanying biological processes taking place in the egg, e.g., fertilization and cleavage.

The micropipet is held at the level of the center of the egg resting on the bottom of the chamber so that the sucked part of the cell surface is not pulled up or down during measurement. The leakage of the pressure around the tip of the micropipet can be checked by carefully observing particles occasionally or deliberately contaminated in the surrounding seawater.

## V. Mechanical Properties of the Cell Surface

### A. Stiffness of the Cell Surface

The stiffness of the cell surface is defined by Mitchison and Swann (1954a) as the slope of the pressure–deformation curve, which is practically linear in both

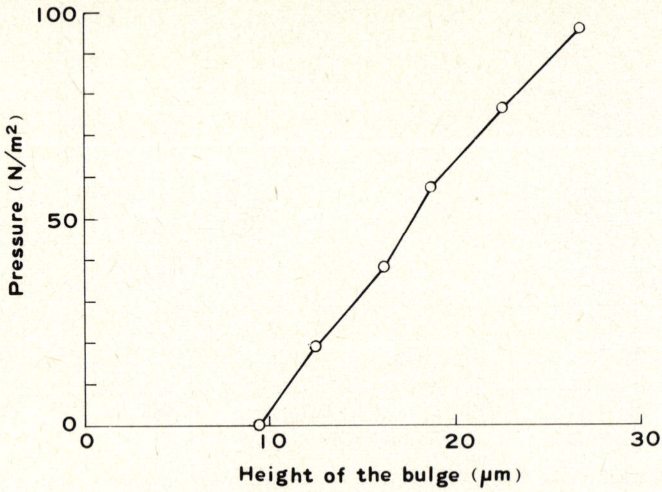


FIG. 4. Negative pressure vs height of the bulge in a starfish oocyte. Starfish oocytes are much stiffer than mature eggs (compare Fig. 3).

sea urchin eggs (Mitchison and Swann, 1954a) and in starfish oocytes (Nakamura and Hiramoto, 1978; Fig. 4). The slope, and therefore the stiffness, vary according to the duration of the applied negative pressure, as mentioned above.

The definition of "stiffness" can be extended to include a nonlinear pressure–deformation property. It is possible to represent the "stiffness" by the magnitude of the negative pressure, obtained from a "constant-deformation run" or by the reciprocal of the height of bulge obtained from a "constant-pressure run," because these values are positively correlated with the resistivity of the cell surface against deformation, provided that the size of the micropipet is the same.

Figure 5 shows the change in negative pressure in a "constant-deformation run" of a sea urchin egg before and during cleavage. The negative pressure change represents the change in the stiffness of the cell surface at cleavage.

## B. Surface Force

As a result of their experiments with sea urchin eggs and with rubber ball models, Mitchison and Swann (1954a) concluded that the stiffness of the cell surface is mainly due to the bending resistance of the cell cortex. This conclusion was, however, criticized by Wolpert (1960) and Hiramoto (1963), who stated that the "stiffness" of the cell surface is due to the force in the direction tangent to the cell surface. This force, which is the sum of the forces generated by the cell cortex, the cell membrane, and any overlying extraneous coats and/or membranes, is called the "surface force" or "tension at the cell surface" (see

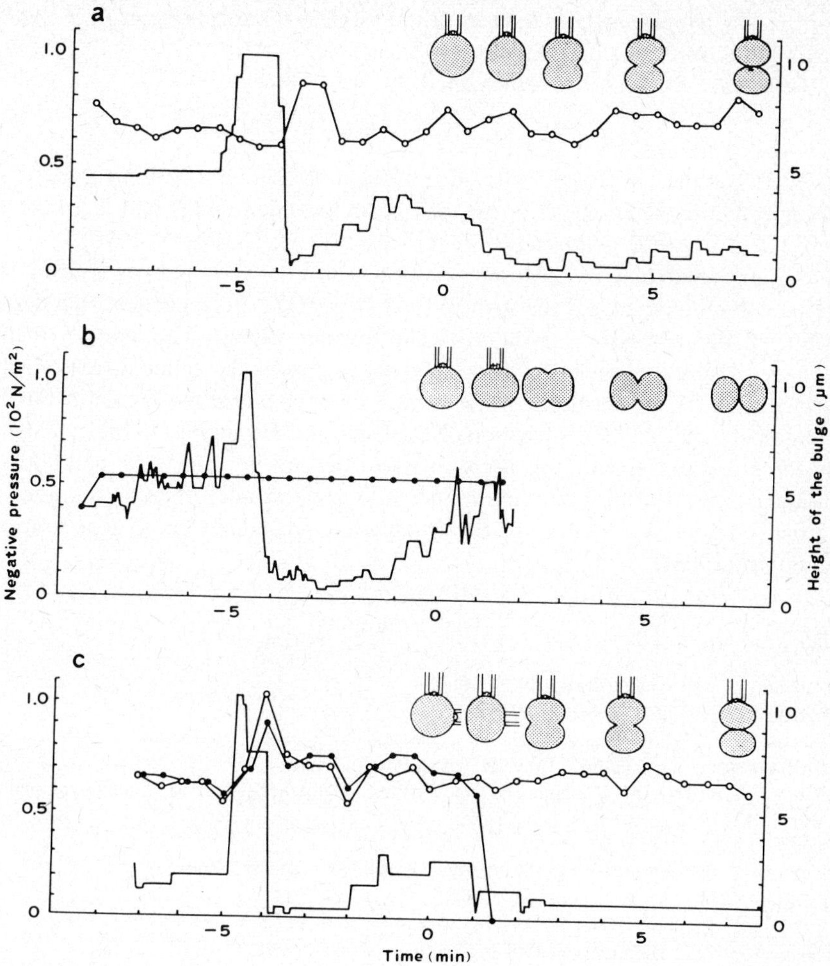


FIG. 5. Stiffness changes of the cell surface in sea urchin eggs before and during cleavage. The stiffness is represented by the negative pressure (continuous lines) required to maintain a constant height of the bulge (○ or ●) at the cell surface in the micropipet. Insets indicate stages of cleavages. (a) The stiffness is measured at the polar surface; (b) the stiffness is measured at the furrow surface; (c) the stiffness is measured at both furrow and polar surfaces simultaneously. ○ and ● are heights of the bulges at the polar surface and at the furrow surface, respectively. The height of the bulge is not maintained absolutely constant, especially after the first peak in (a) and (c), owing to technical difficulties of manual adjustment. The origin of the time scale represents the onset of cleavage. The diameter of the micropipet is 20 μm.

Yoneda, this volume). It has the same dimension as the surface tension of liquid, which is the force per unit length of the surface.

The surface force ($T_0$) is represented by

$$T_0 = S_0\,(x_0^2 + d^2/4)^2/(d^2 - 4x_0^2) \qquad (1)$$

in which $S_0$ is the "stiffness" (the slope of the pressure-deformation curve) when the deformation is small, $d$ is the radius of the micropipet, and $x_0$ is $R - \sqrt{R^2 - d^2/4}$ where $R$ is the radius of the egg (Fig. 1; Hiramoto, 1970).

It is expected that the pressure–deformation cuerve is convex toward the pressure coordinate axis if the surface force is constant irrespective of the deformation of the cell surface (Mitchison and Swann, 1954a). The linearity of the pressure–deformation relation mentioned above may be due to the increase in the surface force by the elasticity of the surface components at the region inside the micropipet. Under this assumption, Nakamura and Hiramoto (1978) calculated the force required to produce per unit strain per unit length of the cell surface (elasticity value) from their measurements of pressure–deformation relations in starfish oocytes. For details of the calculation procedure, see Nakamura and Hiramoto (1978).

## REFERENCES

Hiramoto, Y. (1963). *Exp. Cell Res.* **32,** 59–75.
Hiramoto, Y. (1970). *Biorheology* **6,** 201–234.
Hiramoto, Y. (1979). *In* "Cell Motility: Molecules and Organization" (S. Hatano, H. Ishikawa, and H. Sato, eds.), pp. 653–663. Univ. of Tokyo Press, Tokyo.
Hiramoto, Y. (1981). *In* "Cellular Dynamics: Mitosis/Cytokinesis" (A. M. Zimmerman and A. Forer eds.). Academic Press, New York.
Mitchison, J. M. (1956). *J. Exp. Biol.* **33,** 524–532.
Mitchison, J. M., and Swann, M. M. (1954a). *J. Exp. Biol.* **31,** 443–460.
Mitchison, J. M., and Swann, M. M. (1954b). *J. Exp. Biol.* **31,** 461–472.
Mitchison, J. M., and Swann, M. M. (1955). *J. Exp. Biol.* **32,** 734–750.
Nakamura, S., and Hiramoto, Y. (1978). *Dev. Growth Differ.* **20,** 317–327.
Ohtsubo, M., and Hiramoto, Y. (1985). *Dev. Growth Differ.* **27,** 371–383.
Sawai, T., and Yoneda, M. (1974). *J. Cell Biol.* **60,** 1–7.
Vlès, F. (1933). *Arch. Zool. Exp. Gen.* **75,** 421–463.
Wolpert, L. (1960). *Int. Rev. Cytol.* **10,** 164–216.
Wolpert, L. (1963). *In* "Cell Growth and Cell Division" (J. C. Harris, ed.), pp. 227–298. Academic Press, New York.
Wolpert, L. (1966). *Exp. Cell Res.* **41,** 385–396.

# Chapter 26

# *Determination of Mechanical Properties of the Egg by the Sessile Drop Method*

**YUKIO HIRAMOTO**

*Biological Laboratory*
*Tokyo Institute of Technology*
*Tokyo, Japan, and*
*Department of Cell Biology*
*National Institute for Basic Biology*
*Okazaki, Japan*

**MITSUKI YONEDA**

*Department of Zoology*
*Faculty of Science*
*Kyoto University*
*Kyoto, Japan*

---

# I. Introduction

A liquid drop resting on a horizontal plate (a sessile drop) takes a characteristic form determined by the counterbalance between the surface tension tending to make the drop spherical and gravity tending to flatten the drop. Large spherical cells, e.g., avian, amphibian, and teleost eggs deprived of extraneous shells, layers and/or membranes, take similar forms when they are placed on horizontal planes. It is supposed that the tension of the surface membrane of the egg is analogous to the surface tension of the sessile drop and balances the gravitational force tending to flatten the egg. Therefore, it is expected that the tension at the cell surface (here termed as "surface force," see Yoneda's chapter in this volume) of the spherical cell can be determined from the degree of flattening, if the gravitational force tending to flatten the cell is known. In fact, E. N. Harvey and his co-workers succeeded in determining such surface forces in large eggs (Harvey and Fankhauser, 1933; E. N. Harvey, 1933a).

Vlès (1926) estimated the surface force of the sea urchin egg by comparing the form of the egg observed with a horizontal microscope with the profiles of oil drops whose surface tensions were known. However, his values for the surface force may not be reliable because the deformation of the egg that he observed might have been due to its intrinsic nonsphericity rather than to the extrinsic gravitational force, as pointed out by Harvey (1933a), who stated that sea urchin eggs are too small to be flattened by gravity.

Sea urchin eggs, however, can be flattened when they are placed in a centrifugal field of the order of $10^2$ *g*. Indeed, Hiramoto (1967) determined surface forces in sea urchin eggs by measuring the degree of flattening of the egg observed with a centrifuge microscope.

# II. Measurement of the Form of the Egg

## A. Measurement in a Centrifugal Field

A centrifugal force of the order of $10^2$ *g* to flatten sea urchin eggs is well within the capacity of centrifuges used in the laboratories, but what we need is to observe the form of eggs *during* centrifugation. The centrifuge microscope meets this requirement. Harvey-type centrifuge microscopes (E. N. Harvey, 1933b; E. B. Harvey, 1951) were made by some manufacturers (e.g., Bausch & Lomb, USA, and Struers, Denmark) many years ago, although they are no longer commercially available. A custom-made Harvey-type centrifuge microscope with a continuous observation and recording system was described by Kuroda

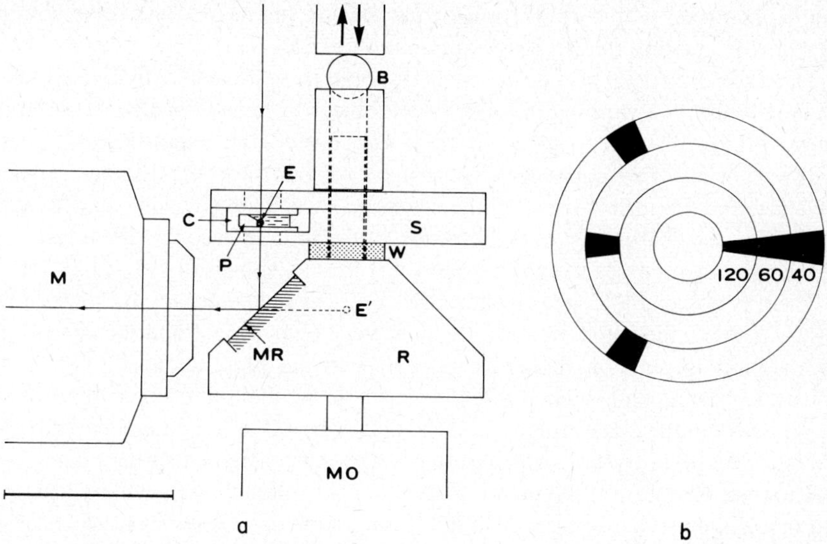


FIG. 1. Brown-type centrifuge microscope. B, Bearing of the rotor axis which can be moved in vertical direction for adjusting the distance between the stage, S, and the mirror, MR; C, centrifuge chamber; E, eggs; E′, image of the egg formed by mirror MR; M, microscope; MO, motor; P, plate supporting the egg; R, rotor; W, rubber washer. (a) Indicates side view of the entire set-up, and (b) indicates top view of the cover of the stage showing stripe patterns. Bar indicates 1 cm.

and Kamiya (1981). These centrifuge microscopes can be used in determining mechanical properties of echinoderm eggs by the sessile drop method.

For those who do not have access to such ready-made centrifuge microscopes, we recommend making a Brown-type centrifuge microscope (Brown, 1940); it can be fabricated in a laboratory machine-shop without difficulty. Figure 1a shows the diagram of a Brown-type centrifuge microscope. It consists of a rotor (R) with a stage (S), a motor (MO), and a microscope for observation (M). A first-surface mirror (MR) is attached to R at 45° to the axis of rotation. The specimen is put in a centrifuge chamber (C) set on the stage above the mirror, as shown in Fig. 1. The egg is supported with a plate (P) set perpendicularly to the direction of the centrifugal field in the chamber. The chamber is made of acrylic plastic and glass. A low-power dissecting microscope is useful in constructing the chamber and setting the egg in the chamber. Extraneous membranes and/or coats, e.g., the jelly coat and fertilization membrane, are removed before experimentation, because they interfere with exact measurement. The specimen is illuminated when the mirror passes in front of the objective lens of a horizontal microscope. Because the centrifuge microscope is constructed so that the mirror forms an image (E′) of the specimen (E) on the axis of the rotor (R), the image can be observed with the microscope focused on the axis of the rotor without appreciably moving during each illumination period.

The rotor is made of acrylic plastic and should be sufficiently small to allow its rotation within the working distance of the objective lens. Objective lenses with long working distances (about 10 mm) and fairly high magnifications (e.g., Nixon 10 or 20×) are commercially available. In our rotor, the distance between the stage and the mirror fixed on the rotor can be adjusted during centrifugation by moving the bearing (B) in the direction of the rotor axis so that the image (E′) of the specimen comes exactly on the rotor axis as shown in Fig. 1a.

There is no need to counterbalance the centrifuge chamber on the stage or the mirror fixed to the rotor, because they are very light weight and the centrifugal acceleration is only on the order of $10^2$ $g$ in the present experiment. It is, however, recommended that a shield with a hole for the microscopic observation be set between the ‘‘centrifuge’’ and the observer and to use an inclined tube for the microscope in order to put observer's eye out of the centrifuge plane.

The rotation speed of the motor should be controllable over a wide range, and the motor's axis should be well fixed during rotation. Centrifugal acceleration ($A$) is calculated from Eq. (1):

$$A = r(2\pi n/60)^2 \tag{1}$$

where $r$ is the distance of the specimen from the rotor axis and $n$ is the rotation speed of the rotor (rpm). The rotation speed is determined by observing a stripe pattern marked on the cover of the stage (Fig. 1b) under stroboscopic illumination at a controlled frequency, because the stripe pattern appears to be stationary when the lighting frequency coincides with the rotation frequency. Alternatively, the rotation speed can be controlled so that the stripe pattern is illuminated by stroboscopic light of a fixed frequency. The number of rotations can be adjusted to divisors of twice the frequency of the ac power source by using a neon lamp or an ordinary fluorescence lamp, because its intensity fluctuates at twice the ac frequency. For example, patterns shown in Fig. 1b appear to be stationary when the number of rotations is 120, 60 (= 120/2) or 40 (= 120/3) per second if the pattern is illuminated by 60-Hz ac lamp.

Stroboscopic flash illumination of the specimen synchronized with the rotation is also recommended to obtain sharp images in observation and photography. The height and the width of the egg are measured using an ocular micrometer. The exact profile of the egg is recorded by photomicrography.

## B. Measurement of an Egg Deformed by Gravity

Exceptionally large echinoderm eggs have been reported for sea cucumbers (cf. Inaba, 1957). In such large eggs, it is possible to determine the surface force by simply measuring the form of the egg placed on a horizontal plate in a trough for observation with a horizontal microscope, because the deformation by gravity is sufficiently large. It is desirable to verify that the observed deformation of the egg is really due to the gravity because some eggs are deformed by other factors, e.g., by packing in the ovaries before shedding. It is preferable to use eggs that have been kept in seawater for a few hours after release from the ovary because they tend to be more spherical.

Eggs are usually placed on a horizontal plate and observed from a horizontal direction. It has been reported that some giant echinoderm eggs are lighter than seawater (e.g., Amemiya and Tsuchiya, 1979). Such eggs are placed under a plate supported horizontally in a trough for observation. Isosmotic sucrose solution or other isosmotic solutions with high density (cf. Section III,C) may be added to the medium in order to increase the deformation of the egg by increasing the density difference between the egg protoplasm and the surrounding medium.

## III. Determination of Mechanical Properties of the Egg

### A. Stiffness of the Egg as a Whole

In order to understand the structural basis of such physiological processes as fertilization or cleavage, it is worthwhile to determine the accompanying stiffness changes. The stiffness of the egg as a whole is inversely related to the degree of flattening under a definite gravitational or centrifugal force. Thus, the height ($H$) of the egg, the ratio of the height ($H$) to the diameter ($2R$) before deformation, or the ratio of the height ($H$) to the width ($W$) may be taken as an estimate of the stiffness of the egg (Fig. 2). The stiffness expressed by these parameters depends on various mechanical properties of the cell components, e.g., the surface force, rigidity of the cortex, intracellular pressure, rigidity of endoplasmic structures, etc. The surface force, which is the same dimension as

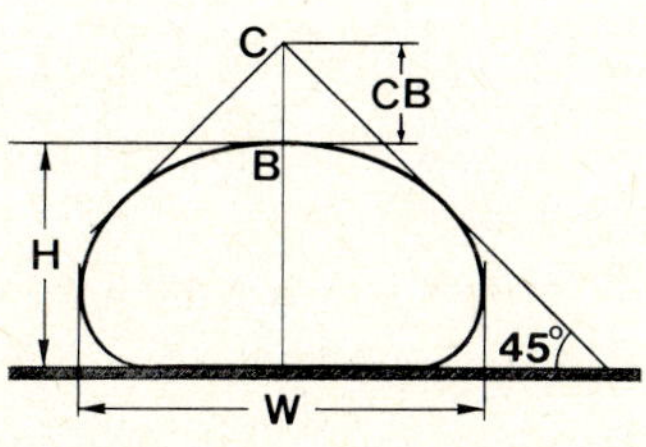


FIG. 2. Liquid drop resting on a horizontal plane (sessile drop). B, Top of the liquid; C, intersecting point of tangents at the drop surface inclining by 45° to the horizontal plane; H, height of the drop; W, width of the drop.

the surrace tension of liquid, can be calculated from the measurement of $H$ or $W$, as mentioned in the next paragraph.

## B. Surface Force

The surface force is determined from the degree of deformation of the egg under a gravitational or centrifugal force, the density difference between the egg protoplasm and the surrounding medium, the size of the egg, and the magnitude of the gravitational or centrifugal acceleration, using relationships obtained from the form of a theoretical sessile drop. In this case, it is postulated that only the force tangent to the cell surface (surface force) is responsible for opposing the force tending to flatten the egg as a function of the density difference between the protoplasm and the surrounding medium. Thus, the stresses due to the deformation of the endoplasm and to the bending of the cortex are neglected. In addition, it is assumed that the surface force is uniform over the entire surface and that the density of the protoplasm is uniform within the egg.

As shown by Hiramoto (1967), the form of the sessile drop is determined by a single dimensionless parameter, $S_1$, that is,

$$S_1 = A(D-D')\,(2R)^2/T \tag{2}$$

where $A$ is the gravitational or centrifugal acceleration, $D-D'$ is the density difference between the drop and the surrounding medium, $2R$ is the diameter of the drop without deformation, and $T$ is the surface tension. Figure 3 shows a series of profiles of a drop of fixed volume for various $S_1$ values obtained by computer drawing (cf. the Appendix). Numerals in Fig. 3 indicate curve numbers in the first column of Table I, in which $S_1$ values corresponding to each curve number are shown in the seventh column. Table I also includes similar dimensionless parameters $S_0 = A(D-D')\,R_1{}^2/T$ and $S_2 = A(D-D')W^2/T$, where $R_1$ is the radius of curvature at the top of the drop and $W$ is the largest diameter [cf. Eqs. (11, 12, and 14)]. $CB/(W/2)$ is a parameter representing the degree of flattening used in Dorsey's method mentioned below.

The first step in determining the surface force is to find $S_1$ or $S_2$ values corresponding to $H/W$, $H/(2R)$, or $CB/(W/2)$ values measured in flattened eggs. The surface force is represented by $A(D-D')\,(2R)^2/S_1$ or $A(D-D')W^2/S_2$, in which $A$ is the centrifugal acceleration [cf. Eqs. (11–14)]. Some examples of the practical procedure are as follows.

### 1. Example 1

Figure 4 shows an unfertilized sea urchin egg in a centrifugal field photographed through a centrifuge microscope recently designed and made by Professor Eiji Kamitsubo of Hitotsubashi University and Dr. Munehiro Kikuyama of

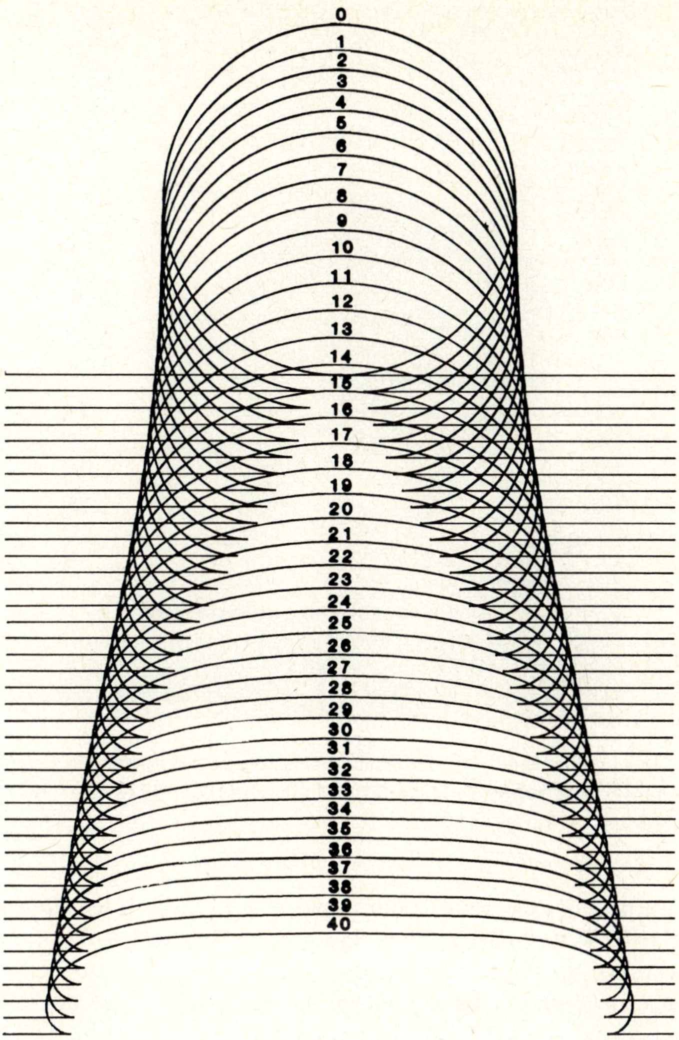


FIG. 3. Theoretical forms of a liquid drop drawn by a computer. Numerals indicate curve numbers shown in Table I.

Niigata College of Pharmacy. The centrifugal acceleration ($A$) is calculated to be $4.0 \times 10^2$ m/sec$^2$ from Eq. (1), because $r$ is 3.65 cm and the rotation speed is 998 rpm. The $H/W$ value measured on the photograph is 0.618, which corresponds to 5.98 for $S_2$ obtained by interpolation from values in Table I. Because the density difference $D-D'$ was determined to be 47 kg/m$^3$ in eggs of the same

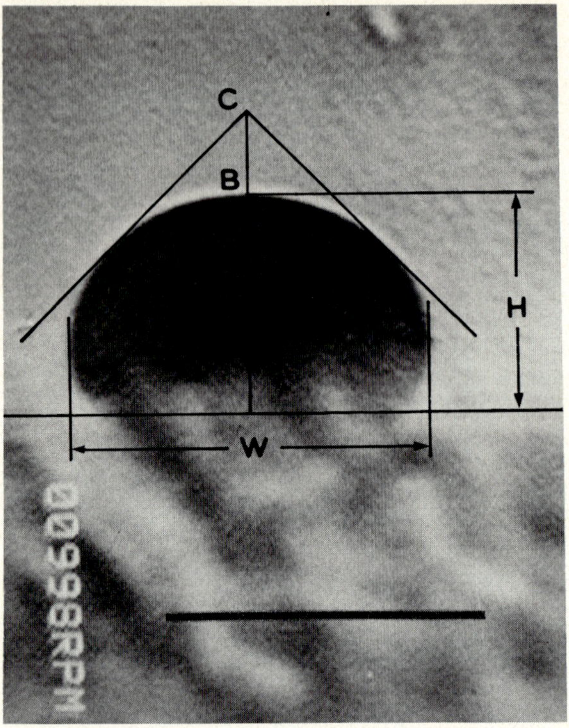


FIG. 4. An unfertilized egg of the sea urchin, *Hemicentrotus pulcherrimus* in a centrifugal field supported with a plate perpendicular to the direction of centrifugal force of $4.0 \times 10^2$ m/sec$^2$. Bar indicates 100 μm.

batch by the methods mentioned below, and $W$ is 111.4 μm, the surface force is calculated to be $(4.0 \times 10^2) \times 47 \times (111.4 \times 10^{-6})^2/5.98 = 3.9 \times 10^{-5}$ N/m (0.039 dyn/cm).

The value of $H/2R$ of the egg shown in Fig. 4 was 0.703 and its diameter before centrifugation was 98 μm. This value corresponds to 4.62 for $S_1$ in Table I. Therefore, the surface force is calculated to be $(4.0 \times 10^2) \times 47 \times (98 \times 10^{-6})^2/4.62 = 3.9 \times 10^{-5}$ N/m, the same as the above value.

## 2. EXAMPLE 2

Another parameter representing the degree of flattening was used by Dorsey (1928) for calculating the surface tension of a sessile drop. His method does not require knowledge about either height ($H$) or initial diameter ($2R$). Consequently, this method is applicable even to conditions where the exact position of

TABLE I

CALCULATED PARAMETERS OF SESSILE DROPS

| Curve | $S_0$ | $H/(2R)$ | $W/(2R)$ | $H/W$ | $CB/(W/2)$ | $S_1$ | $S_2$ |
|---|---|---|---|---|---|---|---|
| 0 | 0.0000 | 1.000 | 1.000 | 1.000 | 0.414 | 0.000 | 0.000 |
| 1 | 0.0313 | 0.973 | 1.005 | 0.969 | 0.416 | 0.123 | 0.124 |
| 2 | 0.0442 | 0.965 | 1.007 | 0.959 | 0.416 | 0.172 | 0.174 |
| 3 | 0.0625 | 0.955 | 1.010 | 0.946 | 0.417 | 0.240 | 0.245 |
| 4 | 0.0884 | 0.942 | 1.013 | 0.930 | 0.419 | 0.335 | 0.344 |
| 5 | 0.1250 | 0.927 | 1.018 | 0.910 | 0.420 | 0.463 | 0.481 |
| 6 | 0.1768 | 0.909 | 1.025 | 0.887 | 0.423 | 0.637 | 0.669 |
| 7 | 0.2500 | 0.887 | 1.033 | 0.859 | 0.426 | 0.869 | 0.927 |
| 8 | 0.3536 | 0.863 | 1.044 | 0.827 | 0.430 | 1.171 | 1.276 |
| 9 | 0.5000 | 0.836 | 1.056 | 0.792 | 0.436 | 1.559 | 1.740 |
| 10 | 0.7071 | 0.808 | 1.072 | 0.754 | 0.443 | 2.047 | 2.350 |
| 11 | 1.0000 | 0.778 | 1.089 | 0.714 | 0.452 | 2.644 | 3.135 |
| 12 | 1.4142 | 0.747 | 1.108 | 0.674 | 0.462 | 3.361 | 4.127 |
| 13 | 2.0000 | 0.716 | 1.129 | 0.634 | 0.474 | 4.200 | 5.356 |
| 14 | 2.8284 | 0.686 | 1.152 | 0.596 | 0.488 | 5.163 | 6.847 |
| 15 | 4.0000 | 0.657 | 1.175 | 0.559 | 0.503 | 6.247 | 8.623 |
| 16 | 5.6569 | 0.629 | 1.199 | 0.525 | 0.520 | 7.446 | 10.70 |
| 17 | 8.0000 | 0.602 | 1.223 | 0.492 | 0.537 | 8.752 | 13.09 |
| 18 | 11.314 | 0.577 | 1.247 | 0.462 | 0.555 | 10.16 | 15.79 |
| 19 | 16.000 | 0.553 | 1.271 | 0.435 | 0.573 | 11.65 | 18.82 |
| 20 | 22.627 | 0.530 | 1.294 | 0.410 | 0.591 | 13.23 | 22.16 |
| 21 | 32.000 | 0.508 | 1.317 | 0.386 | 0.608 | 14.88 | 25.82 |
| 22 | 45.255 | 0.489 | 1.340 | 0.365 | 0.625 | 16.60 | 29.80 |
| 23 | 64.000 | 0.471 | 1.362 | 0.346 | 0.641 | 18.37 | 34.09 |
| 24 | 90.510 | 0.453 | 1.384 | 0.328 | 0.657 | 20.20 | 38.68 |
| 25 | 128.00 | 0.437 | 1.405 | 0.311 | 0.672 | 22.08 | 43.57 |
| 26 | 181.02 | 0.422 | 1.425 | 0.296 | 0.686 | 24.00 | 48.76 |
| 27 | 256.00 | 0.408 | 1.445 | 0.282 | 0.699 | 25.97 | 54.24 |
| 28 | 362.04 | 0.395 | 1.465 | 0.269 | 0.711 | 27.97 | 60.01 |
| 29 | 512.00 | 0.382 | 1.484 | 0.258 | 0.723 | 30.00 | 66.06 |
| 30 | 724.08 | 0.371 | 1.502 | 0.247 | 0.734 | 32.07 | 72.39 |
| 31 | 1024.0 | 0.360 | 1.520 | 0.237 | 0.744 | 34.17 | 78.99 |
| 32 | 1448.2 | 0.349 | 1.538 | 0.227 | 0.754 | 36.30 | 85.88 |
| 33 | 2048.0 | 0.340 | 1.555 | 0.218 | 0.763 | 38.46 | 93.03 |
| 34 | 2896.3 | 0.331 | 1.572 | 0.210 | 0.772 | 40.65 | 100.5 |
| 35 | 4096.0 | 0.322 | 1.589 | 0.203 | 0.780 | 42.85 | 108.2 |
| 36 | 5792.6 | 0.314 | 1.605 | 0.196 | 0.788 | 45.09 | 116.1 |
| 37 | 8192.0 | 0.306 | 1.621 | 0.189 | 0.795 | 47.35 | 124.3 |
| 38 | 11585 | 0.299 | 1.636 | 0.183 | 0.801 | 49.63 | 132.8 |
| 39 | 16384 | 0.292 | 1.651 | 0.177 | 0.808 | 51.93 | 141.6 |
| 40 | 23170 | 0.285 | 1.666 | 0.171 | 0.814 | 54.25 | 150.6 |
| 41 | 32768 | 0.279 | 1.681 | 0.166 | 0.819 | 56.60 | 159.9 |
| 42 | 46341 | 0.273 | 1.695 | 0.161 | 0.825 | 58.97 | 169.4 |

*(continued)*

TABLE I (*Continued*)

| Curve | $S_0$ | $H/(2R)$ | $W/(2R)$ | $H/W$ | $CB/(W/2)$ | $S_1$ | $S_2$ |
|---|---|---|---|---|---|---|---|
| 43 | 65536 | 0.268 | 1.709 | 0.157 | 0.830 | 61.35 | 179.2 |
| 44 | 92682 | 0.262 | 1.723 | 0.152 | 0.835 | 63.76 | 189.3 |
| 45 | 131072 | 0.257 | 1.737 | 0.148 | 0.839 | 66.18 | 199.6 |
| 46 | 185364 | 0.253 | 1.750 | 0.144 | 0.843 | 68.63 | 210.2 |
| 47 | 262144 | 0.248 | 1.763 | 0.141 | 0.847 | 71.09 | 221.0 |
| 48 | 370728 | 0.244 | 1.776 | 0.137 | 0.851 | 73.58 | 232.1 |
| 49 | 524288 | 0.239 | 1.789 | 0.134 | 0.855 | 76.07 | 243.5 |
| 50 | 741455 | 0.235 | 1.802 | 0.131 | 0.858 | 78.59 | 255.1 |

the plate supporting the egg is difficult to determine. Dorsey's method (1928) is to draw a pair of tangents at the drop surface inclined at 45° from the horizontal plane (cf. Fig. 2). The distance ($CB$) between the intersecting point ($C$) of the tangents and the top ($B$) of the drop divided by half a width ($W/2$) is a parameter representing the degree of flattening. The surface force of the egg is calculated from the $S_2$ value corresponding to the $CB/(W/2)$ value determined in an egg under a centrifugal field. In the example shown in Fig. 4, $CB/(W/2)$ is 0.474, which corresponds to 5.356 for $S_2$ in Table I. Because $D-D'$ is 47 kg/m$^3$, $A$ is $4.0 \times 10^2$ m/sec$^2$ and $W$ is 111.4 μm, $T$ is calculated to be $4.4 \times 10^{-5}$ N/m.

In Dorsey's original method (1928), the surface tension is calculated from the equation.

$$T = A(D-D')W^2P(f)/4 \tag{3}$$

where

$$P(\mathrm{f}) = 0.052/f - 0.12268 + 0.048f \tag{4}$$

and

$$f = CB/(W/2) - 0.41421 \tag{5}$$

It was found by checking with a computer that Dorsey's function $P(f)$ is virtually equal to $4/S_2$, as expected by comparing Eq. (3) with the definition of $S_2$.

## 3. Example 3

A still simpler method for estimating the surface force is to find the $S_1$ or $S_2$ value by fitting the profile of the flattened egg with the theoretical curves shown in Fig. 3. The image of the flattened egg is projected at an appropriate magnification onto theoretical curves in Fig. 3, or vice versa, the images of theoretical curves are projected onto the photographic print of the egg in a centrifugal field.

From the number of the theoretical curve that fits the profile of the egg, it is possible to read off $S_1$ and $S_2$ values from Table I.

In the example shown in Fig. 4, the profile of the egg fits between curves 13 and 14 in Fig. 3, which correspond to 5.356 and 6.347 for $S_2$ in Table 1. Thus, the surface force is calculated to be $(3.4 - 4.4) \times 10^{-5}$ N/m by using the $D-D'$ and $W$ (or $2R$) values mentioned above.

The accuracy in determining the surface force by the present method depends on the degree of flattening of the egg. The analysis of the relation between $S_1$ and $H/W$ indicates that the accuracy in determining the surface force is maximal when $H/W$ is around 0.5. Centrifugal acceleration of order of $10^2$ $g$ produces deformation in this range in sea urchin eggs.

## C. Determination of the Density of the Egg

The density difference between the protoplasm and surrounding seawater was reported to be 30 to 60 kg/m$^3$ in various species of sea urchin eggs (Table II). A simple and accurate method to determine the density of the egg protoplasm is by the "isopycnotic method" (cf. Harvey, 1931; Hiramoto, 1954). Eggs are put into various mixtures of seawater and a high-density aqueous solution that is isosmotic with seawater; the eggs are then centrifuged with a conventional centrifuge to find the mixing ratio in which eggs move neither centrifugally nor centripetally. Either 0.73 to 0.75 $M$ sucrose solutions or seawater containing a high-molecular-weight substance such as Percol can be used as the high-density

TABLE II

DENSITY OF SEA URCHIN EGGS

| Species | Density (g/cm$^3$=$10^3$ kg/m$^3$)[a] | References |
|---|---|---|
| *Arbacia punctulata* | | |
| (U)[b] | 1.083–1.084 | Harvey (1931) |
| (F) | 1.074–1.079 | Hiramoto (1967) |
| *Lytechinus variegatus* (F) | 1.057–1.061 | Hiramoto (1967) |
| *Hemicentrotus pulcherrimus* | | |
| (F) | 1.076–1.083 | Hiramoto (1954) |
| (U) | 1.0790–1.0846 | Hiramoto (1954) |
| *Anthocidaris crassispina* (F) | 1.070 | Oshima (1983) |

[a] Density of seawater ranges from 1.024 to 1.028 g/cm$^3$, depending on the locality (cf. E. B. Harvey, 1956). If information is unavailable for the density of seawater, a value of 1.025 g/cm$^3$ may be assumed.

[b] U, Unfertilized egg without jelly; F, fertilized egg without fertilization membrane.

solution. It is desirable to check the sizes of the eggs after they are put into the above solutions, because the density of the egg changes if its size is changed. The densities of seawater and solutions are determined by comparing their weight with the weight of distilled water of the same volume by using the same weighing bottle.

## IV. Limitations of the Present Method and Errors in Measurement

Hiramoto (1967) discussed possible errors and limitations of the sessile drop method in determining the surface force of the sea urchin eggs. He stated that neglecting the rigidity of the endoplasm may not introduce serious errors in surface force determination in unfertilized sea urchin eggs. In fertilized eggs, however, the errors may be larger because structures having considerable rigidity such as the sperm aster or mitotic apparatus are formed in the endoplasm (cf. Hiramoto, 1969).

It is supposed that the density of the protoplasm is practically uniform in echinoderm eggs because eggs do not display any orientations in gravitational fields. A density gradient in the protoplasm due to any stratification caused by centrifugation may be scarcely present because the centrifugal acceleration is small in the present experiment; nevertheless, large structures such as the nucleus and the mitotic apparatus move in the egg protoplasm during prolonged application of centrifugal force (cf. Hiramoto, 1967).

While viewing the egg during centrifugation, the exact location of the plate supporting the egg is difficult to determine owing to the large optical depth of the plate. The contact angle of the egg surface against the supporting plate is not always zero because of the adhesion of the egg surface to the plate, while the theoretical $H$ values in Table I are calculated assuming that the contact angle is zero. The above two facts may introduce errors in determining the surface force from $H/W$ or $H/2R$ determined from an egg in a centrifugal field.

In the methods for determining the surface force from Dorsey's parameter $CB/(W/2)$ and by the curve-fitting mentioned above, no problem is created by adhesion of the egg surface because the forms of the free surface alone are used in the calculation.

## V. Appendix: Derivation of the Profiles of Sessile Drops

Since the sessile drop is axially symmetric along the direction of gravity, the calculation of its contour is reduced to finding a single meridian of the drop, expressed in orthogonal coordinates $X$, $Z$ (Fig. 5). The calculation is based on the Laplace formula which relates the surface tension ($T$) with the internal pressure ($P$) as

$$P = T(1/r_1 + 1/r_2) \tag{6}$$

where $r_1$ and $r_2$ are radii of principal curvatures at a point on the surface. The angle held between the axis of the drop and the line drawn normal to the surface is denoted by $\phi$. By classical theory of differential geometry of a shell of revolution, $r_2$ is equal to $X/\sin\phi$. In Fig. 5, a line segment $ds$ is equal to $dX/\cos\phi$ and also to $r_1\ d\phi$ so that $r_1 = (dX/d\phi)/\cos\phi$. From Eq. (6) we have

$$dX/d\phi = \cos\phi/(P/T - \sin\phi/X) \tag{7}$$

Under the centrifugal or gravitational force the hydrostatic pressure inside the drop increases as an increase in $Z$ (cf. Fig. 5). Thus the excess internal pressure $P$ is a linear function of $Z$ as

$$P(Z) = P_0 + A(D-D')Z \tag{8}$$

where $A$ is centrifugal or gravitational acceleration. If $R_1$ is the radius of curvature at the top ($Z = 0$), from Eq. (6) we have the internal pressure at $Z$=0 as $P_0 = 2T/R_1$. Expressing $X$ and $Z$ in relative values $x$=$X/R_1$ and $z$=$Z/R_1$, and introducing the parameter $S_0 = A(D-D')R_1^2/T$, from Eqs. (7)–(8) we have

$$dx/d\phi = \cos\phi/(S_0\ z + 2 - \sin\phi/x) \tag{9}$$

Since $dz/dx$ is equal to tan $\phi$,

$$dz/d\phi = \sin\phi/(S_0\ z + 2 - \sin\phi/x) \tag{10}$$

The profiles in $X$–$Z$ coordinates are obtained by serial calculations of $\Delta x$ and $\Delta z$ using Eqs. (9)–(10), from $\phi$=0° (top) to $\phi$=180° (bottom), for various values of $S_0$ which define the degree of flattening.

Table I and Fig. 3 presented here are results of such calculations, made by the Runge–Kutta scheme with the mesh size ($\Delta\phi$) of 0.15°. The relative volume ($v$) of the drop, $\pi\sum x^2\Delta z$, was also computed. A trouble encountered was that the

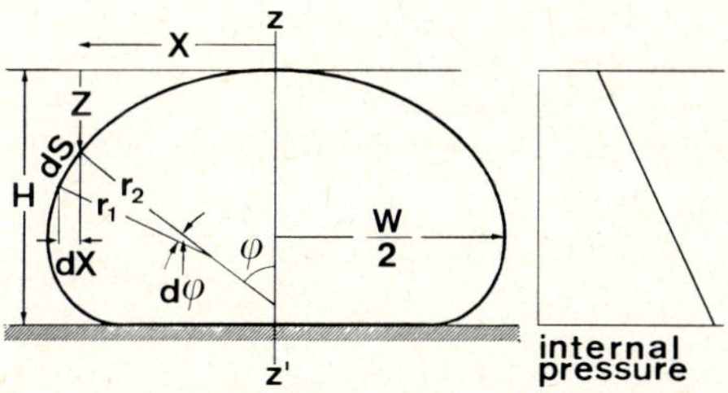


FIG. 5. Diagram of a sessile drop representing various dimensions used in computer calculation.

fourth coefficients in this scheme tends to diverge at the first step ($\phi$=0°) when $S_0$ is very large. This difficulty was mitigated by separately calculating the initial and final parts of $\phi$ (0° to $\Delta\phi/100$ and 180° − $\Delta\phi/100$ to 180°) by proper approximations.

For convenience in practical use of the calculated profiles, the terms $x=X/R_1$ and $z=Z/R_1$ were converted into $X/2R$ and $Z/2R$, respectively, by the relation of $vR_1^3 = 4\pi R^3/3$, where $R$ is the radius of a sphere with the volume same as the drop. Correspondingly, the term $S_0$ was replaced by more convenient terms, $S_1$ and $S_2$, as

$$S_1 = S_0\,(2R/R_1)^2 = A(D-D')(2R)^2/T \tag{11}$$

$$S_2 = S_0\,(W/R_1)^2 = A(D-D')W^2/T \tag{12}$$

Therefore

$$T = A(D-D')(2R)^2/S_1 \tag{13}$$

or

$$T = A(D-D')W^2/S_2 \tag{14}$$

## References


Amemiya, S., and Tsuchiya, T. (1979). *Mar. Biol.* **52,** 93–96.
Brown, R. H. J. (1940). *J. Exp. Biol.* **17,** 317–324.
Dorsey, N. E. (1928). *J. Washington Acad. Sci.* **18,** 505–509.
Harvey, E. B. (1951). *In* "McClung's Handbook of Microscopical Technique" (J. R. McClung, ed.), pp. 586–590. Harper, New York.
Harvey, E. B. (1956). "The American Arbacia and Other Sea Urchins." Princeton Univ. Press, Princeton, New Jersey.
Harvey, E. N. (1931). *Biol. Bull.* **41,** 273–279.
Harvey, E. N. (1933a). *Science* **77,** 430–431.
Harvey, E. N. (1933b). *J. Cell. Comp. Physiol.* **4,** 35–47.
Harvey, E. N., and Fankhauser, G. (1933). *J. Cell. Comp. Physiol.* **3,** 463–475.
Hiramoto, Y. (1954). *Jpn. J. Zool.* **11,** 227–243.
Hiramoto, Y. (1967). *J. Cell. Physiol.* **69,** 219–230.
Hiramoto, Y. (1969). *Exp. Cell Res.* **56,** 209–218.
Inaba, D. (1957). *In* "Invertebrate Embryology" (M. Kume and K. Dans, eds.), p. 224, (English edition translated by J. C. Dan, Nonit, Belgrade).
Kuroda, K., and Kamiya, N. (1981). *Biorheology* **18,** 633–641.
Oshima, N. (1983). *Biol. Bull.* **164,** 483–492.
Vlès, F. (1926). *Arch. Phys. Biol.* **4,** 263–284.

# Index

## F

## G

## H



## I